Rotterdam
JAN 2009

Delft School of Design Series on Architecture and Urbanism

Series Editor
Arie Graafland

Editorial Board
K. Michael Hays (Harvard University, USA)
Ákos Moravánszky (ETH Zürich, Switzerland)
Michael Müller (Bremen University, Germany)
Frank R. Werner (University of Wuppertal, Germany)
Gerd Zimmermann (Bauhaus University, Germany)

Crossover.
Architecture
Urbanism
Technology

Editors **Arie Graafland and Leslie Jaye Kavanaugh**

With contributions by **George Baird, M. Christine Boyer, Joan Busquets, Marisa Carmona, Kees Christiaanse, Wouter Davidts, Adalberto Del Bo, Chris Dercon, Paul Drewe, Leen van Duin, Mick Eekhout, Henk Engel, Hal Foster, Rosina Gomez-Baeza, Arie Graafland, Christoph Grafe, Kari Jormakka, Vittorio Magnago Lampugnani, Lars Lerup, Francesc Magrinyà, Michael Mehaffy, Han Meyer, Ákos Moravánszky, Anne Vernez Moudon, Michael Müller, Kas Oosterhuis, Scott Page and Brian Phillips, Alberto Pérez-Gómez, Stephen Read, Jürgen Rosemann, Nikos A. Salingaros, Clemens Steenbergen and Wouter Reh, Kelly Shannon, Roemer van Toorn, Philip Ursprung, Jan Vogelij, Karel Vollers, Frank R. Werner**

010 Publishers **Rotterdam 2006**

8 List of Contributors & Abstracts

20 Introduction Arie Graafland, Program Director Delft School of Design

30 Culture Scape Holland Photos: Jeroen Musch

Section I: New Instruments of Design – A Matter of Network Thinking

42 Introduction Paul Drewe   **46 The New Charter of Athens 2003: The European Council of Town Planners Vision on the European City of the 21st-Century** Jan Vogelij   **60 Urban Interfaces: Designing In-Between** Scott Page and Brian Phillips   **80 The General Theory of Urbanization of Ildefons Cerdà: An Urban Network Theory** Francesc Magrinyà   **100 Compact City Replaces Sprawl** Nikos A. Salingaros   **118 Codes and the Architecture of Life** Michael Mehaffy

Section II: Representation in Architecture

138 Introduction Arie Graafland   **140 The New Simplicity: A Problem of Representation in Architecture and Town Planning?** Frank R. Werner   **154 Built Images: The Eberswalde Library by Herzog & de Meuron** Phillip Ursprung   **170 After Criticality: The Passion for Extreme Reality in Recent Architecture . . . and Its Limitations** Roemer van Toorn   **184 Poesis doctrinae tamquam somnium** Kari Jormakka

Section III: Design & Engineering: Liquid Design & Engineering in Architecture & Building Technology

210 Introduction Mick Eekhout   **212 Integration and Coordination of Complex 3D 'Blob' Design Processes and Engineering** Mick Eekhout   **240 A New Kind of Building** Kas Oosterhuis   **270 Bending and Folding Glass** Karel Vollers

Section IV: The Urban Question

282 Introduction Jürgen Rosemann   **284 The Artful Reinvention of the Suburban Landscape** Lars Lerup   **294 The City of Tolerant Normality** Vittorio Magnago Lampugnani   **312 The Urban Question in the 21st Century: Epistemological and Spatial Traumas** M. Christine Boyer   **338 Productive Space** Stephen Read   **368 Opportunities of Clustering IBIS Network Research and the Urban Question** Marisa Carmona

Section V: The Museum and the Media

388 Introduction Christoph Grafe   **394 'You can be a museum or you can be modern, but you can't be both'** Chris Dercon   **408 Why Bother (about) Architecture? Contemporary Art, Architecture and the Museum** Wouter Davidts   **424 Order on Display** Michael Müller   **436 Cultural Centres in Europe** Christoph Grafe   **456 ARCO Spatial Project: On Ephemeral Architecture and Art** Rosina Gómez-Baeza   **470 A Little Dictionary of Design Clichés** Hal Foster

Section VI: Urban Compositions in the 21st century

486 Introduction Han Meyer   **494 Urban Composition: City Design in the 21st Century** Joan Busquets   **516 Situations in an Inhabited Landscape** Kees Christiaanse   **530 The Polder as a Design Atlas** Clemens Steenbergen and Wouter Reh   **552 Field Building and Urban Morphology: A Challenge for the Next Generation of Urban Designers** Anne Vernez Moudon   **568 Concluding with Landscape Urbanism Strategies** Kelly Shannon

Section VII: Project & Research

594 Introduction Leen van Duin   **596 Traditional Architecture and the Swinging Market** Leen van Duin   **608 To Create Order out of the Desperate Confusion of Our Time** Adalberto Del Bo   **622 Merz City** Henk Engel

Section VIII: Theory and Praxis

646 Introduction Arie Graafland   **648 'Criticality' and Its Discontents** George Baird   **660 Power Lines** Ákos Moravánszky   **676 Ethics and Poetics in Architectural Praxis** Alberto Pérez-Gómez   **688 On Criticality** Arie Graafland

List of Contributors & Abstracts

George Baird

is Dean of the Faculty of Architecture, Landscape, and Design at the University of Toronto, and partner in the Toronto-based architecture and urban design firm Baird Sampson Neuert Architects. Baird Sampson Neuert is the winner of numerous design awards, including the **Canadian Architect Magazine** awards over many years, and a Governor General's Award for Cloud Gardens Park in 1994. Baird has been the recipient of the Toronto Arts Foundation's Architecture and Design Award (1992) and the da Vinci Medal of the Ontario Association of Architects (2000). He is co-editor (with Charles Jencks) of **Meaning in Architecture** (1969), and (with Mark Lewis) of **Queues Rendezvous, Riots** (1995). He is author of **Alvar Aalto** (1969) and **The Space of Appearance** (1995).

'Criticality' and Its Discontents [p. 648]

Baird elucidates the trajectory of the older exponents of 'critical practice' among whom were Tafuri, Eisenman, and Hays; and opposes them to a newer generation of advocates of 'projective practice': Somol, Whiting, Hickey and Lavin. Nevertheless, Baird encourages us to continue our commitment to questioning. Although a new (anti-)critical stance seems to be emerging in this century, this attitude has yet to be truly tested in practice and has yet to substantiate itself with a supporting body of 'projective theory'.

M. Christine Boyer

is the William R. Kenan Jr. Professor of Architecture and Urbanism, at the School of Architecture, Princeton University. She is the author of **CyberCities: Visual Perception in the Age of Electronic Communication** (Princeton Architectural Press, 1996). In addition, she has written many articles and lectured widely on the topic of urbanism in the 19th and 20th centuries. Boyer received her Phd. and Masters in City Planning from Massachusetts Institute of Technology. She also holds a Masters of Science in Computer and Information Science from the University of Pennsylvania, The Moore School of Electrical Engineering. Boyer was a full-time visiting professor at the DSD from February 2005 until August 2005. She is currently writing a book on Le Corbusier's writings entitled **Le Corbusier: Homme de Lettre**.

The Urban Question in the 21st Century: Epistemological and Spatial Traumas [p. 312]

In 'The Urban Question', Boyer begins with the foundational work of Manuel Castells on Network Theory as applied to urban systems. This critique, although fruitful, was still unable to produce constructive ways of grappling with material practices. For this attempt, Boyer turns to Stefano Boeri and his research network called Multiplicity. In doing so, she enlarges their argument and calls into question the technologies of representation and their subsequent possible interpretations. Yet Boyer acknowledges, ultimately, that the collage of theoretical apparatuses – linguistic paradigms, cartographies, and autopoietic systems theory – are not adequate to the task: the urgent need for an innovative procedure that can not only describe in a new way the conditions of built environment, but lead to improved urban practice.

Joan Busquets

is the first Martin Bucksbaum Professor in Practice of Urban Planning and Design at the Graduate School of Design at Harvard University.

Prior to joining the GSD faculty, Busquets was Professor of Town Planning in the School of Architecture at the Polytechnic University of Barcelona from 1979 until 2002. A world-renowned urban planner, urban designer and architect, Busquets served as Head of Urban Planning for the Barcelona City Council during the formative years, from 1983 to 1989, and in the preparations for the Barcelona Olympics in 1992. In 1969 he was a founding member of the Laboratorio de Urbanismo in Barcelona. Busquets has published many articles and books, including **Barcelona: Evolución urbanistica de una capital compacta** (1992 and 1994), **La urbanización marginal** (1999), **The Old Town of Barcelona: a past with a future** (2003), **Bringing the Harvard Yards to the River** (2004) and **Aleppo: Rehabilitation of the Old City** (2005). He has participated as a visiting professor in London, Urbino, Rotterdam, Rome, Lausanne, Geneva, and at the GSD, as well as on numerous juries for international competitions. Professionally based in Barcelona since the 1970s, Busquets received his degree in architecture (1969) and his doctoral degree (1975) from the University of Barcelona.

Urban Composition: City Design in the 21st Century [p. 494]

In his essay, Busquets outlines ten 'lines of work' in order to come to terms with the complexities of our contemporary urban situation. In his opinion, an over-arching prescriptive ideology is no longer an effective strategy to address the re-formatting of space. In the end, he interrogates the significance of Urbanism as a form of practical knowledge in order to generate more productive urban design practices.

Marisa Carmona

is an Associate Professor in the Department of Urbanism TUDelft and member of the advisory board of the United Nations Global Research Network on Human Settlements. She is coordinator of the International Research network IBIS on Globalisation Urban Form and Governance. She is author and editor of several books, including **The Challenge of Sustainable Cities** (Zed Books, London 1997) and **Strategic Planning and Urban Projects: Responding to Globalisation from 15 Cities** (DUP, Delft).

Opportunities of Clustering IBIS Network Research and the Urban Question [p. 368]

Carmona explicates the work of the IBIS research group, which is a global network composed of 40 universities engaged in research on 25 cities. This collaborative network investigates questions of urbanism and the impact of globalization generally upon urban forms. Specific attention is given to the interconnection between factors of economics, social and governmental forms, new ITC technologies, and the environment.

Kees Christiaanse

studied Architecture at the Delft University of Technology, graduating together with Art Zaaijer in 1988. His graduation project, 'Kavel 25', was realized as part of Christiaanse's urban plan for the housing festival in The Hague. He was awarded the Berlage Flag for this project. Between 1980 and 1989, Christiaanse worked for the Office for Metropolitan Architecture in Rotterdam, becoming a partner in 1983. In 1989, he started his own firm in Rotterdam, lr. Kees Christiaanse

Architects & Planners, which was renamed KCAP Architects & Planners in 2002. In 1990 he founded ASTOC Architects & Planners in Cologne. From 1996 until 2003 he was professor of Architecture and Urban Design at the Berlin University of Technology. Now he is connected to the Swiss Federal Institute of Technology to Zurich. He regularly acts as a jury member for international competitions, and is the author of several publications about architecture and urban design.

Situations in an Inhabited Landscape [p. 516]
The tasks confronting the urban scientist have become more complex than ever. 'What do we as urban planners do', Christiaanse asks, 'when we can no longer control the exact programming, the speed of realization, and the quality and style of the architecture? We certainly don't stop!' Not only the legislation, but also the spatial conditions of Northern Europe form a challenge that increasingly calls for invention and fortitude. New strategies are needed in order to prevent small enclosed, mono-functional atomistic spatial units, or indeed spatial interstices, 'left-over' spaces, becoming disjuncted from a larger infrastructure. Christiaanse discusses strategies regarding transportation infrastructures in order to strengthen the urbanists' influence upon what he calls 'intermediate landscapes'.

Wouter Davidts
is Senior Lecturer in Architectural Theory at the Department of Architecture of Vrije Universiteit Brussel and Postdoctoral Researcher at the Department of Architecture & Urban Planning of Ghent University. In recent years he has published on the museum, contemporary art and architecture in magazines such as **Architecture d'Aujourd'hui, Archis, De Witte Raaf, Kritische Berichten** and **Parachute**, and in several books and exhibition catalogues. In 2000 he co-edited the publication **'B-sites. à propos de la place d'un centre d'art et de recherche à Bruxelles,'** and recently he edited a new volume on the work of the Belgian artist Jan De Cock (Atelier Jan De Cock, 2005). He is preparing a book on museum architecture (A&S/books, 2006). His current scholarly work focuses on post-war artistic attitudes towards architecture and on the architectural nature and identity of the contemporary artist's studio.

Why bother (about) Architecture? Contemporary Art, Architecture and the Museum [p. 408]
Davidts 'bothers'; Davidts worries about the status of museum architecture and its role in redefining, re-framing, or positioning the debate on museums. In recent decades a number of 'attacks' on the institution of the museum and its relevance have centered around the architecture that houses these institutions. Museum architecture, Davidts proposes, occupies a paradoxical and ambiguous position in this debate: being the site of the discourse without being allowed to intervene in the critical questioning of the function of the museum as an institution in society. In the end, he asks: 'Is architecture still an instance to bother, or rather to bother about?'

Adalberto Del Bo
graduated in Architecture in Milan where he is now full professor in Composizione architettonica e urbana at the Facoltà di Architettura Civile del Politecnico di Milano. He has participated with projects and constructions in exhibits and national and international conferences. He has published studies and research in the fields of architecture and urban form with a particular focus on the aspects related to architectural code and to theoretical/practical aspects of design. His design and construction work constitutes the verification of different issues and topics, and the Campus of the Chieti University, realized in collaboration with other architects, was published in **Il Campus Universitario di Chieti**, Electa, Milano, 1997. The Sports Centre on the Campus was awarded the 'European architectural prize for sport facilities' in 1996.

To Create Order out of the Desperate Confusion of our Time [p. 608]
Both the collaboration between Mies van der Rohe and Hilberseimer, and Le Corbusier's six-points are still, according to Del Bo, fruitful sources for architectural design. For him, the modernist masters lived in a critical time that has been neglected by critics and design professionals alike. From analytical knowledge through understanding and seeing, he argues that the contemporary theoretical apparatus is inadequate to the problems confronting architects dealing with large-scale urban aggregates. These concepts he then applies to the current situation in Northern Italy.

Chris Dercon
is an art historian, documentary filmmaker and cultural producer. He was at the end of the 80's a program director of PS1 Museum in New York. In 1990 he became director of Witte de With, Center for Contemporary Art in Rotterdam, as well as a board member of INIVA in London. From 1996 until 2003 he was the director of the Museum Boijmans Van Beuningen in Rotterdam. In 2003 Dercon became director of the Haus der Kunst in Munich.

You can be a museum or you can be modern, but you can't be both [p. 394]
Dercon takes a quote from Gertrude Stein as a starting point in order to speak about what he calls '"re"-factors' regarding museums; that is to say the increasing changes in museum policies and museum practices in the form of renovation, revitalization, refinancing, and reorganization. What would the role of curator be in the contemporary museum that must compete with other means of cultural representation? Indeed, what is the function of the museum, the audience, the museum building, or indeed art itself? In answering these questions, Dercon reminds us that the participants in this discussion must be prepared to look broader afield that its own doors and consider that the museum might pursue an entirely different function than exhibiting and archiving. Yet the question remains: 'Indeed, where are we going to put all this stuff?'

Paul Drewe
is Emeritus Professor of Spatial Planning (since September 1 2005) at the Faculty of Architecture, Delft University of Technology. He is still active as a researcher and consultant both in the Netherlands and abroad. Professor Paul Drewe received his education in economics and sociology in Cologne, Germany. In 1990 he received his Doctors degree of honor in Urbanism, Spatial Planning and Development at the University of Ghent, Belgium. In 1973 he became Professor of Spatial Planning, TU Delft Faculty of Architecture. In 2002 he became Guest Professor in cross-border and inter-regional cooperation, Faculté Pluridisciplinaire de Bayonne/ Anglet/Biarritz, Université de Pau et des Pays de l'Adour.

Leen van Duin

is an architect and since 1994 a professor of Architectural Design at the Faculty of Architecture of the TU Delft. He teaches and does research on design methods and the programming and typology of buildings. From 1979 until 1993 he was a member of De Nijl Architecten, an architecture office that he co-founded. Recently he published **A Hundred Years of Dutch Architecture, 1900–2000**, Amsterdam (SUN) 2003, together with Umberto Barbieri and Willemijn Wilms Floet. Van Duin holds several public and administrative functions. At the moment he is chairman of the Dutch State Examination Committee for Architecture and board member of the EAAE (European Association for Architectural Education).

Traditional Architecture and the Swinging Market [p. 600]

Van Duin offers a polemic against recent trends which offer up architecture to the market, becoming a mere fashion accessory. Instead, he pleads for a return to research on the traditional Dutch City, not in order to copy it, but to gather knowledge pertinent to contemporary concerns.

Mick Eekhout

is Professor of Product Development, Faculty of Architecture, Delft University of Technology and Director of Octatube Space Structures BV. From 1968 to 1973 Mick Eekhout studied Architectural Engineering in Delft. After his graduation (with honors) – supervised by professor Oosterhof and professor Weeber – he worked in an architectural office for two years, before founding his own architectural office where he realized several buildings in eight years' time. In 1982 he founded the company Octatube Space Structures, specialized in three dimensional constructions and structures for the building industry. For over 20 years Octatube has realized many design & build projects, both in the Netherlands and abroad. In 1989 he got his PhD degree (with honors) – supervised by professor Oosterhof and professor Zwarts – with his thesis 'Architecture in Space Structures'. Since 1992 Eekhout is professor of Product Development at the Delft University of Technology, Faculty of Architecture. In 2003 he was the first designer since 1856 to be accepted in the Koninklijke Nederlandse Akademie van Wetenschappen (Royal Dutch Academy of Sciences).

Integration and Coordination of Complex 3D 'Blob' Design Processes and Engineering [p. 212]

Rehearsing a perennial problem between visionary architects and structural engineers who often have the unsung task of executing the vision, Mick Eekhout discusses the challenges and ultimately the rewards of the new complex 'Blob' designs in architecture. In an extensive explication of the design processes between various architects and his own firm, **Octatube Space Structures**, we arrive at various insights as to the lessons to be learned in executing and materializing these exciting new forms. Specifically, the examples of the 'co-designed' Blob designs at the **Floriade** for the province of Noord-Holland realized in April 2002, are taken up at length; one by Kas Oosterhuis of Rotterdam, and the other by Asymptote Architects of New York. In the end, undoubtedly, these new fluid designs not only ask us to think about spatial structures in a challenging way, but also have a remarkable influence on the engineering of structures.

Henk Engel

is an architect and at present co-director of the architecture office De Nijl Architecten in Rotterdam. In 1998 his office had an exhibition on their work in the NAi, which was accompanied by the publication **Als we huizen bouwen, praten en schrijven we** (NAi 1998). Engel is an associate professor of Architectural Design in Delft, and teaches at several Academies of Architecture in the Netherlands. He has been a visiting lecturer in Liverpool, Milan, and Pescara. He has written extensively on various topics concerning modern and urban architecture, and worked on several exhibitions, amongst which are **Frankfurt am Main** (1987), **Colour in Architecture** (1986) and others on the work of De Stijl and J.J.P. Oud.

Merz City [p. 622]

In his discussion of the present direction of research on the Dutch City, Henk Engel appeals to the Dadaist Kurt Schwitters' notion of 'Merz'. 'By definition', Engel states, 'the townscape is not homogeneous but polymorphous, fragmented, full of contrasts and contradictions. A townscape is a collage.' In this way, a productive notion of the city developing over time in connection with other urban conglomerates can be developed in contradistinction to the present notion of the 'Network City'. Mapping and typo-morphogical analysis are important and useful tools, but without representation of the developments over historical time frames, this research is of limited practical utility. Although in the 17th century, urban configurations in the Netherlands could be seen as a hierarchy with Amsterdam at the pinnacle, at present the Randstad agglomeration has to be understood in connection within a network of other cities and towns, both large and small.

Hal Foster

is the Townsend Martin '17 Professor and Chair of the Department of Art & Archaeology at Princeton University; his latest book is **Pop Art** (Phaidon, 2005). As a Professor of Art History at Princeton University, Hal Foster teaches lecture and seminar courses in modernist and contemporary art and criticism; he also teaches in the programs of Media and Modernity and European Cultural Studies. As a recent recipient of Guggenheim and CASVA fellowships, he is at work on a textbook on 20th-century art for Thames and Hudson/Norton, as well as on a project on the problem of the arbitrary in modernist art. His most recent book is **Design and Crime (and Other Diatribes)** from Verso (2002). He continues to write regularly for the **London Review of Books**, the **Los Angeles Times Book Review**, **October**, and the **New Left Review**. His most recent book, written with Rosalind Krauss, Yve-Alain Bois and Benjamin Buchloh, is **Art since 1900**.

A Little Dictionary of Design Clichés [p. 470]

In an instigative polemic, Hal Foster in the spirit of Deleuze's 'L'Abécédaire…', issues a challenge against both the 'self-involvement' of the architectural vanguard' as well as the emerging 'anti-critique' theories of the past few years. He argues for a little moderation – an 'average architecture'; that is, an architecture which is investigative, creative and reflexive.

Rosina Gomez-Baeza

is the Director of ARCO, the International Contemporary Art Fair in Madrid, Spain. Gòmez-Baeza has a Bachelor's Degree in English

Language and Literature from the University of Cambridge, and also studied Education at London Polytechnic University and French History and Civilization at the Catholic University of Paris. In 1979, she joined IFEMA, the Madrid Trade Fair, where she has created and directed nearly 50 different exhibitions. She headed IFEMA's Marketing and Development Department between 1989 and 1999. Since 1986, she has been Director of ARCO, the International Contemporary Art Fair, as well as creator and head of the ARCO Foundation and the Amigos de ARCO Association. Moreover, she is founding Editor-in-Chief of the quarterly magazine **ARCO, ARTE CONTEMPORANEO**. A contributor to different specialized publications and national Spanish newspapers, she is also a member of the Madrid Region's Culture Council, and on the Board of the Galician Centre for Contemporary Art (CGAC) in Santiago de Compostela. Major awards include: Prize for Latin American Co-operation, granted by the International Press Club and Latin American correspondents in Spain (1997); European Distinction for Excellence within the field of culture (2001); 'Juan Antonio Gaya Nuño' Award, from the Madrid Art Critics Association; Cross of the Chevalier de l'Ordre des Arts et des Lettres, from the French Embassy in Madrid (2002); 'Benito Pérez Galdós' Award for International Promotion within the framework of the 2002 edition of City of Madrid's Villa de Madrid Awards; Cross of the Commendatore della Stella della Solidarietà, awarded by the President of the Italian Republic.

ARCO Spatial Project: On Ephemeral Architecture and Art [p. 456]

As viable alternative to exhibitions mounted in museums, Spain's **ARCO** (COntemporary ARt) is a forum for not only the display of contemporary art, but also a site of forward-looking explorations. Housed within a series of large trade-fair buildings, **ARCO** also provides an opportunity for young architectural firms – even student groups – to be commissioned for innovative design. As such, even though the cultural is intermingled with the commercial intentions, **ARCO** fulfils some of the same purposes as museums, the promotion and dissemination of culture.

Arie Graafland

is the Antoni van Leeuwenhoek professor of Architecture Theory, Faculty of Architecture, TU Delft and the Director of the Delft School for Design (doctoral research program). Recent publications include: **Versailles and the Mechanics of Power: The Subjugation of Circe** (2002) and **The Socius of Architecture: Amsterdam, Tokyo, New York** (2000). After a three-year technical education in Rotterdam, he studied sociology and philosophy in Amsterdam. He worked for several years in urban research for the city of Arnhem. In 1978 he started at the department of Urbanism at the faculty. In 1986 he received his PhD in architectural theory. He is the editor of **The Critical Landscape** Series (010 Publishers) and has been co-editor of the **Delft Annual**.

On Criticality [p. 688]

Graafland asks, in this paper grounded in contemporary debates, what new technologies, post-Fordist critical theory, and design theory might mean for architectural practices. In his opinion, in daily practice, critical thinking was never a part of the concerns of an architectural office. Only in academia, where the possibility exists to do scholarship, are the questions of 'criticality' vs 'projective' really relevant. The rest just get on with it. Graafland proposes, instead of these unproductive hard and fast distinctions, a 'reflexive architecture' or an 'architecture of the streets'. In 'On Criticality', he outlines a strategy of doing architecture that attempts to pragmatically intervene in the built environment, whilst at the same time folding in a critique of accepted notions of space and time, contemporary practice, and the socio-economic context under which an architecture must operate.

Christoph Grafe

is an architect and writer based in Amsterdam and London, and Associate Professor of Archtiectural Design/ Interiors at the Faculty of Architecture of TU Delft. A graduate of TU Delft, he worked in practice in Amsterdam and Rotterdam and completed the postgraduate Histories and Theories Programme at the Architectural Association School in London. He has been an editor of OASE journal for architecture since 1993 and has published on post-war European architecture and architectural culture. His current research focuses on the architecture and programmatic aspects of cultural centers in Europe and the architecture of sociability. Christoph Grafe is series editor of the new Routledge series Interior Architecture and is currently preparing the first volume of this series entitled **Bars and Cafés: Living in Public** (forthcoming Autumn 2006).

Cultural Centres in Europe [p. 436]

Grafe historicizes the trajectory of the 'cultural centre' from the 'House of the People' to the present day cultural initiatives. This trajectory is by no means smooth. Up until the 1970's, 'culture' was bundled together in a 'default-solution' under neo-liberal policies in the United Kingdom. By the time of the late-welfare state, under the blank sub-urbanization of the European city, culture had also become bland and leveled-out. The discussion centered around the cultural center was indeed the question of culture itself. What is the 'public', what is the culture – dominant or otherwise – that is to be disseminated, and how to provide broad access to what was generally held to be elitist. Using the South Bank in London as an example, Grafe is able to explicate the agendas of the post-war period in terms of cultural aspirations, to its partial failure to fulfill these expectations, and finally to the discussions about its future. All in all, the Cultural Center in Europe is a sort of paradigmatic case of architecture's engagement with urban policies and neo-liberal municipal politics.

Kari Jormakka

studied architecture at Otaniemi University and Tampere University of Technology, and philosophy at Helsinki University, Finland, receiving a Master's in 1985, a Ph.D. in 1991 and a Habilitation in 1993. After teaching urban design as Assistant in Tampere from 1986 to 1988, he was Asst. Professor of Architecture at the Ohio State University from 1989 to 1995, and then at the University of Illinois in Chicago until 1998. From 1993 to 1997, he also held the position of the Walter Gropius Professor of Architectural Theory and Design at the Bauhaus-Universität Weimar. Since 1998, he has been Ordentlicher Universitätsprofessor of Architectural Theory at the Technical University of Vienna. Author of nine books and some eighty papers on architectural history and theory, his publications include **Genius locomotionis** (Wien: Edition Selene, 2005), **Geschichte der**

Architekturtheorie (Wien: Edition Selene, 2003), **Flying Dutchmen** (Basel: Birkhäuser, 2002), **The Use and Abuse of Paper** (Tampere: Datutop, 1999), **Form & Detail. Henry van de Veldes Bauhaus in Weimar** (Weimar: Bauhaus – Universitätsverlag, 1998), **Heimlich Maneuvres** (Weimar: Verso, 1995) and **Constructing Architecture** (Tampere : TTKK/Ars Magna, 1991). Principal of Wombat Architects since 1989, Jormakka has also realized a dozen buildings in Finland.

Poesis doctrinae tamquam somnium [p. 184]

Jormakka offers a careful exposition of the foundations in ancient Greek philosophy for the problematic of architectural representation. Architects don't make buildings, he states, they make representations of buildings. Instead of uncritically accepting naïve standpoints that new computer software programs are capable of 'accurate' and 'real' representations of architectural objects, Jormakka provides an exegesis of the Platonic Form-copy distinction that underpins all attempts at 'simulation'.

Vittorio Magnago Lampugnani

is Professor of the History of Urban Design, Head of the Network City and Landscape at the Swiss Federal Institute of Technology (ETH) in Zurich and director of an architectural practice in Milan. He has also held positions at academic institutions in Stuttgart, Berlin, Frankfurt, Milan, and Pamplona, as well as at the Universities of Columbia and Harvard. He has curated exhibitions about architecture that have been shown in Germany, France, Italy, and the US; and between 1990 and 1995 was Director of the German Architecture Museum in Frankfurt. From 1986 to 1990 he was deputy editor and from 1990 to 1995 editor of **Domus**. Many of his numerous publications have been translated into English, including **Museum Architecture in Frankfurt 1980–1990** (1990) and **Museums for a New Millennium: Concepts, Projects, Buildings** (with Angeli Sachs, 1999).

The City of Tolerant Normality [p. 294]

According to Lampugnani, the revolution in transportation systems of the 19th to the 20th centuries caused an even more radical transformation of the European city than the telecommunications revolution in the last decades. In this urban explosion, he argues for restraint, modesty, and generosity on the part of architects and planners; as opposed to the hubristic failed utopias and attention-grabbing maneuvers so commonly witnessed today. He rightly states that the discreet design of the background for living is often more difficult and complex than the large media-publicized commissions. Using Marc Augé's term 'non-places', Lampugnani pleas for an architecture of 'silence'.

Lars Lerup

is the William Ward Watkin Professor and Dean at Rice School of Architecture, at Rice University, Houston, Texas. Born in Sweden he holds degrees in Engineering (Sweden), Architecture (UC Berkeley) and Urban Design (GSD, Harvard). He is an Emeritus Professor of Architecture at the University of California, Berkeley and holds a Doctorate **Honoris Causa** in Technology from Lund University, Sweden. Lerup has written several books: **Villa Prima Facie** (1976), **Building the Unfinished** (1977) also published in German, **Planned Assaults** (1987) also published in Chinese, **After the City** (2000)

and forthcoming **Journeys in Alphabet City** (2006), and some sixty essays in international magazines such as **DU, Architese,** and **Werk:** Switzerland; **Lotus** and **Abitare:** Italy; **AD, AA Files:** England; **Places, Architecture Plus, Harvard Architectural Review, Design Book Review,** and **Assemblage:** US; **A&U:** Japan; **SKALA:** Denmark; **Magasin Tessin:** Sweden. Lerup has lectured extensively at universities in the US: Harvard, Princeton, Columbia, University of Washington, University of Miami, UCLA, Sci Arc; in Europe in Barcelona, Zurich, Berlin, Munich, Vienna, Paris, Grenoble, Milan, and Naples; and in New Zealand, Asia, South America and South Africa. **Vasa Orden,** Stockholm, Sweden elected Lerup Swedish American of 2004, and in 2005 Educator of the Year by the Houston Chapter of the AIA. Lerup was invited to give the **Megacities Lecture 2005** in Amsterdam, 2005.

The Artful Reinvention of the Suburban Landscape [p. 284]

Lerup offers a provocative and poetic reading of the Danish artist Olafur Eliasson's work in various installations worldwide. In doing so, he not only brings into **bas-relief** the so-called nature/culture divide and the opposition between object and process, but also the ability (hopefully) of art to confront, to insist, and to affirm the tensions around our common world that for other means often seem ineffectual.

Francesc Magrinyà

was educated as a Civil Engineer (1988) and received his Ph.D. in Urban Planning (ENPC, Paris, 2002). He was the Director of the Exhibit **Mostra Cerdà. Urbs i Territori** (1994). His publications include **La Ingeniería en la evolución de la urbanística** (2002). Francesc Magrinyà is professor at the department of Infrastructure and Territory of the Polytechnical University of Cataluña, Barcelona, Spain. He is a specialist in sustainable Urbanism and human settlements, as much in developed countries as in developing countries.

The General Theory of Urbanization of Ildefons Cerdà: An Urban Network Theory [p. 80]

Magrinyà provides an exposition of the contributions of Ildefons Cerdà to urban planning. Cerdà, the nineteenth century architect who was responsible for the Proposal for the Reform and the Extension of Barcelona, could be considered to be the Spanish founder of the science of urbanization. He provided a comprehensive approach for tackling the problems of the rising industrial city – an approach which took into account all the factors impacting the city – economic, political, as well as technical. His influence in the alignments grid and the chamfered corners on the street grid in the extensions to the old city of Barcelona can still be seen today as organizing the complexity with great flexibility. Because he considered mobility as a chief factor in a successful extension plan, his work in both urban design and theory of urbanism can be just as pertinent today for understanding an integrative approach to urbanism.

Michael Mehaffy

is the Managing Director of the Centre for Environmental Structure-Europe, a new research centre set up by Christopher Alexander in association with the Martin Centre, Cambridge. Previously he was Director of Education for the Prince's Foundation in London, where he established a highly-regarded new education programme of conferences and short courses. He has designed and/or built many structures, including the award-winning new light rail community Orenco Station, in Portland, Oregon. He did graduate work in architecture at the University of California, Berkeley, and in philosophy at the University of Texas at Austin. As an undergraduate he studied 20th-century music and art at California Institute of the Arts.

Codes and the Architecture of Life [p. 118]

In an attempt to address the process-oriented nature of our culture – and also our 'human settlement' patterns – Mehaffy introduces concepts from the most contemporary discussions on evolutionary biology, the processes of morphogenesis. In modernism, specifically in the scientific model developed from Newtonian physics, the complexity of the flux of life was reduced into definable and ultimately mathematically manipulable entities. However, what tools, Mehaffy asks, can now be developed in order to account for the variety of life and the complexity of process? Life is 'messy', but new advances in complexity theory, and biological networks which consider the environment as an interconnected, emergent whole, offer a starting point for the understanding of urban morphologies as morphogenetic living systems.

Han Meyer

is Professor of Theory and Methods of Urban Design, TU Delft since 2001. He graduated as an engineer of architecture, specialized in urbanism in 1979 at the TU Delft Faculty of Architecture. From 1979 to 1990 he worked as an urban planner for the City of Rotterdam. In 1997 he finished his dissertation on **City and Port – Transformations of Port Cities**. He is a member of the Scientific Committee of Europan, member of the Flemish Fund for Scientific Research and member of the Research Advisory Board of the Dutch Spatial Planning Agency (RPB). He has published books concerning the transformations of port cities, the fundamentals of the discipline of urbanism, the development of the Dutch urban block, the Dutch Watercities and the present state-of-the-art of Dutch urbanism.

Urban Compositions in the 21st Century [p. 486]

Han Meyer outlines the course of research into urbanism at TU Delft for the last twenty-five years. The Netherlands, being a land in multiple river deltas largely under sea-level, was always subject to extensive land-management policies and technologies. However, in the last decades a new approach has been developed that takes into consideration the limits of a 'top-down' approach to urban planning. A diverse and multi-disciplinary strategy has been developed incorporating the research in design disciplines of landscape, urban land-use and infrastructural networks in order to reconsider the complex relations between large-scale infrastructure, the 'natural' landscape, and the urban fabric. Consequently, a more flexible and responsive urban science has been developed at TU Delft, concerning 'the transformation of the Randstad Holland from a network of cities to a new, unprecedented and complex **networkcity** or "**Deltametropolis**".

Ákos Moravánszky

is a Professor of the Theory of Architecture at the Institut gta of ETH Zurich. He studied architecture at the Technical University in Budapest between 1969 and 1974 and then worked as an architectural designer in Budapest. From 1977 on he studied art history and historic preservation with a Herder scholarship at the Technical University in Vienna, Austria, where he received his doctorate in 1980. Between 1983 and 1986, he was Editor-in-Chief of the magazine of the Hungarian Union of Architects, '**Magyar Építőművészet**'. From 1986 until 1988 he was a Research Fellow at the Zentralinstitut für Kunstgeschichte in Munich, Germany, with an Alexander von Humboldt Scholarship. Between 1989 and 1991 he was invited to the Getty Center for History of Art and the Humanities in Santa Monica, California as a Research Associate. From 1991 until 1996 he was appointed Visiting Professor at the Massachusetts Institute of Technology (Cambridge, Massachusetts). During the academic term 2003/2004 he was appointed Visiting Professor at the University of Applied Art in Budapest, Hungary as a Szent-Györgyi Fellow. His publications include **Die Architektur der Donaumonarchie** (Budapest: Corvina 1988, Berlin: Ernst&Sohn 1988); **Die Erneuerung der Baukunst: Wege zur Moderne in Mitteleuropa** (Salzburg: Residenz 1988); **Competing Visions: Aesthetic Invention and Social Imagination in Central European Architecture** (Cambridge, Mass.: The MIT Press, 1998); **Spacepieces: Valentin Bearth & Andrea Deplazes** (Luzern: Quart, 2000), **Architekturtheorie im 20. Jahrhundert: Eine kritische Anthologie** (Vienna, New York: Springer, 2003).

Power Lines [p. 660]

Moravánszky begins his discussion of the problem of the representation of the abstract with Alajos Landau, a painter and art teacher in the late 19th century in Budapest. In painting, perspective must be constructed on a two-dimensional surface in order to represent three-dimensional space. Similarly, a mapping occurs in urban planning by projecting a logic onto the landscape and fashioning a vanishing point on the horizon. With the straight lines of modernization, specifically with the architecture of Otto Wagner in Vienna, this projection becomes a tool for representation, a process of turning theory into practice.

Anne Vernez Moudon

is Professor of Architecture, Landscape Architecture, and Urban Design and Planning at the University of Washington, Seattle. She is past-President of the International Seminar on Urban Morphology (ISUF), a Faculty Associate at the Lincoln Institute of Land Policy, in Cambridge, MA; a Fellow of the Urban Land Institute in Washington, D.C.; and a National Advisor to the Robert Wood Johnson Foundation program on Active Living Policy and Environmental Studies. Dr. Moudon holds a B.Arch. (Honors) from the University of California, Berkeley, and a Doctor ès Science from the École Polytechnique Fédérale de Lausanne, Switzerland. Her work focuses on urban form analysis, land monitoring, neighborhood and street design, and non-motorized transportation. Her current research is supported by the Washington State Department of Transportation and the Centers for Disease Control and Prevention. Her published works include **Built for Change: Neighborhood Architecture in San Francisco** (MIT Press 1986), **Public Streets for Public Use** (Columbia University Press 1991), and **Monitoring Land Supply with Geographic Information Systems** (with M. Hubner, John Wiley & Sons, 2000). She has also published several monographs, such as **Master-Planned Communities: Shaping Exurbs in the 1990's** (with B. Wiseman and K.J. Kim, distributed by the APA Bookstore, 1992) and **Urban Design: Reshaping Our Cities** (with W. Attoe,

University of Washington, College of Architecture and Urban Planning, 1995).

Field Building and Urban Morphology: A Challenge for the Next Generation of Urban Designers [p. 552]

At a time when in recent decades, the world's populations, and consequently the need for urban environments have been exponentially exploding, the need for urban design as a profession is greater than ever. Not only specific projects, but also research and analysis of the social processes affecting our cities are required. As such, Vernez Moudon argues for an end to 'the false dichotomization of design as art versus science, and of creativity versus social responsibility'; rather, a multi-disciplinary approach that incorporates design practice with a conceptual framework derived from a broader theoretical concern with socio-economic factors. That is to say, urban design as a field of both practice and research. As a possible methodology, Vernez Moudon suggests the use of 'urban morphologies' in close incorporation with the detail data that are now available through (GIS) Geographic Information Systems.

Michael Müller

is a Professor and Chair for art history and cultural studies at Bremen University. His most important books are: **Die Villa als Herrschaftsarchitektur** (1970); **Die Verdrängung des Ornaments** (1977); **Funktionalität und Moderne** (1984**)**; **Architektur und Avantgarde** (1984); **Schöner Schein** (1987); **Die Macht der Schönheit**.(1995); **Die Straße** (1998); **Die ausgestellte Stadt. Zur Differenz von Ort und Raum** (2005).

Order on Display [p. 424]

For Michael Müller, our cities are only intelligible to us by creating an order that can be seen not only as a social expression of the public, but a collecting of culture as well as an arranging of objects in space. In fact, the very singularity of a place is determined by its ability to order and to objectify. This 'compulsion to arrange' attains its highest abstraction and penultimate historical determination in the arrangement of objects in a museum space. The museum, the ideal of gathering and displaying of knowledge for the edification of mankind, is complicit in this 'compulsion'. Uniting and Showing, the museum is firmly embedded in the theoretical underpinnings of the Age of Enlightenment, yet just as complicit in the acts of consuming, commodification, and accumulating as periodicals, advertisements, and cultural manifestations of our contemporary cities. Through the examples of the exhibition '**Wunderkammer des Abendlandes**' of 1994, and the World Exhibition of 1851 in London, Müller shows that the city itself has become a collection of objects, a system of meaning, a storage facility, and indeed ultimately a consumable commodity.

Kas Oosterhuis

was educated as an architect at the TU Delft, and is since 2000 Professor in Practice at that university. From 1987 to 1989 he was a Unit Master at the AA in London, and in 1994 was Founder and chairman of the Attila Foundation Rotterdam. The **Hyperbody Research Group** [HRG] at the Faculty of Architecture at the Delft University of Technology is directed by him. He is also director of the internationally renowned architectural office ONL [Oosterhuis_Lénárd]. The aim of the HRG is to study interactivity in architecture and to develop practical applications. Programmable buildings will illustrate the paradigm shift from animation towards realtime behavior. Hyperbodies are pro-active building bodies acting in a changing environment.

A New Kind of Building [p. 240]

Whereas traditional building historically relied upon the 'master builder' in a symbiotic relationship between 'design' and 'construct', with modernity and the rise of the profession of the architect, the process of building was specialized into various intermediate phases, each requiring complex strategies of communication. With the groundbreaking research of Kas Oosterhuis, digital techniques are at once returning to a direct 'peer-to-peer' form of communication in the construction process, and at the same time radicalizing production and the execution of building components and procedures. With his architectural firm ONL and his research group **Hyperbody**, at the Delft University of Technology, Oosterhuis proposes a new building paradigm, 'mass-customization' using computer software that he has adapted from gaming. In this way, his 'constructs' truly become 'interactive', both in the sense of the buildings' interaction with an environment, interdisciplinary communication, and with the interface of building and user. People literally effect the constitution of the built environment. Subsequently, Oosterhuis provides evocative and compelling projects that shake the foundations of architecture as a static, eternal, or artificial construct. Or, as he says: 'we can start looking at the world from yet another level'.

Scott Page and Brian Phillips

are the Principals and co-founders of Interface Studio LLC, a collaborative design office focused on the intersection of urban design, architecture and emerging information technology. Page lectures at the University of Pennsylvania's Graduate School of Design, and Phillips is an Adjunct Assistant Professor at Temple University's Tyler School of Art. Interface Studio is based in Philadelphia.

Urban Interfaces: Designing In-Between [p. 60]

Traditionally, urban design was form-based and site-specific, rooted in physical proximity, and defined in time only in a limited way in terms of the logistics of transportation systems. However, in the twenty-first century, the city can no longer be seen as an isolated entity. Multiple layers of dynamic systems interact with the city: networks of communication, global economics, transportation infrastructures, information technologies, virtual/digital spatial experiences. Page and Phillips argue for a concept of 'urban interface' that would take into account not only the place-specific aspects of traditional urban design models, but also fold in the dynamic systems that obviously connect the city in multiple layers, scales and densities of information. In the end, Page and Phillips break down the problem into three areas of concern: questions of convergence, questions of interaction, and questions of place. Undoubtedly, with the new era of globalization and networked communication, the city or urban environment must embrace multiple descriptions and entail a variety of tools.

Alberto Pérez-Gómez

was born in Mexico City in 1949, where he studied architecture and practiced. He did postgraduate work at Cornell University, and was awarded an M.A. and a PhD by the University of Essex (England). He has taught at universities in Mexico, Houston, Syracuse, Toronto, and at London's Architectural Association. In 1983 he became Director of Carleton University's School of Architecture. He has lectured extensively around the world and is the author of numerous articles published in major periodicals and books. He is also co-editor of a well-known series of books entitled **CHORA: Intervals in the Philosophy of Architecture**. In January 1987 he was appointed Bronfman Professor of Architectural History at McGill University, where he chairs the History and Theory division, and is currently Director of Post-Professional (Master's and Doctoral) Programs. His book **Architecture and the Crisis of Modern Science** (MIT Press, 1983) won the Hitchcock Award in 1984. Later books include the erotic narrative theory **Polyphilo or The Dark Forest Revisited** (1992), and more recently (with Louise Pelletier) **Architectural Representation and the Perspective Hinge** (1997), which traces the history and theory of modern European architectural representation. At present, Dr. Pérez-Gómez is engaged in a project tracing the points of convergence between ethics and poetics in architectural history and philosophy.

Ethics and Poetics in Architectural Praxis [p. 676]
Architecture endures, according to Pérez-Gómez, in a poetic manner, absolutely essential for the understanding of what it means to be human historically and culturally. Although he acknowledges that utopian ideals are futile, architecture is a proposition of the future, a project innately optimistic and ethical, calling upon the imagination and poetic responses capable of acting in the present. Architecture is an **event**, Pérez-Gómez emphatically states, an experience beyond words, yet still necessarily ethical in that architecture proposes a 'better future for a polity'.

Stephen Read

is an urbanist and teacher at the Delft University of Technology. He obtained his PhD from Delft in 1996 and completed a fellowship at The Bartlett before starting the Spacelab Research Laboratory of the Contemporary City in 2002. He is author of numerous papers on urban formation and transformation, and of the forthcoming **Urban Life: A Design Guide** (Techne Press, 2006). He is editor of and contributor to **Future City** (Spon Publishers, 2005), and of **Visualizing the Invisible: Towards an Urban Space** (Techne Press, 2006).

Productive Space [p. 338]
Stephen Read proposes that urban space is not only produced, but becomes also itself productive by way of '**difference**' and of '**interface**'. He puts forward a dynamic model of interaction in the urban surface as a way of suggesting how movement, change, and growth may be linked by urban spatio-temporal processes. In contrast to social theoreticians of the post-war period who tended to look upon the city as dialectical, Read emphasizes the role of site and situatedness in conjunction with 'convergent' productive dynamics in setting up what he calls a 'choreography of place'. The micro-social and micro-economic of urban situation in this view becomes the direct product of processes of movement and connectivity. The city is 'made' by us; and then literally turns back upon us, Read says, and 'makes' us situated in return. As an alternative to the dialectic of the city, Read proposes a generative friction productive of everyday social situation in the interfaces of distinct movements in the urban surface.

Wouter Reh

is a landscape architect and has been since 1994 an Associate Professor at the Architecture Faculty of the TU Delft. He has published widely on topics pertaining to landscape architecture, urban design and planning. His dissertation was entitled **Arcadia en Metropolis** (1996) which treated the anatomy of pictorial landscape design such as it developed in the English Romantic garden. In 2003, he published with Steenbergen, **Architectuur en Landschap. Het ontwerpexperiment van de klassieke Europese tuinen en landschappen**, and together with Diederik Aten in 2005, **Zee van Land**. He is presently working with Clemens Steenbergen on a volume of design studies of city parks and urban landscapes provisionally entitled **Urban Landscapes**.

The Polder as a Design Atlas [p. 530]
The Dutch Polder Landscape, that is to say reclaimed land in primarily the Western part of the Netherlands, not only originally provided a logical scheme for the division of agricultural plots and manageable drainage areas, but now also in the era of urban expansion, provides a compositional system that is both flexible and effective. This scheme Steenbergen and Reh term a 'polder grammar'. In this paper are analyses of the theory and technologies employed in three different areas: Watergraafsmeer, Purmer, and Haarlemmermeer. The polder has always been an artifice – a perfect confluence of the natural landscape and 'culture' in the sense of cultivation. Now the polder and its systematic and rational division encompasses once again both 'natural features' and the accommodation necessary between nature, technology and art whilst at the same time preserving the heritage of the Dutch flatlands.

Nikos A. Salingaros

is professor of mathematics at the University of Texas at San Antonio and the author of three books: **Principles of urban structure** (Techne Press, Amsterdam, 2005), **Anti-architecture and Deconstruction** (Umbau-Verlag, Solingen, 2004), and **A Theory of Architecture** (forthcoming from Umbau-Verlag). Dr. Salingaros is a member of INTBAU College of Traditional Practitioners, and is Vice-President of Katarxis Urban Workshops. For his contributions to architecture, he was elected member of the International College of Traditional Practitioners, whose patron is His Royal Highness Charles, The Prince of Wales. Dr. Salingaros is directing student theses at universities around the world, is involved as a consultant on architectural and urban projects, and is a visiting member at the architecture faculties of several universities.

Compact City Replaces Sprawl [p. 100]
Nikos Salingaros provides anti-sprawl polemic. 'Sprawl is a remorseless phenomenon', he says, and it could be said to be just as endemic. Sprawl is motivated by the logic of the automobile and is commercially driven. Pairing himself on the side of Christopher Alexander and Andrés Duany, Salingaros proposes a compact and coherent city integrated into a system of infrastructure, as opposed to either suburban sprawl, or a high density, high-rise urbanism. Ultimately, he argues for a human-oriented urban environment, and who, indeed, could argue with that?

Kelly Shannon

is an Associate Professor at the Department of Architecture, Urban Design and Planning at KU Leuven, Belgium. She holds a Bachelor of Architecture from Carnegie-Mellon University [USA], a Master of Architecture from the Berlage Institute [Netherlands] and a doctorate from KU Leuven [Belgium]. The paper in this volume is taken from her doctorate **Rhetorics and Realities, Addressing Landscape Urbanism, Three Cities in Vietnam.**

Concluding With Landscape Urbanism Strategies [p. 568]

Shannon explicates the conclusions of her extensive doctoral research on issues of urbanization and development in the region of Vietnam. She uses strategies of 'landscape urbanism' in order to generate a 'layered narrative' of the specificity of three Vietnam cities, a narrative that belies not only the urban/rural/ecological divide, but also questions the usefulness of urban theories generated in the West and their pertinence to the situation in Southeast Asia. 'Urban' and 'rural' as defined in traditional urbanism, are inapplicable to the 'hybrid' condition where a simultaneity of conditions coexist in a kind of patchwork. Shannon proposes that case studies derived from areas outside of Europe and North America – multiple world-view narratives – can be used to contribute to a richer and more expansive theory of landscape urbanism.

Clemens M. Steenbergen

is a landscape architect and since 1993 a Professor of Landscape Architecture at the Faculty of Architecture, TU Delft. He has published various books about architecture and landscape architecture, including **Italian Villas and Gardens** (1992). His dissertation of 1990, entitled **De stap over de horizon**, was a full-length study and analysis on the rational design of the Italian Renaissance villa and the formal French Baroque Garden.

The Polder as a Design Atlas [p. 530]

The Dutch Polder Landscape, that is to say reclaimed land in primarily the Western part of the Netherlands, not only originally provided a logical scheme for the division of agricultural plots and manageable drainage areas, but now also in the era of urban expansion, provides a compositional system that is both flexible and effective. This scheme Steenbergen and Reh term a 'polder grammar'. In this paper are analyses of the theory and technologies employed in three different areas: Watergraafsmeer, Purmer, and Haarlemmermeer. The polder has always been an artifice – a perfect confluence of the natural landscape and 'culture' in the sense of cultivation. Now the polder and its systematic and rational division encompasses once again both 'natural features' and the accommodation necessary between nature, technology and art whilst at the same time preserving the heritage of the Dutch flatlands.

Roemer van Toorn

is Head of Projective Theory at the Berlage Institute, Rotterdam and Ph.D. researcher at Delft University of Technology (Berlage Chair). Roemer van Toorn is an architect, critic, photographer, and curator. After graduating from Delft University of Technology in architecture, he published **The Invisible in Architecture** (1994) in collaboration with Ole Bouman. He has been several times the editor of the annual publication **Architecture in the Netherlands**, as well as an advisor of the magazines **Archis** and **Domus**. As author and photographer he also contributes to many other publications. His photography work has been exhibited in Winnipeg, Los Angeles and was part of the exhibition **Cities on the Move** curated by Hou Hanru and Hans-Ulrich Obrist. In October 2004 his photos on the **Society of The And** were exhibited at **Archilab 2004: The Naked City** curated by Bart Lootsma. Currently he is working on a publication as part of his Ph.D research **From Fresh Conservatism to Radical Democracy. Aesthetics as Form of Politics.** Forthcoming is his English/French photobook the **Society of The And**, which includes his photographs and articles supplemented by essays from Stefano Boeri and Bart Lootsma.

After Criticality: The Passion for Extreme Reality in Recent Architecture ... and Its Limitations [p. 170]

The critical philosophy born out of the dissolution of the Second World War, influenced by European Marxism, is no longer seen in some circles of architectural theory as particularly relevant anymore to a world subsumed in global capitalism. The 'new-garde', among whose proponents are Bob Somol, Sarah Whiting, and Roemer van Toorn, argue instead for what they are calling 'projective practice' – an attempt to intercede in a specific context without adjunct ideological apparatuses. Roemer van Toorn, in his essay entitled 'After Criticality', calls for a radical break with the theoretical positions of deconstructivism and 'criticality', and longs for a return to the practice of architecture as a constructive discipline that in itself is inherently critical. Critical theory was 'backward looking'; projective practice 'looks forward' to the possibilities without getting paralysed by aporias. In borrowing Ulrich Beck's idea of a 'second' or 'reflexive' modernity, van Toorn characterizes three types of 'projective practice': 'projective autonomy', 'projective mise-en-scène', and 'projective naturalization'.

Philip Ursprung

is since 2005 Professor for Modern and Contemporary Art at the University of Zurich. He studied art history, history, and German literature in Geneva, Vienna, and Berlin, receiving a Ph.D. from the Freie Universität Berlin in 1993. He wrote his Habilitation at the Swiss Federal Institute of Technology Zurich (ETH Zurich) in 1999. Ursprung has taught art history at the Université de Genève (1992–93), worked with Kurt W. Forster at the Institute for the History and Theory of Architecture at ETH Zurich (1993–99), and served as visiting professor of art history at the Hochschule der Künste Berlin (1999–2001), the University of Basle (2001–02) and the University of Zurich (2002–03). From 2001 to 2005 he was Science Foundation Professor for Art History in the Department of Architecture at ETH Zurich. He was a curator at the Kunsthalle Palazzo, Liestal, Switzerland, from 1990 to 1996. He was a visiting curator at the Museum of Contemporary Art in Basle in 1999. At the Canadian Centre for Architecture in Montréal he curated the exhibition **Herzog & de Meuron: Archéologie de l'imaginaire** (2002–03) and edited the catalogue **Herzog & de Meuron: Natural History** (Baden: Lars Müller, 2002). At the CCA he was co-curator of the exhibition **Out of the Box: Price, Rossi, Stirling, Matta-Clark** (2003–2004). His publications include: **Images: A Picture Book of Architecture** (with Ilka & Andreas Ruby) (Munich: Prestel, 2004), **Pictures of Architecture, Architecture of Pictures, a Conversation by Jacques Herzog and Jeff Wall, moderated by Philip Ursprung** (Vienna: Springer, 2004), **Minimal Architecture** (with Ilka & Andreas Ruby, Angeli Sachs) (Munich: Prestel, 2003), **Grenzen der Kunst: Allan Kaprow und das Happening, Robert Smithson und die Land Art** (Munich: Silke Schreiber, 2003), and **Herzog & de Meuron: Natural History** (editor), exhibition catalogue, Canadian Centre for Architecture, Montreal (Baden, Lars Müller Publishers, 2002) (2nd edition, 2005).

Built Images: The Eberswalde Library by Herzog & de Meuron [p. 154]

In 1999 Kurt Forster, who at the time was the director of the Canadian Centre for Architecture in Montreal, asked Phillip Ursprung to curate an exhibition on Herzog & de Meuron. Ursprung was, however, dubious about the possibility of exhibiting architecture at all — the great paradox being that architecture is a three-dimensional object that is only really experienced in the present historicity and its context, not in photographs and drawings. Yet the experience of Herzog & de Meuron's Eberswalde Library changed his mind. In his essay, Ursprung traces the development of his exhibition concept whilst providing at the same time a critique of the Library, and Herzog & de Meuron's work in general. In the end, the exhibition was an innovative look at the possibility of exhibiting architecture: a 'direct confrontation' with a collection of study models, experiments with materials, and collaborations in progress.

Jan Vogelij

is President of the European Council of Town Planners and Director of Spatial Development at Royal Haskoning. He was educated as a town planner at Delft University of Technology and graduated in 1971. Since then, he has worked as a planning consultant at Zandvoort Urban and Regional Planning Consultants, in the Netherlands as well in several developing countries, where he has been general manager since 1984. The firm merged in 1997 with Royal Haskoning. Vogelij was the chairman of the Netherlands Association of Town Planners, (BNS) from 1989 to 1995, and is presently President of the ECTP, the Association of European Spatial Planners.

The New Charter of Athens 2003: The European Council of Town Planners Vision on the European City of the 21st Century [p. 46]

Vogelij in his contribution to this book, outlines the precepts of The New Charter of Athens 2003 adopted in November 2003 by the European Council of Town Planners, ECTP. Several factors had changed dramatically since the previous Charter of Athens of 1933–42; namely, improved environmental and manufacturing facilities, a service-based economy, concerns for sustainable development, a re-interpretation of participatory democracy, and finally a redistribution demographically in the European population. As such, with the New Charter of Athens 2003, the forward-looking vision of **The Connected City** was proposed, addressing three components of sustainable development: social connectivity, economic connectivity and ecological connectivity. Within this vision, spatial planners may have different roles in the process: as scientists, as designers and visionaries, as political advisors and mediators, and/or as urban managers.

Karel Vollers

received his doctorate in 2001 at the TU Delft with his dissertation 'Twist & Build' on the urban and architectural design of twisted buildings. He accelerated further studies as the post-doctorate head of the 'Blob' research group of the Chair of Product Development within the department of Building Technology, TU Delft.

Bending and Folding Glass [p. 270]

Vollers explains the technology necessary for the execution of complex double-curved glass panels. Innovative design pushes the technology available, and simultaneously, producers respond by developing the processes required for the manufacture of glass surfaces of a complex geometry. Iterative processes are key — both in order to reduce costs-per-unit, and in order to develop the market. The result is a transformation in glass technology, as well as groundbreaking new façade systems.

Frank R. Werner

received his education in architecture at the University of Stuttgart in 1972. Ten years later he became a teacher of building history at the academy of arts in Stuttgart. Since 1994 he is in charge of the Institute for Environmental Organization/Studies at the University of Wuppertal. In 1999 he became Dean of the Architecture Faculty of the University of Wuppertal. He is one of the founding members of the European research group **CoRa** for architectural theory and architectural history. Since 2003 he has been associated with the PhD program of the TU Delft and the Berlage Institute for architectural theory. He has published extensively in periodicals and books.

The New Simplicity: A Problem of Representation in Architecture and Town-Planning? [p. 140]

Werner attempts to bring some nuance into increasingly emergent calls for a 'new simplicity' in architecture. Certainly, this call is understandable in light of the recent decades of exuberant, singular, iconographic, set-piece architecture. Yet to build simply, indeed, to design simply, is in Werner's view neither simple nor facile. 'Simplicity' can be approached either historically/typographically, materially, or programmatically. Therefore, Werner identifies four different modes of influence: the autistic individual building, the 'dead city', the city as the normal case, and the unbroken power of the line. Nevertheless, both in theory and in practice, 'simple building' is an extremely complex undertaking.

Introduction

Arie Graafland, director Delft School of Design

The Delft School of Design (DSD) was instituted in the beginning of 2003 as a laboratory for emerging research and experimentation concerning architecture, urbanism and technologies of construction. The DSD is located at the host institution of the Delft University of Technology, Faculty of Architecture in the Netherlands, and provides training for the doctoral students. Traditionally the Delft Architecture School has developed methodologies primarily grounded in plan analysis, building typologies, construction technologies, and historical research. Yet these approaches are coming more and more under pressure due to the complexities of the contemporary social and technical conditions and their mutual relations.

As a consequence, we began the Delft School of Design, laying out a field for research in the Netherlands, yet encompassing a wide variety of geographical areas – from Tokyo to Mexico City – as an object of study that various PhD candidates and their professors associated with the DSD engaged critically. Two of the longer running projects so far have been *Mapping Urban Complexities* on Tokyo and Shanghai, and the *Space Fighter* project we developed with the Dutch firm MVRDV. At the moment the *Ranking Randstad* project and *Digital Technologies* are under way. As such, the DSD attempts to find new methodologies, evocative ways of doing historical analysis, bridging the gap between analysis and 'projective practice'. This volume, *Crossover*, is the inaugural volume of scholarly research between architects, urbanists, and structural designers either at the DSD, or closely associated with the DSD. These investigations range from theoretical considerations, to historical studies, to urban and architectural practices, to contemporary structural designs. What they all have in common is the emerging 'condition' of architectural knowledge in both the academic context and professional practice.

Inspiration for the DSD came from similar initiatives in the Humanities, one in particular – the Amsterdam School for Cultural Analysis, Theory and Criticism (ASCA), a multi-disciplinary research institute at the University of Amsterdam. Recognizing that the traditional borders defining academic disciplines since the nineteenth century no longer reflected the demands of contemporary scholarship, Mieke Bal and Hent de Vries, both internationally known professors, began ASCA in 1995, with scholars from comparative literature, philosophy, visual culture, religious studies, and film and television studies. At the beginning, there was only a commitment to finding a new methodology. How to actually do cultural analysis was still to be discovered in the doing. Broadly speaking, 'culture' was the object to be studied, and ASCA was organized upon the Salon system, informal meetings bringing scholars from disparate backgrounds together, in order to address a single theme. Fruitful interactions were the result, whilst each scholar brought specific expertise; the whole was greater than the sum of the parts. For us in a technical context of laboratories and other facilities, a Salon might look like a holiday resort. But a closer look in the faculties dealing with architecture, urbanism, industrial design and management, shows that debate and assessment are central to the argument of how and why we should develop new products. Also the majority of the PhD dissertations are dealing with notions that directly relate to society as a whole. No architecture, urbanism, industrial product or management decision is possible without society as its central focus. It was here where I implemented the notion of 'Salon' as a platform for critical assessment into the DSD.

The DSD brings together architects, historians, theoreticians, urbanists, and structural designers. Whilst each has detailed and extensive specialized knowledge in their chosen field of endeavor, the goal is to organize seminars, meetings and debates around various themes found to be mutually provocative. These encounters, whilst respecting the diverse disciplines coming to the table, are productive and generative for fresh approaches to the complex sets of problems engaging professionals, academics, and emerging doctoral scholars today. The DSD provides a platform for the exploration of interconnected frameworks of knowledge, detailed problem analysis, and technical methodologies, both within the Netherlands and in collaboration with other excellent institutions. As such, at its best moments, the DSD is a laboratory for emerging developments, free experiments, and generative approaches. Any attempt that is yet to be discovered is bound not always to succeed. Nevertheless, the most important goal is indeed *the doing*: how to develop new tools to disentangle the problem, how to come to grips with new methods, and how to do critical design theory in the advent of 'global capitalism'. Yet what remains the common ground is, indeed, a commitment to advance the scholarship, a practice that engages (whether tectonic or theoretical), and shared themes worthy of sustained research. This impulse was the original vision of the DSD.

Consequently, the DSD developed a strategy or a field of inquiry in order to map out new means of approaching the complexity of the contemporary urban and architectural conditions and their mutual relations. The current means are getting more and more inadequate to the present complexity in three ways:
- The complexity of the object of study
- The complexity of the condition of knowledge
- The complexity of the necessary methodology

Firstly the traditional approaches are seen to be increasingly inadequate due to the complexities of the object of study. For architects, historians, theoreticians, urbanists, and structural designers, the object of study – whether at a global scale or an intimate construction detail scale – is the relation between their newly developed insights in their specific domains, and the relation to society, or 'client' if you wish. This complexity is in part due to the nature of the information age – we simply know more and are confronted with difficulties as never before. Yet, on the other hand, precisely those same technologies offer us the challenge and possibility to represent our world in unprecedented ways. Our design instruments will need a serious *mutation* to understand even what's going on. For example, in the most recent Report of the United Nations Human Settlements Programme, *The State of the World's Cities 2004/2005*, we find a sketch of pervasive and persistent urban problems. Problems that include growing poverty in many regions, deepening inequality and polarization, widespread corruption, high levels of crime and violence, and deteriorating living conditions with inadequate sanitation, unsafe water, etc. Nonetheless, cities also function as engines of economic growth and an examination of promising practices around the world show examples of low-income communities who mobilize successfully, to improve difficult situations.

In describing the characteristics of today's urban world, the UN Report describes the specific character of globalization. At least three out of the four are well known. Global connections function at a much greater *speed* than ever before, globalization operates at a much *larger scale*, leaving few people unaffected, and third, *the scope* of global connections is much broader, it has multiple dimensions such as economic, technological, political, legal, social and cultural dimensions. Especially the later mentioned characteristic has our attention. The dynamic and often unmediated interactions among numerous global actors create *a new level of complexity* for the relationships between policy, research and design practice. A general trend, especially in East Asia, is the 'informalization' of urban economy, with increasing shares of income earned in unregulated employment. This informalization of labor even occurs in one of the most advanced countries of the Asian world, Japan. Interestingly enough, the UN Report also notes that we need better and more advanced ways to analyze and to describe our contemporary cities. A recent assessment of United States cities put it this way: 'It is becoming a commonplace that established representations of the city and the suburbs do not hold. Our capacity to describe or theorize the social and spatial organization of the contemporary metropolis is manifestly inadequate to what we know of the metropolitan experience.'

One of the major instruments in this field is of course a map. Maps today come in all kinds of varieties. In our present-day situation with GIS systems, a lot has changed. We now measure the earth from space satellites which gives us more precise knowledge of the globe.

But architects and urbanists are no geodesists. They use maps for other purposes. Nor is our perceptual apparatus completely neutral to its observations. Donna Haraway, the American biologist and philosopher writes that the 'eyes' made available in modern technological sciences shatter any idea of passive vision; these prosthetic devices show us that all eyes, including our own organic ones, are *active* perceptual systems, building in translations and specific ways of seeing. There is no unmediated photograph or passive *camera obscura* in scientific accounts of bodies and machines.

What kind of map of the world are we talking about today? What we are confronted with are innumerable seemingly incomplete maps of the world drawn from different space-time 'positions'. Spatiality simultaneously unifies and separates. David Harvey has argued that the multiple windows on the same reality, like the multiple theorizations available to us, can constitute a way of triangulating in on this same reality from multiple perspectives. The technique of conjoining information from different positionalities is a basic principle of all cartographic construction: to make an *accurate* map (representation) of the world we require at the very minimum a procedure of triangulation that moves across multiple points.

Secondly this problem of the object of study in fact leads us to the second problem, the complexity of the condition of architectural and urban knowledge. Now, some twenty years after Jean-François Lyotard's book *La Condition Postmoderne, rapport sur le savoir*, the DSD has redefined its design research in relation to other technical faculties. The central theme of this research is very much in line with Lyotard's notion of the Postmodern Condition, specifically in the context of architecture and urbanism – the '*condition*' of the *means of design* in relation to their effectiveness and social legitimacy. Teachers in many architecture faculties are asking themselves the same questions: 'What am I teaching my students, and why is one solution better than the other?' Obviously it is one of the questions that is most imperative to ask about architectural and urban research and teaching in Delft. Of course, there are always the architectural and urban examples, the standpoints of architects and urban designers, the diverse claims emanating from the different disciplines that are involved, and the legitimizations of the students. But whoever bothers to look beyond this façade will find a great deal of uncertainty. It is fine to discuss examples, but who selects them? Which examples are pertinent and which are not? Only the high profile names?

These questions bring us to the following series of questions: Which instruments are actually valid for use in analyzing a design, and which ones have become obsolete? In other words, can plan analysis, typology, composition, semiotics, linguistics, philosophy, social theory, hermeneutics or phenomenology still help us discover the value of a design? Or are we basically in the position of a postmodern universe where all these disciplines are used to dress up design? We have to realize that most design questions formulated in the Technical Universities have to do with the '*how*' question, the *question of how to make it*. The other question is however the '*why*' question – for more than a century, *the* issue in the social sciences. Since architecture and urbanism are *social constructions of reality*, they have to deal with the '*why*' question also. 'Why' is this design so important, 'why' is it any better than the other one?

In contrast to our other technical faculties, Architecture Faculties are dealing with society *right from the start* – we are *not* only discussing cause and effect here (as might be the case in technical sciences like physics). We deal with a *material concept that has a double connotation, both technical and social*. *Technical* in the sense that materials and procedures of architecture and urbanism themselves form a rich cultural matrix, capable of sustaining dense intellectual argument without much recourse to concepts and language borrowed from other fields; *social* because we are dealing with the social construction of reality. This nexus of the technical and the social might be a '*declaration of position*' for the DSD.

At the various faculties of Architecture, the criteria for answering these questions are not only technical; they are also – in fact primarily – societal. The Faculty of Architecture at the TU Delft distinguishes itself from the rest of the faculties in this regard. We are not suggesting that social relevance is not also important at the other faculties, but perhaps only as an *effect*, not from the outset as 'taste', 'value', 'quality' or 'beauty'; they all inform and form the design. The design process takes place in an open and dynamic system, one which is in continuous flux: that system is society itself. There are only limited opportunities to 'isolate' the system like in technical laboratories, and that is only possible if design takes the changes and dynamics of society into account. The conclusion might be that *architecture and urbanism are, in their first instances, material practices*. Practice here can also mean theoretical practice.

The DSD defines itself, of course, as a scientific or scholarly practice. The construction of the natural sciences, in contrast, aspires to universal application. Buildings, plans, management decisions and industrial products have a far more unique character however. The conditions with which the architect and urban planner and industrial designer have to work, come from *the outside*; programs are determined beyond the control of the individual. Or, as Stan Allen has put it: the practice of architecture tends to be messy, and inconsistent, precisely because it has to negotiate a reality that is itself messy and inconsistent. And maybe, it is better to drop the conventional theory/practice distinction, as Allen has put it, and distinguish broadly between *practices* that are *primarily hermeneutic* – that is devoted to interpretation and analysis of representations, and *material practices* like urbanism, architecture and industrial design. The vector of analysis in *hermeneutic* practices always *points toward the past*, whereas *material practices* analyse the present in order to *project transformations into the future*. In this sense too, the DSD distinguishes itself from the other faculties, where the function of the 'laboratory' is fundamentally different.

The DSD *does* support the idea that a fruitful approach is possible by using concepts developed from cultural studies and reflective design practices. We do not wish to be entangled in pointless arguments about the dichotomies relating to humanities and technical sciences. We believe we have learned from the German *Positivismuss Streit;* we have learned from Kuhn's *paradigms*. We certainly do not adhere to any form of scientific 'mimicry', we do not borrow 'method' as used in natural sciences (in an isolated and repeatable laboratory context) to fulfill tasks in a dynamic social cultural context. We would like to state clearly that we see architectural and urban design as a complicated process of analysis of a large and increasing range of requirements, social and cultural as well as economic and technological, the outcome of which can be understood through a precise study, but which cannot be predetermined by whatever stable method given in advance. Nor have I ever discovered any architectural office in the Netherlands that uses 'method' as in the sciences in its daily practice. Our position of course is a *difficult one* in a Technical University. This 'messiness' will be one of the *risks* of our research and teaching program. Unfortunately, the constant threat of *'utilization'* parameters and the illusions to find fully objective methods which can be instrumental in design activities defined by technical laboratory research is the impending real threat. Some might even consider this ambiguity a *'weakness'* in our program. However, we believe the DSD is on the right track. We actually believe it is our *strength* not to address 'technical mimicry', indeed, not to address questions of instrumentality *alone*.

Thirdly the problem of the object of study and the problem of the condition of architectural and urban knowledge, leads us inevitably to address the complexity of the necessary *methodology* in order to grapple with the first two complexities. Initially, what we must do is look at the present-day curricula at other (Technical) Universities and assess how architectural and urban thinking is going to cope with its technical contexts. What we need is to analyze our current social and technical condition and the requirements arising from this. But that is not possible without also characterizing the problems connected with the theories of, for instance, Manfredo Tafuri and the present school of ideology critique of, for instance, Fredric Jameson. This case is especially true where it concerns the consequences of these analyses and the specific situation of design training. The faculties of Architecture, as I see it, are acutely affected in the seeming antithesis of social and political sciences and their future oriented design work. Apart from the fact that social sciences can maintain a critical position with regard to their social object, architecture faculties are at the same time *constructing*, that is, designing and future oriented. The most important characteristic of critical analysis is that it is retrospective – historical analysis explains why something happened the way it did. In order to achieve such an analysis, the narrative must be shut down. The German philosopher Hegel wrote that the owl of Minerva, representing our knowledge, takes its flight at dusk. In fact it always arrives too late – certainly for the planner and the architect who are left standing empty-handed. Through the rigorous separation of historical critique and design practice as witnessed in, say, Manfredo Tafuri's *Architecture and Utopia*, the theory is saved, but the planner or designer is left with nothing. Tafuri's approach in *Theories and History of Architecture*, where he writes that one cannot 'anticipate' class architecture (an architecture for the liberated society), also had a crippling impact. He believed that the only recourse was to introduce 'class criticism into architecture'. Fredric Jameson's critique of this mode of thinking, which he relates to Theodor Adorno's 'negative dialectic' and Roland Barthes' 'degree zero of writing', illuminates at least a number of problems of such analyses. This rigorous analytical history must in turn be bought by the stoic renunciation of action and value, according to Jameson. As I see it, that inspires Jameson to the disputable assumption that the position of Tafuri and that of the postmodern is the same – the idea that in fact there is nothing to be done, that no fundamental change is possible anymore within global capitalism. One leads to a self-conscious stoicism, the other to a post-modern relax, as he puts it.

But we do find ourselves in the era of postmodernism. Let me give you an example that is close by. The situation within architectural training is complex. The competition of the various social, political and aesthetic ideas – following Lyotard I call these *types of discourse* (*genres de discours*) – is large. All these species of discourse are essentially heterogeneous – one cannot replace the other. The departments, faculty bodies, and boards who must administer this field, apportion it and pay for it from fixed amounts of money, yet find it harder and harder to make balanced decisions. Whereas the Beaux Arts, or Russian Vkhutemas could still have a dominant idea about the profession, this is no longer possible today. There is no general language which leads to control of these different so called 'language games' and the financial claims attached to them. There is no meta-rule, no simple, ultimate principle like 'quality of training', no recent purposeful discourse policy for the final decision makers, no functionalist maxim of 'form follows function' with which to encompass them. Internal conflicts can still be settled, but not the external ones between the various systems. In Technical Universities the idea of a simple machine technology is no longer decisive; instead technical universities deal with a knowledge technology in the sense of symbol interpretation. The actuality of this situation – the heterogeneity of the discourses – precludes a harmonious arrangement. While this condition has been analyzed many times, one striking example comes to mind. In the 1985 exhibition *Les Immateriaux* in the Centre

Pompidou in Paris, thirty authors were asked to communicate with one another using personal computers connected to a central memory. The various writers, philosophers, physicists, biologists, and linguists were not only asked for their own contributions, but to react to the contributions of others. Central to this event was the question of the influence of modern technology on changes in writing and thinking. The results were utterly worthless. All the authors shielded their own discourse and there was no communication at all.

To conclude the current design means are inadequate to contemporary complexity in three ways: the complexity of the object of study, the complexity of the condition of architectural and urban knowledge, and the complexity of the necessary methodology. In the DSD, architectural, urban and constructional understanding is not only developed on a theoretical level, but also questions the status of this knowledge itself, as well as attempting to develop the necessary tools with which to even speak about this body of knowledge. Such is the 'condition' of architecture, urbanism, and structural design. As a laboratory for emerging research and experimentation, the DSD is not only interested in understanding the world, but also in *improving* it by concrete interventions, reconciling theory with practice. This vision is the birthright of the Delft School of Design.

Culture Scape Holland

Photos: Jeroen Musch

Section 1:

New Instruments of Design — A Matter of Network Thinking

Talking about 'new' instruments of design implies the existence of 'old' instruments, cherished by practitioners, teachers, and students of urbanism, including architects acting as urban designers.

A New Charter of Athens has been published in 2003, seventy years after its predecessor. The Athens Charter of 1933 had provided the dogma of the mainstream until today, so-called zonal urbanism. Over the years, important changes have occurred in every conceivable domain, changes that are challenging the future of cities. However, the dogma provided by the Athens Charter of 1933 cannot cope with these changes because it is based on two false premises:

A it is desirable to concentrate functions in giant packages;
B the geometry within each package is homogeneous.

However, a city contains so many complex functions that it is impossible to isolate them, let alone concentrate them, so that imposing a simple geometry on urban form inhibits the human activities that generate living cities.[1]

Moreover, giant homogeneous packages – combined with automobile dependence – underlie today's by-and-large unsolved, urban mobility problem. As to the adoption of 'new' instruments of design, one may take a pessimistic view following Dorothy Parker: 'One cannot teach old dogma new tricks.' Take for example the Les Halles design competition in Paris in 2004. The competition was meant to repair the errors of an urban intervention from 30 years ago. The solutions proposed by four famous architects, acting as urbanists, hardly differ from the competitions held in the years 1964 to 1974. Only the styles have changed (which is what architects sell).[2]

The contributions that follow in this session are all from network thinkers who try to refute Dorothy Parker. Ever since Dupuy unearthed network urbanism, we know that some new tricks may also be time-honored, old tricks invented by Alexander, Lynch, Cerdà, and Jacobs among others. However, a new network urbanism is called for today, because 'geographical scales are dilating to the edge of infinity; technology is developing at ever faster speeds; and with the liberalization of the network utilities, monopolies are giving way to almost uncontrollable competition.'[3]

What better way of starting a debate about new instruments of design than presenting the New Charter of Athens. Jan Vogelij has played an active part in preparing it: as president of the European Council of Town Planners as well as one of the co-authors. Vogelij provides background information on the process leading to the New Charter. He highlights its content and focuses, by way of conclusion, on the commitments or roles of spatial planners. However, 'The Charter' is keyword-rich but content-light. It contains the right 'sound bites' such as sustainable development and its three components; that is social, economic, and environmental.[4] Yet these abstract concepts still must be translated into new instruments of design in order to become effective and to reassert the grounding powers of urbanism. How does 'connectivity', another keyword of the New Charter, relate to the new network urbanism?

For teachers and students of urbanism, the Charter contains an important message. Spatial planners should play different roles in the urban development process: scientist, urban designer and visionary, political advisor and mediator, as well as urban manager. Yet we may ask: Do our schools of urbanism properly prepare students for such a broad spectrum of roles? What is urbanism for the future?[5]

Technology is developing at ever faster speeds, in particular the Information and Communication Technologies (ICT). ICT is one of the reasons why a new network urbanism is needed. According to Scott Page and Brian Phillips, ICT is urban design's 'blind field'. The users or forms of community are also blind in a way, with the traditional emphasis on physical urban space. These two American practitioners and teachers advocate urban design as interface design, focusing on the relationships between reorganized forms of community, physical urban space and ICT. Implementation is illustrated by the project of revitalizing a distressed community in North Philadelphia. Urban design as interface is an alternative to physical development as the primary tool to revitalize the area in question, practiced over ten years to no avail. Nevertheless, three questions remain for further exploration:

- **convergence** of physical and digital flows and their nodes (as in the Hudson County Cyberdistrict proposal of Scott and Phillips);
- **interaction** or participatory design (to bring in users at an early stage of the design process has proven to be key to innovation, that is new products and services.[6] How about urban design involving in a similar way local community members?
- **place**, that is convergence and interaction augmenting and amplifying how individuals perceive place (thus customizing non-place ICT).

Following, the question addressed to Francesc Magrinyà is whether Cerdà can make a contribution to today's debate about new instruments of design. Dupuy, unearthing the network urbanism in 1991, classifies Cerdà as one of the prominent network thinkers who has emphasized strongly the topology and kinetics, and to a lesser extent, the adaptation of networks. Cerdà's work comprises both theoretical reflections and urban planning instruments in action. The scene is set by the city of Barcelona and its extension through three consecutive projects (1855, 1859 and 1863). According to Magrinyà, the urban planning instruments developed by Cerdà are based on a number of principles: independence of the individual within the home; independence of the home within the city; independence of different kinds of movement on urban roadways; and finally, 'ruralize the urban, urbanize the rural'.[7]

Cerdà's most important tool is the 'ways-interways' concept. 'With mobility increasingly the determining factor, buildings and building plots can no longer determine urban form, but must fit in the "interways" ensemble. The Extension grid shows how Cerdà's city block can absorb different types of building and site adjustment'.[8] Cerdà practiced in the second half of the 19th century what indeed the European Council of Town Planners advocated for planners in the 21st century in the New Charter of Athens. Cerdà certainly has combined the four roles in Barcelona's urban

---

1 Salingaros, N.A.; 'Complexity and Urban Coherence', in N.A. Salingaros, **Principles of Urban Structure** (Amsterdam: Techne Press, 2005) p. 101.
2 cf. Drewe, P.; 'Welke stedebouwkunde voor de toekomst?' {What urbanism for the future?} **Atlantis**, 16, no. 2: (2005) pp. 39–43.
3 Dupuy, G.; 'Networks and Urban Planning: Evolution of a Two-way Relationship', in E.D. Hulsbergen, I. T. Klaasen and I. Kriens (eds) **Shifting Sense, Looking Back to the Future in Spatial Planning** (Amsterdam: Techne Press, 2005) pp. 125–129.
4 This mirrors, incidentally, the European Spatial Development Perspective.
5 Drewe, **op. cit.**
6 See Musso, P., Ponthou, L. and Seulliet, E.; **Fabriquer le futur. L'imaginaire au service de l'innovation** (Paris: Village Mondial, 2005).
7 cf. Francesc Magrinyà, this volume.
8 cf. Francesc Magrinyà, this volume.

development process[9] – take as an example, only the role of visionary. Already in 1867, Cerdà wrote: 'Today everything is movement, everything is expansion, everything is communicativeness.'[10]

Nikos Salingaros' treatise on design methods, emergence, and collective intelligence is part of a larger treatise on principles of urban structure and connects to a criticism of contemporary architecture.[11] Salingaros points out the importance of adaptive design following a Darwinian process. If a complexity is required that exceeds the intelligence of individual human beings, then the traditional built environment can serve as a source of collective intelligence. This approach is a matter of combined, adaptive top-down and bottom-up design methods. The latter may be participatory as in Scott and Phillips. Yet non-adaptive design methods – in widespread use for some time – are criticized as 'simplistic image-driven typologies', not adapted to human needs.

Salingaros once again links urban design to computer science. After having pointed out the 'pattern language' of software design, he is now using sorting algorithms as an analogy for design – a far cry from non-adaptive design. Nikos Salingaros is an applied mathematician and physicist, looking from the outside in, on urban design. He can teach us new tricks similar to Barabasi, another physicist researching complex networks in a multidisciplinary way. Salingaros has extended the approach to complex networks known as cities. Consequently, the time has come to look from the inside out.

Michael Mehaffy's contribution is on the same wavelength as Salingaros' article. Mehaffy, who was with the Prince's Foundation in the United Kingdom, starts from a 'new scientific understanding of the structure of nature, and in particular of living systems – the science of the behaviour of small, rule-based iterations, or algorithms – cellular automata and the like, and the so-called "emergent" patterns they create'. The application of the 'new science' started as early as 1961 with Jane Jacobs pleading for organized complexity. Mehaffy, in fact, recommends re-reading the last chapter of *The Death and Life of Great American Cities* entitled 'The Kind of Problem a City Is'.[12] The keyword in Mehaffy's article is 'code', inspired by the work of Christopher Alexander. After the pioneering pattern language in the 1970's, Alexander recently has explored what he calls the 'nature of order'. Mehaffy describes a ten-step process or cycle of handling codes in line with the 'new science'. The process includes fifteen so-called structure-preserving transformations, based on the fifteen properties that Alexander has identified in natural morphologies. Codes also play an important part in the New Urbanism movement (and its participatory feature of stakeholder collaboration in the design process). But it still remains to be seen whether the New Urbanism – although obviously successful in practice – really is in line with the 'new science'.

While Koolhaas complains about the 'massive crater in our understanding of modernity and modernization'[13], the new sciences – according to Mehaffy – imply the dawn of a new modernity that we are only just beginning to understand. However, the proof of the pudding is as always in the eating. This also holds for the work of the Prince's Foundation: http://www.princes-foundation.org.

Unfortunately, our final contribution in the debate about new instruments of design from Bonfiglioli has not been made available. It is, however, an important one: urban time policies, in particular the work of LABSAT at the Politenico di Milano. Urban time policies in Italy are not only a matter of research but also a matter of design and implementation. Instruments of design center around the so-called chronotopes: 'A chronotope is a physical time configuration: 1. constructed historically; 2. inhabited permanently/temporarily by a mix of changing populations; 3. breathes according to open/closed cycles of local functions, both permanent and ephemeral; 4. is used by inhabitants according to phantasmagorical designs of timetables, calendars and presence cycles; 5. is embedded into multi-scalar nets of person, goods and information mobility'.[14]

In 1969, Hagerstrand, the founder of the space-time model, presented a paper to the European Congress of the Regional Science Association in Copenhagen entitled 'What about people in regional science?' Now we might ask: Does not the question 'What about people?' also hold for conventional urban design? Does not asking 'What about time?' bring back people in urban design? This is most clearly proven by urban time policies. The chronotopic analysis reveals echoes of the past, present rhythms and the simulated presence of the future. What better way of answering Lynch's time-honored question: 'What time is this place?'

Network thinking is the common thread amongst the different contributions in this section. However, no royal road to innovation exists in urban design. The diversity of the contributions makes for fascinating avenues of research. The experience of the Athens Charter of 1933 shows, if nothing else, that the assumed existence of a royal road only leads to dogma. If 'the kind of problem a city is' presses for organized complexity, then Ashby's Law of Requisite Variety holds: 'only variety can destroy variety'.

9 cf. Vogelij.
10 cf. Francesc Magrinyà, this volume.
11 Salingaros, N.A.; **op. cit.** and Salingaros, N.A. **et al.; Anti-architecture and Deconstruction** (Solingen: Umbau-Verlag, 2005).
12 Jacobs, Jane; **Death and Life of the Great American Cities** (New York: Random House, 1981).
13 Rem Koolhaas, 'Whatever Happened to Urbanism' in **S,M,L,XL** (New York: Rizzoli, 1995).
14 Bonfiglioli, S.; (2005) **The City of Time and the Culture of Planning**, unpublished manuscript; see also Drewe, P.; 'Time in Urban Planning and Design', in E.D. Hulsbergen, I. T. Klaasen and I. Kriens (eds); **Shifting Sense, Looking Back to the Future in Spatial Planning** (Amsterdam: Techne Press, 2005) pp. 197–211.

The New Charter of Athens 2003: The ECTP's Vision on the European City of the 21st Century

Jan Vogelij

In November 2003, the European Council of Town Planners, ECTP, launched The New Charter of Athens 2003 at an international conference in Lisbon, at which about 400 spatial planners, officials, and scholars participated. In this chapter the New Charter of Athens is discussed, addressing questions like: What was the reason for formulating such a new charter? How can the new charter be positioned in relation to the previous Charter of Athens? What about the content and the vision behind this new charter? What does it mean for the commitments of spatial planners in their daily practice?

Background   It was the European Council of Town Planners (ECTP) taking the initiative to formulate the New Charter of Athens. The ECTP is the European umbrella organisation for professional spatial planners, uniting the national associations of planners throughout Europe: in total 25 associations in 23 European countries with about 26000 individual members. Planners are active in all countries of Europe, on the local level, the regional level, the level of the national states and also on the European level. They are employed by both the public authorities and private consultancies.

Although there is a large variety in the ways that planning is practised in the European countries, there are also big similarities in the problems planners are facing and in the interests of individual planners. A common aspect is the relative weak position of the planning disciplines in the public debate. Although the importance of sound spatial planning is generally acknowledged, the role and aims of spatial planners are not always well understood. The weakness of their position is on the one hand related to the character of the subject of spatial planning that includes large scale and long term spatial developments, that are both difficult to communicate to the large public. On the other hand the relative weak position is related to the position of planners in the spatial development processes. Planners are advising public authorities who in the end are responsible and planners are practicing (relatively more than other disciplines) integrated approaches and team work which make their individual contributions less visible. Both aspects of the position of planners make them relatively unknown for the public. Anyhow, planners are less visible in comparison to architects with their clear physical results. The results of sound planning work tend to be perceived as quite normal and taken for granted. Only if large mistakes are made during planning activities, or if spatial planning was even not involved at all, the lack of planning quality is noticed.

Since the European Community was developing and legislation in many different fields was prepared, influencing the conditions in which professionals are operating, the need to promote the interest of spatial planners was extended from the national to the European level. The ECTP started as a so called 'comité de liaison' between a group of national associations of planners, which in 1985 was reorganised as a vzw under Belgian law into the actual European Council of Town Planners.

From its start on, the ECTP concentrated on two items. First the need for an agreement between the national associations about the essential characteristics of the planning profession. The resulting document was named the common core (of the planning profession). Secondly the ECTP agreed on the requirements for the education of planning professionals. These basic documents are still part of the ECTP's constitution, although the developments in modern society, as well as the unification process in Europe, demand a constant critical review.

The Need for a New Charter of Athens Although this is seldom recognised, a big part of non discussed, common understanding about urban planning is related to the original Charter of Athens (1933–1942). Almost everybody involved in spatial developments is referring to the four distinct functions: living, working, recreation and mobility. Also most legal systems on spatial planning are implicitly assuming that those are the four functions with which planners are able to spatially organise the territory. In several European countries the Charter of Athens is explicitly used in the education of spatial planners, whereas in other countries like the UK, the Charter of Athens was hardly known among planners.

The Charter of Athens was the result of a series of meetings of the Congrès International d'Architecture Moderne, CIAM. Especially, the boat trip on the Patras of 1933, that started in Marseille and ended unforeseen in Piraeus near Athens provided an important step in those discussions. The group of modernist architects, among which names like Le Corbusier and Giedion, discussed under chairman Van Eesteren, principles of planning modern cities.

It was an Exciting Era Modern sciences developed fast and offered new great and promising possibilities. Not only in the field of construction but more especially in mechanics and industrialisation new opportunities developed. Enlarging of scales and the concentration of polluting activities posed limitations for the possibilities to combine functions in a territory. Large scale production should be separated from other, quieter human activities. Mass traffic using the up-coming motor cars must be separated from slower, old fashioned ways of transportation. The city could be considered like a set of functions each located in its own position in the urban machine.

The distinction of four functions with different requirements for their functioning was introduced in the Charter of Athens in a non rigid way. The text refers to the importance of mixing functions and the qualities of the combination of activities. Nevertheless, although this was clearly not according to the intentions of the authors, eventually the simplistic interpretation of those planning principles became quite common. In planning practice it became usual to dedicate specific areas to specific destinations. Residential areas were located separated from industrial areas, that were separated again from recreational areas, even if the specific activities in reality did not hinder each other. Many extensions of European cities as well as new towns, built after World War 2, reflect this mechanistic interpretation of the principles of the Charter of Athens.

The ECTP, reflecting on the main sources of inspiration of the profession, noticed that a review of the principles of planning was needed. Acknowledging that consistency of principles is an important quality of planning, the developments in society since the text of the Charter of Athens was finalised by Le Corbusier, have been so strong, that the original Charter of Athens did not fit anymore with the needs of the actual society.
The developments in society that are relevant for this revision are covering a wide range of fields:
First, new techniques resulted in less polluting and less dangerous production and mobility.
Second, modern economy is strongly characterised by service activities.
Third, general awareness of limited natural resources lead to broadly adhering to the objective of sustainable development.
Four, actual interpretation of democratic principles includes a strong public participation in decision making processes.
Five, demographic developments result in the confrontation of different groups with sometimes conflicting interests.

These big changes in modern society since the period that the Charter of Athens has been drafted, created societies as well as urban and regional problems, that cannot anymore be addressed in a technocratic approach. The scope in which actual spatial developments must be addressed, is much wider than that of the modernistic architects.
Nowadays it is about the combination of economic, social, cultural and environmental aspects in complex spatial planning processes instead of about only architecture.

The question arose, whether the name of the new document resulting from the discussions as organised by the ECTP, should refer to the Charter of Athens. This question was relevant since it was clear that the content of the new document would differ completely from that of the old one. The question was especially relevant for the many planners that were fighting, some for a long time already, against the negative influence of the application of the Charter's principles. Two main arguments resulted in the decision to name the new document The New Charter of Athens 2003.

By referring to the old Charter of Athens it is demonstrated that views evolve, according to developments in society, without necessarily completely denying the value of previous steps. It gives due respect to the notion of continuity in long term spatial developments: the Charter of Athens was the appropriate answer to the situation of that moment in history. By adding the year of revision, new revisions can be considered if that is felt necessary. The second argument for referring to the (old) Charter was that the new name clearly indicates a replacement. It makes clear that The New Charter of Athens 2003 comes in the place of the old one.

The Need for a Common Orientation It was not only the notion that the content of the old Charter of Athens was not relevant for the actual spatial developments that brought the ECTP to starting its revision, the need for a new, common orientation in the profession was also large. More than many other professions, spatial planning is a profession that is carried out differently in the different European countries. Legislation differs per country, decision making processes differ per country, the problems to be addressed are different and should be approached in a location specific way. In short, spatial planning is, and should be, intensively embedded in the national, regional and local cultures.

That situation results in large differences in the perception of planners of European policies with regard to the profession and spatial developments. Therefore it is quite complex to come to common standpoints needed to defend and promote the discipline. This adds to the relatively quantitative weakness of the profession in the European arena. In order to create a common orientation, the ECTP decided to focus the discussion on building a common vision on the European city of the 21st century. As in actual planning processes, discussions on a wished future, not dedicating all energy to the analysis of problems and difficulties, but searching for the opportunities offered by long term developments, may result at least in an answer to the question what would we like our future to be?

Of course these discussions should avoid resulting in unrealistic dreams, but planners should help local and regional societies to identify their ambitions and to elaborate them for gradual realisation. In that sense, working on the New Charter of Athens 2003 was a tool for the ECTP to create coherence within the community of European spatial planners which, as a result of its cultural embeddings tends to be strongly segregated. The process towards an agreement between all European planning associations involved, resulting in the launch of 2003, was long. In 1996 a working group started, drafting texts and submitting those to the council and the member associations. Large discussions revealed great differences in the views and problems of actual planning. National hypes were sometimes promoted as general professional issues for the whole of Europe. Different views on the influence of environmental aspects on spatial development questioned the identity of the planning discipline. An agreement with the character of a compromise was reached in 1998, when the version of that moment was presented in a dedicated conference in Athens.

Discussions Continued However   National associations of countries more advanced with integrating sustainability in planning practice did not really accept the document as sufficiently future oriented, and also discussions with students during the third Biennial of Towns and Town Planners in Herne, Germany resulted in a critical rejection. The ECTP decided to continue the process and composed a new working group presided over by the Portuguese professor in planning Paulo Correia. New texts were drafted, starting with a partial revision, but ending up in a complete revision of the Charter. After first presentations and discussions in the General Assembly of the council, the drafts were again sent for comments to all individual members of the national associations in two rounds, with discussions during General Assemblies in between.

The structure became clearer, sustainability was applied as an integrative notion, actual and future economic, social and ecological issues were addressed, and spatial developments according to the notions of networks and urban and regional specialisation were integrated. Also the final chapter discussing the variety of roles spatial planners can play, was offering important clarity. In November 2003 the text, accepted and welcomed by the ECTP's General Assembly was launched under the title The New Charter of Athens 2003, the ECTP's vision for the European city of the 21st century.

The Content of the New Charter of Athens 2003   The vision of the ECTP on the European city of the 21st century is summarised as THE CONNECTED CITY. The notion of connectivity is stressing the contrast with internal as well as external segregations which are related to actual problems of cities and citizens. It emphasises the objective of the city as a whole, undivided entity which is intensively interrelated to its environment, the region and wider networks of cities, by a variety of tactile, functional and virtual/informational connective mechanisms. It also stresses the importance of connection through time, where the urban identity connects the future with the past.

It is widely acknowledged that the influence of spatial planning is limited, spatial planners are part of complex decision making processes with many other actors. Urban and regional design are important but relative small parts of these processes. At the same time it is acknowledged that the influence may be most effective if the planners convey clear objectives and visions for desirable future situations. If those visions imply interests of relevant parts of society, and are supported by those and other actors, synergy might result.

The ECTP's vision is therefore quite comprehensive and may be considered to be ambitious but that does not mean that the European spatial planners are naive about the magnitude of their influence. The fact that meeting specific objectives in reality is extremely difficult does not have to include that they must not be aimed at any more. The notion of the connected city has been elaborated in the three components of sustainable development: social connectivity, economic connectivity and ecological connectivity.

Social Connectivity Social connectivity covers a wide range of social aspects of cities. It is a pledge for supporting coherence in urban societies. The involvement of as many groups of the population as possible in the urban development processes is important, the urban offer should meet the population's demand in its variety. Intensive involvement of all groups is not possible, some only want to consume the city on a commercial basis.

The city as a part and a symbol for someone's identity gets an increasing importance, especially since cities absorb so many immigrants. In some cases, as a consequence, the notion of identity has grown as part of a defensive reaction of people that do not feel safe. But, almost without exception, everywhere urban culture and identity resulted from a history of foreign groups immigrating, occupying or trading in a specific location. Therefore, when referring to a specific urban culture or identity we have to describe a location specific mix of different cultures, which is continuously developing. Newcomers are not absorbed in a static existing culture, but they are reacting dynamically, together towards a new cultural identity. In that sense, multicultural richness offers new exciting opportunities for the future as it did in history.

Another demographic aspect relevant for the future of our cities is the gentrification of the population. Not only because this poses stronger demands with regard to the solidarity between generations, an essential component of sustainability, but also because it will require specific services and facilities and it may jeopardise future urban prosperity. The concept of the connected city implies the need to formulate different, city specific balances for social cohesion.

Economic Connectivity Also in the future, every urban entity will experience a tension between its local characteristics and the influences of global commerce. Since the whole world develops into a single market, even the most remote locality is under pressure of globalisation. Although this often is mainly perceived as a threat, cities should orientate to the opportunities in their future development.

The unification of Europe, in which national states tend to become less important, results in a relatively increased importance of the regions. This also applies for the economy of the regions. Where globalisation stands roughly for standardisation and homogenisation of products, regionalisation stands for regional specificity and quality products. Where globalisation stands for modernity, regionalisation may stand for traditional aspects. In that sense, new interpretations of traditional aspects may be important for the future. Economic activities as basis for the city's prosperity, require connections to the outside world, not only physical connections for the exchange of goods, but especially the exchange of information will be essential.

The European Spatial Development Perspective, ESDP, adopted in 1999 in Potsdam by the Ministers responsible for spatial development, promotes the application of urban and regional networks. Cities are part of different networks, physical as well as virtual and thematic networks. The economic connectivity of cities as promoted by the New Charter of Athens 2003, implies intensive connections with their regional environment as well as with the other cities in relevant networks.

Cities can on the one hand be centres, symbolising the identity of their region, on the other hand they should search for their specific position in relation to the other cities in the relevant networks. Explicitly identifying the city's position in networks clarifies their roles in chains of the production and trading of certain products or services.
The interrelations in networks consist of co-operation, and individual cities are complementary to each other. This requires specialisation on the basis of existing, sometimes historic qualities. The competitive advantages of each city depend on their relative strengths, in comparison with other cities. The more the specialisation of economic activities relates to specific, distinct qualities, the better the city's position might be. Economic connectivity, aiming at urban prosperity, argues for articulating the diversity of European cities.

Environmental Connectivity   Sustainable development is a generally supported objective, promoted by a variety of European programmes. Although, the environmental aspect of sustainability was an important starting point, in European policies it is nowadays embedded in the triangle of social, economic and ecological aspects. The city considered as an ecosystem, looking for a balance between input and output of resources, is strongly recommended by the report of the European experts group on the urban environment: European Sustainable Cities of 1996. The use of energy, the production of waste and the management of water should be considered in the context of wider systems. This requires environmental connectivity.

Next to that, environmental problems, resulting from polluting industrial activities and transportation, have caused severe health problems. Polluted cities will be considerably less successful in attracting new economic activities or residents. Several European Directives about air quality and noise, integrated water management and soil quality address those problems. Reduction in the use of fossil energy, other resources and water and the production of non recycled waste, are areas of concern and investments, also in the future. Apart from the environmental concerns that are essential for healthy cities, the environmental connectivity as promoted by the New Charter of Athens 2003, implies sound connections with landscapes and open spaces.

Enhanced efforts in building the Pan European Ecological Network of which the Natura 2000 programme is an essential activity, is needed to support Europe's natural qualities. But also the possibilities for recreation of the population that are offered when cities are well connected to the region are essential spatial qualities. This includes improved connections to the cultural aspects of the surrounding landscapes as well as to the ecological values of the natural areas.

Urban Differentiation and Spatial Planning If the concept of the connected city is elaborated in the way as meant here, the result will be an increased differentiation in the variety of cities. Every city as such will contain different internal functional and thematic networks, based on their historic qualities. And at the same time, every city will be part of wider networks in which it should articulate its distinct qualities and define its position.

The task of spatial planners will include the identification of those qualities with which a city distinguishes itself from others. Those qualities should be formulated explicitly and accepted as opportunities for future developments. Moreover, spatial planners should focus on enhancing the specific spatial qualities in their urban design activities. A revival of urban design is hoped for in the New Charter of Athens 2003. Responsible authorities should decide for city specific balances in the social and demographic aspects, for city specific balances between globalisation and regionalisation in the urban economies, as well for city specific connections to the regional cultural landscapes and natural areas. Altogether this will result in increased differentiation of cities.

Commitments of Planners One of the main messages of the New Charter of Athens 2003 is that the strategic capabilities of the planning discipline should be utilised in a better way for the sake of the future of the European cities. Spatial planners can be helpful in exploring opportunities, in showing solutions developed elsewhere, in designing new solutions, in mediating between interests, in communicating about the processes and informing stakeholders about limitations and possibilities. Their attention should focus however on the relation of individual problems and solutions with the wider spatial contexts and with the developments in time, avoiding segregated *ad hoc* solutions. Spatial planners may be involved in spatial developments in different roles in the process, as scientists, as designers and visionaries, as political advisor and mediator and as urban manager. Some planners combine two or more of these roles, others concentrate on one of these roles:

The spatial planner as a SCIENTIST should be committed to inform planning processes with knowledge and results of scientific analyses. He or she should also take the responsibility to be aware of the most recent relevant information and to maintain adequate databases.

The spatial planner as an urban DESIGNER and VISIONARY should be committed to utilise his or her creativity to search for optimal solutions within the given context. These activities focus on creating attractive environments and public spaces and on results that are improvements of the city as a whole. Relating interventions to the local culture and identity in specific ways is thereby essential.

The spatial planner as a political ADVISOR and MEDIATOR concentrates his or her activities on balancing interests of the relevant stakeholders. Hereby the general interest of the local community should be identified and promoted. For the mediator, the creativity of a designer is helpful to find new possible solutions for conflicting interests.

The spatial planner as an urban MANAGER is in the position to organise processes in such a way that strategic considerations can be utilised in the municipality's interest. This position may also enable him or her to involve colleague spatial planners playing one of the other roles, at the right moment in the spatial development processes.

The ECTP has initiated a working party that takes the above described roles of spatial planners as a starting point for considering the requirements of planning education. It is envisaged that this will result in a review of the existing list of requirements of the education of planners that will be discussed with the AESOP. The ECTP aims at the utilisation of that outcome in a European Platform for the professional qualification of spatial planners.

Conclusions The New Charter of Athens 2003 indicates some important aspects of the direction that the spatial planning discipline in Europe should take, according to the national associations of spatial planners that are united in the ECTP. It relates the work of spatial planners to the EU policy objectives with regard to spatial development as has been adopted in the European Spatial Development Perspective. Within that policy and its elaborations, local and regional planners can offer important contributions to identify those specific local and regional qualities that should be enhanced to establish stronger positions of the cities in their various networks.

The fact that the European Commission decided to take the notion of *territorial cohesion* as the third pillar of its cohesion policy for the development of Europe is of great importance for the planning discipline. Territorial Cohesion, based on networks of complementary cities in combination with the ecological main structure will enhance the variety of cities. Spatial planners, interpreting local and regional qualities for future development, may thus offer important contributions to the development of Europe.

The New Charter of Athens 2003 (NCA)

The ECTP's vision for cities in the 21st Century

Jan Vogelij
President of European Council

Content of the Presentation

The European Council of Town Planners

1. Positioning of NCA
2. The vision of NCA
3. Background of NCA
4. Commitment of planners
5. Conclusions

The European Council of Town Planners

The European Council of Town Planners

- **Comprises**
  - 23 European National Associations of Spatial Planners
  - Together about 26.000 planners
- **Consulting body to European Institutions**
- **Organises**
  - Biennials for Towns & Town Planners
  - European Planning Awards
  - Conferences
  - Publications

The European Council of Town Planners

Positioning of NCA 1/5

1.1 Historical background

- **Charter of Athens 1933-1942 (CIAM)**
  - City as a functioning machine, applying modern technology (mobility!)
- **New Charter of Athens (ECTP)**
  - First version in 1998, second adopted version in 2003
  - Next to new technology, stronger orientations on social, economic, ecological aspects.

The European Council of Town Planners

Positioning of NCA 1/5

1.2 Charter as a tool

- Spatial planning requires co-operation between planners and other professionals
- Spatial planning as an activity in society requiring support and consensus

Both demand for a certain degree of common orientation.

The European Council of Town Planners

Positioning of NCA 1/5

1.3 Target

The NCA is addressed to professional planners throughout Europe and those concerned with the planning process:
- to direct their actions by a common vision
- for greater coherence, when building networks of cities, connected through time, scales and sectors

The European Council of Town Planners

The Vision 2/5

2.1 The connected city

- Variety of connective mechanisms tactile, functional, information, virtua
- Connecting through time character / identity connects future and past

The European Council of Town Planners

The Vision 2/5

2.2 Social connectivity

- Social balance
- Involvement
- Multi cultural richness
- Connections between generations
- Social identity
- Mobility, facilities, services

The European Council of Town Planners

The connected city

social connectivity

The Vision 2/5

2.3 Economic connectivity

- Globalisation / regionalisation
- Competitive advantages
- City networking: complementary and co-operation
- Economic diversity

The Vision 2/5

2.5 Aspects of the connected city

- Part of and consisting of meaningful networks
- Urban and regional design
- Different solutions, cultural richness, enhanced variety

The connected city at different levels

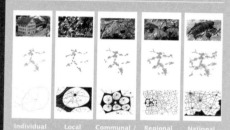

Individual | Local | Communal / Municipal | Regional | National

Centralism
Strong hierarchy of the cities, central orientation in politics and infrastructure

Network / poly-c... as a s...
Poly-c... and ne... at inte... and tra...

Background 3/5

3.2 Trends

- **Social and political changes**
  Europeanization, cultural contrasts (immigrants), multi-cultural cities, diverse services and products, planning & management of cities (deregulation and privatisation, PPP, city marketing, public participation)

- **Economic and technological changes**
  Speed of technological development, e-commerce, e-business, footloose economy, international competition (specialisation and co-operation)

The changes

Social and political changes

Economic & technical changes

www. ... .com

- **Environmental** ...
  Urbanization, expanding ... pollution and wasteful c... resources, contamination ...

- **Urban changes** ...
  past: city – never only one ... development, variety of ur... planning techniques as we... future: world-wide commu... generating new urban form...

Considered with issu... and spatial planners...

Commitments of planners 4/5

4.2 Positioning spatial planners

- Spatial planning is essential teamwork
- Transdisciplinary involving different professionals

But, the spatial planners approach specifically distinguishes by stronger orientation on
- spatial context and
- continuity in time

Commitments of planners 4/5

4.3 Spatial planners in strategic roles

- as a scientist
- as a designer and visionary
- as a political advisor and mediator
- as an urban manager

sometimes combined, sometimes separated

designer and visionary

scientist

The Vision 2/5

2.4 Environmental connectivity

- Input output
- Healthy cities
- Nature landscape open space
- Energy

The European Council of Town Planners

The connected city

environmental connectivity

Example 2 – Germany

Spatial polarisation | Large regions and polycentric networks | Compacted corridors with new spatial functions

Background 3/5

3.1 Issues and challenges

- Developments affecting cities
- Challenges to the cities
- To be addressed by the vision

The European Council of Town Planners

The changes

Environmental changes

Urban changes

Commitments of planners 4/5

4.1 Values and objectives for professional planners

when
- elaborating and implementing the vision
and
- advising politicians and the public

The European Council of Town Planners

Conclusions 5/5

The European Council of Town Planners wishes to enhance the strength of the profession reinforcing

- self-confidence
- cohesion and
- solidarity among planners

Gracias por su atención.

The European Council of Town Planners

Urban Interfaces: Designing In-Between

Scott Page and Brian Phillips

The buildings, roads, and squares that served as key forums for communication in the traditional city are now but one layer in a network of infrastructures which facilitate individual and community interaction within urban space. Information technology provides an expanding realm of tools and platforms to understand, perceive, and participate with the city. This paper focuses on the relationships between reorganized forms of community, physical urban space, and information technology as a new territory for urban design. We propose that these three elements can be conceptualized as an integrated approach to linking the computer and the city in more intimate ways.

Urban Design's 'Blind Field'

'It's as though the creative process is no longer contained within an individual skull, if indeed it ever was. Everything, today, is to some extent the reflection of something else.' – William Gibson[1]

Rethinking the designers' approach to the city has become a mandate. While advances like the telephone and automobile clearly had wide-ranging impacts on the twentieth-century city, the recent wave of information technology promises to provide many more. Most of us have become intimately familiar with the hardware and software associated with information networks including mobile phones, laptop computers, wireless Internet, GPS systems, and personal digital assistants. As indicated by the proliferation of on-line shopping, dating, and web-based news services, these tools are radically impacting the ways we get information and form our social networks.

The early literature on the emerging information city came out of economic, social science and philosophical circles by the latter half of the twentieth century (see Castells, Lefebvre, Virilio, etc.). A new wave of globalization, facilitated by enhancements in transportation and communications, led to numerous descriptions of how our lives would change due to these new real-time, networked phenomena. These observations became the impetus for designers to recalibrate their approaches to the city. From the Smithson's *Team X Primer*, to the Situationists' *Naked City*, to Banham's *Four Ecologies of Los Angeles*, the modern city became the scaffold upon which to build new theories of urbanism from the perspectives of media, speed, and personal perception.

In recent years, many urban theorists and practitioners have drawn attention to the fact that cities are as much about how they are organized and experienced through information technologies and networked communication infrastructures as they are physical form. Only a portion of what we know or experience in cities is through direct physical contact as an increasing number of sources are available to download and browse urban information. As expressed by Mark

Wigley, 'Cities are experienced in terms of images. Visitors bathe in images before going anywhere – scrutinizing guidebooks, web-sites, business brochures, videos, airline magazines, friends' snapshots, and so on – then project these images onto the place, trying to match what they see with what they expected to see... Physical form is at best a prop for launching or modulating streams of images.'[2]

Lev Manovich further expands on this intertwining of culture and digital technology in stating: 'Cultural categories and concepts are substituted, on the level of meaning and/or language, by new ones that derive from the computer's ontology, epistemology, and pragmatics. New media thus acts as a forerunner of this more general process of cultural reconceptualization.'[3] While this phenomenon is of particular interest today, the symbiosis between technology and culture has always shown itself in the forms of cities. Baroque urban planning with its rational geometries and strong vistas were a clear reflection of the larger cultural frameworks of the period. Today, information technology lies between the physical city and the people who inhabit and use it, promoting an expanded array of opportunities for individual and community expression.

But as the volume of information being created and distributed exponentially grows, finding opportunities for expression is increasingly the product of how we access, filter and use information. This trend invites all of us to edit – in fact, positions editing as a fundamental state of being – as our sources of information and criteria for truth and meaning become of utmost importance. With information overload also comes enhanced tools for editing made possible by technology. Web browser software, PDAs, cell phones and other devices have spawned on-line user groups, list-serves, community networks and collaborative software. The Microsoft Windows operating system acts as a personal filter for software programs and personal files where large amounts of data are stored in discrete and navigable folders. The filter limits the amount and type of information available from any single perspective. The collective integration of these tools and platforms has helped to filter the 'data glut', enabling each individual to personally browse particular subject matters.

We increasingly *cut and paste* the world. With digitalization, the reassemblage of existing media has become a popular mode of production as exemplified by reality television, music sampling, and Adobe Photoshop. High-powered processors and large data storage capacities have shifted the editing room to the personal desktop or laptop. Everyone is contributing to a particular 'edition' of the city at any given moment – a perpetual, collective process of editing, publishing and revising that guides our perceptions of the built environment.

Urban designers are already in many ways editors. Adobe Photoshop, for instance, has had a powerful role on the architectural profession in terms of how we conceptualize and represent the world. Our role in projects is in part our ability to collect and re-present information to stakeholders guiding them through the complex nature of urban spaces. Designs too have taken form as a blending from other projects and readily available 'products'. Catalogues of products from lights and benches to windows are the components that designers are assembling. Glossy design publications are often the first place a designer goes for inspirational direction at the beginning of a project, drawing consciously and sub-consciously on the latest trends.

While information technology is facilitating new cultural practices and design processes, they are misunderstood in the design of urban environments, playing a diminished and often superficial role from the overall conceptualization to implementation. Anne Beamish discusses the problems often associated with community technology initiatives. She noted that: 'Project goals are often too vague, simplistic, and unrealistically utopian. Participants are frequently seen as passive consumers of information rather than active producers. Information is emphasized rather than the communication side of the technology.'[4]

Urban centers have always been important devices of communication. Mumford writes, 'The ability to transmit in symbolic forms and human patterns a representative portion of a culture is the great mark of a city.'[5] We are interested in how this evolving context of the urban environment might impact the way we design cities. The overall rise in importance of various media and specifically various means of production, consumption, and editing mandate a reflective response from those engaged in designing and planning cities. Salingaros said in a recent keynote address, 'If we can get over the ideological blinders imposed on the world by otherwise well-meaning, but false ideas about 'modernity', then we can begin to understand how the urban fabric forms itself and changes dynamically.'[6] This 'blind field'[7], as Lefebvre would call it, is apparent within current urban design practice as its traditional techniques remain undaunted in the face of widespread changes in the ways in which we engage the city and form social networks.

The Community and the Individual

Traditionally, the fundamental meaning of community – a structure that lies at the very foundation of urbanism – is based upon physical proximity. It is an expression of locality brought about through careful attention to the urban realm as a forum for that interaction to take place. The impact of information technology on the ways that we define the idea of *community* is crucial to the manner in which we think about

---

1 Gibson, William. **Pattern Recognition.** New York: G. P. Putam's & Sons, 2003. p. 68.
2 Brouwer, Joke and Arjen Mulder, eds. 'TransUrbanism.' **Transurbanism.** Rotterdam: NAi Publishing, 2002. p. 104.
3 Manovich, Lev. **The Language of New Media.** Cambridge, MA: MIT Press, 2001. p. 47.
4 Schon, Donald A., Bish Sanyal, and William J. Mitchell, eds. 'Approaches to Community Computing: Bringing Technology to Low-Income Groups.' **High Technology and Low-Income Communities.** Cambridge, MA: MIT Press. 1999. p. 351.
5 Mumford, Lewis. **The City in History.** San Diego, CA: Harvest, 1989. p. 93.
6 Salingaros, Nikos. 'Connecting the Fractal City.' **Keynote speech, 5th Biennial of towns and town planners in Europe:** Barcelona, 2003.
7 Lefebvre, Henri. **The Urban Revolution.** Minneapolis: University of Minnesota Press, 2003. p. 29.

urban design. Historically, physical networks have played an important role in the cultivation of civic life. Streets have always been a space for community building as people interface with friends, neighbors, and strangers from their porches, stoops and sidewalks. Public space has traditionally been at the center of civic life. The Ramblas in Barcelona, Times Square in New York, and Trafalgar Square in London all provide invitations for individuals to abandon the isolation of their private lives and enter into a collective realm — a place of exposure and unpredictability.

Public spaces were (and still are) often connected to a city's transportation network. The great railroad palaces of the nineteenth century were not only transfer points between the railroad car and the city itself, but also intense concentrations of commerce, entertainment, and cultural diversity. These connection points between different networks (in this case, the city fabric and the railroad network) were celebrated zones of public life and community building. With the introduction of portable, personal digital telecommunications devices into our daily lives, the last few decades have ushered in a radical transformation of this link between public space and community. Portals between key networks have expanded to include not only the train station and telephone booth but also the home/office and ultimately the body through cell phones, PDAs, and laptop computers. This trend will continue to proliferate as the devices become smaller, faster, and more affordable. Paul Goldberger observed how cell phones remove the pedestrian from the street environment creating, in his view, a less public city.[8] While such a trend may seem problematic at first, it is important to seek what new opportunities these trends provide for the public realm. Change has always been and always will be a defining characteristic of urban life. New tools simply provide more opportunities to amplify our interests and activities. Therefore, it is not surprising that the impacts of technology on urban life today are extremely varied.

People are using digital telecommunications to become members of innumerable networked communities. The staggering number and increasing diversity of online users yields an equally wide-ranging body of issues from which to form communities. Technology has facilitated tools to find, join and participate in these communities from any particular location. We create our own webs of interaction that transcend spatial constraints, self-selecting our identities. This is a modern version of Richard Sennett's concerns relating to the expressed desire to create community.[9] By actively seeking a community, one seeks others like him or herself thus eradicating different perspectives. Software has facilitated this process providing us with tools to easily manipulate our frame of view. A recent New York Times article explains that for many young Americans, 'because of the way they've trained themselves to use media, they never have to be exposed to an idea, an artist, or anything that they did not select for themselves.'[10] The same has been noted in the recent use of the Internet for American political campaigns. The 'cyber-balkanization' has all but ruled out debate, accelerating the ability to participate in forums that already confirm their own individual viewpoints.[11] These forums and Blogs are monologues for gathering and coordinating like-minded people.

But while our information culture has in many ways facilitated online and, at times, homogenous communities, place-based communities are also finding new modes of connectivity and representation through expanded use of information technology. Social and community networking both from within local community-based services and activities, as well as those at larger urban, regional, and national scales of organization are being facilitated increasingly through the use of new technology. Ten years ago it would have been nearly unthinkable for a small community organization to participate and benefit from national organizations of peer groups that are present today. The ability to share not only their experiences, but also to actively participate through a consulting network of services is a significant asset to grassroots organizations.

Furthermore, much of what happens online has subsequent impacts in physical space. In combination with the Internet, the cell phone is having a revolutionary effect on connections between virtual and physical spaces. Various websites now coordinate meet-ups in physical space among like-minded people organized around a theme of shared interest (from cancer support groups, to single parents, to Gothic literature). Dodgeball.com maintains databases of individuals and their friends and then allows real-time searching via cell phone of their locations to find a friend of a friend at a nearby nightspot.[12] Bot-Fighters, an on-line game played with mobile phones, uses locations within Stockholm as one part of the playing field. Digital 'post-it' notes and graffiti – programs that enable posts of text to be uploaded and downloaded at specific locations via cell phone – are marketed and designed for use by the individual but enhance our connection to a particular place by providing information previously unavailable in real time. With the increased use of location-awareness technology and the expanding functions of handheld digital tools, our digital and physical worlds are intertwining, providing a connection to each other and potentially a place that brings forward expanded notions of community and the individual.

Urban design conventionally operates at the scale of a physical community, emphasizing the needs of a district over that of a specific individual. Modern computer interfaces, however, were designed as personal environments, concentrating on the needs of the individual. The recognition of the individual, organized by information technology, has gone so far as to impact modern marketing where 'rather than pushing the same objects/information to a mass audience, marketing now

8 Goldberger, Paul. 'Disconnected Urbanism.' Metropolis Magazine (November 2003). http://www.metropolismag.com/html/content_1103/obj/index.html.
9 Sennett, Richard. **The Fall of Public Man**. New York: W.W. Norton & Company Inc, 1976.
10 Leland, John. 'Beyond File Sharing, a Nation of Copiers.' New York Times. 14, Sept. 2003.
11 Harmon, Amy. 'Politics of the Web: Meet, Greet, Segregate, Meet Again.' The New York Times. 25 Jan. 2004.
12 See http://www.dodgeball.com.

tries to target each individual separately … every visitor to a website automatically gets her own custom version of the site created on the fly from a database.'[13] With the help of these databases, the actions of many individuals can be amplified to have significant impacts socially, economically and physically. Stephen Johnson and many others refer to these complex relationships as a type of self-organizing system. Johnson states, 'A city is a kind of pattern amplifying machine: its neighborhoods are a way of measuring and expressing the repeated behavior of larger collectives … you don't need regulations and city planners deliberately creating these structures. All you need are thousands of individuals and a few simple rules of interaction.'[14]

Interestingly, the individual perceptions of people were critical to the *Situationist International* urban experiments in the 1950's and 60's. Their maps and collages reflected their interest in how each individual's unique worldview and subjectivities could provide a new way of editing and reading the city. Although the Dutch architect Constant attempted to formalize a city based on play and exploration, other Situationists emphasized propaganda and social experiments as the main conduit for altering people's view of the city. The city the Situationists were engaged with, however, is vastly different from the urban issues and opportunities facing us today. The new *derive* is an evening of web-surfing and in some cases, surfing the web and the city simultaneously through portable technologies. Organizing the actions of multiple individuals into collective action is easier than at any time in history as evidenced by flash mobs, political rallies and other events coordinated, planned and implemented in real-time. We must question the relevance of the physical community as the sole basis for urban design as top down systems are supplemented with bottom up networks augmented by information technologies.

Interface
'… software is constantly in a state of "in-between". It's a whole series of instructions that lie in the interstices of everyday life.' – Nigel Thrift[15]

Throughout history great urban spaces have acted as interfaces – zones where a locus of activity plays out due to the connection between infrastructure, program and people. On the Ramblas in Barcelona the sheer scale and extent of people and activity is facilitated by its location, negotiating the edges of very different neighborhoods as it stretches from the waterfront to Placa de Catalyuna, a major transportation hub. This input of activity and circulation creates an audience to which a collection of street performers, shops, and restaurants cater. The surface of the Ramblas itself becomes an interface upon which these dynamics are expressed through commerce and spectacle.

Lev Manovich describes how the modern notion of interface from its early inceptions in painting and cinema has consistently provided a lens through which to edit and represent the world.[16] Interfaces today inscribe a frame around how we interact with our environments not just in entertainment but in all aspects of our lives. Although typically conceived of as a screen, the interface is both digital and cultural, drawing elements related to the new language of data processing and memory storage and those of our established social conventions. Interfaces are the primary interpreters of our global societies. As stated by Stephen Johnson, 'As our machines are increasingly jacked into global networks of information, it becomes more and more difficult to imagine the datascape at our fingertips, to picture all the complexity in our mind's eye – the way city dwellers, in the sociologist Kevin Lynch's phrase, "cognitively map" their real-world environs.'[17] As the amount of available data expands, so too does our reliance on tools to navigate this context.

Urban Design as Interface Design

Merriam-Webster defines interface as, 'a surface forming a common boundary of two bodies, spaces, or phases; the place at which independent and often unrelated systems meet and act on or communicate with each other.' This zone has always been the territory of media artists who have sought new interpretations of the relationship between technology, culture and space. Jenny Holtzer's LCD projections have emphasized specific spaces, using technology to question specific cultural perceptions about our interactions, other people and the city. Krzysztof Wodiczko uses city space as a canvas to spur debate on issues that are often beyond the traditional tools of urban design. Their work makes the viewer 'an active reader of message rather than a passive contemplator of the aesthetic.'[18]

The idea of interface provides a way of seeing new territory for urban design which integrates – a view which sees the link between the city and the computer as a membrane – a flexible, mutable, design concept which can accommodate the complex dynamics associated with urban processes including elements of social, political, infrastructural and physical form. From this perspective, we generally refer to an urban interface as a medium through which input and output are gathered and represented to impact urban flow, perception and physical space. At a basic level, we argue that this includes addressing both the hardware and software of the city. The physical embodiment of the city can be thought of as hardware – its buildings, its streets, its subways, its telecom infrastructure, its parks. The software is a combination of the social, political, and economic structures that capitalize on the physical city. Traditionally, we might argue that urban software was written and implemented through a hierarchical structure – by

---

13 Manovich, Lev. **The Language of New Media**. Cambridge, MA: MIT Press, 2001. p. 42.
14 Johnson, Stephen. **Emergence: The Connected Lives of Ants, Brains, Cities and Software**. New York: Touchstone, 2001. pp. 40–41.
15 Thrift, Nigel. 'Software Writing Cities.' Address to Information and the Urban Future Conference at NYU (February 2001). http://www.informationcity.org/events/feb26/thrift-presentation/Thrift.pdf. opening page.
16 Manovich, Lev. **The Language of New Media**. Cambridge, MA: MIT Press, 2001.
17 Johnson, Stephen. **Interface Culture: How New Technology Transforms the Way We Create and Communicate**. San Francisco: HarperEdge, 1997. p. 18.
18 Karasov, Deborah. 'Urban Counter-Images: Community Activism Meets Public Art.' **Imaging the City**. Eds. Lawrence J. Vale & Sam Brass Warner, Jr. New Brunswick: Center for Urban Policy Research, 2001. p. 339.

decree as mayors and city planners implemented master plans with coordinated physical and policy directives. Today, information and communications technologies are expanding and enhancing the importance of urban software — of the networks that connect ideas, information, and people.

'Interface' as a concept is already frequently used in urban projects. As a metaphor for bringing people together, the term is used to describe design concepts for buildings, parks and infrastructure networks. Several finalists to design a High School in Perth Amboy, NJ use the term interface explicitly, as a means of providing a type of communication system between the new school and the larger context. The winning entry, by John Ronan Architects, is described as, 'the interface between the community and its high school … a hybrid institution which functions simultaneously as school and civic cultural center, blurring the boundary between the community and its institutions. The high school is herein defined not as a "building," but as the sum total of three superimposed systems.'[19]

'Interface' has also been used quite literally, to visibly project the role of technology onto space through large projections and screens. Times Square represents the pinnacle of this appliqué, but numerous examples are evident including Shibuya Station in Tokyo, and the vault shaped video screen roof of Fremont Street in Las Vegas. Several years ago, in our own work, we proposed to elevate the visibility of technology in our work in Newark, NJ where large screens showing real-time activity in downtown were proposed along the State's main highway — Interstate 95. Anthony Townsend documents the increasing ease in integrating these technologies in cities as the cost of diode technology decreases and the physical impacts are better understood.[20]

But given some of the challenges many neighborhoods face, a broader approach to developing urban interfaces is needed to make connections to services, markets and employment opportunities available elsewhere. The use of the term as a metaphor to better understand an urban problem or the direct application of digital interfaces in public space has no doubt provided benefits to both designers and users. We believe there are opportunities for bringing added relevance to these positions by understanding some of the key components traditional interfaces offer. The following discussion is by no means comprehensive, but captures three initial observations about how this idea may be explored in practice.

Interface and Convergence

Cities operate at multiple scales — they establish overlapping relationships between their immediate territories, metropolitan zones, regional metroplexes, national boundaries, and global networks. The twenty-first century city is particularly susceptible to great contrasts in physical, economic, and social conditions as global influences interact with very local ones. We use the concept of the 'Interface' as a means to seek new 'sites' that exist between the multiple scales of the city. Traditionally, urban design has considered a project site to be a defined area of real estate. Are there other types of sites from which to design cities? Amin and Thrift write, 'places … are best thought of not so much as enduring sites but as *moments of encounter.*'[21]

Modern interfaces for our computers and cell phones, for instance, enable these moments of encounter by facilitating close connections between an individual user, information, and larger communities – converging data into one location. We can use them for personal, one-on-one interactions through email or as a platform to globally reach customers. The flexibility engendered in interfaces allows us to bridge multiple scales of interaction with a click of the mouse. The reach that we choose to have often encompasses both local and global interactions. We believe that urban interfaces represent the potential for an urbanism that is simultaneously compact and distributed. Uses and amenities typically confined to one space or site can be accessed at locations throughout an urban area through digital networks. Programs operated by institutions, corporations and other organizations will be relevant to one another and to multiple communities despite geographic proximity. Visible public access points through LCDs, screens and kiosks, mobile training centers, web portals and new physical development have the potential to be digitally networked, providing multiple means for individuals to experience and learn from the urban environment.

The question for designers is threefold: the nature of the specific programs and services, their adaptation to specific contexts, and where investments should be targeted. The objective would be to organize a strategic framework that determines key positions within the urban fabric that overlap with critical nodes of physical and digital flows. Our recent work in Hudson County, NJ approached these locations as potential 'seeds' that collect and distribute networked services, engaging local users and the physical context while creating and enhancing relationships with organizations, programs and neighborhoods beyond their immediate boundaries.

Interface and Interaction In the sixties, avant-garde artists such as Nam June Paik and Fluxus experimented with closed circuit television installations. These artistic expressions brought new conceptions of interactivity to the museum and subsequently impacted the mainstream nearly a decade later through what became referred to as community-based public-access programming. These initiatives were founded to promote local communication and create stronger ties between people and their neighborhoods. The technology had its limits however, enabling only one method of 'plugging in' – through living rooms. Television also could not

---

19 See http://www.trenton.k12.nj.us/comp/PerthAmboyRon.html.
20 Townsend, Anthony. 'Digitally Mediated Urban Space: New Lessons for Design.' Praxis 6 (March 2004): 100–105.
21 Amin, Ash and Nigel Thrift. **Cities: Reimagining the Urban.** Oxford: Blackwell, 2002. p. 30.

customize its output to specific groups or neighborhoods on the fly. In essence, it runs off of one main platform too broad to capture the diversity of the audience and one that few people have access to for expressing their views.

Issues related to user participation are being explored in recent computer design fields called *participatory design*. Participatory design seeks to include potential users in the initial design of software application. Traditionally in urban design, public meetings and, recently, project web sites enable local residents or stakeholders to voice their opinions. The records of these comments are often meeting minutes or scrolling posts on-line. The flexibility with which technology is engendered, however, enables other opportunities. Public participation GIS, for instance, enables residents a stronger voice in local planning initiatives by transferring their local knowledge directly into a GIS database. A digital bulletin board anchored in public space puts the production of neighborhood news and information into the hands of individual users, adding value to our understanding of space from multiple perspectives.

Similarly, some designers have used interactive systems to enhance public space through participation. Rafael Lozano-Hemmer's Vectorial Elevations enables users to input parameters online that impact the position and direction of large lights within a square in Mexico. Lozano-Hemmer's work makes major strides in promoting interaction and enabling the activities of people to directly impact the quality and nature of urban space. The opportunity lies in combining this type of physical innovation with more common education and social service initiatives operated by community groups. We believe that urban design will evolve to not just plan for the activities of designated groups or city organizations, but actively seek ways of promoting local productivity that have a visible impact on the physical fabric of cities. Local business and arts initiatives – tapping the untapped creativity of local residents – are all too often a secondary priority even though innovative programs exist in various cities. RecycleArt in Brussels, for instance, is a community driven organization dedicated to local productivity. Shared resources such as a metal working and wood shop combined with education and outreach programs have enabled local residents to produce art, street furniture and play equipment for the neighborhood. They are active participants in a long process to physically alter the image and meaning of the area. The output is ongoing, formed by the participation of local community members without the need for urban designers guiding their actions.

Interface and Place Typically, many interfaces are viewed as non-place specific. A computer can be virtually anywhere and personalized despite place. But there is a symbolic role for technology – one that expresses unique local conditions and represents the infrastructure that is often hidden in physical space. Urban designers are well suited to sensitively capture these conditions and bring digital initiatives to the public realm. In a recent class taught at NYU, students were asked to design software that is grounded within a local community's needs and social organization. The instructor, Clay Shirky, describes this type of application as 'situated software',[22] which embraces the idea of non-scaleable applications (those which are not dumbed-down to address the widest possible audience). In other words, the software of the city, the interface, the means through which we manipulate the forces at play, are contextualized. They are tailored to a specific group of individuals that share a common space unique from anywhere else. We believe urban interfaces are rooted in this customization. They are territories for communication between people that balance the needs of the community with those of the individual.

Seoul's *Digital Media City* uses technology in the public realm for multiple purposes, from enhancing public access to measuring and presenting the activity of the area in real time. Even simpler initiatives are targeted toward similar results. Neighborhood marketing programs are often focused, in part, on gathering local information and increasing the awareness of programs, events and the environment itself (historic architecture walking tours for instance). An urban interface would bundle programs, services, community relations and the necessary physical wiring (or unwiring in the case of wi-fi development) to reinforce place-based initiatives. It would provide a unique digital network that expresses and represents the unique characteristics of each neighborhood while enabling opportunities to access services and information beyond its physical boundaries. At the same time, place would be a key means of accessing and upgrading this digital network. Place, as well as data streams, can be used to edit information with the net effect of creating more of a fabric between people and place. From this standpoint an urban interface is a publicly accessible filter for strengthening local information.

Customizing information technology can further local knowledge by storing histories, perceptions and events in publicly accessible databases. Much like a computer which organizes file systems and email by date, storing past exchanges in the CPU, neighborhoods have a storage capacity that can be enhanced. In the past, the storage capacity of neighborhoods resulted from the social networks and stories relayed gradually over time. Today, cities are rife with physical reminders of past decisions and events – like weathering, graffiti, and historical markers. Locales like Shibuya demonstrate one particular and extreme version of recording events and displaying information for public consumption but other initiatives are working at a much smaller scale and focusing on the collective interaction by individuals. Urban Tapestries, for instance, is developing programs for cell phones that enable multiple people to author 'location-specific' content. As stated on their website, 'It enables a community's collective memory to grow organically, allowing ordinary

22 Shirky, Clay. 'Clay Shirky's Writings on the Internet. Economics & Culture, Media & Community, Open Source.' http://www.shirky.com/writings/situated_software.html.

citizens to embed social knowledge in the new wireless landscape of the city.'[23] Framing a place through customized programs and platforms requires collective participation that actively shapes its content and role in the urban environment. This reflects recent thinking in the design of online environments that emphasizes user cooperation, community-based memory and icons of physical space to guide users.

A Proposal for North Philadelphia

These underlying concepts related to urban interfaces have driven our project to revitalize the distressed APM community in North Philadelphia. The APM community is profoundly disconnected from resources and assets within adjacent neighborhoods. Its isolation is furthered by the racial tensions and the extensive vacancy that have long plagued the area. Despite over 10 years of active physical revitalization by the local community development corporation (CDC), the neighborhood remains one of the most daunting challenges in Philadelphia. To further combat the negative trends that have impacted the area for decades, the local CDC has actively worked with us in developing an alternative urban design strategy that moves beyond physical development as the primary tool to revitalize the area.

The project seeks to create a new interface between Temple University and the neighborhood, inhabiting a stretch of vacant land around a train station that isolates both. Our objective was to seek new ways of instilling communication between different sets of people, empowering residents to gain deeper understandings of their own and neighboring communities by integrating digital information into physical space. The plan emphasizes two critical social and physical connections: between Temple University and the APM area, and between residents within APM. To facilitate this interaction, an idea of 'permeability' guided the urban design approach. Vertical and horizontal connections were emphasized as buildings, landscape and information technology were jointly designed to visually and socially bridge the station. A derelict and elevated viaduct will be transformed into a nursery to support local greening efforts and a mix of community uses; retail and housing are distributed around the station to ensure clear visual corridors between Temple and APM.

A major component of the physical program developed through a concurrent initiative focused in information technology undertaken by the authors is a proposal for a 'media station' at the base of the train station and the new development within APM. Occupying the most critical location in the area, the station will provide a centralized facility for education, communication, and the arts through digital technology. It seeks to build connections to services, programs and neighborhoods across the city and instill a sense of local productivity whereby local residents create, produce and broadcast art that will have physical impacts within the community. A plaza to publicly express these activities extends from the rail station to the media station, providing a prominent gateway for the area. To reinforce the potential value of this initiative, the authors are working with Temple University to create a proposed North Philadelphia network of shared services and partnerships supported by Temple faculty.

Initial proposals seek to create a point-to-point, real-time linkage between the rail station and the local supermarket – the two most prominent local public spaces which serve very different populations. This early initiative proposes to use off-the-shelf technology to bridge *local* space, instilling communication and interaction among very different cultural groups that already share a common physical place. It seeks to use communication to bridge and interpret the characteristics, problems and opportunities in the community from different perspectives.

Operating as a 'mood ring' for the neighborhood, two large interactive screens are designed to download information about the neighborhood and surrounding areas, communicate via instant messaging, view real-time video from each location in physical space, and post comments and news on a digital bulletin board. Real-time monitoring enables the screens to change color according to the level of input and activity. Linked to a web presence accessed through wi-fi 'hotspots' at each location, the screens are a platform for the local production of information that is intended to shape, alter and inform the resulting physical program and its relevance within the community as the individual projects are designed and developed. By re-presenting the neighborhood through the perceptions of different user groups, information technology is used to reinforce place and interaction.

Summary The language of information technologies and that of our cultural interactions are mutually reinforcing. While the contemporary computer interface has emerged from traditional cultural conventions, how we view and experience the world is framed in part by our interface culture. We have come to rely on information previously unavailable, but now made accessible through web-based resources such as real estate services, e-government and news agencies, to name a few. Our online worlds are impacting our physical ones, presenting new territories for urbanism.

We argue that if communication lies at the heart of the city's purpose, the changing nature of how we communicate is of key concern to thinking about city design. The form-based traditions of urban design were rooted in a city where social and economic relationships were built on proximity and vernacular traditions. By contrast, the globally networked metropolis is reliant on a complex web of relationships that are local and global, as well as physical and virtual.

23 See http://www.urbantapestries.net.

With new tools, we can envision new urban interfaces that enable physical proposals to be supported, augmented, and transformed by an integrated approach of digital technology and software customized to the needs of the neighborhood. Our intent is to uncover how the linkages between physical and digital space can be manipulated to produce, enhance or intensify urban environments — to reinvigorate the design project that is the city.

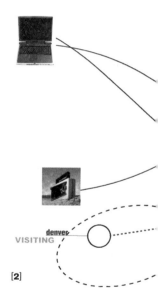

[2]

1 Strategic sites for physical and digital initiatives in Hudson County, NJ (Work completed by the authors while with Wallace Roberts & Todd LLC).
2 Editing the 'data glut': a personal derive encompassing physical and digital connections.

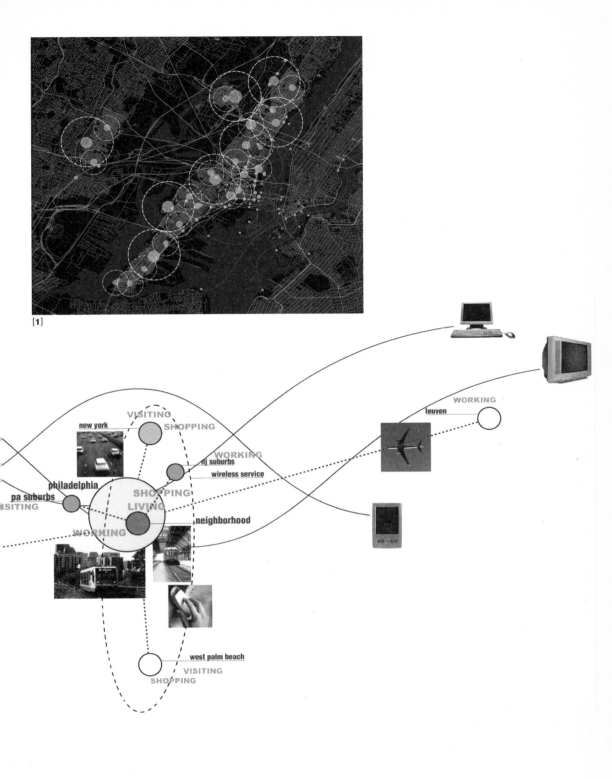

 Temple University
 APM

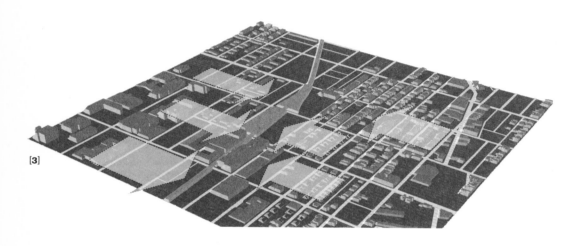
[3]

[4]

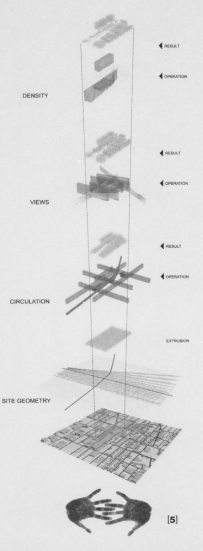

DENSITY
◀ RESULT
◀ OPERATION

VIEWS
◀ RESULT
◀ OPERATION

CIRCULATION
◀ RESULT
◀ OPERATION

SITE GEOMETRY
EXTRUSION

[5]

3 North Philadelphia: The rail station area isolates Temple University and APM despite their close physical proximity.
4 A reading of the context formed the urban response.
5 Initial massing insertion of potential development adjacent to the station.

RESIDENTIAL

OFFICE

OPEN SPACE

COMMUNITY SPACE

COMMERCIAL

[6]

6 Programmatic approach.
7 Windows into APM defined by new development.
8 Conceptual site plan.
9 View from the rail station platform illustrating a digital bulletin board on the rail platform, LCD displays and an elevated tree nursery.
10 Preliminary program for the 'media station' — using information technology to meet local needs.

[8]

[7]

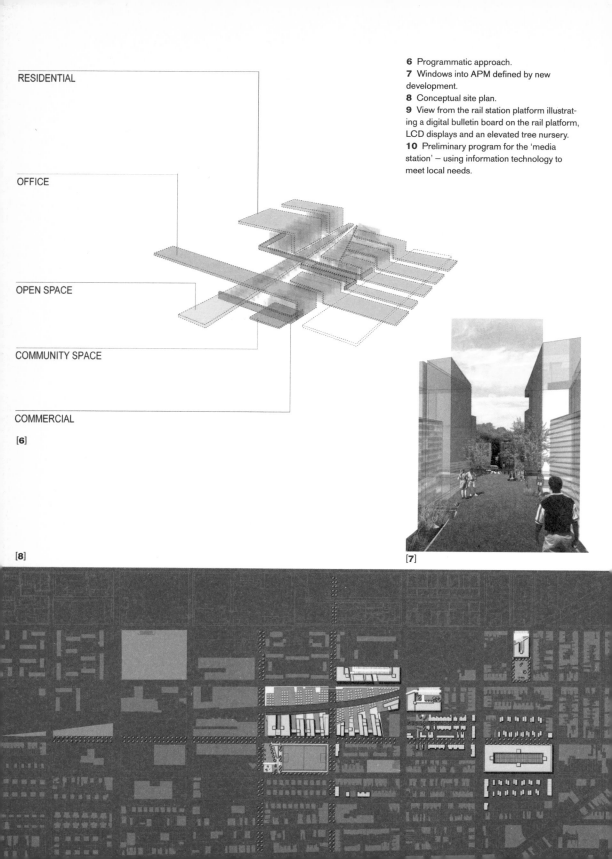

	Tools	**Production**
Performing Arts	Theater Lighting / Sound Writing Production Live Drama	*Dramatic Productions* *Dance Productions* *Children's Theater* *Film Screening*
Music	Music Recording Engineering Production Marketing Internships	*Holiday Concerts* *Music Performance*
Visual Arts	Painting Photography Graphic Design Web Design Animation Textiles	*Public Art* *Merchandise* *Web Design Consulting* *Craft Shows*
Life Skills	Basic Computer Skills GED / Prep / Testing Job Readiness Budget / Credit Counseling Homeownership Prep Parenting Counseling After School Programs Work Skills Certificate	*Adult Education* *Literacy Programs*
Economic Development	Resume Services Job Counseling Retail Entrepreneurship Training Library Shared Facilities	*Workforce Preparation* *Job Creation* *Housing Investment* *Small Business Assistance* *Business Incubators*
Community Building	Community Room Outdoor Plaza Planning Meetings Technology Resident Council	*Local Information* *Art Exhibitions* *Community Events*

The General Theory of Urbanization of Ildefons Cerdà: An Urban Network Theory

Francesc Magrinyà

Cerdà produced the first real treatise on urbanism. His *General Theory of Urbanization* (Cerdà, 1867) and its application to the Proposal for the Reform and Extension of Barcelona embody his comprehensive approach for industrial cities, which rested on five pillars: technical, legal, economic, administrative, and political. He considered the essential requirements of the new industrial city: '*Today everything is movement, everything is expansion, everything is communicativeness*' (Cerdà, 1867). For this reason, urbanization was to be based on mobility and networks. Rather than imposing a specific urban form, he offered a set of parameters to articulate an urban fabric based on complexity and to enable urbanization based on freedom. This is best summed up in his maxim, '*Ruralize the urban, urbanize the rural*' (Cerdà, 1867), where ruralized urbanization represents individual independence harmonized with enjoyment of social life.

1 Cerdà and His Times

The industrial revolution began to hasten change in the fabric of European cities by the mid-nineteenth century. Barcelona's unique situation at the time goes a long way to explaining Ildefons Cerdà and his work (Magrinyà, 1999). In 1714, the walled medieval city of Barcelona was declared a military post. Building outside the walls was banned, so when factories began to use steam power they still had to be built inside the medieval fabric of the city (1832). Until 1854, population density grew to extremely high rates. Real estate prices thus increased to levels that were unbearable for the working class, and a number of strikes ensued (1855), which were even noted by Engels himself. Faced with this untenable situation, the city's earliest urban planning demands were voiced in *¡Abajo las murallas!* (Down with the Walls) by the hygienist P. Monlau (1841); they never ceased until the military enclosure was torn down (1854).

Cerdà was a full participant in the city's transformation. His training at the School of Highway Engineers (i.e. civil engineers) in Madrid (1835-41) and the influence of the French Ponts et Chaussées school largely explains the origins of his urban approach. In 1854, he was elected to the Barcelona Council. He entered the fray by writing his *Monografía de la clase obrera* (Monograph on the Working Class) in 1856, an exhaustive analysis of Barcelona and its inhabitants in the tradition of early nineteenth-century sociological treatises, which were veritable cross-sections of the working class such as *Les ouvriers européens* by Le Play (1855).

At that time, formal consideration of urban issues harkened back to the early nineteenth-century utopian movements (following Owen, Fourier, and Cabet, among others), and had reached Spain only through isolated, scattered experiences. Cerdà was of the next generation, a generation of social reformers reluctant to pursue utopian solutions ungrounded in reality, seeking to adapt cities to the new needs for transport and expansion. Development was now driven by the new, railway-based mobility, as well as the needs defined by the 'hygienists'. Cerdà realized the impact of the railway when he first saw a train in Nimes (1844) and later with Spain's first peninsular railway, the Barcelona-Mataró line (1848).

Furthermore, ship transport was forming great expectations, especially among the followers of Saint-Simon. Their focal point was the opening of the Suez Canal (1859), designed by Ferdinand de Lesseps, a contemporary of Cerdà who had been the French consul in Barcelona in 1841.

Just as railways were beginning to spread, major reforms began in Europe's great metropolises. Paxton's project for London (1855), Haussman's boulevards in Paris (1852-70), and the Vienna Ring (1858) were the chief examples. Cerdà took in all these recent technological and urban developments, but he also salvaged the Spanish tradition in the founding of new settlements. This tradition, first developed as a result of Christian reconquest of the Iberian Peninsula and subsequently adapted to new towns founded in Latin America, was encoded in Philip II's *Ley de Indias* (Indies Act) (1573). Thus cities like Buenos Aires and Havana also became reference models in his proposals.

Cerdà's concept was a holistic, complete city. It entailed both reforming and extending the city in a locomotive-based urbanization process. But alongside the actual project, under the influence of positivism, he also set out to establish a new '*Science of Urbanization*', as he called it. Thus, his chief work, the *General Theory of Urbanization* (1867) and its practical application, the Project for the Reform and Extension of Barcelona, became the first modern treatise on urban planning, as pointed out by F. Choay in *La règle et le modèle* (Choay, 1980). His Project for the Barcelona Extension (1859) was the beginning of formal urban planning practice in Spain. By 1864, the first *Ley de Ensanches* (Extensions Act) already referred to *ensanches* in Barcelona, Madrid, Bilbao, and San Sebastian. Many other Spanish cities were to follow, so that *ensanches* became the preferential form of urban development in Spain well into the twentieth century, whereas the rest of Europe had no urban planning legislation until the work of Sitte and Baumeister in Germany in the 1870's.

2 The Origin of His Projects

Cerdà's early engineering work involved designing a number of roads. In 1844, on a trip to southern France, he was gripped by his first sight of a train, four years before the first railway line was built in Spain (the Barcelona-Mataró railway, 1848). In view of this new means of transport, which together with the telegraph was to generate a whole new civilization, Cerdà had an insight: mobility and communications would bring about the transformation of towns and landscapes. In 1849 he inherited his family's wealth and resigned from the Civil Engineering Corps in order to devote his entire life to developing a new discipline: the science of urbanization.

In 1854, during a brief Progressive rule in Spain (1854-56), the government authorized demolition of Barcelona's walls. In 1855, the government commissioned from Cerdà a topographical survey map of the outskirts. With his *Plano topográfico de los alrededores de Barcelona*, he seized the opportunity to submit a preliminary extension project. However, it was not until December 1858, when Barcelona ceased to be a military fortress, that he was officially commissioned to draw up his *Proyecto de reforma y ensanche de Barcelona* (Proposal for the Reform and Extension of Barcelona), produced in 1859 with final approval obtained in 1860. Later still, in 1861, he was to draw up a preliminary project for the reform of Madrid's inner city, entitled *Teoria de viabilidad urbana* (Theory of Urban Viability), in which he developed a theory of city reform.

Between 1860 and 1866, he was actively involved in the development of the Barcelona Extension (*Eixample*), as a technical expert for the Civil Government, as city councilor, and as technical director of a construction company, *El Fomento del Ensanche de Barcelona*. In 1863, he drew up his *Anteproyecto de Docks de Barcelona* (Preliminary Proposal for the Barcelona Docks), a chance to revise his Extension project. Cerdà had to bring his technical experience and political activism to bear in order to implement his urban planning schemes for the Barcelona Extension.

3 The 1855 Preliminary Extension Project of Barcelona: Housing Models and Urban Services

In 1855, Cerdà was commissioned to draw up a *Topographical Plan of the Outskirts of the City of Barcelona*. He drew it at a scale of 1:5000 with contour lines at one-meter intervals (one meter = 3.1 feet), which was incredibly advanced and precise for his time. He submitted this survey map along with a *Memorandum for the Preliminary Proposal for the Barcelona Extension* (hereafter MAEB, from the Spanish initials). Unfortunately, the overall design plan has been lost, but his *Atlas del MAEB* (found in 1988), which was attached to the memorandum, provides some idea of the main features of his first proposal for the Barcelona Extension.

3/1 A Metropolitan Project

The Extension became a metropolitan project, as it was to connect the surrounding towns to the city and increase available land tenfold. To begin with, Cerdà designed a great perimeter drainage sewer (*ramblar colector*) to intercept the streams that ran through the Extension area so as to avert flooding (Magrinyà, 1996a). In his Preliminary Project for the Extension, Cerdà designed a network of 35-meter-wide streets following the principle that different forms of locomotion should move independently. His street cross-section provided separate 'lanes' for plain pedestrians, load-carrying pedestrians, carriages, and railways. He also designed service alleys under which to run a service conduit, linked to adjacent buildings, grouping sewage, water, and gas pipes as well as telegraph lines (fig.1 and fig.2).

3/2 Housing Models

Cerdà designed eight types of model housing – four for the bourgeoisie and four for the working class. In his view, the ideal home was a 20 × 20 meter (62 feet by 62 feet) detached house with a palace-like structure; he took this ideal, superimposed it and juxtaposed it, and came up with smaller bourgeois housing units down to his fourth-order terraced housing, which was to become the standard for housing blocks in his 1859 project. Worker housing was designed in the shape of galleries built around a central courtyard; there were three different types of unit for different-sized families and a fourth type for single workers. He arrived at this minimal housing unit by meeting construction and hygiene requirements and keeping the cost affordable for workers.

3/3 A Garden City Proposal

Cerdà grouped his city blocks (as opposed to housing blocks) by combining different types of bourgeois and working-class housing in various arrays. He looked to city blocks in New York, Edinburgh, and especially London which had examples that were closest to the ideas he was developing at this early stage. The roadway structure of this early project has not been accurately identified. However, a number of scenarios have been worked out to provide perspective views of the garden-city-like urban fabric initially designed by Cerdà for the Barcelona Extension.

4 The 1859 Proposal for the Reform and Extension of Barcelona: The Integrative City Plan

With the experience of the 1855 Preliminary Project, Cerdà was way ahead of others in 1859, and drew up his 1859 Proposal for the Reform and Extension of Barcelona in just three months. He did not merely draw up a general plan including a memorandum to explain the project; he actually seized this opportunity to present the underlying theory that had been applied to this particular project: the Theory of City Construction.

4/1 An Industrial City Project

Cerdà's plan consisted of an open, egalitarian grid-based city (Magrinyà, 1996b), the opposite of a radial scheme. The basic roadway structure consisted of 20-meter-wide streets for the urban grid, plus a number of 50-meter-wide 'transcendental ways', as he called them, or main thoroughfares to link the city to the outside world and to give structure to the overall plan (fig.1). The underlying theoretical model is borne out by analysis of the plan: there was an equitable distribution of services and public facilities consisting of a total 60 × 20 city block arrangement divided into 3 sectors (20 × 20 city blocks each), 12 districts (10 × 10 city blocks) and 48 'neighborhoods' (5 × 5 city blocks). There was to be a community center for each 'neighborhood', a market for each district, a suburban park for every two districts, and a hospital for every sector.

4/2 The City Block Model

The city block arrangement consisted of

two blocks of housing. Housing units were based on the fourth-order bourgeois housing developed in 1855, which ensured good ventilation and sunlight. Cerdà worked out the size of the street block using an equation that included the following variables: street width, façade length, building depth, certain features of the 1855 housing unit, and number of square meters per resident. These modular requirements were then turned into a number of arrangements in which the two equal-length housing blocks on a city block were parallel, in an L-shape or in a T-shape. These arrangements allowed for a public garden to be set up on every city block, as well as private gardens for ground-floor apartments. This is what Cerdà termed the *vía-intervías* (the way-interways), and it was the way to follow his maxim, 'Ruralize the urban,' parallel to 'urbanizing the rural.'

4/3 The Chamfered Grid

One of Cerdà's chief contributions to urban planning was his chamfering of the corners on the street grid to ensure continuity of movement. He cut off the corner of each block, so as to create a small square at every intersection. This created an alternation of street lengths and octagonal plazas with 20-meter sides, at 133-meter intervals (113 meters for buildings and gardens plus 20 meters for street width; one meter = 3.1 feet). Intersection design assumed that intersections would attract and concentrate many public street-based occupations encouraged by mobility. He listed over 90 functions that could be performed in these plazas – trade, exchange, guarding, etc. Thanks to these open spaces, Barcelona has managed to survive the gridlock caused by cars in urban centers.

5 The 1863 Preliminary Project for the Barcelona Docks: Adapting for the Railway

In his *Teoría General de la Urbanización* (1867) Cerdà established the basis of a theory of the urbanization of the locomotive, but his first attempt to develop a solution in terms of a project was his 1863 Preliminary Docks Project (Magrinyà, 1996c). In Nîmes in 1844, Cerdà was deeply struck by the novelty of trains: *'I saw, for the first time, the way railways work, and the thought first came to me of studying the transcendental influence they will exert on urbanization as they become widespread.'* The following paragraph from one of the reports on the contents of his Preliminary Project conveys its central point: *'The idea that the locomotive might penetrate into a town devoted to the service of movement on land and sea, and do it in such a way as to extend through every district, cut through every street-block, pass close to every house, and even enter the houses to provide its inestimable services throughout.'*

The compilation of all the materials regarding the 1863 Preliminary Project for the Docks, the text of which unfortunately has not been found, has allowed us to recreate something close to the proposal with which Cerdà attempted to solve the problem of adapting the City to the introduction of the railway. Early in the process of developing the Extension, Cerdà was compelled to revise his 1859 Proposal according to the Final Approval Decree issued in 1860. This he did, and he included his 1863 Revised Proposal in a Preliminary Proposal for the Barcelona Docks. The main development in this revision was the ultimate inclusion of the railway within the street grid, in his own words 'the urbanization or domestication of the locomotive'.

5/1 A Proposal Adapted To Introducing Railways

Cerdà set out to create a distributed railway network and design street-blocks specifically to hold the factories, warehouses, shops, and housing that would develop in the new industrial city. Linking shipping to overland transport became the chief consideration. His new 'locomotive urbanization' solution for trains to run through the Extension involved a circle railway line connected to a 'station of stations' operating as an intermodal exchange center, and three parallel railway lines running through the city with platforms connected to houses (fig.3 and fig.4).

5/2 The Railway Block

Cerdà's proposals for Barcelona were inspired by an 1844 articulated train design by the French engineer Arnoux which reduced the turning radius considerably. Cerdà had already designed a railway project around this particular train system in 1856. Extending the railway lines throughout the city led him to adapt the grid to the new needs that emerged. He designed what we have termed 'railway-blocks' by grouping four city blocks, each with buildings arrayed in an L-shape, with the semi-underground railway running through the overall square formed by the buildings. Streets became specialized according to the means of locomotion, so that those assigned to the railway were kept separate from the rest and at a lower level. The new buildings were linked to the railway stations and platforms on two levels (underground and ground floor); they were designed to hold warehouses, workshops, shops, and homes (fig.5).

5/3 Densification of the Street-Block and New Construction Proposals

The Final Approval Decree (1860) entailed enlarging the urbanized area, allowing higher buildings, and higher construction density. The city block that consisted of two parallel blocks of buildings in the 1859 Extension project now became a city block built up on three sides, in a U-shape. As technical director of a private-enterprise development company called *El Fomento del Ensanche de Barcelona* – one of the organizations related to the construction of the Extension – Cerdà designed a solution to the densification requirement – a grouping of two city blocks, with buildings in facing U-shapes and a central garden area, in which the buildings on the chamfered corners were designed to allow access to the central courtyard from the street (Gimeno & Magrinyà, 1996). Analysis of the 1863 Revised Project shows the new dominant building layouts. Cerdà redesigned the city blocks to form linear groupings of facing U-shaped buildings; groupings of four city blocks built in L-shapes facing the center, to form the railway blocks; and large, monumental groupings, as well.

The link between shipping and overland transport became central to his plan. Cerdà found a solution for the introduction of the steam engine in a *circle line* and a *three-line axis* built through the Extension. Around this three-line axis, Cerdà designed groupings of four street-blocks forming a 2 × 2 square using the same street alignments as the 1859 project; this grouped shape we have termed a 'railway-block'. Each of these groupings or railway-blocks had a railway running through its central axis below street level. This combination provided the railway-block with the same outer frontage as the 1859 project, and combined 'ordinary' traffic (stagecoach and tramway) with 'perfected' traffic (railway) on different levels.

Introducing railways involved adapting the Extension grid to new needs. Furthermore, two of the housing storeys – the ground floor and basement – were adapted to link with the (road)ways and the railways, and thus became the interconnection for these different types of ways. By introducing the third dimension into his scheme, Cerdà had to plan for compatibility of the service grids, especially the sewers, and we have inferred that he devised a sewer plan for Barcelona.

6 The Evolution of Cerdà's Urban Planning Proposals

6/1 Three City Models

Cerdà's earliest proposal, the 1855 Preliminary Proposal for the Extension, was shaped by a study of housing, the introduction of urban services, and an eye for the needs of new means of transport. All in all, it is a typical garden city proposal. The second proposal, contained in the 1859 Extension Proposal, is much more sophisticated, including a general layout that structured the city into neighborhoods, districts, and sectors. Facilities relevant to the new industrial city were all enmeshed within this structure. The basic unit was a square, chamfer-cornered street-block with sides 113 meters long. Finally, in the 1863 Preliminary Project for the Docks, Cerdà designed a revised version of the 1859 proposal in which the overriding consideration was the final inclusion of the railway in the street grid, as well as increased building density entailing three blocks of buildings on every city block instead of two; he even designed a housing project on this basis (fig.1).

6/2 Evolution of the Way-Interways Concept

At city block level, we note that Cerdà's first proposal groups together the three elements of urban fabric – buildings, gardens, and streets. The urban fabric is structured by groupings of different types of bourgeois and working-class housing, standard streets of 35 meters wide, and a service conduit for new urban services. In his 1859 proposal, Cerdà struck new theoretical ground and defined the 'way-interways' (the city block and the street all around it) as the minimal unit. He proposed two detached blocks of buildings and a central garden area on every city block, with chamfered corners, 20-meter streets, and octagonal plazas. In the 1863 Revised Project, the 'way-interways' idea was adapted to a new means of locomotion – the railway – which required specialized *ways*. New street-block groupings now termed 'railway blocks' comprised housing with direct links to train platforms below street level.

6/3 Three Different Urban Fabric Proposals

Focusing on a particular area in which the old city merged with the Extension helps illustrate the evolution of Cerdà's designs. The alignments drawn in the first project show a dominant pattern of rectangular city blocks with detached housing following vertical lines parallel to the maximum slope line for the sewage system. The 1859 proposal had matured into the present alignment grid with 20-meter-wide streets and 113-meter-long city blocks predominantly built up on two parallel sides. The 1863 revised proposal preserves the alignments in the preceding scheme, but the dominant street-block patterns are the so-called railway blocks or city blocks built up on three sides.

7 Cerdà's Urban Planning Instruments for Implementing the Extension

Cerdà played an active role in the building of the Extension in several capacities: 1860-1865, as technical advisor to the State; 1863-1865, as city councilor; and 1863-1865, as technical director of a company called *El Fomento del Ensanche de Barcelona*, which dealt in real estate and housing development (Gimeno & Magrinyà, 1996). Analysis of Cerdà's actions during this period has cast light on the instruments he used in order to guide his plan to completion.

7/1 Flexibility of the Extension Grid

One of the essential traits of Cerdà's project is its outstanding combination of order-creating potential and great flexibility. He defined a system of alignments able to accommodate many different urban models, whatever system might be used to group buildings on city blocks. Thus, he used the same basic grid in two separate instances to propose two completely different systems of streets and city block groupings, both perfectly suited to the predefined needs for mobility and habitability. Overall, this shows the adaptability and flexibility of the grid. The Alignments and Gradients Plan proposed by Cerdà remained in place until 1953, when the *Plan comarcal de Barcelona* (Regional Plan of Barcelona) came into force.

7/2 A System for Site Adjustment and Compensation

In 1861, Cerdà published a leaflet – *Cuatro palabras sobre el Ensanche dirigidas al público de Barcelona* (Four words concerning the Extension of Barcelona, directed to the people of Barcelona) – which considered all the potential adjustment situations that might develop when rural real estate was made urban. It was the origin of what is now known as a 'site adjustment system' (fig.6). This technique was essential to the successful implementation of Cerdà's theories. In those early years, as technical director in charge of the Extension, he had almost absolute control over its implementation. For every landowner wishing to build, Cerdà drew up a geometric site-adjustment plan on which he drew the area that could be used and the buildable surface area, according to the alignments in the project. This information was also faithfully copied onto a set of 28 larger-scale land adjustment plans called '*planos particularios*,' which gave him total control over the course of Extension building (Gimeno & Magrinyà, 1996).

7/3 Elements of Construction That Allowed the Project to Prevail

An infinite number of plans in the history of urban planning have remained mere paper. Cerdà's great virtue was his ability to impose his proposal and secure implementation of its essential points. In the early days, three issues were central to the effectuation of his plan. First, and fundamentally, his redesign of the linkages between the Extension and the old city, which appears in his 1865 Proposal for Alignments for the Tracts of the Former City Wall. Second, his managing to retain the chamfered corners in the early years preserved the fundamental virtues of the project. Last, because the central stretch of the *Gran Via*, a 50-meter-wide street, was built in the early years, the Extension was finally built according to his grid layout instead of a radial layout.

Present-day Barcelona is the product of a number of operations that have determined its urban morphology. Cerdà was directly involved in the 1850's and 1860's; his Extension project remained in force until 1953. However, his Building Ordinances and his Economic Plan to finance the work were never approved. For almost a century, a succession of building ordinances regularly increased the original building density. Yet today, the value of the Extension still lies in the preservation of the basic features in Cerdà's project.

8 Cerdà's Theory

Cerdà's *General Theory* is a paradigmatic instance of the break with utopian approaches and the implementation of modern urbanism as a discipline: '*Nothing can or should be offered the present generation, with its spirit of practical positivism, if after thorough reasoning it fails to satisfy all the necessary conditions for immediate, rapid implementation. The famous planners and utopians of the sixteenth, seventeenth, and even the eighteenth centuries would quite rightly be a laughing stock in our own time.*' (*Pensamiento económico*, 1860) (Cerdà, 1861).

8/1 Cerdà, the Founder of a New Discipline: the Science of Urbanization

By developing the concept of urbanization and writing his *General Theory of Urbanization* (1867), a summary of his earlier writings on theory, Cerdà became the founder of urban planning as a science, which he termed 'the science of urbanization'. Along with his generation, he was aware of the significance of the harnessing of steam for industry and transport: '*I understood that the harnessing of steam as a driving force had signaled the end of one era for humanity and the beginning of another.*' The current relevance of his *Theory* rests on his ability to perceive 'the demands of the new civilization, with movement and communicativeness as its distinguishing traits', which led him to proclaim '*today everything is movement, everything is expansion, everything is communicativeness.*' Therefore, Cerdà established a modern conception of urbanism, but he also read the process of urbanization as a function of the development of networks for mobility and communication – a precursor of network urbanism (Magrinyà, 2002).

8/2 Urban Planning Instruments Contributed by Cerdà

The central ideas in Cerdà's work are reflected in the following principles:
- Independence of the individual within the home
- Independence of the home within the urbs
- Independence of different kinds of movement on urban roadways (fig.7)
- 'Ruralize the urban, urbanize the rural'

Cerdà felt that the main objective was to achieve '*the harmony that reigns in ruralized urbanization between independence of the family and enjoyment of sociability*' (*Teoría General de la Urbanización*, 1867). He put his ideas to work in his Proposal for the Reform and Extension of Barcelona.

Cerdà proposed an integrated view of urbanism relying on five bases: technical, legal, economic, administrative, and political. He even stated: '*It seems impossible that there might be a professional whose heart does not tremble at drawing the first lines of the plan for a city, knowing as he must that these lines determine the material and moral future of countless families.*' With his comprehensive, integrative approach to urbanism, not only did he design the overall layout; he also defined the rights and duties of landowners and of the government, as the legal basis of the city; established rules and mechanisms for financing urbanization work and sharing out the burden and the profit, as its economic basis; drew up building ordinances following a set of general administrative principles; and his principle for political action was to harmonize the desirable with the achievable, allowing trade-offs as a form of transition to the ideal (Soria, 1996).

Foremost among the theoretical tools devised by Cerdà was his 'way-interways' concept. It derives from his view of urbanization as a balance of stasis and mobility, and explains change in urban layouts as a result of evolution in transport networks and urban services. Putting this new concept to work, Cerdà defined a system of alignments that enabled straightforward site adjustment and compensation for the transformation of rural properties into urban land. By using his Alignments Plan for the Barcelona Extension (1861) and defining the 'elements of construction' (housing models, urban services, types of street, etc.), he had the requisite parameters to bring order into complex urban fabric so as to develop urban practice based on individual freedom – its legacy, the modern *Eixample/Ensanche* in Barcelona (fig.8) (Magrinyà 1996d).

9 The Extension, A Model Urban Fabric, an Example of Urban Network Theory

In the case of the Barcelona Extension, the square grid is the result of applying the general principles of Cerdà's General Theory of Urbanization. Independence of the individual within the home and of the home within the city (or 'urb') led to alignments that allowed for detached construction, a feature of rationalistic urbanism, which evolved into the present-day Extension city block. The principle of independence for each means of locomotion led to a standard 20-meter-wide street, perfectly capable of absorbing the different means of transport. Cerdà designed street intersections on the principle of continuity of movement, a critical requirement for traffic. The result is the chamfer-cornered street-blocks typical of the Barcelona Extension, which ensure adequate mobility. Finally, city block inner courtyards with a central public garden area should implement the principle of 'urbanizing the rural, ruralizing the urban' (Magrinyà 2002).

9/1 The Alignments Grid: Structuring Complexity

The Barcelona Extension includes a number of features which, while allowing great latitude, give structure to the construction complexity typical of urban fabric. This insight comes into clear focus with aerial photographs showing a sector of the modern *Eixample* and part of the old city (fig.9). Overall, we have a simple yet complex right-angled grid with octagonal plazas at every intersection. Cerdà's 'transcendental ways' give overall structure to the grid and connect it to the land beyond. Finally, the features that articulated the *city* block as a whole allow the elements of stasis (buildings, plot, parcel) and those of mobility to come together. Thus, the grid preserves a high degree of flexibility and adaptability, demonstrating how it can order the complexity of an urban ensemble (Magrinyà, 1998).

9/2 The Relevance of the Way-Interways to Building and Site Adjustment

Mobility systems prove the relevance of the way-interways concept. With mobility increasingly the determining factor, buildings and building plots can no longer determine urban form, but must fit in the interways ensemble. The Extension grid shows how Cerdà's *city* block can absorb different types of building and site adjustment. City blocks ranging from the standard *Eixample city* block as it stood in the late nineteenth century, to the new city blocks in the Olympic Village, through such arrangements as the *Mercat de Sant Antoni* (St. Anthony's Market), the *Fàbrica Batlló* (The Batlló Factory) industrial site, or modern patterns of occupation, all show the relevance of Cerdà's 'way-interways' as a determining element of the new balance between stasis and mobility. Furthermore, the grid has managed to absorb and adapt to the various, changing forms of architecture practiced in the Extension over the years, which explain its special nature as an architecturally and urbanistically qualified city.

For Cerdà, the introduction of steam power for locomotion meant a radical change in mobility systems, but his design only came into its own with the arrival of the motorcar. Even so, the revolutionary nature of his plan is apparent, because it was the first attempt to create a theory of urbanism and because it integrated all the service and transport grids (Magrinyà, 2002). This makes it an essential reference for the new concept of network urbanism (Dupuy, 1991) which accords a central role to mobility and communications.

Bibliography

Cerdà, I. (1859), 'Teoria de la construcción de las ciudades'; In: AAVV (1991), **Cerdà y Barcelona**, vol.I, INAP & Ajuntament de Barcelona, Madrid.

Cerdà, I. (1861), 'Teoria de la viabilidad urbana'; In: AAVV (1991), **Cerdà y Madrid**, vol.II, MOPT & Ayuntamiento de Madrid, Madrid.

Cerdà, I. (1867), **Teoría General de la Urbanización**, Imp. Española, Madrid, 2 vol.; In: Estapé, F. (1971), **Teoría general de la urbanización**, vol.I & II, Instituto de Estudios Fiscales, Madrid.

Choay, F. (1980), **La règle et le modèle: Sur la théorie de l'architecture et de l'urbanisme**, Seuil, Paris.

Dupuy, G. (1991), **Urbanisme des réseaux, théories et méthodes**, Armand Colin, Paris.

Gimeno, Eva & Magrinyà, Francesc (1996), 'Cerdà's Part in the Building of the Extensions'; In: Magrinyà, F. & Tarragó, S., **Cerda Urbs i Territori. Plannning Beyond The Urban**, Ed. Electa, Barcelona, pp.167-188.

Magrinyà, Francesc (1996a), 'Service Infrastructures in Cerda's Urban Planning Proposals'; In: Magrinyà, F. & Tarragó, S., **Cerda Urbs i Territori. Plannning Beyond The Urban**, Electa, Barcelona, pp.189-204.

Magrinyà, Francesc (1996b), 'Way-Interways: A New Concept Proposed by Cerdà'; In: Magrinyà, F. & Tarragó, S., **Cerda Urbs i Territori. Plannning Beyond The Urban**, Electa, Barcelona, pp.205-224.

Magrinyà, Francesc (1996c), 'The 1863 Preliminary Docks Project: a proposal for the railway urbanization of Barcelona'; In: Magrinyà, F. & Tarragó, S., **Cerda Urbs i Territori. Plannning Beyond The Urban**, Barcelona, Electa 1996, pp.225-254.

Magrinyà, Francesc (1996d), 'Les propositions urbanistiques de Cerdà pour Barcelone: une pensée de l'urbanisme des réseaux', **Flux**, n°23, January-March, 5-20.

Magrinyà, Francesc (1998), 'Urbanismo de redes y planeamiento urbano', **Revista Obra Pública. Ingeniería y Territorio**, n°43, Urbanismo II, 48-57.

Magrinyà, Francesc (1999), 'Las influencias recibidas y proyectadas por Cerdà', **Ciudad y Territorio. Estudios Territoriales**, Vol. XXXI, n°119-120, 95-117.

Magrinyà, Francesc (2002), **La theorie urbanistique de Cerdà et son application à l'»Ensanche» de Barcelone: une genèse d'urbanisme de réseaux**, Ecole Nationale des Ponts et Chaussées (enpc) – Director Gabriel. Dupuy.

Soria Y Puig, Arturo (1996), **Las cinco bases de la teoría general de la urbanización**, Barcelona, Ed. Electa, 447 pp..

[1]

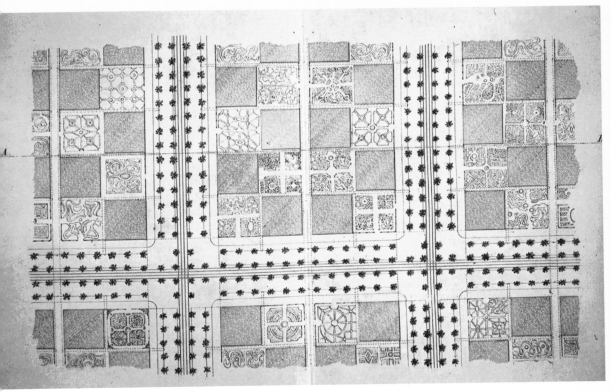

1 Grouping of street-blocks in the 1855 Preliminary Project (a), in the 1859 Extension Project (b), and in the 1863 Preliminary Project for the Barcelona Docks developed in 1863 (c) (Magrinyà & Gassull, 1994).
2 Elongated street-blocks grouped along the sewer line proposed in the 1855 Extension Plan (**Atlas del Anteproyecto de Ensanche de Barcelona, 1855**).

[3]

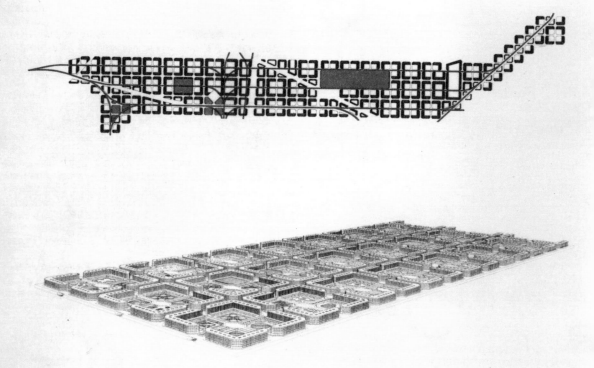

[4]

3 The Preliminary Project for the Barcelona Docks developed in the 1863 Teoría del Enlace de las Vías Marítimo Terrestres (Theory of the Linkage of Landways and Seaways) (Magrinyà, 1994).
4 Three-line trenched railway axis proposed by Cerdà for the introduction of the railway within the Extension (above). 2x2 street-block grouping, stacking dwellings, workshops and factories, and the railway (below) (our own production).
5 Conjectural railway street-block, connecting railways and housing, proposed by Cerdà (Magrinyà, 1994).

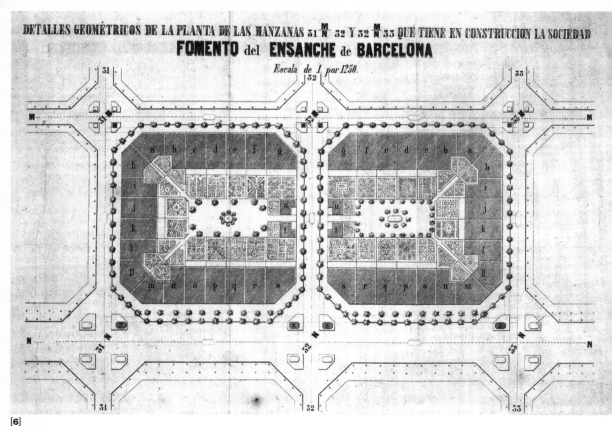

[6]

[7]

6 Brochure for the El Fomento del Ensanche de Barcelona company, of which Cerdà was Technical Director from 1863 to 1866 (Caixa Cerdà, Arxiu Historia de la Ciutat de Barcelona).

7 Street cross-section defined following the criterion of independence of means of locomotion (Atlas of the Preliminary Extension Project of Barcelona, 1855).

8 Different grouping of street-blocks proposed by Cerdà or built during the initial stages (Magrinyà, 1994).

9 Aerial photographs showing a sector of the modern **Eixample** and part of the old city (Cerdà Urbs i Territori, 1996).

AGRUPACIONS INTERVIÀRIES DEL PROJECTE DE 1859

AGRUPACIONS INTERVIÀRIES DE LA REELABORACIÓ DE 1863

DIFERENTS PROJECTES D'AGRUPACIONS INTERVIÀRIES

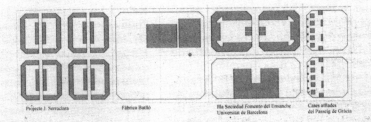

Projecte J. Serraclara — Fàbrica Batlló — Illa Sociedad Fomento del Ensanche Universitat de Barcelona — Cases aïllades del Passeig de Gràcia

[8] [9]

Compact City Replaces Sprawl

Nikos A. Salingaros

The compact, geometrically integrated city can and should replace suburban sprawl as the dominant development pattern in the future. This approach to urban planning and design is well established among proponents of the New Urbanist and Smart Growth movements. However, the more radical scenario I propose in this paper is that the compact city should also replace the high-rise, ultra-high-density megacity model. I will present arguments for the compact city from both directions, criticizing both conventional suburbia and the hyper-intensity of the urban core. A radical intervention is required on the part of concerned urbanists. We need to rethink the positioning of individual buildings to form a coherent urban fabric, as well as the role of thoroughfares, parking, and urban spaces. New zoning codes based on the rural-to-urban Transect and the form of the built environment are now available to assure predictable densities and mixed use for the compact city.

1 Introduction

Sprawl is a remorseless phenomenon. We see it covering more and more of the earth's surface, whether it is in the form of *favelas*[1] invading the countryside in the developing world, or as monotonous subdivisions in the United States. Nevertheless, the city of tomorrow (actually, in many parts of the world, the city of today) has a low-rise, compact human scale. If the government does not forbid it (or cannot control it), *favelas* eventually condense to define compact urban regions, but the same organizing process cannot occur in subdivisions because of anti-urban zoning. A *favela* can become living urban fabric, whereas its high-priced US analog remains dead. The difference is in the connectivity.

Suburban sprawl has become a self-generating, self-fulfilling 'machine' that produces an enormous amount of mechanical movement, but is not conducive to natural human actions and needs. Sprawl persists because vehicles define a now-familiar self-perpetuating entity: the auto-dependent landscape. Cars enable sprawl, and sprawl needs cars. This suburban 'machine' now circumvents its human creators and feeds in directly to the globalized economy. Yet it wastes untold amounts of time and resources, while trapping those without cars in their homes.

High-rise apartment and office towers are equally unsustainable. The serious threat of high-energy costs makes both ultra-high-density environments based on skyscrapers, and low-density suburban sprawl no longer feasible. Ultra-high-density urbanism creates more problems than it solves, in the form of energy reliance that draws on the resources of an enormous surrounding region and shortsightedly depends on an uninterrupted supply of cheap oil. Our only alternative is the smaller-scale, compact city, ideally surrounded by and close to agricultural lands for local food supply. We should produce viable settlements at optimal densities for the human scale, just as body tissue has a compact structure at an optimal density. This approach can be achieved through thoughtful planning and the appropriate codes.

1 A term developed in Brazil meaning 'shanty town' [**ed**.].

Urbanism once meant dense city living for humans, but anti-urban forces have (literally) driven people out to the opposite condition: low-density suburban sprawl. The correct solution is not formless sprawl, however, but an intermediate density low-rise compact city that is geometrically integrated. The huge commercial success of postwar suburban growth (a low-density phenomenon) took place because it harnessed genuine and powerful socio-economic forces. It also generated and fed some of those forces by means of clever media manipulation and advertising. Those same forces can be channeled to build a better environment for human beings – the compact, geometrically integrated city – so as to make an urban environment adaptable as much as possible. Suggestions for achieving that on a theoretical level are offered in (Salingaros, 2005a).

There is nothing wrong with either high density or low density per se, as long as it is well integrated with other densities and is in the right place (not too much of the same thing). People in the past several decades seem to have bought into the false notion of geometrical uniformity, which goes back to the now discredited 1933 Charter of Athens (Salingaros, 2005b). That document introduced notions that turned out to be catastrophic for cities, such as separating functions into single-use zoning, the false 'economy of scale', and also seductive but toxic images of ultra-high-density skyscrapers, vast open plazas, and uniform housing developments. It gave planners the idea of disintegrating the city into non-interacting components, or at best, ones that interact with each other only at tremendous cost and inconvenience; the opposite of a geometrically integrated city.

2 Andrés Duany and the Transect

Even the best theoretical urbanism is close to useless without changes in our zoning codes, however. The existing codes, more than anything, determine the pattern of urbanism. The planner-architect Andrés Duany and his partner Elizabeth Plater-Zyberk are at the forefront of efforts to reform these codes. They coded and designed the highly successful New Urbanist community of Seaside, Florida in the mid-1980's. The momentum from Seaside propelled traditional town planning again into the mainstream of planning options.[2] Duany and his colleagues have built numerous New Urbanist projects around the world, and in each case they work closely with the local government to adopt codes based on urban form instead of the separation of uses. Without a form-based code, one cannot predictably plan a human-scale community. Duany will not work for a community that wants to rebuild itself, but which stubbornly retains its postwar anti-urban codes. He has found out from experience that it leads to time-consuming and irresolvable conflicts.

Using a very pragmatic approach to urban form, Duany classifies different zones according to a Transect (i.e. a cross-section of a continuum) of the built environment, according to intensity and density of urban components. He then proposes that communities ensure their desired urban character by adopting written codes that prescribe it. In Transect planning there are six zones, but the three zones T3 (Sub-Urban), T4 (Urban General), and T5 (Urban Center) contain the areas that we would identify with a compact, walkable, mixed-use village or city neighborhood.[3] Unfortunately, the single-use zoning of the past sixty years has made such compact patterns illegal. (Note that, as explained below, Sub-Urban is not the same thing as suburban.)

I propose that a compact T3/T4/T5 city or town begin to substitute for suburban sprawl everywhere around the world. The compact city is sustainable, whereas both sprawl and the high-rise megacity are not. The Transect codes are ready for immediate use, and should therefore be adopted by government agencies. The 'low-density city' we now see erasing farmland is not a city: it feeds off and depletes a vast region that it keeps at a distance, so the functioning city is much larger, has a higher net density than first appears, and is ultimately unsustainable.

3 The Three Urban Transect Zones of the Compact City

Transect Zone T3 allows single houses on large lots, with a looser road network than in the higher zones. A Transect-based code limits the density to maintain a relatively rural character. Still, there would be walkable street connectivity to the denser Zones, so that residents are not isolated and forced to use cars for all their daily needs. Thus, T3 is part of the compact city, not estranged from it. (Country houses, on the other hand, would be part of T2, the Rural Zone, which is by definition outside the city.) The T3 Zone may be the same density as the dreary suburban tract houses we see in sprawl – technically referred to as Conventional Suburban Development (CSD) – but other key design elements in the new codes ensure much more housing diversity, walkability, and connectivity.

Transect Zone T4 is the denser Urban General Zone, with houses closer to each other and to the sidewalk. More mixed use is permitted, with corner stores and restaurants within walking distance of most houses. As soon as the density permits, therefore, the mixing of functions is actively encouraged by the Transect-based codes.

Finally, Transect Zone T5 is the Urban Center, thoroughly mixing commercial uses with housing. This is analogous to the neighborhood center or small-town Main Street in early twentieth-century America, as well as the traditional European village. Transect-based zoning supports the compact city from both of the critical standpoints identified earlier, for it also prevents the erection of high-rise buildings and vast parking lots, whose expanse and density destroy the desired human-scaled character of T5.[4] Other important details, such as sharp curb radii and narrow

---

2 Duany, Andrés & Plater-Zyberk, Elizabeth; **Smart Code**, Version 6.4, <www.dpz.com>, Miami (2005). and Duany, Andrés; Plater-Zyberk, Elizabeth; Speck, Jeff; **Suburban Nation**, (New York: North Point Press, 2000).
3 Duany & Plater-Zyberk; (2005) **op. cit.**
4 The height limit in the Duany Plater-Zyberk Transect-based Smart Code is three storeys for T3, four for T4, and six for T5.

streets, help to calm traffic. The urban geometry in these Transect Zones is entirely different from that of sprawl (Conventional Suburban Development): roads and buildings correspond more to the compact small town found at the turn of the last century. Suburban sprawl, on the other hand, is neither a low-density CITY nor true country living – in pretending to be both, it accomplishes neither. The correct Transect codes ensure that the complex urban morphology necessary to support the city for people will not disintegrate into disconnected sprawl.

One crucial point of the Transect is that the three zones T3, T4, and T5 connect to and adjoin each other. Each one is kept by its own code from changing wildly, yet each one needs the other two next to it. Suburbia without an urban center requires constant driving, while a downtown without a healthy mix of uses is dead after business hours.[5] The codes prevent the repetition of one single zone over a wide area, thus preventing the monoculture of sprawl.

Theoretical work[6] based upon earlier work by Christopher Alexander[7] supports Duany and Plater-Zyberk's practical prescriptions with fundamental arguments about urban form and structure. New Urbanist solutions also draw upon the neo-traditional notions of Léon Krier.[8] The same approaches will, of course, also work for the Urban Core (T6), as well as for Natural and Rural Zones (T1 and T2), and the appropriate Transect-based codes apply to those densities as well. Nevertheless, here my topic is the compact city, a human-scaled city to replace both sprawl and the high-rise megacity. The compact city, therefore, involves only the medium-density zones of T3, T4, and T5.

4 Sprawl is Driven by the Car

Sprawl exists only because it is an outgrowth of car activities. In turn, this automobile dependence generates urban geometries that accommodate cars first and pedestrians second. These are the wrong priorities for a healthy life, especially for those who cannot drive: the young, the old, and the poor. The sustainable compact city must be designed for the pedestrian first.

People have been encouraged by the automobile industry and by government agencies promoting the automobile industry to indulge in an impossible and destructive fantasy of inappropriate urban types. In practical terms, sprawl comes about from misunderstanding urban morphology. The needs of the car automatically generate an urban morphology appropriate to the car. Sprawl relies totally on the automobile, and thus follows the dendritic (treelike) geometry of roads. A dendritic geometry is good for the automobile, but is inappropriate for human beings. Sprawl occurs when buildings are erected with no regard or understanding of which connective geometries encourage walking. Suburban sprawl grows uncontrollably, generated by anti-urban zoning codes that achieve the opposite geometry to what human beings need.

Complex urban fabric means condensation, connectivity, and mixing; the opposite of homogeneity.[9] And yet, most postwar planning has deliberately spread a homogeneous, amorphous structure over the earth, replacing healthy urban fabric in existing compact cities. Monoculture displaces and stretches its vital connections to complementary nodes, making the functioning city (a much larger entity that encompasses the entire commuting distance) tremendously wasteful of both time and energy.

With the wrong codes in place almost everywhere today, roads in fact determine the geometry of urban settlements. Let's examine what happens when the government builds a road to connect two towns. A road in the countryside attracts new buildings along its length, thus linking each building with that particular road and with nothing else. But human beings do not link to a road: they link to work, school, church, medical facilities, etc. Clustering is supposed to occur among linked human activities, and not strictly between houses and a road. It's the wrong linking, and it destroys the meaning of a city.

The solution is obvious to some of us. Zoning codes should prevent the dendritic growth of buildings along roads, and instead promote an urban geometry that concentrates human connections inward to focus on local urban nodes. Transect-based zoning has the correct zoning codes that do this, replacing anti-urban zoning codes that allow the unrestrained growth of the auto-dependent landscape.

5 Laws, Regulations, and the Democratic Ideal

I have proposed Transect-based zoning to regulate the development of urban areas of different density. It may appear to a reader that this represents a rather strict set of regulations. The notion of regulations runs counter to our utopian conception of civic freedom, and may cause strong protests if not revolution. In the case of Transect zoning, however, I am simply advocating a REPLACEMENT of very rigid zoning codes that already exist, governing the geometry of buildings and roads. Most people are woefully unaware of how tightly the built environment is now controlled by existing codes on planner's books. They have been sold the false image of 'suburban freedom'. In fact, Transect-based zoning provides MORE choices for development than does current single-use zoning.

---

[5] Salingaros, Nikos A.; 'Towards a New Urban Philosophy: The Case of Athens', in **Shifting Sense — Looking Back to the Future in Spatial Planning**, edited by Edward Hulsbergen, Ina Klaasen & I. Kriens (Amsterdam, Holland: Techne Press, 2005) pp. 235–250.
[6] Salingaros, Nikos A.; **Principles of Urban Structure** (Amsterdam, Holland: Techne Press, 2005).
[7] Alexander, Christopher; Ishikawa, S.; Silverstein, M.; Jacobson, M.; Fiksdahl-King, I.; & Angel, S.; **A Pattern Language** (New York: Oxford University Press, 1977).
[8] Krier, Léon; **Architecture: Choice or Fate** (Windsor, England: Andreas Papadakis Publisher, 1998).
[9] Salingaros, **Principles of Urban Structure, op. cit.**

Another misconception about Transect zoning and the New Urbanism is that it places severe restrictions on cars. It merely changes the geometry of how they move and where they park. True, in the compact city, the movement of cars is calmed, and parking is no longer dominant and obvious in front of buildings. But cars are not banned, and parking is adequate.

Still, for a variety of reasons, including energy costs and population growth, car use must be curtailed over time. Unfortunately, the immensely powerful car industry has successfully coupled the idea of personal 'freedom' with a car purchase, and it has been almost impossible to convince people to reduce car use. They don't see that giving unlimited 'freedom' to the car has to be paid for by the destruction of a city, and of their own human environment. One's car today represents something almost inviolate – a right of ownership and object of fetish all at the same time. It is going to be very difficult to educate people on this point.

6 The Auto-dependent Landscape Self-generates

The auto-dependent landscape consists of the road surface, parking, and all areas devoted to the care and feeding of vehicles, such as gasoline stations, garages, muffler shops, tire stores, hubcap stores, car dealerships, parts stores, car washes, automotive junkyards, etc. Shopping areas and restaurants take the form of drive-ins or malls set back in a sea of parking. In this way sprawl is a self-generating system with mechanisms for spreading and enlarging itself. In the auto-dependent landscape – occupying more than half the urban surface in many regions – vehicles no longer serve simply as a means of human transportation, but as ends in themselves.

Since the auto-dependent landscape feeds on and generates much of the world's economy, it is not feasible to simply eliminate it. Many countries' industries and economic base depend on producing cars and parts, or petroleum and petroleum products. Global wars are fought over the petroleum supply. At the same time, the auto-dependent landscape is changing the earth and human civilization, so it has to be contained. What is good for General Motors is no longer good for America, to turn around an old American slogan. Car-related activities within a city are still essential for our economies, but they must be kept on the proper geographic scale. The great planning fallacy in our times is trying to mix up (instead of carefully interface) the auto-dependent landscape with the city for people: all that happens is that the former takes over the latter.

Most important, vehicular speed must be calmed. The highways of the auto-dependent landscape are designed to maximize a smooth and fast flow of traffic, without any consideration of human beings outside a car. Those same principles of speed maximization at the expense of pedestrian physical and psychological well-being have been automatically applied to all roads inside the urban fabric, making it anti-urban in the process. My book *Principles of Urban Structure*[10] offers rules that reestablish the city for people by giving pedestrians priority over cars. Those rules rely on earlier work by Christopher Alexander, published as *A Pattern Language*[11] more than twenty-five years ago.

Despite numerous, well-documented presentations of energy/oil depletion issues, people remain blissfully unconcerned about their car-dependent lifestyle. They trust the transnational oil companies to continue providing them with affordable gasoline until the end of time. Gasoline will certainly be available – at market price, whatever that may be in the future. I do not add my voice to the doomsayers predicting the end of petroleum, but unsustainable urban and suburban morphologies will simply become too expensive to survive. The compact, small-scale city is sustainable, whereas ultra-high-density skyscrapers and suburban sprawl are not.

7 Sprawl is also Driven by Commercial Forces

The dream of owning an isolated country villa surrounded by forest draws people out to suburbia, and cheap land draws developers there. At the same time, lower rents and taxes draw business there, following residential growth. But because the form of suburbia is already established by single-use zoning, businesses must locate away from residential areas, and they must locate where there is enough drive-by traffic to sustain them. Since developers and builders have made fortunes out of selling this defective geometry, they simply keep building what they have done for decades. Government perpetuates sprawl by building roads and infrastructure in an anti-urban pattern.

Because business in sprawl depends on attracting the drive-by customer, then, it must announce to all drivers that there is ample free parking everywhere. Thus we have the shopping mall surrounded by a vast parking lot; the office tower in the middle of farmland surrounded by its parking lot; the university campus in the middle of nowhere surrounded by its parking lots, and so on. Urban morphology is determined in most places by highways and parking lots. Again, the priorities are exactly backwards. Thoroughfares and parking lots should conform to a compact urban structure, not the other way around.

The geometry of commercial nodes is generally oriented outwards toward high-speed arterials to attract drivers. Current zoning makes sure that it cannot be oriented toward residential neighborhoods. That must change with new Transect-based zoning. When a community adopts such a zoning code, there will be assigned Transect zones as described above and structured so that stores, schools, churches, and parks are within walking distance of homes. Density increases as T-Zones get higher, but never to the extent of the high-rise megacity that depends precariously

10 ibid.
11 Alexander **et al.**, **op. cit.**

upon a vast energy grid. In a Transect-based code, mixed use is allowed in all T-Zones, and the design of streets favors the pedestrian. The first priority is to get rid of the parking lot in front of a store, narrow the streets, and provide a wide sidewalk.[12] On-street parking is fine; as is parking behind, below, and above the store.[13] Parking garages must have liner stores with windows, so that the pedestrian does not walk past blank walls or rows of cars. People are more likely to walk if there are pleasant things to look at on the way.

Sustainable compact cities in place all around the world are now being destroyed by the introduction of anti-urban components. Not only are skyscrapers proliferating as symbols of modernity, but so are more modest typologies that profit one person while slowly degrading the entire city. In Latin America and Europe, for example, a new corner store typology copied from the United States erases the sidewalk and gives it over to parking. If this goes on (along with adopting other similar typologies from the auto-dependent landscape), that will unbalance societies that have depended on a human-scale urban morphology for so long.

Transect-based zoning codes limit the number of storeys in the compact city to three in zone T3, four in zone T4, and six in zone T5. This places a ceiling that protects the urban fabric from the negative consequences of high-rise construction. These problems include: the office tower (which generates traffic congestion for the entire region during rush hour); the residential tower (which generates strongly negative social forces);[14] and the giant parking lot that comes as part of either of these (and which erases the human environment precisely where it ought to be intensified). High-rise buildings don't belong in a compact city. Genuine high-density, high-rise city centers do exist, as coded for in Transect Zone T6, the Urban Core. Examples include the downtown Loop in Chicago, Manhattan, Hong Kong, and Sydney. But I do not foresee a future for new T6 Cores, so I have confined the compact city to a T5 maximum density and six-story height limit.

It is a great pity to see cities in the developing world self-destruct as they try to imitate the images of dysfunctional western cities (to them, symbols of power and progress). Cities in southeast Asia and China that had been working fairly well up until recently, such as Bangkok and Shanghai, have in one bold step ruined their traditional connective geometry. Their mistakes include building megatowers, then widening streets and building a maze of expressways to serve the new ultra-high-density nodes. For their entire future, those cities are condemned to be choked by traffic.

8 Low Speed Encourages Urban Life

The compact city is a LOW-SPEED city. This feature has to be guaranteed by narrow streets and a special low-speed geometry. Planning has for several decades concentrated upon increasing vehicular traffic flow. This has diminished the livability of cities and urban regions. To rebuild a living environment for people, we need to reverse almost all the traffic-boosting planning measures implemented since the end of World War II – that is, rewrite the traffic codes. Roads inside the compact city should not be built to accommodate fast vehicular traffic. Cars should go slowly inside this region. The physical road surface and width will force them to. Transect-based planning calls for thoroughfare design to respond to the context of the T-Zone, not the other way around.

The key is to permit internal access everywhere for large vehicles such as fire trucks, delivery trucks, and ambulances, but in the immediate vicinity of a house cluster around an urban space, all the roads should be woonerven, the Dutch model of very low-speed roads shared with pedestrians.[15] Here we may use narrow roads with occasionally semifinished surfaces. We have forever confused ACCESS with SPEED. Today, fire departments refuse to cooperate with urbanists, insisting on an overwide paved thoroughfare everywhere. The reason is that fire chiefs want to be able to make a U-turn in one of their giant fire engines anywhere along any road.

The compact city mixes shared civic spaces with concentrated arrangements of structures. It defines a highly-organized complex system, in which each component supports and is connected to the whole. A city for people consists of buildings of local character and specific function that contribute to the immersive context of their Transect Zone. This is the opposite of modern 'generic' building types, which are strictly utilitarian and connect only to the parking lot. Fixated on fast speed, governments or developers spend much of their money on paving wide roads and vast parking lots, neglecting the design of urban space. When building a low-speed parking ribbon (described in the following Section), parking costs should be the last priority, thus permitting gravel, and brick/grass surfaces. Such surfaces slow cars down.

Urban space is supported by the geometry of surrounding buildings.[16] Buildings should attach themselves to those spaces, and not to the road. A compact city is defined by internal cohesion achieved via a centripetal (center-supporting) arrangement, versus a centrifugal (directing away from the center) arrangement. Buildings are connected via a network of paths into clusters. A number of buildings should define a cluster perceived by a pedestrian as accessible (a low-speed setting). By contrast, buildings in suburban sprawl are outward-looking and connect to nodes in the far distance, but not to each other (a high-speed setting). There are rarely any local connections in a monofunctional region.

12 Sucher, David; **City Comforts: How to Build an Urban Village** (Seattle, Washington: City Comforts Inc., 2003).
13 ibid.
14 See Alexander **et al.**, **op. cit.**; and Salingaros, **Principles of Urban Structure, op. cit.**
15 Gehl, Jan; **Life Between Buildings: Using Public Space** (Copenhagen: Arkitektens Forlag, 1996).
16 Salingaros, **Principles of Urban Structure, op. cit.**

Sidewalks and all pedestrian paths must be protected from unnecessary changes of level, and any other discontinuities.[17] Cars on the other hand, don't get tired, so their path can easily go around people and pedestrian nodes. Again, that slows them down (anathema to today's traffic engineers!). Pedestrian paths should be laid out to connect urban nodes, and to reinforce a connected complex of urban spaces.[18] A parking ribbon can be designed to snake around buildings and pedestrian urban spaces – not the other way around.

9 Car-Pedestrian Interactions and the Parking Ribbon

The compact city is a city for people, but it still accommodates cars and trucks. However, surface parking lots interrupt the urban structure and sense of an outdoor 'room'; they are dangerous and exhausting for pedestrians, and visually destroy any pleasant walking. They also create runoff from impervious surface, encouraging flooding.

Instead of taking over a vast open area, parking should occur in a ribbon of intentionally constrained road: I am proposing a radically different parking geometry, to be generated by new zoning codes. A parking 'lot,' then, is just another road, not an open space. These long and narrow parking ribbons will branch into each other, assuming a networked form just like urban streets. A maximum dimension of about two car lengths will be stipulated for the width of any parking ribbon, accommodating only one side of head-in or diagonal parking. Parking ribbons don't need to be straight, but can be made to fill up otherwise useless narrow spaces.

Furthermore, pedestrians should be given priority when crossing an existing large parking lot. This means building a raised footpath, sometimes covered by a canopy, and also giving it a distinct color coding for visual separation. Giant, uniform parking lots are hostile to human beings and essentially anti-urban. They can be reformatted into parking ribbons by building other structures inside them. Inserting sections of water-permeable surface into giant parking lots will also solve the serious problem of flooding from storm run-off. Such infill solutions can be written into a new code.

On-street curbside parking (either parallel, or diagonal) should be encouraged in the public frontage, but banned from the private frontage, between the sidewalk and building face.[19] On-street parking actually helps pedestrians feel safer on the sidewalk by providing a buffer between them and moving traffic. Sidewalks are not used if there exists a psychological fear from nearby cars and trucks; vehicular traffic parallel to pedestrian flow can be tolerated only if it flows at a certain distance from people. Adjusting the maximum speed of a road (not by speed limit signs, but by its narrowness and road surface) to tolerable limits also achieves this symbiosis. For slightly faster urban traffic, an excellent thoroughfare type to accommodate both car traffic and safe sidewalks is the boulevard, traditionally designed with low-speed 'slip roads' and parking on the sides.

Parking ribbons already exist in traditional urbanism: as curbside parking on slow-moving roads; and on the sides of a fast-moving boulevard. Most parking garages are indeed wound-up parking ribbons. What I'm suggesting is that ALL parking should conform to the ribbon geometry. A parking lot should never again be confused with an urban space, and cars should never be allowed to take over an urban space.

Another solution is to have orthogonal flow for pedestrians and vehicles (working simultaneously with protected parallel flow). Their intersection must be non-threatening. The two distinct flows cross frequently at places that are protected for pedestrians. In this way, the two flows do not compete except at crossing points. Introducing a row of bollards saves many situations where pedestrians are physically threatened by vehicles. An amalgamation of pedestrian paths defines a usable urban space. This must be strongly protected from vehicular traffic. Any paved space that children might use for play must be absolutely safe from traffic. I discuss all these points at length elsewhere.[20]

10 Beyond the Transect with Christopher Alexander

Where do the Transect-based codes come from? They are a result of thinking how to create an environment conducive to human life, obtained by comparing present-day with older successful environments the world over. They ultimately depend on traditional solutions, such as those collected in Christopher Alexander's *A Pattern Language*. The Transect's value lies in structuring a proven form of compact, traditional urbanism in a way that can be used within the existing planning bureaucracy. As Andrés Duany has so often expressed, he wants to use the system to introduce radical changes without waiting to change the system itself. He calls the Transect-based Smart Code a 'plug' into the existing power grid used to working in terms of zoning.

There is another approach. Alexander's new book *The Nature of Order*[21] is the most important analysis of architecture and urbanism published in the last several decades. Alexander advocates a complete replacement of current planning philosophy, because the existing manner of doing things is so fundamentally antihuman. That may be difficult to implement immediately, but the future of cities does depend upon ultimately applying Alexander's understanding on how urban form is generated, and how it evolves by adapting to human needs. My own work[22] has been profoundly influenced by Alexander.

17 Gehl, **op. cit.**
18 Salingaros, **Principles of Urban Structure, op. cit.**
19 Sucher, **op. cit.**
20 Salingaros, **Principles of Urban Structure, op. cit.** and Salingaros, 'Towards a New Urban Philosophy: The Case of Athens', **op. cit.**
21 Alexander, Christopher; **A Vision of a Living World: The Nature of Order**, Book 3 (Berkeley, California: The Center for Environmental Structure, 2005).
22 Salingaros, **Principles of Urban Structure, op. cit.** and Salingaros, 'Towards a New Urban Philosophy: The Case of Athens', **op. cit.**

Alexander describes his adaptive design process, giving examples to show urbanists how to tailor it to their own particular project.[23] I will not attempt to summarize his extensive results here, but only wish to point out a key finding. Living urban regions have a certain rough percentage of areas devoted to pedestrians-green-buildings-cars as 17%-29%-27%-27%. Contrast this to a majority of today's urban regions, which typically have the percentage distribution as 2%-28%-23%-47%. Alexander describes in great detail the succession of geometrical steps that can be taken to convert one type of urban region into another. His approach is to do this one step at a time, and it is eminently practical.

The result is what all of us (Alexander, Duany, Krier, Plater-Zyberk, and myself) want: a human-oriented urban environment. At the same time, Alexander presents a theory of urban evolution, which could be steered either towards a living city, or towards an anti-urban landscape for cars. The point is to recognize the fundamental mechanisms and forces that push towards either goal, and to channel them to what we want. Most important, we should recognize what we really want, since many people (including prominent urbanists) really do want to sacrifice urban life to the auto-dependent landscape, even though they may not openly admit it.

Alexander's understanding of urban processes probes far deeper than the Transect. Duany and Plater-Zyberk have learned from Alexander, but want to effect immediate improvements. The simplest expedient is a change in zoning codes, such as the Transect-based Smart Code. Today's urban environment is so fragmented, degraded, and antihuman that such code reform is urgently needed. Once healthy urban fabric begins to grow again, then people can see the advantages of a human-scale built environment. They could apply Alexander's ideas to generate vital urban regions once again. Anyone who dismisses the New Urbanism as superficial, or as simply a 'style', needs to read Alexander to really understand urban form.

And yet, I must point out a fundamental difference. Alexander is convinced that genuine urban unfolding – the process of sequential adaptation that generates living environments – is not possible within current planning practice. He fears that the system is not only misaligned, but is also too rigid to accommodate living processes. The new Transect-based codes, significant as they are in improving an abysmal situation, are not flexible enough, according to Alexander, precisely because they work within the present planning system. Since changing a vast and established bureaucracy is next to impossible, Alexander proposes going around the system. These points raise serious tactical questions.

Defining urban character as inherent in the Transect has begun to reestablish an urban structure that can engender a new urban citizen. The Transect, however, is just a beginning: in addition to these sectional prescriptive codes, urbanists must extend their logic to multiple scales and work through a knowledge of urban adaptive processes.[24]

11 Some Contradictions

There are several contradictions I feel I need to discuss. First, the limitations of working with a system of permits and construction that is deeply flawed, threaten to neutralize any code-based way of building cities. Alexander emphasizes that living cities can only come about from an adaptive PROCESS, i.e., building and adjusting urban form step-by-step. This is not easily reconcilable with the present mainstream professional culture. It is, however, the way that traditional building and self-built settlements arose for millennia.

Alexander's fear is that any system that builds cities without a truly adaptive process will never achieve the intense degree of life seen and felt in cities of the past. That is not the aim of the present code-based system, which instead uses the existing bureaucracy to limit such an evolution of urban form. The gradual evolution of cities, akin to the evolution of individual organisms and ecosystems, is now illegal. What is allowed is a large-scale intervention, regardless if it is catastrophic or nearly so. Planners still cling to the myth of an 'economy of scale'.[25]

The second contradiction is that a majority of people go along with anti-urban sprawl and high-rise construction without complaining. It is hardly possible to discuss issues of urban form with a contemporary society that has become desensitized through its addiction to technology. Growing up in suburbia with the false notion of unlimited freedom has distanced people from truly human environments. People who enjoy eating junk food in their parked car; who love the ear-damaging loudness of commercial movie theaters and rock concerts; who own a 'Home Entertainment System' (a monster television/stereo with subwoofer) and another subwoofer in their car, are not going to value the pleasures of a traditional environment – it only reminds them of a pre-technological past.

In the present atmosphere, I see Transect-based codes as the best entry-point for bringing a human environment back to our cities. I have discussed these issues with commercial developers, who insist that they are not setting urban typologies: they are only providing what the market wants, working within the existing codes. Clearly, our society has to learn to appreciate good urbanism before Alexander's work and my own can begin to be applied to cities. The Transect will certainly help to move society in that direction.

23 Alexander, **A Vision of a Living World: The Nature of Order**; **op. cit.**
24 Alexander, **A Vision of a Living World: The Nature of Order**; **op. cit.** and Salingaros, **Principles of Urban Structure, op. cit.**
25 Salingaros, **Principles of Urban Structure, op. cit.**

Alexander would prefer for codes to be optional and voluntary: accepted by ordinary people on the basis of understanding and sensitivity, and not imposed by law. Duany, on the other hand, is suspicious of media-induced fear and manipulative marketing; those forces push people to reject connectivity and to want to live in monocultures.

The third contradiction is that human-scaled cities must be market-driven and implemented by legislation, but people don't seem to be ready to do what is required. Any hope for a positive change must come from an educated society that demands good urbanism instead of its 'junk food equivalent'. Enough popular support has to build up to pressure elected officials to make the necessary changes in urban codes. Those who need it the most – the young, the old, and the poor – are either not educated about city form, or have no influence. New Urbanist ideas have been embraced by upper-income groups simply because of their higher level of education. That is not because of any particular attraction between the compact city and any particular socioeconomic class.

Ultimately, the most disadvantaged classes of society can least afford the expense of sprawl, yet only those who are better educated see the reality of a human-scale urban environment.

The fourth contradiction is the institutionalization of sprawl. In addition to planning codes, sprawl has been adopted as an unshakeable standard by insurance companies and financial institutions. They are reluctant to finance or insure the compact city, but will automatically help to build sprawl because all their offices and agents have been doing this for decades. That mindset is permanently fixed to the extent that even when natural disasters wipe out vast areas of sprawl, the bureaucracy does not permit them being rebuilt as compact city. An opportunity to finally get rid of anti-urban patterns and to reconfigure our cities is thus missed. All the discussion about wasting time in commuting, and wasting one's salary on gasoline seems to be for nothing, if it will not influence rebuilding when an opportunity presents itself. This may be interpreted alternately as the bureaucracy doing the 'safe' thing; or as criminal willfulness.

12 Conclusion This essay put forward a radical idea shared by many urbanists today: that the ultra-high-density city is outdated. There are essential differences with other authors, however. Unlike some of my colleagues who abandon any urban principles out of frustration, I condemn suburban sprawl and high-rise buildings as equally unworkable. Supporting Andrés Duany and Elizabeth Plater-Zyberk, I proposed a 'new' ordered urban form: the compact city. This new urban typology looks remarkably like the old geometry of small-town and village living, so it is really a return to traditional urbanism. Where it is radical is that it requires a complete rewriting of the zoning codes. That is essential, since theoretical urbanism is ineffective if the present anti-urban codes remain unchanged.

This essay also contained an implicit condemnation of planners and designers who refuse to distinguish between good and bad urbanism, or to offer any workable solutions. That is the equivalent of doctors refusing to diagnose and cure patients, deciding to give an equal chance to the microbes. Prominent designers talk about the urban condition, labeling the disconnection of our cities (and civilization) as a new, exciting phenomenon: a natural evolution (instead of extinction) of the city. They also accept, without question, the massive destruction of traditional urbanism taking place in China and the developing world as 'inevitable progress'. Urbanists have a responsibility to intervene; they cannot be neutral observers. From now on, the world can only rely on pragmatic urbanists who are willing to tackle practical issues to create compact cities for humans.

Acknowledgments

I am greatly indebted to Sandy Sorlien, without whose active participation this paper could never have been written. Thanks to Michael Mehaffy for useful advice; to Andrés Duany for support; and to Christopher Alexander for sharing an unpublished letter from himself to Andrés Duany on the differences between their respective approaches to urban form.

References

Christopher Alexander (2005) **A Vision of a Living World: The Nature of Order**, Book 3, The Center for Environmental Structure, Berkeley, California.

Christopher Alexander, S. Ishikawa, M. Silverstein, M. Jacobson, I. Fiksdahl-King & S. Angel (1977) **A Pattern Language**, Oxford University Press, New York.

Andrés Duany & Elizabeth Plater-Zyberk (2005) **Smart Code**, Version 6.4, <www.dpz.com>, Miami, Florida.

Andrés Duany, Elizabeth Plater-Zyberk & Jeff Speck (2000) **Suburban Nation**, North Point Press, New York.

Jan Gehl (1996) **Life Between Buildings: Using Public Space**, Arkitektens Forlag, Copenhagen.

Léon Krier (1998) **Architecture: Choice or Fate**, Andreas Papadakis Publisher, Windsor, England.

Nikos A. Salingaros (2005a) **Principles of Urban Structure**, Techne Press, Amsterdam, Holland.

Nikos A. Salingaros (2005b) 'Towards a New Urban Philosophy: The Case of Athens', in: **Shifting Sense – Looking Back to the Future in Spatial Planning**, edited by Edward Hulsbergen, Ina Klaasen & I. Kriens, Techne Press, Amsterdam, Holland, pages 265-280.

David Sucher (2003) **City Comforts: How to Build an Urban Village**, City Comforts Inc., Seattle, Washington.

Codes and the Architecture of Life

Michael Mehaffy

The history of modern city-making has been the implementation of 'rational' organisation schemes. From Haussmann to Howard to Le Corbusier and beyond, the problem of cities was seen as one of reductive engineering schemes, seeking to isolate smoothly-functioning mechanical parts in place of 'messy' organic conditions.

But recent complexity science – particularly the astonishing progress in the understanding of living systems – reveals that such 'messiness' had much more sophisticated structure than we realised. We now see that the schemes of segregated zoning and their kin have created radically simple urban morphologies lacking a critical threshold of connectivity and iterative complexity, leading to a state of dysfunction and environmental crisis for the modern city and its sprawling suburbs.

This crisis is deeply structural and cannot be addressed with technological bandages or wildly imaginative architecture. We can however identify processes that can regenerate the needed structural complexity. Abundant examples come from anthropology and sociology, which document a complex adaptive 'collective intelligence' in human traditional pattern-making.

These new insights suggest how a new class of 'generative' codes may restore the essential connectivity and dynamic interactivity of a healthy urban system. This paper discusses recent work in the area, including the 'structure-preserving transformations' of Christopher Alexander.

We live in a technological civilisation that is highly coded. There are zoning codes, building codes, engineering codes, legal codes, codes of professional conduct. Moreover, there is a vast system of interlocking regulations and practices that themselves function as coded processes, regulating and sometimes promoting the iterative generation of structures within our civilisation.

Thanks to recent advances in complexity science and in game theory, we are beginning to understand much more about the ways that such codes work iteratively to generate form. Coupled with historic studies of coding approaches, such an understanding has allowed the development of a number of new, more refined approaches to urban codes, as alternatives to the relatively crude regimes at present. This paper discusses several of the more recent approaches, and their implications for urban form.

Let us consider first the historic backdrop of the current period of urban design practice. As many commentators have noted, the old twentieth-century model of industrial design and construction is arguably in a state of prolonged crisis. In his essay 'What Ever Happened To Urbanism?' the architect Rem Koolhaas described it this way:

Modernism's alchemistic promise – to transform quantity into quality through abstraction and repetition – has been a failure, a hoax: magic that didn't work. Its ideas, aesthetics, strategies are finished. Together, all attempts to make a new beginning have only discredited the idea of a new beginning. A collective shame in the wake of this fiasco has left a massive crater in our understanding of modernity and modernization.[1]

As Koolhaas notes, we are still mired in a twentieth-century mechanical understanding of nature that still largely guides our acts of planning and building. One may say that his response as a designer has been merely to play about in the bottom of the crater – to reassemble bits of 'junkspace' from the modernist bits of rubble. But we suggest a more radical idea: that the most recent science in fact shows us a path out of that crater – a path that holds intriguing and exciting possibilities.

This path is to be found in the revolutionary new scientific understanding of the structure of nature, and in particular of living systems: the science of organised complexity, networks, fractals, and so-called 'strange attractors'. This new science also describes the behaviour of small, rule-based iterations, or algorithms – cellular automata and the like, and the highly complex and ordered 'emergent' patterns that they create.

Perhaps the most familiar of such systems is of course the DNA code of life itself, made from just four molecules. Through the complex interactions of millions of combinations of these four molecules, and through their expression into protein structures, an adaptive, morphogenetic process acts over long periods of time to produce the astonishingly varied patterns and intricate wonders of life itself.

We suggest that there are evident lessons to be found here for intentional human morphogenesis as well, and for the morphogenesis of cities and other settlements.

These insights form the basis of what the architectural theorist and designer Charles Jencks has called the 'new paradigm' in architecture.[2] We believe he is on the right track, although we suggest that he doesn't grasp the real revolutionary implications of it. In fact we suggest that scientists and artists of urban and architectural form have only barely begun to scratch the surface.

There is reason to believe that this new science – this new understanding of the structure and the organisation of things – may in time revolutionise our world, just as the old science did previously. Markets, institutions, the very structure of civilisation itself may well be transformed, in important and necessary ways. Indeed we would argue that it has already begun to do so. Moreover, these changes hold out the promise of deeper understanding, and reform, of the horrific mistakes of the early industrial period. The severity of these mistakes increasingly appears to threaten our welfare, our prosperity and even our very future. Therefore, this is much more than just an academic discussion of abstract ideas.

Charles Jencks makes a very interesting case for the application of concepts of the 'new science' to the representations of an exuberantly sculptural new architecture, and to what he describes as its 'enigmatic signifiers'. That is without doubt a very fascinating realm of connected ideas to explore in art.

However, it is precisely that – a realm of *ideas*, not the realm of nature and of connected natural structure *to which the ideas refer*. That is, this architecture is *about* complexity, but not necessarily manifesting genuine emergent *properties* of complexity. For example, Jencks' own fascinating landscapes of strange attractors aren't *really* strange attractors, but forms based on scientific *diagrams* of strange attractors. It is as though one were to create houses made of blueprints of remarkable buildings rather than making the kinds of buildings that the blueprints described.

This is an interesting and perhaps quite lovely artistic idea, to be sure – but it is not the thing about which the idea was generated. It is a form of art that is of course quite far abstracted from life. It turns out that this question of abstraction is a critical one in the current discussion, to which we will return.

Let us be clear, so as to avoid a false charge of illiberalism, that the celebration of ideas – the *adventure* of ideas – is a profound and worthy goal of architecture, and of all the arts. But architecture has to do something else of course, unique among the arts: it has to serve as *the place where we live*. Buildings and cities, unlike paintings or sculptures, must serve as the connective fabric of human life. Whatever ideas we may signify and celebrate in our architecture, nonetheless we must somehow account for the fact that these structures will shape our use of resources, and our patterns of interaction with each other, and the patterns of activity and change on the earth, in a way that no sculpture or painting or piece of music ever need do. It continually shapes and conditions the *emergent* structures of human behaviour – for better or for worse.

Some contemporary artists respond to such concerns by saying that we have no alternative but to remain in the realm of art, and that we should stop trying to change the world – indeed, that it is dangerous even to try. That is because our understanding of nature, like any 'narrative' about the world, is always socially constructed; and therefore, to act upon that understanding is merely to

---

1 Rem Koolhaas, 'What Ever Happened to Urbanism?' (in **S,M,L,XL**, Rizzoli, 1995).
2 Charles Jencks, **The New Paradigm in Architecture** (New Haven: Yale University Press, 2002).

replace an existing political hegemony with our own equally arbitrary version. Therefore, the only noble thing we can do is to free ourselves and others with the truth – and that truth is that our understanding of nature is socially constructed.

I would agree that our understanding of nature is always *conditioned* by social construction; and to the extent to which that social construction is done by a privileged elite, it is a perfectly valid function of art to explore this truth and perhaps to 'deconstruct' it.

But the lessons of the new science are clear: whatever we think of it, the vast complex structure of nature is real enough, and no mere 'narrative'. Death, extinction, and human-made disaster, are real phenomena, not social constructions. Nature may well remain beyond our full comprehension, and our knowledge of it may indeed be forever *conditioned by* social construction. But that is not to say that it is beyond our ability to intelligently interact, and learn to produce better or worse outcomes for human beings.[3]

This, we would argue, is nothing other than the continuing phenomenon of emergent 'collective intelligence' in the human species. It is the level at which our new understanding of nature begins to transform our understanding of the nature of human processes and human patterns – human culture, technology and art.

And it is on this level that things start to get very complex and very interesting indeed.

The question then becomes: what does the new science tell us about the complex structure of human life, in the form of settlement patterns, economic processes, social patterns, interaction with complex ecosystems, and sustainable development? What does it say about urban pattern, urban morphogenesis and architectural morphogenesis?

And what tools might there be to manipulate these complex phenomena to richer human ends, in art and in life? Can we learn something from genetic coding, for example, about the astounding variety and the robust success of life? Can we devise codes based upon similar morphogenetic principles?
One can perhaps begin to see where such a discussion can take us.

Rem Koolhaas' paper quoted earlier, 'What Ever Happened to Urbanism?', poses a question that is on many people's minds right now. Charles Jencks confessed his concern about what he called the 'basket of icons' that seems to be dominating architectural practice today, in place of coherent urbanism. There is a near-exclusive focus upon the radical creativity in each building act, and yet no means to produce a larger connective system that is more than the sum of its parts. One may say that we are still mired in the crater.

George Ferguson, president of the Royal Institute of British Architects, has expressed this same concern about what he calls 'iconitis'. It is widely acknowledged that the vast majority of buildings are unacceptably inferior, and the civic spaces around them are often quite horrible. Nor is this, it seems increasingly, a sustainable state of affairs. Ferguson wants a new emphasis on coherent urbanism, and new tools and practices to address it. The Prince's Foundation has been working with the RIBA on this, and George Ferguson has spoken on coding as a tool to that end at the Foundation.

Now the 'new science' of organised complexity has already begun to be applied to urbanism. Students of urbanism will recall that as far back as the early 1960's, seminal thinkers like the American Jane Jacobs already recognised the powerful implications of complexity science in challenging the then-existing basis of architecture and planning – the rigid modernism that Koolhaas has challenged – and arguing a persuasive case for reform.

The last chapter of Jacobs' *The Death and Life of Great American Cities*[4] is a classic and definitive piece on the subject, called 'The kind of problem a city is'. In it she talks remarkably lucidly and presciently about the history of scientific thought and the way it has shaped human action, and in particular the way it has shaped how we think about and act upon cities.

She described how modern science accelerated, around the time of Newton, when it mastered so-called two-variable problems. In physics, the laws of motion, for example, are two-variable problems. In urban design, one may equate, for example, two variables such as the numbers of dwelling units to supportable quantities of retail square footage.

But in the early twentieth century, Jacobs notes, something interesting began to happen: through statistics and probability we learned to manage very large numbers, with myriad variables interacting. The interesting thing discovered was that one could manage those phenomena as statistical averages without knowing much about the actual interactions.

This statistical science powered into the phenomenal technological power of the industrial revolution of that period. Much of our industry and the prodigious output of 20th-century modern-

---

[3] Such a view of things is essential if we are to avoid being trapped forever in our own abstractions, in a kind of philosophical 'hall of mirrors' – or as Wittgenstein put it, if we let our intelligence become bewitched by our language.
[4] Jane Jacobs, **The Death and Life of Great American Cities** (NY: Random House, 2002).

ity was rooted in these powerful new statistical methods. And indeed, Jacobs points out that the early ideas of Le Corbusier and others, and the later ideas of planners – often to this day – rely upon this notion of large statistical populations.

This progression in science was matched by a progression in city planning: from, say, the rigidly formal, 'rational' plans of Haussmann, or of Ebenezer Howard and his neatly segregated Garden City plans, through to the more statistically informed plans of Le Corbusier, implemented around the world by technocrats such as Robert Moses and others.

In either case the problem of cities was seen as one of devising reductive engineering schemes, seeking to isolate smoothly-functioning mechanical parts in place of 'messy' organic conditions. In the scientific world-view of the day, this was understood to be advancement and modernisation. The two-variable engineering was not eliminated, but was supplanted with the statistical mechanics operating on large numbers. The newer science was added to the old, and transformed it.

Meanwhile, the biological sciences had reached a dead end with the statistical science of so-called 'disorganised complexity' and had to come to terms with the emergent phenomenon called 'organised complexity' – the area in the middle, between simple two-variable problems and vast numbers of variables. Biologically speaking, we now understand that it is in this theoretical 'middle ground' that we can explain many of life's complex processes – where a number of variables are 'interrelated into an organic whole'.

It was clear even then that the problems of the human environment were in many respects emergent problems of 'organized complexity'. But Jacobs pointed out how the planning and architecture professions were still at that time, 1962, mind you, mired in the old scientific world-view. She says:

> Today's plans show little if any perceptible progress in comparison with plans devised a generation ago. In transportation, either regional or local, nothing is offered which was not already offered and popularized in 1938 in the General Motors diorama at the New York World's Fair, and before that by Le Corbusier. In some respects, there is outright retrogression...[5]

It appears, then, that Koolhaas' 'crater of modernity' has been around for a while...

Jacobs' ideas for planning reform have gained wide currency if not yet very deep implementation. The so-called New Urbanism movement in the US is a planning reform movement that is dedicated to implementing many of Jacobs' insights. But one may argue that the ideas on *process* have not yet been fully taken up, particularly at the scale of individual buildings. The conditions can be too rigid, or the designers resort to a literal copying of historical precedent and historical style with insufficient adaptation to novel conditions. Critics maintain that the result is still a rather lifeless human place, and rather too much like its modernist predecessors in that respect.

In other cases the architecture can be a dazzling sculptural idea that only exists in a narrow conceptual range of expression, and does not function as a complex connective structure in the larger system at all. Such structures then become severed objects inserted into a foreign ecosystem. As we've noted, this is true even for the sculptural objects that *signify*, but do not yet *embody*, insights of the new science.

Often the problem with these object-buildings is most severe at finer scales, such as the crucial pedestrian experience, where the detailing drops away in a mess of ugly concrete or poorly weathering blank surfaces. Or there is a severe failure at the level of the emergent urban pattern, which is poorly-ordered, left-over space – what planners call 'SLOAP', or 'space left over after planning'.

Clearly many conventional economic processes leave us with the same kind of problem, or an even worse one. One has only to look at the pattern of sprawling monocultural suburbs around the globe to recognise a lack of differentiation and intelligent adaptation, and to identify a great deficiency in the necessary level of stable or sustainable urbanism. There may be an emergent pattern in these complex processes, but it is nothing we want as human beings. It is rather a Frankensteinian version of Le Corbusier's Radiant City, a mechanical plan based on the 'old science' but metastasised in the new global economic reality into a pervasive and unwelcome form.

So let us consider this transformation now occurring in the sciences, and its contribution to a new and deeper understanding of 'the kind of problem a city is,' in Jacobs' memorable phrase. What can we learn from the newest insights of the new science, both about the *structure* of complex human settlements, and about the *process* of their morphogenesis? What about the cultural processes, the economic and technocratic processes that we are going to have to account for?

It turns out that the new science tells us a great deal indeed. And like a biological problem, in which a living whole cannot be treated as a mere collection of parts, we find that the problems of the built environment are interconnected, complex and emergent.

5 ibid.

But saying that the built environment is not merely a collection of parts is not to say that it cannot be acted upon. It is rather that the actions will not be the usual manipulations of objects from the old paradigm.

It has long been understood that one can code or zone in a top-down way. For example, one can say that offices go in one area, residences in another and so on. One can specify geometric parameters for form. But can one write a bottom-up code – a code that delivers the emergent complexity of nature? And can one do it in a way that produces a livelier, more desirable outcome? The new complexity science suggests that we can, and begins to show us how. (And, interestingly, it shows us how humans once did this inadvertently in antiquity, and how we can now do it more deliberately.)

While nature is vastly complex, it turns out that many of its processes are surprisingly simple. In their operation, the interaction of sheer vast numbers creates patterns and ripples of astonishing, mind-numbing complexity. Yet the emergent structure still has great coherence, because it is adaptive over time. Often that simple process functions as a kind of code.

For example, life itself arises from a code, as noted earlier: the genetic sequence of only four molecules that comprises the genome. The astonishing variety and beauty of life is coded, in what is at its core a remarkably simple way. As we study it and begin to tease apart its secrets, we have found to our surprise that even the entire genome is simpler than we had expected – that much of the complexity of the organism emerges in the *process* of stepwise differentiation of tissues.

So let us repeat: the individual steps can be quite simple, but the emergent pattern is often vastly complex. Furthermore, the complexity of that pattern emerges over time, in a *process*. It cannot arise from a single structural blueprint, or a single act, or a single conception. This is a crucial point.

Another important trait of the code of life is that it is collective. That is, DNA that has become well-adapted in a previous environment is imported and combined, often through a sexual process. Then it is expressed in a morphogenetic process, and then adapted and selected out by the environment. In this way, a bottom-up process – the emergent results of a somewhat random combination of genetic material – is combined with a top-down process, the selection pressures of the environment, and collected and stored in new genetic material. This 'collective intelligence' is another important lesson about the processes of nature.

As noted at the outset, we humans use codes too, and we store our collective intelligence in them just as nature does. Clearly codes for the built environment have been around for a very long time – both literal building codes, and also the larger economic and political rule-based processes. We now see that they are very much like the algorithms of cellular automata, in which many actors take many adaptive steps in relation to one another.

What is interesting about the last several hundred years is that as our technology has become more complex, the regimes with which we have ordered our built environment have paradoxically become radically simpler and more weighted toward the top-down side of things. As a result the morphologies of these structures have become much more rigid and less adaptive – with woefully unsuccessful and even disastrous results.

Meanwhile, a number of reform movements have arisen to develop better codes and better algorithms, to improve the liveability and the sustainability of the built environment. I mentioned the New Urbanism movement, which does explicitly use collective intelligence in the form of stakeholder collaboration in the design process, the charrette – and often the result of that charrette is a form-based code. The New Urbanism also uses the collective intelligence of regional design patterns, documented in builder pattern books, which, linked to a regenerated 'building culture', aims to produce again a satisfactory level of 'good ordinary' buildings necessary for successful urbanism.

One of the most recent and influential of these new coding approaches is a generation of codes based upon the concept of a 'transect' spanning between the most intensive human uses and the most pristine wilderness. Andrés Duany has been working on this with his brother Douglas, and we have been collaborating with them on several aspects of this work.

The work of Christopher Alexander is more explicitly dynamic, and builds upon his earlier pioneering work in the complexity of morphogenesis and computer synthesis of design, influenced by, and perhaps influencing, the pioneer of complexity Herbert Simon and others.

Alexander's work over the years has delved deeper into some fascinating areas of metaphysics – some philosophers find him in very good company with Whitehead, Bergson and a number of other philosophers of process – but it is important to note that Alexander's scientific approach to architecture has never left him. In fact his work on pattern languages has been taken up as a major new initiative in computer science.[6]

---

[6] Alexander's PhD thesis, **Notes on the Synthesis of Form,** is considered a classic in the field, and his later work **A Pattern Language** is said to be the bestselling treatise on architecture of all time. His paper 'A City is not a Tree' was also a landmark in understanding the importance of network structure in the genesis of form. **A Pattern Language** was intended as a methodology for generating such a networked form, while **The Nature of Order** describes in more detail the geometries that such a form must take, and the processes of its transformation and adaptation. The notion that such highly theoretical ideas have no practical application is belied by the keen interest of developers of 'Pattern Language Software', a form of object-oriented programming (and of course by Alexander's own prodigious work). Will Wright, creator of the computer game **The Sims**, has acknowledged Alexander's influence on that popular game series.

Ever the logician, Alexander wants to know how nature achieves the kind of order that we see everywhere around us. His book *The Nature of Order* is a magnum opus inquiring into that question, and he comes up with three remarkable answers.[7]

One is the familiar lesson from biology: that nature does not invent from whole cloth, but adapts and transforms from existing conditions, in a series of step-wise transformations. He calls these 'structure-preserving transformations'.

Second, it is of the essence that these transformations are not re-arrangements of a collection of parts, but rather, *transformations of wholes*. That is, one cannot abstract the parts from the context without losing the essential connective structure of the entire system. Trying to do so is like clutching at a piece of jelly: everything slips through one's fingers.

And yet there are ways to better understand the structure of these wholes, and indeed to engage and manipulate them, in a new generation of dynamic codes. Alexander the mathematician uses detailed mathematical formulas to describe this.

The third conclusion is a much more radical idea, although it will be familiar to more advanced students of complexity science: that these structures are a class of actual emergent *life*. This is a notion of life that is directly connected to our own lives as human beings, in much the same way that, say, an ecosystem can be seen as a kind of living extension of an organism. We can also recognise it as connected to ourselves *a priori*, intuitively, and not in any abstract way. For Alexander, our own perception of value and meaning as living creatures is identifiable in these structures, and can no longer be excluded.

This is not unfamiliar territory for the new science, as it finds itself going into deep questions about life and consciousness in fields like neuroscience, cognitive psychology and others. It seems we are at the point where progress is only possible if we address the metaphysical aspects of experience. Often the answers we then find are astonishing, and powerful.

It follows from Alexander's structural observations that we can construct processes that amplify the life in these structures in the built environment. We can develop adaptive, iterative, rule-based algorithms to do this, very much as nature does.

This offers us the basis for a new generation of so-called 'dynamic' codes.

We have been working with Alexander on several projects employing these kinds of dynamic coding processes in ecologically sensitive settings, including a large community in Oregon. The early results are very intriguing and encouraging. We are also working now on an initiative to develop a pilot project for such a coding scheme in the UK. There are enormous practical issues to be worked out, of course – what Alexander calls 'massive process difficulties'. Again, one cannot simply re-arrange the parts of the system, but must make gradual, even imperceptible transformations of the whole – a daunting task to be sure.

But there is also what we would argue as a very robust and well-informed theoretical approach to these problems, and while the solutions are becoming evident, they will take time to implement. The intriguing thing is the way in which such a dynamic code can in principle be inserted into the existing technocracy, rather like a virus inserting its DNA into an existing cell, and transforming the function of the cell's own existing DNA.

Let us briefly review the characteristics of such a code. (A much more comprehensive description is laid out in the four volumes of *The Nature of Order*.)

Such a code has the following features:

1 It specifies, in some way, a defined, step-wise, generative process.
2 It specifies that in that process, human beings will take certain rule-based actions, in combination with evaluations based upon feeling, and in adaptation to what has come before.
3 At each step, it acts upon the then-existing condition *as a whole*.
4 At each step, it identifies the weakest parts of the structure and acts to improve and amplify them.
5 At each step, it may apply previously-coded solutions and patterns, *and adapt them to the novel conditions*.
6 At each step, it differentiates the space according to a scheme of 'centres'.
7 The centres are differentiated via 15 'structure-preserving transformations' (see below).
8 *Infrastructure follows.* As with the morphogenesis of organisms, where the tissues come first, and the veins and ducts follow, the human patterns and human spaces come first, and then roads, sewers and the like follow – not the reverse.
9 Similarly, *visual expression follows*. The human patterns and spaces come first, and then the visual ideas and 'signifiers' follow – not the reverse. Otherwise we are simply making people live in disconnected sculptures.
10 At the end of each cycle, the result is evaluated and the cycle is repeated.

7 Christopher Alexander, **The Nature of Order**, Center for Environmental Structure, 2002.

The 15 'structure-preserving transformations' are based on 15 properties that Alexander has identified in natural morphologies. They are properties that have very much to do with what we might call the 'architecture of complexity', the structure of centres and fields within a system of wholes. Again, it is worth remembering that Alexander was one of the early pioneers of complexity theory in the late 1950's and 1960's, and this is very much an extension of that work.

Let us summarise these 15 properties here. Note that although some of these names may sound rather mystical, they are in fact grounded in the most specific structural theories – as one might expect from a Cambridge-educated mathematician.

1. Levels of scale.
2. Strong centres.
3. Boundaries.
4. Alternating Repetition.
5. Positive Space.
6. Good shape.
7. Local Symmetries.
8. Deep Interlock and Ambiguity.
9. Contrast.
10. Gradients.
11. Roughness.
12. Echoes.
13. The Void.
14. Simplicity and inner calm.
15. Not-separateness.

The salient observation is that these properties are seen over and over again throughout nature, and throughout the cultures of human beings across time – but they are conspicuously missing from our own technological age.

Why is this? We suggest that this is because humans are still in the infancy of a technological age, and captivated by a particular class of elementary technological morphology. The danger is that this morphology is still tied to a destructive and unsustainable *morphogenesis* – the force of the old modernity, the 'crater' that Koolhaas spoke of. That technological morphology threatens to destroy the richness and even the sustainability of nature, and the collective intelligence of human traditional culture, turning everything into a simulacrum of culture. Even the architectural arts become little more than 'product styling' within a sea of junkspace.

But let us consider that just as the old modernity was profoundly shaped by the old mechanical science, the new sciences imply the dawn of a New Modernity that we are only beginning to understand. This insight holds out the stunning theoretical possibility of a new synthesis of science and architecture and, in time, an organic fusion between technology and human culture. Instead of the old paradigm of object-manipulation and mechanical replication, we may learn the powerful morphogenetic processes of life itself. We may learn to make our own 'seeds' to produce astonishing richness, variety and quality, containing the adaptive morphologies and the ecological coherence of natural systems.

Let us consider, therefore, that before us lies the next great modern advancement. This is the revolutionary promise of the new science: the path out of the old 'crater of modernity'. It does indeed offer us a New Paradigm: a more humane, more advanced kind of 'New Modernity'.

[1]

[2]

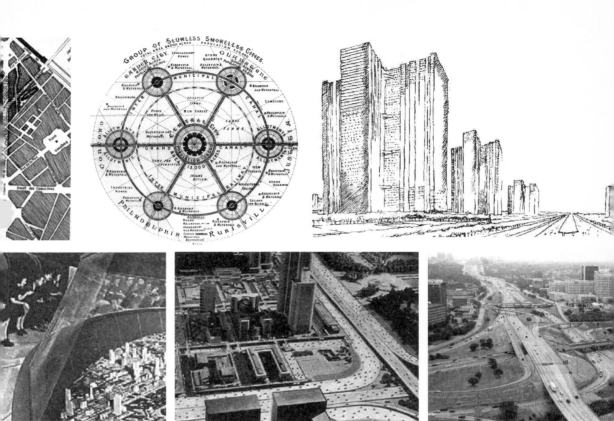

1 'The new sciences of algorithms, fractal structures and strange attractors are offering startling new insights into complex natural processes and their essential geometries.'
2 'The forms of the early 21st century, in spite of occasional borrowings from the dramatic geometries of complexity science, are still largely rooted in the insights of the early 20th century.'
3 (Beginning top left:) 'Rational segregation in Haussmann's Paris, Howard's Garden Cities and Le Corbusier's Radiant City reached its zenith in the intoxicating vision of the 1938 World's Fair "Futurama" exhibition, becoming the basis for later freeway suburbs around the world.'

4 'SLOAP: Space Left Over After Planning, the result of a segregating mechanical process.'

5 'Four stages of many over 900 years in Venice, demonstrating a "Structure-Preserving Transformation". Each of many steps maintains a complete and whole structure.'

6 '15 generic properties of geometric form arising from natural processes.'

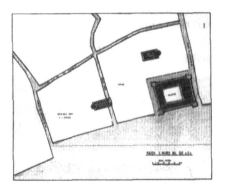

About 560 A.D.

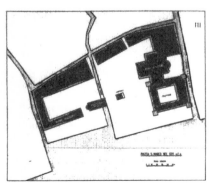

[5] *About 1160* A.D.

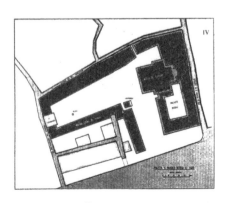

About 900 A.D.

About 1300 A.D.

[4]

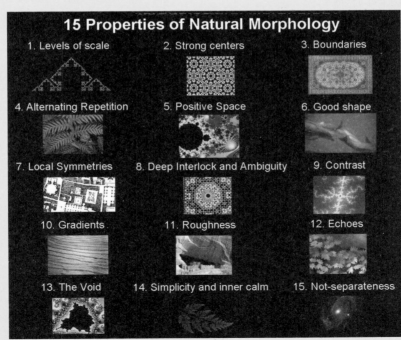

[6]

Section II:

Represen-
tation
Architecture

Drawing necessarily precedes building. Drawing is identified with abstract speculation and geometry, and in turn, with social formation: 'Once codified, architecture tends to imitate pre existing architectures, but what does it originally imitate?', Stan Allen asks. Karsten Harries underscores the artificiality of Alberti's construction that his representation of space does violence to the way we actually experience things. Alberti assumes monocular vision and a flat earth; indeed, 'his perspective offers itself as a figure of Cartesian method'. Classical architecture imitates nature in the form of the primitive hut, but it does so only through a highly abstract and idealized geometrical mediation – 'the stories of origin return to a void. The desire for stable origins always turns up empty'.

Diana Agrest suggests that Modern architecture has abandoned these ideas. With the development of modern architecture, several important changes occur. The perception of image becomes part of a different system of thought: no longer is an object related to the problem of representation or imitation; it becomes a mental construct. The classical language of architecture is negated in favor of a more abstract, 'non-representational', visual, formal system. Not only have styles been banished but also the need for figuration. Similarly, Arie Graafland once suggested that 'contemporary architectural culture is characterized by an increasing interest in architectural drawings… What has changed most radically is the character of the drawings themselves. Until recently there was an almost direct relation between the design drawing and its actual execution in built form. But nowadays this link has become attenuated, and the designs have become "aesthetic".'

How do contemporary architects like Diller & Scofidio, among others, deal directly with this critically abstract, yet practically concrete phenomenon? What exactly has changed in the understanding of, and application through, representation in architecture today? The scholars in this section bring forth both their knowledge and experience towards a position on how research might continue to address representation as both a historical issue and a pressing issue of the day.

Frank Werner takes on the problem of representation head-on in his essay, 'The New Simplicity: A Problem of Representation in Architecture and Town-planning?' where he develops a concept of 'Simplicity'. Yet simplicity is anything but simple, in either theory or practice; rather it is perhaps the most difficult. Subsequently, when a building is built, what does it represent? Phillip Ursprung takes up the question with regard to an exhibition on Herzog & de Meuron that he was asked to curate. The consequent frisson between the built work as opposed to the instruments of architectural representation – drawings, models, material investigations, graphic design, etc. – provide the opportunity for Ursprung to ask in this essay about the status of architectural experience. Nevertheless, instruments of architectural representation cannot be taken to be static ways of seeing. Roemer van Toorn in his essay entitled 'After Criticality: The Passion for Extreme Reality in Recent Architecture … and Its Limitations', explicates the unfolding of contemporary practices that engage tools far beyond what was possible even twenty years ago. Partly because these new firms see things differently, so to speak, and partly because the means of representation allow us to see things differently, an urge to move forward unencumbered by the past and its historical/theoretical weight has gained momentum, calling itself 'Projective Practice'. Finally, sometimes one must look to the past (in this case the very distant past) in order to go forward. Kari Jormakka in his essay, '*Poesis doctrinae tamquam somnium*' in explicating the Platonic form/copy distinction, allows us to understand what is at issue in problems of representation. In the end, this section of papers dealing with problems of representation still must ask: 'what are we representing?'

The New Simplicity: A Problem of Representation in Architecture and Town Planning?

Frank R. Werner

It makes no sense inventing something unless it is going to be an improvement.
Adolf Loos

To 'build simply', one might suppose, is the easiest thing in the world. But the history of architecture and design tells us a different story. The simple artefact can be anything from lofty to trivial, moralistic to destructive, rigorous and reduced to formless. Conversely however, the complex artefact may be either attention-grabbing or lofty, informative or moralistic, meaning-laden or reduced and rigorous. The dilemma is evident. To build simply, to design simply is, both in theory and practice, an extremely complex undertaking.

This can be seen most clearly in discussions focusing on the role of the so-called 'new simplicity' in the context of architectonic representation, which many critics see as currently undergoing a crisis of both form and quality. Thus the architectural semiotician Claus Dreyer recently wrote:

> A cursory examination of some spectacular architectural works of recent years results in a finding that would be easy to supplement and support on the basis of further examples: the crisis of representation is evident. There is no common 'language' in architecture through which common experience, ideas, hopes, values, traditions and conventions could be expressed, just as there is no common ground in society for these issues. There are only a few outstanding individuals with a great artistic potential and almost God-like reputation who have the opportunity to articulate their 'private language' by unique means, and to present it to the public. The work of translating and interpreting this 'language' for the general public takes place mainly in the professional and popular media, where it occasionally comes dangerously close to product advertising. Under these conditions, the communicative correspondence between employer, architect, user, observer, and society as a whole amounts to little more than a coincidence or a stroke of luck. In a complex, multi-cultural, compartmentalized, and media-oriented society, this state of affairs seems impossible to circumvent: everyone must and can seek the representational context in architecture that suits him and is limited only by lack of information or financial means. At any rate, this freedom of choice must be understood as an achievment and an opportunity. Never before in history have so many different forms of architectural representation been in competition. The monotony of classic international modernism is a thing of the past. Today interested people can put together their own (even potentially virtual) 'universe of architectural discourse'. At the same time, there is a possibility for the emergence of a new architectonic paradigm that may develop from the close proximity of the different 'languages'. It could represent a world culture growing closer, without losing the regional varieties and 'dialects'.[1]

---

1 Dreyer, Claus; 'The Crisis of Representation in Contemporary Architecture', in **Semiotica**, no. 143 / 1–4 (2003) p. 180.

Simple building as a dialect of architectonic representation? It would be easy to agree if the topos of simplicity had not been so repeatedly instrumentalized in its recent history.

Under German National Socialism, simplicity developed into a theoretical and practical program of architecture marked by a double standard from which – at least in Germany – it still suffers today. Sailing under the xenophobic flag of the conservative *Heimatschutz* movement (literally 'home protection' movement), and drawing on the folksy, retrogressive 'cultural works' of Paul Schultze-Naumburg for example, the exponents of the 'blood and soil' ideology promoted ultra-simple, regionally rooted architectural stereotypes that inevitably ended in a sort of synthetic regionalism. The vision was of a uniform single-family house throughout the entire Reich from Peenemünde to the Obersalzberg – though in reality family house design was criminally neglected by the Nazis in favour of the great buildings for the state. Goethe's summer-house in Weimar was prescribed as a sort of archetypical ur-matrix for party architects, as it was for all the potential builders and developers of the 'Thousand Year Reich'. This attitude gave birth to innumerable family homes and building developments, for example the notorious Kochenhofsiedlung built in 1933 in Stuttgart as a model housing development, a compact realization of synthetic, regionally accented simplicity imbued with the National Socialist urban vision. It was intended as a counter-blast to the nearby Weißenhofsiedlung, and destined to enter architectural history under that disreputable banner. From this point in time at the very latest, any conception of a simple architecture based in 'local native tradition' seemed, at least in Germany, to be contaminated with associations of folksy retrogressive nationalism and hostility to any form of innovation.

However, this instrumentalization of simple building and shaping should not mislead us into thinking that simplicity of form is *ipso facto* reactionary. A glance at the history of architecture shows that every age had its 'new simplicity', although the reasons for this were always different. Thus in 1788 an anonymous author published one of the most important treatises of late baroque classicism, entitling it *Inquiries into the Character of Buildings; on the Relation between Architecture and the Fine Arts and on the Effects that These Should Produce.*[2] Like a golden thread the epithet 'simple' runs through this enlightened tract, which concentrates entirely on the effect of buildings on people. Simplicity, the author enjoins, should be the keynote of plan, silhouette, interior, furnishings, decor and choice of materials, and the details that constitute this simplicity should be executed with the greatest care. The simple making of simple objects has indeed always been a concern of men and women, albeit in different social contexts and subject to varying connotations and readings. It is almost with melancholy that one thinks of cultures such as Japan, where the 'simple house' – and still more so its sensually ascetic interior appointments – has for centuries been conceived as a 'man-made philosophical continuum', a 'thought-construct' whose world-view transcends time and space in the sureness of its forms and its unquestioned sense of obligation.³

Topical interest, both national and international, in 'simple' building and design focuses less, however, on history than it does on problems of ecological economy, and here it is the example of young international architecture – including a number of Austrian, German and Swiss practitioners – that is now setting the pace. And setting it in a direction that contrasts markedly, one must say, with the navel-gazing architectural banalities of the new Berlin or the no less irrelevant debate on architectonics that has turned that city into a minor battlefield.

Kenneth Frampton has taken a very clear stance on these matters. He believes one can find 'in the tradition of new building [architects who "build simply",] ... following the distinction made by Max Bill between concrete architecture and concrete art'.⁴ This reminds one of Adolf Loos's remark: 'The house is conservative, the work of art revolutionary. That is why people love houses and hate art'.⁵ Both statements – Frampton's as well as Loos's – indicate an awareness of the divide between the object made for use and the object made for art, between the pragmatic and the transformed. Both accept the functional and reject the artistic. And it is this awareness that informs the work of architects like Guyer and Gigon, Diener and Zumthor. Frampton also counts Steven Holl among this group, for he opposes 'the fetish of making materials and details into an end in themselves ... especially when this is divorced from content or context'⁶ – a judgement Holl himself corroborates when he writes: 'for me, material is just a tool to express a concept. If there's no concept, then the result is uninteresting... Material is the flesh giving form and space to the concept. The danger in this whole discussion is that the material could become an end in itself. That would be rather pathetic'.⁷

If, however, one looks more closely at Zumthor's thermal baths in Vals for example – a building that has been called both 'archaic' and 'atavistic' – the danger of such clear distinctions soon becomes evident. For Zumthor's approach in this building is neither programatically 'simple' (in the sense of functional) nor conceptually 'economical', let alone ecological. What it represents is the sculpturally transcendent product of a sensualist aesthetic.

A second group of 'simple builders', harshly criticized by Frampton⁸, comprises architects who 'appear to have succumbed to the hallucinatory effects of the media'. He is thinking of design

2 Anonymous; **Untersuchungen über den Charakter der Gebäude, Faksimile-Neudruck der Ausgabe** (Leipzig, 1788) with an introduction by Hanno-Walter Kruft (Nördlingen, 1986).
3 cf. Yoshida, Tetsuro; **Das japanische Wohnhaus** (Tübingen: Wasmuth, 1954).
4 Frampton, Kenneth; 'In the Nature of Materials: A Note on the State of Things', in **Daidalos** No. 65, **Magic Of Materials II** (Berlin, 1995), p. 12.
5 Loos, Adolf; **Architektur**, 1910.
6 Frampton, **op. cit.**
7 Holl, Steven; in **Daidalos**, no. 65 **Magic Of Materials II** (Berlin, 1995) p. 16.
8 Frampton, **op. cit.**

teams like Herzog and de Meuron or Sumi and Burkhalter. This criterion also seems to me highly questionable, as I hope to show later. A third group, among whom he includes John Pawson, is that of the protagonists of 'zero degree architecture', as he calls it, though he means the term positively enough. This minimalist approach to building responds, in his opinion, to the 'escalating cacophony of the modern world with a palpable stillness – a *beinahe nichts* – in which the grain of a single material is the only allowable figure'.[9] To avoid misunderstandings it must be added that architects like Pawson and Chipperfield, or kindred designers, have nothing whatever in common with reactionary social attitudes.

Have we not for years now, not to say decades, been experiencing a *de facto* renaissance of objects-as-archetypes – clever, even sophisticated archetypes based on Aldo Rossi's, Robert Venturi's or Christopher Alexander's concept of the simple underlying forms of all design? What is genuinely new on the other hand is the approach to the concept of materiality and its appropriate expression. For modern composite and bonded materials simply have too many qualities to be expressed simultaneously in any artefact. No underlying form can penetrate all these strata, and materiality breaks away as a result (as in many timber or concrete buildings), taking on independent existence and value and becoming itself a guarantor of 'presence' and immediacy.[10] Martin Steinmann comments: 'Things show their presence, inscribed in their materiality. Conversely, materiality itself shows its presence.'[11] Hans Frei sees in this:

> The disjunction beween inner and outer, revamped in terms of the indifference of material to form and form to construction. Spatial construction, it follows, can be left to the general contractor. At the very least it becomes a subordinate issue, far removed from the complexity of surface proper to the facade. It is this complexity, now, that constitutes the real mystery of architectonic creativity.[12]

Minimalist architecture draws its inspiration here from high-tech materials; *arte povera* architecture and *arte povera* design from the theories of Joseph Beuys, according to which 'poor materials' have to be enhanced and charged with mental energy. The question in both cases is no longer what role a particular material is intended to play within a construction but what optical effect it can achieve. The new motto, according to Christian Sumi, is therefore to 'forget about the material, look at its effect'. Like certain designers, he says:

> We must look for the magic of our materials beyond the limits of semantic definition. For us the materials with which we work are magic in themselves and our task is to differentiate between and exploit their sensual qualities.

The concept of aura is over-used, but it is important here. If we are hesitant to elaborate on the subject, it is out of carefulness and respect.[13]

Catalogues of building materials and ready-made products seem the immediate source of inspiration for this type of 'simple building', which frequently resorts to ready-made industrial or craft products for its façades: industrial glass for a museum extension in Winterthur (Gigon & Guyer), traditional wooden shingles for a chapel in Sumvigt-bei-Disentis (Zumthor), cast iron manhole-covers for a residential and commercial block in the centre of Basle (Herzog & de Meuron). And not infrequently these ready-made products are transformed very simply, even archaically, into something quite new in appearance. 'In this way Herzog & de Meuron transform the image of a transformer coil into a signal box at the main railway station in Basle and a pile of wood in a timber-yard into the layered façade of the first Ricola factory'.[14] The object seems to be to create architecture by metamorphosis out of the banal images of everyday. Or a small six-storey office comes into being in the area behind the central station in Basle, a quality product from the renowned design team of Diener & Diener 'crazily enough not decked out with the usual self-referential features of architecture as architecture', as Klaus-Jürgen Bauer remarks.[15] The iron oxide that has formed on the concrete façade creates what Martin Steinmann calls 'an impression of poverty appropriate to the streets behind the station where the rust from passing trains alters every façade'.[16] A modest building plays consciously with banality, which seems to constitute a central aspect of its claim to quality. Up to now, however, it has not been customary to use the term banal in a positive sense of architecture or design products. The emphatic poverty of the façade stands, it must be added, in conscious contrast to elegant details, like the bronze frames around the windows of this Basle building. For, as Steinmann notes, 'a building must hold meanings in balance and suspense'.[17]

How does 'simple building' approach the question of total form and its expression? Two different tracks are evident. One sets out to create a 'strong form' – and here the visible construction is not identical with the actual construction. This latter is wrapped and condensed in a form determined not by constructional principles but by 'criteria drawn from the psychology of perception.'[18] The focus is on mass and tectonics, sometimes overemphasized, sometimes questioned.

9 ibid.
10 Frei, Hans; 'Neuerdings Einfachheit' in **Bundesamt für Kultur** (edited volume) **Minimal Tradition, Max Bill und die 'einfache' Architektur 1942–1996** (Baden, 1996) p. 122.
11 Steinmann, Martin; **'Form und Ausdruck'** in Ackermann, Mathias (ed.); **Morger & Degelo Kommunales Wohnhaus 1993** (Basel, 1994) p. 8.
12 Frei, Hans; **op. cit.**
13 Sumi, Christian; 'Positive Indifference' in **Daidalos** No. 56, **Magic of Materials II** (Berlin, 1995) pp. 26 **ff.**
14 ibid.
15 Bauer, Klaus-Jürgen; **Minima Aesthetica, Banalität als Subversive Strategier der Architektur** (Weimar, 1997) p. 32.
16 Steinmann, Martin; **'Neue Architektur in der Schweiz'** in Magistrat Linz (ed.); **Bauart** (Linz, 1990) p. 82.
17 ibid.
18 Frei, Hans; **op. cit.**, p. 123.

Architects and designers approach the issue of form 'by encapsulating the real construction in order to dramatize the visible construction'.[19] It is a trend recognizable from the time of Mies van der Rohe's celebrated buttress for the Pavilion of the German Reich at the Barcelona World Fair of 1928-29, later developed for the Haus Tugendhat in Brünn (1930), or again from his world-famous 'corner-solution' for Alumni Memorial Hall (1945-46) on the I.I.T. site in Chicago.

The second variant is what Hans Frei[20] calls the creation of a 'specific form' intrinsically linking function and surface. The façade sets out in this case to reflect the specifics of the brief; 'materials serve as repositories of meaning', as they can readily be seen to do in the copper-laminated signal-box of the Wolf goods-yard in Basle. As if the copper wrapping were a sort of armour protecting the electronic control systems inside the building from the electromagnetic smog outside. Herzog & de Meuron have, according to Frei, succeeded in 'translating the mental energy of the "Feuerstelle" installation which, as Beuys' assistants, they created for the Basle Museum of Contemporary Art, into the electronic age'.[21]

In both these cases the principle of simplicity consists in structuring the façade to be 'as viewer-friendly an interface as possible, both expressively powerful and readily understood'. The outer skin of a building becomes in this way a 'locus of resistance against pure fiction, a negotiating-place for the truth of vision.'[22] Against this – or complementing it – stands the clear influence of an aesthetics of economy whose aim is to establish the constructive unity of interior and exterior, the economic interdependence of structure and form and ecological adaptation to the (local) *status quo*. Thus in certain areas of Europe, wooden buildings, rigorous in line and 'clean' in structure, have assumed the status of a 'journeyman's entrée into the regional architectural scene'.[23] They demonstrate the rejection of all that is modern and exalted on the one hand, or imported and alien on the other. Their vision is directed to what, though banal and everyday, is nevertheless functional and effective, and this vision has brought with it a new and subversive invigoration of the simple, traditional purpose-designed building. A very practical side of this is that such buildings have always been cheaper and more environmentally friendly than any of the unloved modern implants. 'Out of need an ethical principle has developed'[24] which architects in certain regions (notably the Swiss cantons of Vorarlberg and Tessin) have been able to incorporate almost seamlessly into a striking program of 'new old-simplicity'.

If these various, heterogeneous strategies for 'building simply' finally meet, it is at the level of perception. Martin Tschanz is right when he sees precisely in 'these buildings that appear so simple, even bleak at first … a deeper, sensual multiplicity of meaning'[25] – often a greater multiplicity than postmodern architecture was able to achieve. The result, not infrequently, is 'effects of impressive beauty that can be enjoyed without any previous knowledge'.[26] Despite our enjoyment of these effects, however, it cannot be a matter of indifference whether this architecture aspires to a 'minima aesthetica', to 'banality as subversive strategy', or whether its practitioners have consciously embarked on a program of 'added meaning', rejecting the idea of transforming the physical artefact into an autonomous aesthetic object (a project in which classical modernity also failed) in favour of placing simple constructs in a precisely shaped cultural and political frame.

The question whether the 'new simplicity' has affected contemporary urban planning – and if so how and where – cannot easily be answered. So many conflicting claims can be made – city centres versus suburbs, national strategies versus global, places versus non-places, reconstruction versus deconstruction – that no definitive statement is possible. Nevertheless, on the basis of what I have been saying here, four different modes of influence can be at least concretely described:

1 **The autistic individual building** – viz. the building that stands in an urban context and makes a significant statement but is unable or unwilling to enter into any sort of dialogue with its context or the men and women who use it.

2 **The 'dead city'** (after a metaphor of Ernst Jünger's and Jean Baudrillard's). If simplicity in an urban context allows itself to be instrumentalized by monolithic energies – either by untrammelled capital and power or by the *idée fixe* of urban totality to be realized by main force (cf. Hausmann's Paris or Stimman's Berlin) – then the sum of simplicities (in themselves acceptable) will inevitably add up to blank domesticity, petrification and the exclusion of anything approaching presence.

3 **The city as the normal case.** Simplicity can achieve great things in an urban context when it sees itself calmly and modestly as a complement to the urban whole and its users, a regenerative ingredient within the spaces of the contemporary city, furthering and fulfilling thought and emancipation.

4 **The unbroken power of the line.** Building in lines, suburban terraces, residential rows, seems to me, despite the objection that it is merely a reanimated relic of classical modernism, perhaps the most fruitful statement of simple building currently available. It is open to amelioration and refinement both at the ecological and constructional levels as well as to quantum leaps of improvement in the design of interior living space. Where this happens it continues to represent almost the only building project able to halt the breakdown of settlement patterns and even create compelling local frameworks for global players as well as supporting the generation of new forms of suburban socialization. As the Smithsons stated:

19 ibid.
20 ibid.
21 ibid.
22 Frei, Hans; **op. cit.**, p. 126.
23 ibid.
24 Bauer, Klaus-Jürgen; **op. cit.**, p. 28.
25 Tschanz, Martin; 'Gentle Perversions' in **Daidalos** no. 56, **Magic Of Materials I** (Berlin, 1995) pp. 88ff.
26 ibid.

At a time when our scale of values was still determined by the church and the monarchy, and later by local government and banks, it was a case of erecting buildings which proclaimed a message of power. Now that we are being influenced simultaneously by many different factors, the time for any kind of rhetoric in individual buildings is past.[27]

No one today, however, will subscribe to this statement of Alison and Peter Smithson's from the 70's. Even the simplest present-day architecture is anything but speechless. But the best examples of simple building and design speak a language that cannot be thrust into the corner occupied by restorers and academic adepts of an eternal yesterday. They are neither mirrors of banality nor models of reactionary social philosophy. 'The simple', to speak with Hans Frei, 'is a formula for processes that incorporate as much as they can'[28] – of presence to and in the present, he means. What the simple entirely excludes is any sort of 'ism', of new fashion. For this would, for better or worse, be the end of simple building and simple design, the end of a final optimistic option in whatever remains to us of design culture.

To build simply, to design simply, one might suppose, is the easiest thing in the world. In reality it is the most arduous, the most time-consuming and last but not least the worst paid of any task that one can nowadays assume. For that reason we should cultivate it, intellectually as well as in practice, rather than deliver it up to the empty phrases of populist argument.

27 Smithson, Alison and Peter; **Without Rhetoric: An Architectural Aesthetic** (London, 1973) p. 12.
28 Frei, Hans; **op. cit.**

1 Office Building, Basle, Switzerland, 1986-1988, Northwestfacade, Arch.: Diener & Diener.

2 Japanese house.
3 Standardized Buildings for Railway Technology, various locations throughout Switzerland, 1995-2001. Plans and elevations. Arch.: Morger & Degelo.

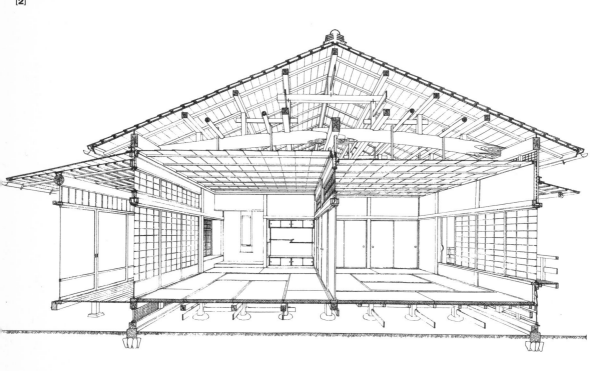

[3]

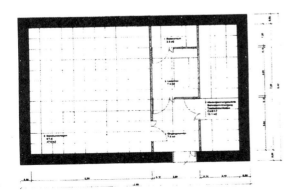
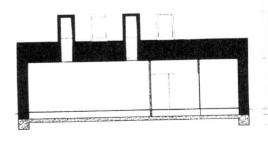

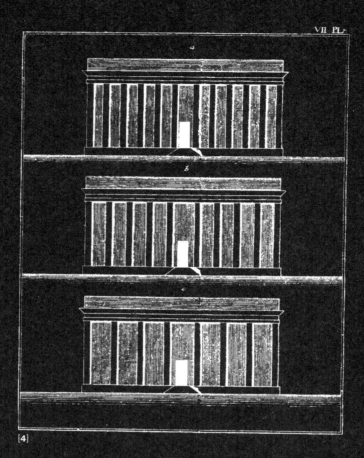

4 'Building with different characters', Illustration from: Anonymous, **Untersuchungen über den Charakter der Gebäude,** Leipzig 1788, Plate VII.
5 Thermal Baths, Vals, Switzerland, 1990-1996. Sketch of section through indoor pool. Arch.: P. Zumthor.
6 Art Museum, Bregenz, Austria, 1990-1997. Exhibition Space. Arch.: P. Zumthor.

[5]   [6]

Built Images: The Eberswalde Library by Herzog & de Meuron

Philip Ursprung

When Kurt Forster, then the director of the Canadian Centre for Architecture in Montreal, asked me in 1999 to curate an exhibition on Herzog & de Meuron, my reaction was lukewarm.[1] On the one hand, I could not resist such an offer; on the other hand, I had never been a fan of architecture exhibitions. More accustomed to curating and visiting art exhibitions, I had always been disappointed by architecture exhibitions. I found it frustrating to view an abundance of documentary photographs, plans, and models referring to something that either stands elsewhere or no longer exists. I always left such exhibitions with a guilty feeling, namely, that I should be where the actual buildings are or at least where they were developed – in the studio of the architects, for example. Of course, I realized that the problem resides in the very nature of architecture, which cannot be exhibited as such. Exhibiting architecture is a paradox – as challenging to architecture as the exhibition in the public realm is to art.

Furthermore, at the time I was not yet a real fan of Herzog & de Meuron. I had witnessed the prodigious growth of the office during the 1990's, after the early success of buildings such as the Blue House in Oberwil, Switzerland (1980), the Ricola Storage Building in Laufen (1987), and the Stone House Tavole, Italy (1988), which led to world-famous projects such as the Sammlung Goetz in Munich (1992), the Dominus Winery in Yountville, California (1998), and the Tate Modern in London (2000). But I had difficulties identifying with their architecture. I rejected the reductionist, formalist, and minimalist attitudes with which they were associated.[2] To me, something was missing in their buildings. But I could not yet say what it was.

Then, in early 2000, I went to see a small building that had been completed in 1999: the library of the Fachhochschule Eberswalde.[3] (ill.1) The city of Eberswalde is located an hour north of Berlin in the state of Brandenburg, which once belonged to the German Democratic Republic. The old industrial city, which was heavily damaged by bombing during World War II, is now infamous for its high unemployment rates and its skinhead gangs. It seems surprising to find cutting-edge architecture in such a place. However, the higher education system in the east was renovated after the reunification of Germany.[4] The moment I saw the building, everything I had thought about the architects changed. Instead of an abstract volume, I was confronted with figurative images depicting people, airplanes, houses, and animals. Everything that I had missed in the architecture of Herzog & de Meuron was here. What had been pieces of a puzzle became an entity. I felt a shock, a kind of a visual rupture, which only happens, to me at least, when a work of art or a building changes the way I see the world. For a moment I was uncertain whether this building was real or I was dreaming it. As if to relieve me of my doubts, two children approached and gently touched the façade with their fingers. (ill. 2)

1 See Herzog & de Meuron, **Natural History**, ed. Philip Ursprung, exh. cat., Canadian Centre for Architecture, Montreal (Baden: Lars Müller Publishers, 2002).
2 See Ilka and Andreas Ruby, Angeli Sachs, and Philip Ursprung, **Minimal Architecture** (Munich: Prestel, 2003).
3 See **Herzog & de Meuron, Eberswalde Library**, with essays by Gerhard Mack and Valeria Liebermann (London: Architectural Association, AA Publications, 2000).
4 Another such building by Herzog & de Meuron is the Informations-, Kommunikations- und Medienzentrum (IKMZ), the library of the Brandenburgische Technische Universität Cottbus, Germany (2005).

The experience of the Eberswalde Library radically changed my view of Herzog & de Meuron and was instrumental in my conception of the exhibition; from a skeptical and distanced observer, I turned into an enthusiastic admirer of their projects. Why did this building strike me as something different? First of all, it was not at all what I had expected from looking at photographs, where it looks massive, almost like a bunker. When I first perceived it, it seemed small and fragile; its grayish green color and its size distinguished it from its context in a way that made the whole scene look artificial. It was as if someone had punched a hole in the continuum of space and time. The urge to touch the façade that my students felt when we returned later that year shows how much the project is about the limitations of mere visuality.

Of course, I am not the first one to observe this strange quality in projects by Herzog & de Meuron. Terence Riley, in an essay written in 1995, had remarked on a 'ghostlike' quality in their projects.[5] And the architects themselves speak of their buildings, as 'ruptures in the urban web which may at first sight even seem violent.'[6] Among the photographic representations of the library only Thomas Ruff's electronically mounted work of art *Bibliothek Eberswalde* (2000) does justice to the project. (ill. 3) Two transparent, 'ghostlike' students are passing by. The image deals with the problem of depicting the building, its surroundings, and people simultaneously on one representational level.

While I was taking pictures, I thought of Robert Smithson's 1967 essay 'The Monuments of Passaic,' in which he describes stopping at a bridge and encountering his first 'monument'. He writes: 'Noon-day sunshine cinema-ized the site, turning the bridge and the river into an over-exposed picture. Photographing it with my Instamatic 400 was like photographing a photograph. The sun became a monstrous light bulb that projected a detached series of "stills" through my Instamatic into my eye.'[7]

In fact, the library's façade can be read as a series of film stills. The motifs are repeated horizontally. One could, of course, establish a genealogy of buildings with decorated façades and refer, for example, to the richly decorated library of the *Universidad Nacional Autónoma de México*, or the *sgraffito*-decorated houses in the Engadine Valley in the Swiss Alps that Herzog & de Meuron admire. One could also discuss the complex relation between architecture and film, for example, the issue of the duration of architecture as it is visualized in Andy Warhol's eight-hour film *Empire* (1964). But I am less interested in the typology or iconographic genealogy of the building than in the way it allows us to discuss general matters of imagery and representation. As I came closer, I realized that the images are not painted onto the surface but are inscribed by means of a specially developed silk-screen technique into the concrete panels and silkscreened by means of enamel paint onto the glass panels. They are blow-ups of newspaper photographs; the dot grids are clearly visible. (ill. 4) But whenever I tried to focus on a detail of the motif, I stumbled over the materiality of the panel. It is impossible to make an abstraction of the physical support while reading the image, and it is impossible to make an abstraction of the image while looking at the material. Like a half-forgotten person in a painting by Gerhard Richter, the motifs are constantly vanishing.

Soaked with Images

Eberswalde goes furthest in Herzog & de Meuron's intention to 'soak the façade with images' as Jacques Herzog puts it.[8] In fact, this 'soaking with images' is a crucial theme in their work. It goes back to an unrealized project for a Greek Orthodox Church in Zurich (competition 1989). They first employed it in the concrete panels used for the façade of the Pfaffenholz Sports Center in Saint Louis, France (1993), then in the motif of a plant by the German photographer Karl Blossfeldt used in the polycarbonate panels of the Ricola-Europe SA Production and Storage Building in Mulhouse, France (1993), and it culminated – from my point of view – in the Eberswalde Library. The complex relationship between buildings and images is a core theme in the recent history of architecture.[9] It challenges traditional sets of value and it is a challenge to traditional conceptions of spatiality. Herzog & de Meuron, standing at the forefront of this tendency arouse much criticism from colleagues. In conversations with colleagues, I often hear that Herzog & de Meuron are 'only' interested in images. Word is that Wolf Prix said of the Signal Box, Auf dem Wolf, Basel (1994) that any student could do that because it is just wrapping a surface around a core. And word is that Peter Eisenman once called them 'Swiss Fascists', perhaps because of the authority expressed by their images and the monolithic shape of their buildings. Being the most radical case of a combination of images and architecture, Eberswalde Library is, of course, particularly provocative. Furthermore, the reason for the disconnect between the façade and the interior is obvious. Because of budget cuts, a local architecture firm was hired to design the interior. However, I found only one published critique. In an article in *AV*, published in 1999, Rafael Moneo criticized the exclusive attention paid to the façade of the library, contrasting this with what he sees as a successful structure, namely, the

---

5 Terence Riley, 'Light Construction', in **Light Construction**, exh. cat., Museum of Modern Art, New York (New York: Abrams, 1995) p. 11.

6 Herzog & de Meuron, in **Arch+** 129–30, special issue 'Herzog & de Meuron: Minimalism and Ornament' (December 1995) pp. 23.

7 Robert Smithson, 'A Tour of the Monuments of Passaic, New Jersey' (first published in **Artforum** (December 1967) in idem, **The Collected Writings**, ed. Jack Flam (Berkeley: University of California Press, 1996) pp. 68–74, esp. 70.

8 'In the project for a Greek Orthodox Church... we combined, in fact we virtually soaked the image, the icon, which is the very essence of this religion with the wall of the church that the images sort of vanishes, dissolves in the spatiality. It dissolves into something total because it is so present—like a tattoo that is so visible that it turns into something abstract, "non-image"-like.' From: 'Discussion between Jacques Herzog and Bernhard Bürgi, Basel, 8 November, 1990,' in **H & deM, Eine Ausstellung in den vier Räumen des Kunstvereins München mit Beiträgen von Helmut Federle und Enrique Fontanilles sowie einem Gespräch Bernhard Bürgi-Jacques Herzog**, exh. cat. (Kunstverein München, 1991) pp. 8–20, esp. 15 [my translation].

9 See Ilka and Andreas Ruby and Philip Ursprung, **Images: A Picture Book of Architecture** (Munich: Prestel, 2004).

Dominus Winery which he praises for the 'exaltation and celebration of material, of a material that doesn't need form'.[10] (ill. 5)

Moneo is missing something in Eberswalde, which he is not missing in the Dominus Winery. For me, it is precisely the reverse. Is the winery less 'hollow' than the library? Is Moneo interested in sculptural, physical spaces, where I am interested in mental, historical spaces? Is Moneo an essentialist who is nostalgic for self-referentiality – in other words, the idea that the meaning of architecture resides in the reflection of its media? Is he interested in 'depth', where I am interested in 'depthlessness' (a term coined by Fredric Jameson)? Is it about the difference between an absolute, modernist space, and a coded postmodernist space, 'historically specific, and rooted in conventions' as Charles Jencks put it.[11]

I would argue that Herzog & de Meuron are developing a new spatiality. They adhere neither to modernist, nor to postmodernist conceptions of space. In fact, their practice is a challenge to the very dualism between modernist and postmodernist spatiality. Their projects make clear how much these spaces have in common and how much they depend on each other to produce a coherent system of representation. The element that most obviously links the spatiality of, for example, Mies van der Rohe with the spatiality of Robert Venturi – to give two typical examples of modernist and postmodernist architects – is the fact that their buildings are highly photogenic. (ill. 6) Although they work with different architectural forms, they both presuppose space as an empty medium, which can be 'filled' either with objects or with signs.

Such a space functions like a free-trade zone in which things, visual codes, or meaning can be multiplied and shifted around at will. This is the neutral space of diagrams that fascinates Peter Eisenman, the historical space of typological reference, which fascinated Aldo Rossi, Herzog & de Meuron's teacher, and the dynamic space of the flow of globalized capital that fascinates Rem Koolhaas. All these various kinds of space are particularly open to mediation and they translate smoothly onto the two-dimensional plane of photography. In her striking analysis of the Barcelona Pavilion, Claire Zimmermann has shown how Mies's architecture seems to find its fulfillment on the plane of black-and-white photography.[12] This translation works particularly well for any kind of architecture that deals with the play of transparency and opacity, of dissimulation and revelation, of reflection, distortion, and 'blurring' – a play that has fascinated architects from Mies to Diller + Scofidio.

The Legacy of Crystal Palace I would argue that there is an archetype for modernist and postmodernist space: Joseph Paxton's Crystal Palace. As the American literary historian Thomas Richards has shown, the Great Exhibition of 1851 established a specifically capitalist system of representation, which is commonly referred to as a 'spectacle'.[13] Paxton's palace created a space that turned consumer objects into works of art, thereby marking the beginning of what Jean Baudrillard called a 'phenomenology of consumption'.[14] As Richards states, the Crystal Palace was factory, museum, market, station, and theater in one. This space was perfect and could not be improved upon in any way. As people said at the time, it could only be repeated as an endless sequence of Crystal Palaces.

And this prophecy was fulfilled. Today, most architecture still stands in the shadow of the Crystal Palace. Shopping malls, buildings on university campuses, corporate headquarters and the buildings of political institutions still follow the modernist ideology of transparency and promise that, merely by establishing eye contact with what is available, people will have a share in consumption, politics, or knowledge. As much as ever, they operate as a game of concealing and revealing, illusion and disillusion, transparency and opacity. Most of these buildings are dominated by a fundamentally naturalistic approach which goes back to Paxton, the successful greenhouse-builder, who virtually built the Crystal Palace around a row of old elm trees in Hyde Park; by its articulation of gravity, sunlight, materials it seeks to evoke that same natural world that it has ousted. Furthermore, the space of spectacle allows maximum continuity and flexibility. In the hands of Mies van der Rohe, the headquarters of Bacardi Rum can become a museum. In the hands of Rem Koolhaas and Bruce Mau, the Kunsthal Rotterdam is translated into a movie script and a chapter of their book *S, M, L, XL*. In the hands of Diller + Scofidio, water turns into a cloud – Blur (2002) – which turns into a landmark.

But projects by Herzog & de Meuron resist such translations and multiplications. Their architecture does not function as a stage set for an aging praxis of representation but instead operates as if the buildings themselves were exhibits in a larger, as yet unfinished exhibition. None of their projects could ever be described, like the Crystal Palace, as palace, station, and museum in one. The names they give their buildings make this clear: Blue House, Wooden House, House for an Art Collector, House for a Veterinary Surgeon, House in Leymen and so on – until some of the earlier projects reappear in a kind of architectural *déjà-vu* in the Rue des Suisses Apartment Buildings, Paris (2000). On Herzog & de Meuron's romanticist stage – and by 'romanticist' I mean the structure of a fragmentary, ever-changing work – they appear like actors in an ongoing play. Again, Smithson comes to mind, as the one who once wrote: 'The artist should be an actor

10 Rafael Moneo, 'In Celebration of Matter,' in 'Herzog & de Meuron 1980–2000,' special issue of **AV** 77 (1999): 25.
11 Charles Jencks, **The Language of Post-Modern Architecture** (New York: Rizzoli, 1977) p. 118.
12 Claire Zimmermann, 'Photography, Architecture, Abstraction,' paper presented at College Art Association, Ninety-first Annual Conference, New York, February 19–22, 2003.
13 Thomas Richards, **The Commodity Culture of Victorian England: Advertising and Spectacle, 1851–1914** (Stanford: Stanford University Press, 1990) p. 58.
14 See Jean Baudrillard, **The System of Objects**, trans. James Benedict (London: Verso, 1996) and Jean Baudrillard, **The Consumer Society: Myths and Structures** (London: Sage, 1998).

who refuses to act.'[15] Could it be, one might ironically ask, that photography steals the soul of these buildings?

Of course it would be naive to say that the projects of Herzog & de Meuron do not reproduce well. It is obvious that their work too is usually perceived by means of mass-reproduced images. It would be equally naive to suggest that they want to resist the gravitational field of the spectacle. I would argue that architecture cannot completely escape the Crystal Palace's capitalist logic of representation and produce an anti-spectacular space. Nevertheless, I would make distinctions. I would interpret projects such as those by Koolhaas or Diller + Scofidio as a constructive critique of spectacle, one that aims at improving the object under critique while retaining its basic set of values and updating a traditional system of representation which has lost some of its appeal since the days of the Great Exhibition. I see Koolhaas and Diller + Scofidio as the legitimate offspring of the Crystal Palace.

Monumental Space Herzog & de Meuron, by contrast, aim – at least from my perspective – at an alternative to the dominant spatiality. It is rewarding to compare their agenda to Henri Lefebvre's critique of 'abstract space' in his book *The Production of Space* (1974). The book is an attack against mechanistic ideologies. It is against modernism and postmodernism, structuralism and post-structuralism, against Sartre and Foucault and Derrida. The sheer amount of negation in his book – and its internal contradictions – recalls Herzog & de Meuron's strategy of constant opposition and negation. And Lefebvre's longing for a 'unitary theory', where physical, mental, and social spaces can be conceived not as bits and pieces of knowledge but as a theoretical unity, applies readily to the architectural practice of Herzog & de Meuron.[16] Lefebvre's unitary theory deals with the crucial question of monumental space, which offered, as he puts it 'each member of a society an image of that membership, an image of his or her social visage'.[17] For Lefebvre, modernist architecture has failed to produce monumental spaces such as a Greek theater, where, in his words 'space, music, choruses, masks … converge with language and actors.'[18] He poses the question: 'How could the contradiction between building and monument be overcome and surpassed?'[19] To him, one of the reasons for the failure of contemporary architecture to produce modernist spaces is the fact that most (modernist) façades disappear in the abstracted space where 'all aspects of an object could be considered simultaneously.'[20]

In my view, this is precisely what distinguishes the Eberswalde Library from the Dominus Winery and this leads us back to my disagreement with Moneo's critique. Eberswalde cannot be viewed from every aspect. But, seen from certain angles, it offers certain people an 'image of their social visage'. In Eberswalde,

I would argue, Herzog & de Meuron have overcome the contradiction between building and monument. At the interface between the individual building and the urban space they condense the abstract space into a substance – to use another of Lefebvre's terms – which knits together the physical, the mental, and the social space. Another word for substance is 'image' – what Henri Bergson describes in his book *Matter and Memory* as 'a certain existence which is more than that which the idealist calls a presentation, but less than that which the realist calls a thing – an existence placed half-way between the "thing" and the "representation".'[21] For Herzog & de Meuron, I would argue, there is no such thing as an empty space. The world is already full of images.

I thus don't conceive the façade of Eberswalde Library as a screen onto which images are projected through the medium of abstract space, but rather as a dam soaked by the flood of images that the phenomena of the world, our observation, our memory, and our imagination are constantly producing. For Herzog & de Meuron, unlike Aldo Rossi, for example, this dam, or sieve, is not a neutral, contemporary device that frames images from the past but is instead informed and transformed by and through the very materiality of the images, transforming them in turn as well. This might explain why Herzog & de Meuron are so popular, since many people are immediately affected by their architecture. And it explains their preference for storage buildings, museums, libraries, wineries, or simple attics – spaces that are already soaked with images and thus absorb further images more easily.

But the Eberswalde Library is also radical because of the way it represents, or embodies, historical space. In Eberswalde, where as in all eastern Germany the air is thick with Western iconoclasm and repressed historical memory, the private and the public, the past and the future merge. Like all of Herzog & de Meuron's projects, the library is highly site-specific. It obviously refers to the eastern European typology of prefabricated concrete buildings, which are often decorated with images.[22] And it refers to the complex visual memory of eastern Germany.

In 1994, the architects commissioned Thomas Ruff to select the motifs for the façade. He chose them from his archive of newspaper images: 'a "diary" of photographs collected between 1981 and 1991 from newspaper articles on a variety of subjects, such as politics, economics, culture, science, or history.'[23] As he puts it: 'They represent a cross section of the issues that concerned people in

15 Robert Smithson, 'A Refutation of Historical Humanism' (written 1966–67, unpublished during his lifetime), in idem, **The Collected Writings** (note 7), pp. 336–37, esp. p. 337.
16 Henri Lefebvre, **The Production of Space**, trans. Donald Nicholson-Smith (Oxford, UK: Blackwell, 1991) p. 11.
17 ibid., p. 220.
18 ibid., p. 222.
19 ibid., p. 223.
20 ibid., p. 125.
21 Henri Bergson, **Matter and Memory**, trans. Nancy Margaret Paul and W. Scott Palmer (London: George Allen and Unwin, 1911) pp. xi–xii.
22 See Peter Guth, **Wände der Verheissung: Zur Geschichte der architekturbezogenen Kunst in der DDR** (Leipzig: Thom Verlag, 1995).
23 'Library of the Eberswalde Technical School, Project Description of the Façade by Thomas

Germany during this period.'[24] The motifs relate more or less directly to the issues of the library, of learning, of German history, of architecture, and art. Ruff selected as he puts it, '12 different pictures to visualize knowledge and the resulting awareness.'[25] On the bottom of the building there are 'Girls on Grass Roof' – what Ruff in his iconographic program calls 'starting with the wrong thing: putting a picture at the bottom that ought to be on the top. In the tradition of Flaubert's *"Dictionnaire des idées reçues"*.'[26] From bottom to top, the motifs are described as: 'Airplane', 'Model Railroad', 'Venus and Cupid', 'Bernauer Strasse', 'Reunification', 'Vanitas', 'Haus am Horn', 'Old Palazzo', 'Students in a Library', 'Stag Beetle', and 'Girls on Grass Roof'.

The central image links the building to a specific historical moment and is inseparably related to the traumatic separation of Germany, namely, the photograph of Bernauer Strasse, in Berlin. As the Wall was erected in 1961, some inhabitants fled to the west over the façade of their homes. For the collective memory of West Germans, this picture is an icon. But when Ruff submitted it to the local authorities, they, having grown up in the East, did not recognize it and wondered why he was interested in people escaping from a burning building. When they learned about the meaning of the image they asked that a more triumphant image be added, one depicting the reunification of Germany after the fall of the Berlin Wall in November 1989.

The Exhibition   But how, to get back to the initial question, can one exhibit these issues? As mentioned above, my encounter with Eberswalde was crucial for my concept because it shifted my view on the work of Herzog & de Meuron: it opened my eyes to their interest in historical space, and it also made clear that photography does not adequately reproduce most of their buildings. Of the documentary photographs I discovered in the Herzog & de Meuron archives, the only shot that looked convincing was of concrete plates in a storage building. (ill. 7) It inspired the whole exhibition, which was conceived as a puzzle that would invite viewers to reassemble it in their own way. We established three rules for the exhibition. The first rule: No documentation, that is, no documentary photography, no plans, and no small-scale models of existing buildings – what Jacques Herzog ironically calls 'Bonsai architecture'. What then should we display, if not documentation? When I visited the studio of Herzog & de Meuron, I was fascinated by the hundreds of models standing around. They are conceptual models, which literally go from hand to hand to test ideas and to facilitate the design process in an office with two hundred collaborators.

I wanted to reproduce this experience for the visitors of the exhibition, namely, the direct confrontation with the mass of models. The exhibition should produce an atmosphere of entering an unorganized archive or an old attic. It should provoke the urge to rearrange, a playful urge one might experience in old natural history museums, not yet tidied up by eager curators. I wanted to pretend to be an archaeologist from the future who discovers hundreds of models in the archive of Herzog & de Meuron without knowing what they mean. Like a curator of a nineteenth-century natural history museum, who arranged the bones of dinosaurs according to size, I wanted to arrange the models according to formal and morphologic criteria. In addition, they should be compared with works of art, with objects of all kinds emerging from the collections of the various venues we had planned for our exhibition.

The second rule was: Each of the 1,200 objects of the exhibition would have a label. The things do not speak for themselves. I wanted to escape the modernistic fetishization of objects and show that meaning is not inherent but produced. By identifying the objects and interpreting them from a personal point of view, I wanted to show the visitors that this was just one of various possible interpretations and that they themselves were invited to participate in the production and dissemination of meaning. The labels should allow the viewers to know as much as the curator and the architects. At the same time there was no vantage point from which the entire project could be grasped and surveyed. The exhibition itself proved not to be very photogenic. Just as we enjoy the visit to a natural history museum because it demonstrates the contingency of any system of classification, I wanted to demonstrate that neither the curator nor the architects know the meaning completely.

This was directly connected to the third rule: No wall texts. The museum as an institution would remain silent. This had nothing to do with 'institutional critique' – the museum enabled the exhibition. Rather, I wanted to demonstrate that the museum cannot speak from offstage, and that the curator does not have to identify with the museum as an institution.

The key space of the exhibition revolved around Eberswalde. It included some of the panels from the façade, the original images from Ruff's archive, and various other items which relate to the issue, from vintage prints by Karl Blossfeldt that had provided the floral motif used on the façade of the Ricola Mulhouse to photography on porcelain. (ill. 8)

In producing the exhibition, I wanted to reflect the mechanisms of architectural exhibitions and produce an alternative way or representation. My aim was not to fix the meaning of the architecture of Herzog & de Meuron. Rather I was trying to locate their work in a broader context of history and of contemporary visual culture. By combining the various forms and layers of representation in the work of Herzog & de Meuron with other realms of visual culture, I saw more and more clearly why Eberswalde was so fascinating to me as a historian.

Ruff,' trans. Catherine Schelbert, Düsseldorf, August, 1999, unpublished typescript, Archive of Herzog & de Meuron.
24 ibid.
25 ibid.
26 ibid.

Even today, the division and reunification of Germany is a repressed collective memory. There is no public monument to represent this recent history. No work of art, it seems, can perform this function at the moment. The Eberswalde Library proposed an alternative kind of monument: By amalgamating the building with images it produces a new kind of spatiality – a spatiality that articulates historical space and intertwines it with the present. It recalls Lefebvre's definition of monumental space as something that is not simple but highly complex. In his words: 'Not texts, but texture.'[27] What intrigued me in Eberswalde, and what I wanted to articulate in the exhibition, was the way the past is incorporated but not sentimentally evoked. It is represented as something that disappears for good, irreversibly, but is nevertheless intertwined with the present. It disrupts the historicity and the self-referentiality of modernist and postmodernist space and opens up a new chapter for architecture.

27 Lefebvre, **The Production of Space** (note 16), 222.

[1]

[2]

1 Herzog & de Meuron, Library, Fachhochschule Eberswalde, Eberswalde, Germany, 1999. Photo: Philip Ursprung, 2000.
2 Herzog & de Meuron, Library, Fachhochschule Eberswalde, Eberswalde, Germany, 1999, detail. Photo: Philip Ursprung, 2000.

3 Herzog & de Meuron, Library, Fachhochschule Eberswalde, Eberswalde, Germany, 1999, detail. Photo: Philip Ursprung, 2000.
4 Storage building for concrete panels used for the Eberswalde Library (unknown photographer, c. 1999: archive of Herzog & de Meuron, Basle).
5 Ludwig Mies van der Rohe, Hermann Lange House, Krefeld, Germany, 1930, detail of the garden terrace. Photo: Philip Ursprung, 1999.
6 Herzog & de Meuron, Library, Fachhochschule Eberswalde, Eberswalde, Germany, 1999, detail. Photo: Philip Ursprung, 2001.

[5]

[6]

7 Thomas Ruff, **Bibliothek Eberswalde**, 2000.
8 Herzog & de Meuron: Archaeology of the Mind, exhibition at the Canadian Centre for Architecture, Montreal, 2002–3. Photo: Michel Legendre, copyright CCA.

After Criticality:
The Passion for Extreme Reality
in Recent Architecture ...
and Its Limitations*

Roemer van Toorn

Reality through critical lenses.

'Look at me! I'm critical! Read me!' A lot of Western (great) criticism draws the research (and its readers) away from experience and pushes them toward the side of deconstruction or 'criticality'. Foucault and the Frankfurter School theorists accord a paramount place to ideology and culture critique, but minimize the possibility of emergent or alternative conscious-

* An earlier version of this text was previously published: Roemer van Toorn, 'Nooit meer dromen? / No more dreams?', **Architecture in the Netherlands Yearbook**, 2003-04, Rotterdam, 2004, pp. 133-141; extended version: Roemer van Toorn, 'No more dreams?', **Harvard Design Magazine 21: Rising Ambitions, Expanding Terrain**, Fall/Winter 2004, pp. 22-31.

ness allied to emergent and alternative phenomena and groups within the dominant society. The problem with the correct ideas of criticality is that they conform to dominant meanings or established passwords; that it is always ideas that verify something, even if this something is yet to come. Trapped in 'winner loses', Jameson notes[1] that the more Foucault wins by portraying society as corrupt the more he loses in so far as his critical voice of refusal becomes increasingly paralyzed. There is the theoretical insistance, against Foucault, of a guaranteed insufficiency in the dominant culture against which it is possible to mount an attack. Raymond Williams says that 'however dominant a social system may be, the very meaning of its domination involves a limitation or selection of the activities it covers, so that by definition it cannot exhaust all social experience, which therefore always potentially contains space for alternative intentions which are not yet articulated as a social institution or even project.'[2] What seems guarded against in this approach from Williams is immediacy, the unknown, that untreated bolus of direct experience, experiences that cannot be reflected as a whole. The criticality of Foucault and Adorno looks backward, is armed with prior theory but what you could call the projective attitude of Williams and for instance Gramsci is not one which comes armed with prior theory, but rather one which helps formulate new problems or suggests new concepts. The very act of doing entails a commitment to the future, more particularly, a commitment to appearing in, making a contribution to, or in various other ways forming and affecting the future. In the light of this Sarah Whiting, Bob Somol and myself,

Libeskind, Jewish Museum, Trauma Architecture as ad, Billboard Berlin 2004.

instead of critical architecture, propose the term Projective. Why the word *projective*? Because it includes the term *project* – that is, it is more about an approach, a strategy, than a product. The projective looks forward [projects], unlike criticality, which always looks backwards.

Criticality in architecture rests like critical theory on a self-affirming system of theoretical and ideological convictions. 'Look at me! I'm critical! Read me!' Criticality in architecture proceeds from a preconceived legibility.[3] It is an architecture that brooks no alternative interpretations. Unless the critical theory and vision are legible in the object, the object fails. One form of critical architecture – exemplified by the work of Peter Eisenman, Daniel Libeskind, and Diller + Scofidio – offers comments within architectural/social discourse and avoids looking for any alternatives in reality. The Frank House by Eisenman, for example, forces the couple living in it to think about the psychology of their cohabitation by placing a slot in the floor between their beds. Critical Regionalism in Europe, Asia, and Australia – exemplified by the works of Ando, Hertzberger, Siza, and Murcutt – tries, out of disgust with contemporary society, to overcome estrangement, commodification, and the destruction of nature.[4] Critical Regionalism does not strive to make difficult or playful comments on society as Eisenman or Tschumi do, but invests in alternative spaces far from the wild city of late capitalism. It hopes to locate moments of authenticity – to calm the mind and the body – in order to survive in our runaway world. While critical architecture deconstructs the discourse of architecture, demystifies the status quo, or locates alternative worlds in the margin, it believes that constructing liberating realities in the center of society is impossible.

The Projective In contrast to both the criticality of deconstruction and Critical Regionalism, Sarah Whiting, Robert Somol and myself are interested in 'projective practices' which aim to engage realities found in specific local contexts. Instead of hanging ideological prejudices on built form (derived from knowing the future to come or from negative critique against reification (*verdinglichung*)), the architectural project must be rendered capable of functioning interactively. It thereby undermines representation. The projective unmasks representation. In the words of Scott Lash, 'Representation is by definition monological, it is the fixed creation of a subject. Presentation, like play, is dialogical, it opens up and involves the playing off of one another of playmates. When a jazz band improvises it is like play. So is football when it's working, when a team is really knocking the ball around, creating openings, running with the ball, moving into

1 Jameson, Fredric; 'Postmodernism, or the Cultural Logic of Late Capitalism', **New Left Review** 146, pp. 53–92.
2 Interview with Raymond Williams 'Politics and letters' with **New Left Review**, 1979 Verso, 1981.
3 Various observations on criticism versus the projective are very clearly set out by Somol and Whiting in Whiting, Sarah and Somol, Robert; 'Notes around the Doppler Effect and Other Moods of Modernisms', in **Perspecta: The Yale Architectural Journal** 33, 'Mining Autonomy', 2002, pp. 72–77.
4 See Kenneth Frampton, 'Towards a Critical Regionalism: Six Points for an Architecture of Resistance,' in **The Anti-Aesthetic: Essays on Postmodern Culture**, Hal Foster, ed. (New York: New Press, 1999) and Liane Lefaivre, Alexander Tzonis, **Critical Regionalism: Architecture and Identity in a Globalized World** (Prestel USA, 2003).

space.'⁵ According to Lash, emancipation does not come about through an ideal dialogue in this case but through an aesthetic creation – as in jazz, for example. It does not happen through any transcendental aesthetic subjectivity, not again by representation as in criticality, but by presentation, to be understood as 'performance' (situated freedom). With a projective practice the distancing of critical theory is replaced by a curatorial attitude. By systematically researching reality as found with the help of diagrams and other analytical measures, all kinds of latent beauties, forces, and unknown possibilities can be brought to the surface.⁶ Preferable, it seems to me, is a projective practice that operates with and within society at large and that sets a collective and public agenda in direct communication with modernization. The victimology (pity science) of critical theory leaves no running room for plausible readings capable of completing a project in the mundane context of the everyday including that of alienation and commodification. Estrangement must not be thought of as something to overcome, but as a position from within which new horizons can open.

Although the urban, capitalist, and modern everyday is pushing towards increased homogeneity in daily life, the irreconcilable disjunctions born in a postindustrial city full of anachronistic interstices make it impossible to think of modernization as only negative. Michel de Certeau's work confirmed the impossibility of a full colonization of everyday life by late capitalism and stressed that potential alternatives are always available, since individuals and institutions arrange resources and choose methods through particular creative arrangements. Often critical experts and intellectuals prefer to think of themselves as outside everyday life. Convinced that it is corrupt, they attempt to evade it. They use rhetorical language, meta-language, or autonomous language – to paraphrase Henri Lefebvre – as permanent substitutes for experience, allowing them to ignore the mediocrity of their own condition. Critical practices reject and react unsubtly to the positive things that have been achieved in contemporary society, such as the vitality of much popular culture, including its hedonism, luxury, and laughter.

So what is it instead of criticality that the projective should produce but has so far failed to produce?

Addiction to Extreme Realism In many Dutch practices the architect waits and sees in the process of creation where information leads him or her. Much of the strange shapes of recent Dutch architecture can be attributed to the devotion to the diagram, and the authorial absolution it grants. By taking traditional Dutch pragmatism to absurd, deadpan extremes, the designer generates new, wholly unexpected forms. Some of Droog Design embodies this absurdist-hyper-rationalism. The designer simply continues to apply the system until the form appears in all its strangeness. The touchstone here is not subjective vision

Model of the Dutch Pavilion by MVRDV in Hanover with former Prime Minister Wim Kok.

but an addiction to extreme realism, a realism that is intended to show no theoretical or political mediation, a kind of degree zero of the political, without thought about the consequences of the social construction it would lead to in reality. The extreme realities the projective is obliged to deal with are the cyborg; the information society; the global migration of money, people, and imagination; shopping; fashion; media; leisure; and the coincidence of the enormous effectiveness and absolute abstraction of digitization. In other words, this practice brings to its extreme the consequences of the processes of commodification, alienation, and estrangement that constitute the contemporary motor of modernity. According to projective practices, involvement, even complicity with given conditions, rather than aloofness, is more productive than dreaming of a new world. The paternalistic 'we know best' attitude that has long hindered critical architecture is a thing of the past.

From my perspective, we can see three basic types in many recent realized projects in the Netherlands, types that display 'projective autonomy', 'projective mise-en-scène', and 'projective naturalization'. The projective autonomy confines itself primary to models of geometry. Projective

---

5 Scott Lash, 'Difference or Sociality?', Lecture for conference 'The Theory of the Image', Van Eyck Academy, Maastricht, 1995.
6 All data regarding location, program, use, and infrastructure as well as the economy, politics, art, fashion, the media, the everyday, technology, typology, and materials that might conceivably help to advance a specific 'found' reality are documented in diagrammatic form, especially charts and graphs. Of course, ideology is implicit in the science of measurement and the way the hidden qualities of reality are communicated. Most projective practices are, however, not aware of this ideological dimension. In addition they are ideologically 'smooth' because the veil of fashion and style hides the many contradictions through the deployment of the design. For more information on the ideological dimension of contemporary Dutch architecture see my article 'Fresh Conservatism: Landscapes of Normality', in Ibelings, Hans (ed.); **Artificial Landscape: Contemporary Architecture, Urbanism** (Rotterdam: NAi Publishers, 2000).

'Projective autonomy': Claus & Kaan, Forensic Institute, Ypenburg, 2004.

mise-en-scène and projective naturalization, by contrast, experiment with architecture as infrastructure. Projective autonomy tries to restore contact with the user and the contemplator through passive experience, while projective mise-en-scène and projective naturalization seek interaction. While projective autonomy is interested in form – what the aesthetic by its own means is able to communicate – the projective mise-en-scène seeks the creation of theatrical situations, and projective naturalization seeks strictly instrumental and operational systems.

Projective Autonomy The architecture of Claus & Kaan (and Martin Pawson and other minimalists) reveals what I am calling 'projective autonomy'. The meticulously crafted forms characteristic of their projective strategy offer comfort and reassurance. Projective autonomy revolves around the self-sufficiency of tasteful, subdued form, which, notwithstanding the vicissitudes of life or passing dreams, is in theory capable of enduring for centuries. For Claus & Kaan, the organizing principal is not typology but the typographic autonomy of a building. Just as the typographer selects his typeface and searches for the most appropriate spacing, so Claus & Kaan deal in a craftsmanly and repetitive manner with windows, columns, doors, façade panels, and volumes. They pursue a conventional architecture that inspires confidence and eschews controversy, that is about mass, boxy volumes, light, beauty, and style.⁷ The abstract language and meticulous detailing lend their buildings a self-satisfied, stylish gloss. Minimal chic glosses over vulgarities with its abstract perfection.

Projective mise-en-scène   In the projective mise-en-scène favored by MVRDV and NL, the user becomes an actor invited to take an active part in the theater choreographed by the architects. In these projective practices, projects are not to be contemplated; rather they throw reality forward through the help of scenarios inspired by the theatrical programs theze architects write based upon the data they find within 'extreme reality'. They turn life into an optimistic and cheerful play that generates new solutions while making jokes about our constantly mutating reality. MVRDV translates the program into a carefully choreographed spatial experience that incorporates the user into science fictions hidden in the everyday.

While NL makes jokes and develops a trendy lifestyle, MVRDV looks for new spatial concepts capable of giving our deregulated society the best imaginable and spectacular shape. In projective mise-en-scène, it is not the autonomous force of the type, of chic minimalism, that is given free rein – as in projective autonomy – but the daydreams alive in society. Objects are not important as things in a projective mise-en-scène; they are there to be used as a screen onto which fragments of our extreme reality can be projected. Instead of continuing to hide the more than sixteen million pigs in thousands of pitch-roofed bioindustry barns spread over the picturesque countryside of the Netherlands, MVRDV proposes that it is more efficient and animal-friendly to house pigs in high-rise flats in the harbor of Rotterdam. Suddenly – without any value judgement – the facts that there are more pigs than people in the Netherlands and that pigs can be happy in high-rises with a view, looks plausible.

As in the social sciences, objects are seen as the carriers of everyday culture and lifestyle. The architecture is a co-producer in the embodiment of cultural and social meaning. The shock effect of the surreal and pragmatic mise-en-scène – like the Benetton billboards by Olivier Toscani with an AIDS patient dying in a living room – will immediately grab attention. But whether this bewildering realistic mode of representation is interested in either a better world or in exposing our Brave New World remains uncertain. The fables that lie hidden in the everyday are made visible by MVRDV's opportunistic imagination and make users into leading actors, as in the Dutch pavilion in Hanover.

Like Steven Spielberg, architects must provide new representations that everyone can enjoy.⁸

---

7 'We do not believe in designing aesthetic objects with complicated forms that can only be built through craftsmanship. Instead, we use standard industrial materials, spans, and constructions: ordinary products and ordinary techniques.' **cf.** Claus & Kaan, '109 Provisional Attempts to Address Six Simple and Hard Questions About What Architects Do Today and Where Their Profession Might Go Tomorrow' in Sigler, Jennifer and van Toorn, Roemer; **Hunch** 6/7, Berlage Report (Rotterdam: Episode Publishers, 2003) p. 140.
8 Winy Maas mentioned Spielberg in a call to architects during the presentation of a research studio at the Berlage Institute, March 2, 2004.

Entertainment first confronts you even with the dystopias (e.g., sixteen million stacked pigs), then guarantees a happy ending by glossing them over with 'pragmatic solutions' ensuring conformity. The attitude is the putatively cool 'Whatever.' As long as it generates difference.

Foreign Office Architects, Yokohama Terminals.

Projective Naturalization While the Projective mise-en-scène is busily projecting meaning onto things, it forgets that materials and structures can themselves convey meaning, can be sensitive and active, and can activate processes in both the eye and the body. That performative capacity is at the heart of practices that follow the route of what could be called 'projective naturalization'. In the Netherlands, projective naturalizations have been developed by, among others, Oosterhuis.nl, UN Studio, Maurice Nio, and NOX Architekten.

Projective naturalization is not about signs, messages, codes, programs, or collages of ideas projected onto an object, but about technologies that allow matter to be performative. Architect Lars Spuybroek of NOX is not interested in technology as a way of regulating functions and comfort. He sees it as a destabilizing force whose function is to fulfill our craving for the accidental by providing a variety of potentialities and events. What geology, biology, and even history have taught the architects of projective naturalization is that mutable processes generate far more intelligent, refined, and complex systems than ready-made ideas ever can.[9] This non-conventional architecture comprehends many shapes and schools.[10] What these manifestations have in common with nature is that the shapes they produce exhibit similarities with the structures, processes, and shapes of biology. A façade is not simply a shell, but a skin with depth that changes in response to activity, light, temperature, and sometimes even emotions.

A blobbish interactive 'D-tower' designed by NOX is connected to a website at which the city's inhabitants can record responses to a questionnaire, designed and written by artist Q.S. Serafijn, about their everyday emotions: hate, love, happiness, and fear. The answers are graphed in different 'landscapes' on the website that show the valleys and peaks of emotions for each of the city's postal codes. The four emotions are represented by green, red, blue, and yellow, and determine the colors of the lamps illuminating the tower. Each night, driving through the town of Doetinchem, one can see which emotion is most deeply felt that day.

A host of measurable data and technologies gives rise to a sophisticated metabolism that, as in Foreign Office Architects' Yokohama Terminal, channels the flows of people, cars, ships, and information like blood cells through and near the organism of the building. The project tries to function without obstacles or other complications and avoids communicating cultural meaning through shock, as does the work of MVRDV.

It is not ideology but the (wished for) instinct of artificial organisms that ensures that complex processes are operating appropriately. Buildings are intended to function like bodies without heads following complex biomechanical logic. When Foreign Office Architects exhibited their Yokohama Terminal at the Venice Biennale, they showed sections of a body scan parallel to the terminal, suggesting that the logic of a building should resemble the body's.

In contrast to projective mises-en-scène, projective naturalizations are not interested in projecting scenarios onto objects related to society, religion, power, politics, globalization, or individuals. Projective naturalizations possess a super-functionality that revolves around movement, self-organization, and interactivity.[11] Projective naturalizations are about modulating precise and local decisions from a mechanistic perspective interested in self-organizing systems that allow flows of consensus to follow their different trajectories. While concentrating on organic abstractions, projective naturalizations totally neglect the fact that every appropria-

---

9 See also Manuel DeLanda, **A Thousand Years of Nonlinear History** (New York: Zone Books, 2000) and **Intensive Science and Virtual Philosophy** (London and New York: Continuum, 2002).

10 The pavilions by Oosterhuis.nl and Asymptote in the Floriade Park in **Architecture in the Netherlands. Yearbook 2002–2003**, Anne Hoogewoning, Roemer Van Toorn, Piet Vollaard, Arthur Wortmann (eds) (Rotterdam: NAi Publishers, 2003) pp. 38–40 and the saltwater pavilion by Kas Oosterhuis and the freshwater pavilion by Lars Spuybroek at Neeltje Jans (in **Architecture in the Netherlands: Yearbook 1997–1998**, Hans Van Dijk, Hans Ibelings, Bart Lootsma, Ton Verstegen (eds) (Rotterdam: NAi Publishers, 1998) pp. 42–47.

11 Projective naturalizations also aspire to be operational. See also what Koolhaas has to say about this in a reaction to the manifesto of Van Berkel, Kwinter, Zaera-Polo, and Lynn during the 1997 **Anyhow** conference in Rotterdam: 'They had fresh and new ambitions and postures — antisemantic, purely operational — represented in virtuoso computer (in)animation. I remember being critical of their claim, then, that they had gone beyond form to sheer performance, and their claim that they had gone beyond the semantic into the purely instrumental and strictly operational. What I find (still) baffling is their hostility to the semantic. Semiotics is more triumphant than ever—as evidenced, for example, in the corporate world or in branding—and the semantic critique may be more useful than ever.' Rem Koolhaas, in 'Spot Check: A Conversation between Rem Koolhaas and Sarah Whiting', **Assemblage** 40, December 1999, p. 46. See also Felicity D. Scott, 'Involuntary Prisoners of Architecture', **October 106**, Fall 2003, pp. 75–101.

tion of a project depends on narratives of use – is about the interaction between social behavior and a given objective condition. What projective naturalizations tend to forget is that our social actions and behavior, not our biological bodies, constitute our identities.[12]

Larger Ambitions   Breaking with criticality, a passion for extreme reality and a return to what architecture as a discipline is capable of projecting are essential to make the most of the many possibilities inherent in our 'second modernity.'[13] Instead of predicting the future, we have to be attentive to the unknown knocking at the door. Projective practices also demonstrate that the question is not whether architecture should participate in late capitalism. That is a given. But what form this relationship with the market realism should take is an ethical and political question that cannot be curated only in pragmatic, technical, or aesthetic terms. The projective practices described here create spaces cut from the same cloth as the garments of the ruling systems. As such they confine themselves to forms of comfort enjoyed in particular by the global middle class. Apart from fear of confrontation with the unknown, the chief concerns of this middle class are the smooth processes that guarantee its rights to power, individualism, career, identity, luxury, amusement, consuming, and the infrastructure that makes all this possible.

This totalitarianism of difference, of individual rights – celebrated as the 'multitude' of neoliberalism – overlooks the fact that it is essential to pay attention to the collective interests of the world population (including that of the transnational middle class). Instead of the paradigm of difference, we should vivify a paradigm of sameness and collective responsibility. Culture is now all about diversity, flexibility, and the search for permanent novelty and effect that a project initiates, about how an object can relate to the market as an open supposedly neutral platform. This is a strategy without political ideals, without political or socio-historical awareness, that is in danger of becoming the victim of a dictatorship of aesthetics, technology, and the pragmatism of the blindly onrushing global economy. Instead of taking responsibility for the design, instead of having the courage to steer flows in a certain direction, the ethical and political consequences arising from the design decisions are left to the market[14], and the architect retreats into the givens of his discipline. In that way, all three projective practices described here are formalistic and politically correct.

Copulation of Clichés   It's a politics of what I call Fresh Conservatism, the one of Neoliberalism, or a market realism (design as crime), that embraces the many transformations of our reflexive modernization with no regard for the consequences of the many subversive (thus fresh) lifestyles they design. While the freshness of their design generates endless differences on the cultural level, they generate a new homogenisation on the level of the political, a culture of sprawl for a new global middle class.

The positive thing about projective practices is that in the making of a project, under the influence of the material, the economy, the construction, the form, the program, the specific context, and with the help of architectural knowledge, projections can be tested and developed. In the very act of walking, projective practices create their paths. Not a priori ideas, but the intelligence of a certain condition is used to the full. In the making of work, reality projects itself.

Smooth Space, Anti-Utopian

What these show projective practices fail to see, however, is that utopias – as a principle of hope – are necessary in order to develop in a project a perspective that reaches beyond the status quo. I am not suggesting that utopias should be realized, but that such utopian dreams provide frames of reference for political action. Utopian dreams also enable us to make a detached diagnosis of the present. This moment of exile from the addiction to extreme reality could make us aware of our own inevitable and implicit value judgements, of the fact that excluding political and social direction itself sets a political and social direction. It is the interaction between the moment of utopia with reality that could help a projective practice develop a new social perspective. What should fascinate projective practice is how it might inflect capitalism towards democracy. The only problem is that so far almost nobody has been prepared to rethink the now-eroded concept of democracy or to carry out research into what democracy could mean today in spatial terms.

Talking about democracy is simultaneously a taboo and a fetish. We treat the word *democracy* as a palliative that relieves us from having to think hard about its realization. If we were to dream about new formations of democracy in space and time, we would develop visions that shake off the current political ennui, the blind pursuit of the market, and our incessant navel-gazing.

At times, the practice of Rem Koolhaas (although he refuses to talk about it) seems to experiment with new notions of democracy in space. Alongside the three projective practices mentioned there are also 'projective juxtapositions', in which the permanent crisis of late capitalism is a source of inspiration. Projective juxtapositions are characterized by an indefinable critical

---

12 I am always surprised when Van Berkel & Bos (UN Studio) show their 'Manimal' metaphor for a new architectural practice—an image hybridizing a lion, a snake, and a human, and only talk about the process of generating the Manimal but never about its cultural, ideological, and symbolic implications. For them it's all about form and not how social practices of use unlock such a metaphor. It would not surprise me if Hollywood cast this Manimal in a horror film.
13 The idea of a 'second' or 'reflexive' modernity was first developed in Ulrich Beck's **Risikogesellschaft: auf dem Weg in eine andere Moderne** (Frankfurt am Main: Suhrkamp, 1986) (**Risk Society: Towards a New Modernity** (London; Newbury Park, CA: Sage Publications, 1992).
14 For the role of the market in architecture see also 'Lost in Paradise' in **Architecture in the Netherlands: Yearbook 2001–2002**, by Anne Hoogewoning, Piet Vollaard, Roemer Van Toorn (eds) (Rotterdam: NAi Publishers, 2002) and 'Propaganda' in **Architecture in the Netherlands: Yearbook 2002–2003**. Op. cit.

'OMA meets Mies', National Gallery, Berlin.

detachment that continually places the program and with it the organization of society in a state of crisis. In projective juxtapositions – such as the ones of OMA – a project never reaches a conclusion but instead provokes a never-ending subjective interpretation and inhabitation. The early projective juxtapositions of OMA were a vessel to experiment with new freedoms, as for example the Kunsthal resisting the current idea that a museum needs to be a temple with quasi-neutral white exhibition spaces. There a projective juxtaposition is combined with what Immanuel Wallerstein calls Utopistics. With Utopistics Wallerstein is not referring to a progressivism that already knows what is to come, but is pleading for a science that seriously assesses liberating historical alternatives – what best possible path for a far (and uncertain) future can be followed. Reassessing Utopistic examples – which proved successful in creating freedom in the past – can help in the creation of new situations of freedom.

Such an approach can be found in OMA's Seattle library, which to a large extent reworks the public library of Hans Sharoun in Berlin. When utopistics are combined with a projective juxtaposition, we come close to what I am after. But the OMA experiments with Prada and the Guggenheim in Las Vegas went no further than a projective mise-en-scène which Salvador Dalí would have loved: 'It is not necessary for the public to know whether I am joking or whether I am serious, just as it is not necessary for me to know it myself.'[15]

---

[15] Salvador Dali, in **Diary of a Genius**, London, Creation Books, 1964.

Poesis doctrinae tamquam somnium

Kari Jormakka

'Poetry is as a dream of learning, a thing sweet and varied, and that would be thought to have in it something divine, a character which dreams likewise affect. But now it is time for me to awake, and rising above the earth, to wing my way through the clear air of philosophy and the sciences.'
Francis Bacon, *De Dignitate et Augmentis Scientiarum (Advancement of Learning)*, 1623

In an oft-quoted passage in Plato's *Sophist*, the main protagonist, a Stranger from Elea, asks his respondent, the young Theatetus: 'Do we not make one house by the art of building, and another by the art of drawing, which is a sort of dream created by man to those who are awake?'[1] This passage has always reminded me of Bob Dylan's classic *Talkin' World War Three Blues* which describes his nightmare circa 1962 of being the only survivor in New York after a nuclear war. After waking up, the protagonist lies on the analyst's bench, discussing his dream, and, after a complex psychoanalytic dialectic, concludes: 'Part of the people may be alright all the time and all the people may be alright part of the time, but all the people cannot be alright all of the time. I think Abraham Lincoln said that. I'll let you be in my dream if I can be in yours. I said that.'[2] Dylan's irony, of course, is in that one cannot really let other people into one's dream: your dream is your private world. Then what does Plato mean by calling a drawing of a building a dream (*oneiros*) created by man to those who are awake?

Reality In a banal sense, architectural drawings are indeed often daydreams for those who are awake in that they represent wishes that may never be fulfilled. In a more fundamental but equally banal sense, Plato generally contrasts dreams with reality. The house as built, has real being, while the house that is merely drawn is not real in the same sense. However plausible this view may be, it brings up a problem for architects.

There is a widespread and persistent tradition according to which architecture is about real buildings, the reality of construction, and materials.[3] If this standpoint is correct, it is ironic that architects – who claim to produce architecture – actually produce only representations of buildings, visual or verbal.

As early as the late thirteenth century, Nicolas de Biard complained that 'the master masons … do not work: nevertheless they receive the greater fees, as do many modern churchmen… Some work with words only. Observe: in these large buildings there is wont to be one chief master who orders matters only by word, rarely or never putting his hand to the task, but nevertheless receiv-

[1] Plato; **Sophist** 266c. Unless otherwise indicated, the translations are from **The Dialogues of Plato**, translated by Benjamin Jowett (Chicago/London/Toronto: Encyclopaedia Britannica, inc. 1952).
[2] Dylan, Bob; 'Talkin' World War III Blues' song lyric published in **Bob Dylan: Lyrics 1962–2001** (New York: Simon & Schuster, 2004) pp. 64–65.
[3] For example, Peter Zumthor argues that **'Die Wirklichkeit der Architektur ist das Konkrete, das Form-, Masse- und Raumgewordene, ihr Körper. Es gibt keine Idee, ausser in den Dingen.'** Consequently, he rejects the theory of architecture as a thought, as a **Denkform**, a view that he imputes to Herzog & de Meuron. See Zumthor, Peter; **'Der harte Kern der Schönheit'.** In Moravánszky, Ákos; **Architekturtheorie im 20. Jahrhundert.** (Vienna/New York: Springer, 2003) pp. 252, 254.

ing higher wages than the others. … they themselves do nothing.'[4] While architects produce representations or likenesses of buildings and images of architecture, they themselves seldom make the buildings, although they do claim the honor for having created them.

Speaking through the mouth of the Stranger, Plato divides craft (*techne*) into acquiring (*ktetike*), and producing or making (*poietike*); further, making is divided into the making of originals of real things (*autopoietike*) and the making of images (*eidolopoiike*).[5] Here Plato does not use the usual term *eikon* for an image or likeness, but rather the word *eidolon*, which connotes a false image or a counterfeit. He adopts the same terminology in his attack against the arts in the tenth book of the *Republic*, where he also talks about the image as *phainomenon* or *phantasma*, with connotations of apparitions and magic.[6]

The making of images is further divided into the making of likenesses (*eikastike*) and the making of phantasms or semblances (*phantastike*). A genuine likeness is one that conforms to the original in all three dimensions and whose every part has the same color as the original.[7] To make a likeness (*eikon*) on this definition is to make a perfect replica.[8] In painting and sculpture, however, it is typical that one does not make a likeness but a semblance, an appearance. For Plato, picture-making is twice-removed from the truth. He gives the example of tables and beds. The Form itself is not the creation of the craftsman; rather, the carpenter makes 'not what is, but something which is like what is, but is not that'.[9] The painter, on the other hand, makes a likeness of the bed, not of the Form. Painters and also poets make false images of excellence (*eidola aretes*) which deceive children and fools.[10]

A painting, in Plato's theory, is not the imitation of things as they are, but rather as they appear.[11] He claims that in large works of sculpture and painting, there needs to be a certain degree of deception. To counter-act the effects of perspective, the upper part of a large statue needs to be made proportionally larger than the lower part which is closer to the viewer.[12] There is really no alternative: a painter cannot imitate a Form directly, because Forms do not have physical form.[13]

Who then makes the Forms, if not the carpenter? Socrates hesitantly suggests that Forms might come from the gods, but we need not give this suggestion too much importance, since the Forms are timeless and then do not really need ever to have been created, any more than we think of universals – such as blueness or triangularity – of having been created by a god.[14] Making images, by contrast, is not much of an achievement according to Plato. Everyone can do it with a mirror: 'Carry a mirror with you everywhere; you will then quickly make the sun and things in the heavens, the earth as quickly, yourself and the other living creatures, manufactured articles, plants, and all that was mentioned just now. Yes, he said, I could make them appear, but I could not make them as they truly are.'[15] A mirror image or a drawing of a house is and is not a house; the image is poised between the realms of being and non-being. The Stranger argues:

> When we speak of 'things which are not' are we not attributing plurality to not-being? … But, on the other hand, when we say 'what is not', do we not attribute unity? … Nevertheless, we maintain that you may not and ought not to attribute being to not-being? … Do you see, then, that not-being in itself can neither be spoken, uttered, or thought, but that it is unthinkable, unutterable, unspeakable, indescribable?[16]

The Stranger is fully aware that he has just predicated something of not-being, and this must be contradiction: we cannot say that not-being *is* unthinkable, or anything else. To further complicate matters, he asks for a definition of an image. Theaetetus, who is described as the likeness of Socrates, describes an image as 'something fashioned in the likeness of the true'.[17] However, the Stranger then asks if this something is to be some other true thing and the young man answers: 'Certainly not another true thing, but only a resemblance.'[18] The Stranger concludes: 'And you mean by true that which really is? … And the not true is that which is the opposite of the true? A resemblance, then, is not really real, if, as you say, not true? … Then what we call an image is in reality really unreal.'[19] Theaetetus exclaims: 'Not-being does appear to have become twisted together with being, by some such weaving, and it is very strange.'[20] The clever Stranger has demonstrated that an image is not real, it has no real being, and yet it exists in some sense.

4 Frisch, Teresa Grace (ed.); **Gothic Art 1140–1450. Sources and documents** (Englewood Cliffs, N.J.: Prentice-Hall, 1971) p. 55. Plato also notes that 'the master-builder does not work himself, but is the ruler of workmen'. See **Statesman** 259d.
5 Plato; Sophist, 219a–c, 264c–265b; see also 235b–236c, 266d–268d.
6 Plato; Republic, 598b.
7 **Sophist**, 235d-e.
8 Janaway, Christopher; **Images of Excellence. Plato's Critique of the Arts** (Oxford: Clarendon Press, 1998) p. 170.
9 Plato; **Republic**, 597a.
10 Plato; **Republic** 598b–c, 600e.
11 Plato; **Republic**, 598a.
12 Plato; **Sophist**, 236a.
13 cf. Plato; **Statesman**, 285e–286b, but also consider **Republic** 500e3–4, where Plato says, perhaps metaphorically, that no city can be happy which is not designed by artists who imitate the divine model.
14 Plato; **Republic**, 597b.
15 Plato; **Republic**, 596d–e.
16 Plato; **Sophist** 238b.
17 Plato; **Sophist** 240a. In Plato's **Theatetus**, 114d–e, Socrates invites Theaetetus to join him so that he 'may see the reflection of myself in your face, for Theodorus says that we are alike'. In the **Statesman** 258a, Socrates says that both Theaetetus and Young Socrates can 'be said to be in some way related to me; for the one…has the cut of my ugly face, the other is called by my own name.'
18 Plato; **Sophist** 240b.
19 Plato; **Sophist** 240b–c.
20 Plato; **Sophist**, 240b–d.

One of the problems in Plato's account of images has been pointed out by Arthur C. Danto.[21] This kind of imitation theory assumes that the medium is logically invisible. The aim of imitation is to conceal from the viewer the fact that it is an imitation. And yet the painting of the bed is clearly not just 'of the bed', it is also a painting, for example a flat surface with paint on it. The picture of a bed may be a good painting as a painting, or a bad one. While the drawing of a house is just an image of the house, and as such as unreal as a house, it is as real as a drawing. To see what the drawing itself is, we have to take a look at different architectural representations.

Representations   In Plato's time, Greek architects used few or no drawings to create the paradigms of Western architecture, such as the Parthenon. Some researchers speculate that ground plans might have made their first appearance in the second century when Hermogenes 'dedicated at the temple of Athena' and also 'executed' something called a *hypographe*, possibly a plan carved in stone. Drawings were not really necessary because the proportions and basic principles of Doric, Ionic and Corinthian temple architecture were fixed by written texts and the dimensions of a temple could be determined from below to above while the building was under construction. There were, however, several different methods of representing buildings. The most important were called *syngraphai* or 'specifications', of which a few examples have survived. In the case with the naval arsenal of Piraeus by the architect Philon of Eleusis in 340 B.C., the description is clear enough that a reconstruction of the building has been possible. A *syngraphe* set out the general lines of the building amplified by a good deal of information about the way it was built. Instead of making references to drawings, the necessary details are conveyed by precise measurements and by conventional technical terms, such as 'triglyph' or 'Ionic cornice'. In some cases, the specifications include explanations of the design, which are presumably intended for the clients rather than for the builders. In the description of the Piraeus arsenal, for example, it is said that the columns are to be placed so as to leave a twenty-foot wide passage for the public through the middle of the arsenal. Plato's 'city of words', his plan for an ideal city, as presented in the Republic, could be seen as an extended *syngraphe*.[22]

The *syngraphai* could also contain much information of a purely technical nature, but if special instructions about a particular detail were needed, the architect would, during construction, supply a *paradeigma* or an *anagrapheus*.[23] It is not clear what the *anagrapheus* was. It might have been a template, used particularly for blocks with complex mouldings which could be specified by a two-dimensional profile. The *paradeigma* is better known: it was a full-scale model made of wood, stucco, or clay, used for a three-dimensional representation of small elements like triglyphs or capitals and for the description of colored or carved ornaments. Such models raise the problem of the architect's role in design; there is no evidence of these specimens made by architects but there are such made by other craftsmen. While the overall design of the temple was more or less fixed and its size related to the column as a module, the details determined by *paradeigmas* were a major variable in design. This fact would seem to make the craftsmen more like today's artists than the architects of the temple. However, the status assigned to the details represented in *paradeigma* was not equal to the general structure, as indicated in the verbal specifications.

George Hersey sets the *syngraphai* in the context of Vitruvius' emphatic remarks that architecture consists of *fabrica et rationacione* or practice and theory, and that in all matters there are two things: the signifier and the signified.[24] In Hersey's reading, the *syngraphai* are the signified and the building is the signifier. This view privileges the specifications and in fact identifies them with the architectural work proper. He claims that in commissioning a building, the patron only had to utter a signified, such as 'I want an eustyle Ionic pseudodipteral octastyle temple with a module of two feet and a three-step stylobate.' This statement, with its various subsidiary formulas, is the temple completely translated into an algorithm. In this sense, the written document containing the *syngraphai* and the building are two particular embodiments of a universal, the architectural work which in essence is a principle of geometrical or numerical order.[25]

Plenitude   The *syngraphai* represent the building without looking like it; representation is here not based on resemblance but on the codification of the elements. But modern architects are seldom satisfied with such verbal descriptions; they impede creativity and leave a lot of the final result undefined. Pictures, it is claimed, can multiply the informational content a thousand-fold. However, our standard architectural drawings, whether on paper or computer screen, also suffer from a host of limitations. Orthographic projections and axonometric images do not show very accurately how the building would appear in reality, and while perspectives may give a more realistic image (with certain restrictions), they distort the shapes and sizes of the building so that they cannot be used for construction purposes. Faced with this dilemma, Wilfried Wang declares that 'there is simply no way to reduce three-dimensional architecture to two dimensions.'[26]

Hence, it seems more reasonable to resort to models for they are as three-dimensional as the buildings they represent. Furthermore, models do not limit the spectator to one or more views

21 Danto, Arthur C.; **The Transfiguration of the Commonplace** (Cambridge, MA: Harvard University Press, 1981) p. 151.
22 **Republic**, 59 2a10–b3. Plato also uses the word '**paradeigma**'.
23 Coulton, J. J.; **Ancient Greek Architects at Work** (Ithaca, N.Y.: Cornell University Press, 1977) pp. 52–57.
24 Vitruvius Pollio, Marcus; **The Ten Books on Architecture**. Tr. Morris Hicky Morgan (N.Y.: Dover, 1960) I,i,1 ; I, i,15.
25 Hersey, George L.; **Pythagorean Palaces** (Ithaca, N.Y.: Cornell University Press, 1976) p. 25.
26 Wang, Wilfried; 'A Plea for Durability and Unobtrusiveness', in **Overload.** HdA Dokumente zur Architektur 7. Ed. Roland Ritter and Bernd Knaller-Vlay (Graz: Haus der Architektur, 1996) p. 20.

designated by the architect but allow one to look around. However, it is obvious that not even models reproduce too many features of the building. The materials are different so that the tactile values (and possibly the tectonic ones as well) of 'critical regionalism', for example, would be misrepresented. Models are usually quite small, so that the effects of size have to be imagined. Neither are models nearly as heavy or noisy as the real thing; moreover, they smell different. It is not clear, then, that models resemble the building more realistically than the standard drawings do.

All of the standard representations from drawings to models and computer simulations of architecture perpetuate what Juhani Pallasmaa calls the 'visual bias' of architectural theory, the allegedly harmful tendency to see architecture as a visual phenomenon. In line with this thinking, Wang remarks that:

> … human beings are endowed with more than the sense of vision to perceive their environment. The synthesis producing the overall impression of a room or a form not only depends on seeing but on our other senses as well. Even architects do not exclusively rely on their … vision when walking through an underground garage: they also make use of their senses of hearing and smelling.[27]

Already Plato pointed out this problem in his attack against images. He claimed that 'mimetic art is far removed from truth, and that is why it can make everything, because it touches only a small part of each thing.'[28] Several writers respond to this critique of pictures by concluding that a responsible designer should use many different methods of representation in order to bring out many different conceptual distinctions, interests and ideas.[29] By combining different representations that each have their own limitations, a designer might approach the complete description of the sensory effects of an environment.

With such an infinitely sensitive method of representation it should also be possible to demonstrate what Wang calls 'a fact'; namely, that 'every place … is unique.'[30] If everything can be taken into account, any two places or any two real things will turn out to be different from any other. However, there are some problems in applying this Leibnizian doctrine. No description of a real thing can ever be complete without also being infinitely long, and thus feasible only to gods and demiurges. Moreover, such an indiscriminate principle entails also that no real thing maintains its identity over time. This notion goes back to Heraclitus, who according to Plato, 'says that everything moves on and that nothing is at rest; and, comparing existing things to the flow of a river, he says that you could not step into the same river twice' – because the river is no longer the same and you are also not the same.[31] In contrast to ancient Greek speculation about identity, Eddy M. Zemach has argued that there are no identity conditions in general. Rather, the individuation of things is value-bound in that objects of thought are constituted in relation to particular interests. Things are classified as 'same' or 'different' for a purpose; this purpose validates some perspectives and invalidates others.[32]

From this perspective, then, complete descriptions are not only impossible but also uninteresting as representations. Everything that exists in the real world necessarily exists in a multi-sensory setting, but not everything in the situation is equally relevant. When I attend a musical performance I have to be in some space which will have a certain temperature, lighting, smell, etc. Yet, these are no properties of the musical work: the music is not considered to have changed if the temperature in the next performance were two degrees lower.

Resemblance   A representation of an object can never duplicate it in every respect without being that thing itself. Still, it might seem reasonable to say that for something to be a representation of another, there must be a partial resemblance between the two things. This would seem to establish a relation of similarity between two physical objects. There are however, several good reasons to reject the idea that representation would be an objective condition of resemblance between material things.

Firstly, while resemblance is a reflexive relation, representation is not. A person resembles his or her passport photograph as much as the picture resembles the person; yet the picture represents the person and not vice versa. Secondly, representation is intentional. Twins look alike but do not represent each other. A photograph of one of them is not a representation of the other; neither does a photograph of one represent a photograph of the other nor another photograph of the first. Thirdly, there are an infinite number of resemblances between any two physical objects. If representation were based on resemblance then every thing would be a representation of everything else. A photograph of the Barcelona Pavilion shares with the original a rectangular shape, a shiny surface, artifactuality and so on.

27 ibid., p. 22.
28 Plato; **Republic**, 598b.
29 Lehtonen, Hilkka; **Perspektiivejä arkkitehtisuunnitelmien esityskäytäntöihin**. Väitöskirja (Espoo: TKK/YTK, 1994) p. 34. Hume, David; **An Enquiry Concerning Human Understanding** (Indianapolis: Bobbs-Merrill, 1955) Sec. II, p. 26.
30 Wang, **op. cit.** p. 22.
31 Plato; **Cratylus** 402a.
32 Zemach, Eddy M.; 'No Identification without Evaluation.' **British Journal of Aesthetics 26**, 1986, pp. 239–251. In another text, Zemach postulates that all things (not just all works of art) are types and that types are particulars that have physical existence. Rejecting Aristotelian predication and accepting only proper names and identity, Zemach maintains that the sentence, 'Plato is white' contains two names, 'Plato' and 'white' which are claimed to be identical at a certain index of time, space and a possible world. Things occur at various indices and are thus types even though they are both physical and particulars. Even fictional entities exist in the same sense as other things, thanks to the inclusion of the reference to a possible world in the definition of the index. A number of counterintuitive conclusions follow from this position, including the possibility that the same physical thing exists at the same time in different places. For Zemach, Rembrandt's **Night Watch** is not only the painting at the **Rijksmuseum**; a photograph of that painting in a book can also be that very same thing. In some context, Gerrit Lundens' copy of the **Night Watch** in the National Gallery in London is also Rembrandt's work. See Zemach, Eddy M.; **Real Beauty** (State Park, Pa: Pennsylvania State University Press, 1997) especially pp. 140–166.

From the infinity of resemblance follows that only some features of a representational object are relevant. A photograph resembles another photograph more than anything but it represents that which is shown in the picture. Neither does every property in an architectural drawing denote something. Unless one possesses a code which identifies the relevant properties, there is no way to relate two physical objects with infinite amount of properties, a drawing and a house, to one another. In this sense, it is clear the iconic signs are conventional in the same sense as indexical signs, and similar to symbols. This observation was not made first by Nelson Goodman or Umberto Eco but already St. Augustine. Talking about dance, the Church father stresses the conventionality of iconic signs: 'it is true that everyone seeks a certain verisimilitude in making signs so that these signs, in so far as is possible, may resemble the things that they signify. But since one thing may resemble another in a great variety of ways, signs are not valid among men except by common consent.'[33]

Essence Instead of basing representation on resemblance, Rudolf Arnheim suggests that:

> ... the usefulness of photographic reproductions depends on how much of the original essence is preserved. A black-and-white slide of a painting by Poussin retains much of the pattern of shapes on which the composition is based, whereas the kind of work conceived primarily in color relations, as for example Josef Albers's *Homage to the Square*, becomes pointless ornament or, in the work of some other colorists, a meaningless conglomeration of patches. Reduction, therefore, leads to falsification when indispensable aspects of the work are suppressed. Otherwise, the abstracted image can even help to clarify certain aspects that are less noticeable in the undiminished composition.[34]

Arnheim goes as far as to claim that 'a reproduction should be treated as the work to the extent to which it conveys essential qualities of the work.'[35] This is presumably the reason why there are not only photographs but also plan and section drawings in books on architecture: photographs cannot capture the essentials of architecture to the same extent that they can show paintings, for example. A few critics, such as Bruno Zevi and G. E. Kidder Smith, have actually argued against the use of photographs as the critic's substitute for the direct perception of buildings.[36] Still, both of them have also used photographs in their books on architecture, amended with other representations.

To determine which representation of architecture is accurate we apparently have to decide what is essential in architecture, for as the Stranger asks, 'can he who does not know what the exact object is which is imitated, ever know whether the resemblance is truthfully executed?'[37] The question to ask is which qualities are necessary to reproduce in representations. It is true that models are not as heavy as buildings, but the real question is what the model is a representation of – a building, or, for example, the architectural object embodied in the material thing. The difference is crucial. Although, as a building, the Farnsworth House is quite heavy, its architecture is extremely light. Architectural representations differ from other representations of buildings in that they present some or all of its architecturally relevant properties. In short, an architectural representation is accurate to the extent that it represents the building as a work of architecture.

There are two major problems with this idea. One concerns the difficulty of determining the essence of such things as architecture, a difficulty exacerbated by the possible dependence of essence on the representational models available. Another problem is brought up by Ludwig Wittgenstein. He writes:

> I draw a few dashes with a pencil and paper, and then ask: 'Who is this?' and get the answer: 'It is Napoleon.' We have never been taught to call these marks, 'Napoleon.' … No one would say: 'This is the same as that' in one sense. But on the other hand, we say: 'That's Napoleon.' On one peculiar balance we say: 'This is the same as that.' On one balance the audience easily distinguishes between the face of the actor and the face of Lloyd George. All have learnt the use of '='. And suddenly they use it in a peculiar way. They say: 'This is Lloyd George' although in another sense there is no similarity.[38]

Neither does the representation reproduce the essence of Lloyd George in any Greek sense of the word. There are cases when just naming the object or denoting the referent seems to be enough. We could perhaps say that names are expressions that are taken to denote objects directly while descriptions are used to get more information about the objects.

Identity To be acceptable, a representation has to function as a name, to denote and to preserve the work's identity, but not necessarily its essence. It depends not only on the work but also on the context that is required to preserve the work's identity. In some cases, something can represent another thing without reproducing its essence in any way. Rather, it only has to pick out an object from among a number of objects.

We can perhaps formulate the criterion for the correctness or adequacy of a representation by applying Aristotle's definition of a definition: *species est genus et differentia*. A representation is

33 Augustine; **On Christian Doctrine**. Tr. D. W. Robertson, Jr. (New York: Bobbs-Merrill / The Liberal Arts Press, Inc., 1958), ii, 24, p. 61.
34 Dutton, Denis; **The Forger's Art** (Berkeley: University of California Press, 1983) p. 239.
35 **ibid**., p. 238.
36 Bonta, Juan Pablo; **Architecture and its Interpretation** (London: Lund Humphries, 1979) pp. 146–147.
37 Plato; **Laws** 668e.
38 Wittgenstein, Ludwig; **Lectures and Conversations on Aesthetics, Psychology and Religious Belief.** Ed. Cyril Barrett (Berkeley: University of California Press, 1966) p. 32.

adequate if it correctly picks out the entity among others in a particular universe. A picture on a restroom door is a correct representation if it distinguishes between men and women while a portrait is correct when it identifies a particular person among human beings.[39] What determines the class of objects to be considered or the universe of discourse? One class is formed by the medium of expression: we have to know how the medium affects the outcome. Consider the famous portrait of Adolf Loos by Oskar Kokoschka. Even though Loos probably never had such a multi-colored skin, the painting can still be truthful and point to Loos, rather than any other person. Since the style of the painter is very stable and instantly recognizable, we can easily bracket the constant elements in all mature works by Kokoschka and infer what the person portrayed would look like in another medium of representation – say, in a Modigliani painting, or an ink drawing by Picasso, or a photograph by Brassaï. As Goodman says, representation is not a matter of any constant or absolute relationship between a picture and its object, but a matter of a relationship between the system of representation employed in the picture and the standard system of perception.[40]

Another determinant is the point that is being made. When children are playing, a broom may represent a horse if the possibility of riding is what matters. If one wishes to discuss the anatomical differences and similarities between the leg of man and that of a horse, as in the notebooks of Leonardo, it is not a good idea to draw a broom. Once the intention of the discourse is set, one chooses from all objects available the one which best reproduces the relevant qualities of the original, as it needs to be reproduced in a particular discourse.

In addition to the author, the art-historical context also influences what is deemed relevant in an object. Since what is relevant in the object (and therefore necessary to show in representations) partly depends on the historical context and the narrative, one chooses what is necessary to project on the historical material. For example, Nikolaus Pevsner can legitimately use illustrations which hide the ornaments on the window frames of Sullivan's Carson, Pirie, & Scott building, or the *entasis* of the brick pilasters in Gropius' Faguswerk; these aspects are considered retrograde and do not fit the historian's progressivist narrative.[41] Consequently, these aspects are not deemed relevant to the representation.

Original   If we try to apply Plato's initial remark to architectural drawings, and not just any picture that shows a house, we arrive at a further complication. The problem is that a likeness is an imitation of an original being. However, the drawings by architects represent buildings that are only becoming, if that. Hence, these drawings cannot be called imitations, since there is in the material world as yet nothing to copy. Plato admits that there can be a meaningful image without it being a representation of any really existing particular. Thus, he says, a painter who has painted a model (*paradeigma*) of a perfectly beautiful man should not be any the worse even if he was unable to show that such a man could ever have existed.⁴² However, if there is nothing to imitate, when there is no 'original', where does the image come from?

While Plato can evoke the concept of the idea as an archetype, Finnish philosophers Simo Säätelä and Christian Burman attempt to explain how an architectural drawing could mean anything at all without resorting to such notions. They start with the assumption that the meaning of architectural drawings is based on correspondence either with reality or with the architect's intentions.⁴³ Since at the time of the design process the building does not yet exist, they conclude that drawings do not refer to anything empirical but correspond to the intentions of the architect. Simultaneously, they also propose the independent claim that drawings get their meaning from the correspondence to the author's intentions.

This argument seems to be based on a simple mistake of confusing the sense and the reference of a word, as distinguished by Gottfried Frege. The expressions 'morning star' and 'evening star' refer to the same object, namely the planet Venus, but they are different in their sense. Säätelä and Burman may be correct in claiming that either buildings or the architect's intentions are the referents of architectural drawings, but this does not entail that buildings or intentions would need to constitute their sense. We would need to ask if it is necessary to know the referent to understand the meaning of a statement or a sign. Perhaps we could agree that we understand what signs mean if we know to which real objects they refer. However, with the help of a suitable translation manual, it is often possible to make a translation from one language to another with less than complete understanding. In an architectural drawing, there can be a sign for the material concrete; we check it in a legend and tell the builders to get some 'concrete' without having any idea what kind of material concrete is. Indeed, Plato says in the *Republic* that a painter can make a likeness of a cobbler though he understands nothing of cobbling.⁴⁴

Säätelä and Burman continue their argumentation by claiming that drawings which refer to the author's intentions would be meaningless. Since intentions, rather like dreams, are private, only the author could have access to both the drawings and to their source of meaning. If Wittgenstein's argument against private languages is correct, there can be no private rules. The author of the drawing could never check the meaning of the drawings; he could have no more than a feeling of

39 There is no necessity to apply one dimension consistently in a given class of objects. One selects a prototype aspect of each entity but the prototype status might be independent of the context of the particular representation.
40 Goodman, Nelson; **Languages of Art** (Indianapolis: The Bobbs-Merrill Company, Inc., 1968) p. 38.
41 For a criticism of Pevsner, see Bonta (1979).
42 Republic, 472d4–7. Likewise, the ideal city is no worse for existing as idea only. **Republic**, 592a10–b3; cf. 598b8–c1.
43 Burman, Christian and Säätelä, Simo; 'On Understanding Architectural Drawings,' in **Describing Man-Made Structures. Aesthetically Qualified Physical Environment and New Planning Techniques** (Helsinki: Technical Research Centre of Finland, Research Notes 1240, 1991).
44 Plato; **Republic**, 601a.

correctness. Of course, an objection could be made that the situation would not change even if the rules were public; the speaker would still have to trust his feelings that he has correctly matched the public rules and the meanings whether he is judging his own linguistic behavior or somebody else's. We can only talk about 'using language correctly' as opposed to 'thinking that one is using language correctly' if there is a transcendental subject that can escape the epistemological difficulties that plague the original solitary speaker. If such a transcendental subject exists, then a match between the intention of a solitary speaker with the expressions is possible since the transcendental subject would be able to compare the intentions and expressions with public usage. If no such subject exists, there is never an independent way of comparing my intentional meanings with public rules. Indeed, I could always misrepresent my own past intentions to myself. Nevertheless, the two authors conclude that a drawing is not a model of a pre-existing mental image but that intentions are constituted by the drawings.

This conclusion may indeed hold for the reconstructions of the artist's or architect's intentions *ex post facto* by a critic or a historian, but we cannot dismiss intentionality as easily. In his argument against the doctrine of intentionalism, or the view that the work of art means that which the artist intended, philosopher Max Black points out that if he intends to draw a horse but happens to make something that looks more like a picture of a cow, people would only conclude that he has failed in his intention to draw a horse.[45] Black is undoubtedly correct in suggesting that the intention of the artist does not determine the meaning of the drawing, but his example suggests something that intentions do manage to determine. While the philosopher's intention to draw a horse does not guarantee that the drawing actually represents a horse, it does seem enough to make the piece of paper with lines on it into a drawing, rather than a notation for a musical work or a performance of a play (which might or might not represent a horse in a different way). While the artist's intention cannot determine the meaning of an artwork, it does entail the category to which the work belongs, to use Kendall Walton's terminology.[46]

The question remains: from where does the sense of the drawing come? Nelson Goodman solves the problem of signs with no referents in a simple way by suggesting that 'a picture must denote a man to represent him, but need not denote anything to be a man-representation.'[47] He proposes that since, for example, unicorns do not exist, unicorn pictures do not denote anything, yet the term 'unicorn-pictures' denotes them. In a similar vein, we might suggest that a plan does not denote the building but that the building denotes the plan.

However, it seems obvious that unicorn-pictures are not so much like words without a referent than they are like sentences which do not describe any state of affairs in the world. It is possible to construct understandable sentences which do not derive their meaningfulness from a previous, conventional association with a referent but from grammatical rules which apply in a number of unspecified cases. In an analogical sense, a unicorn picture is read the same way as a picture of a horse and consequently understood as a picture of an animal which happens not to actually exist. It is a combination of simpler signs for a horn, a mane, and so on, and even on a more primitive level it utilizes a number of representational combinations concerning colors, lines, etc. A technical term in an ancient Greek *syngraphe* functions in exactly the same way. Even a simple line in an architectural design drawing gets its meaning from how similar lines in similar drawings have been interpreted before, not from the intentions of the architect. In a plan, for example, a line is to be understood as a vertical difference but not as any specifically shaped wall. The architect draws the particular line because he or she intends the structure (the real thing) indicated with it, not because he or she intends the line (as part of the image) in itself.

Likewise, the correctness of a design drawing, which Säätelä and Burman find problematic, is a quality comparable with the grammaticality of a sentence rather than its propositional content. Just as we can assess the grammatical correctness of a sentence without simultaneously checking its truth value, we can also assess the correctness of a standard architectural representation without comparing it with a building or the intention of the architect. A design drawing is akin to fictional writing, not to writing in a private language. The rules, which determine both the meaning and the correctness of an architectural drawing, have to do with the kind of architectural or art criticism and history, as well as the institution of architecture in the practical sense. At the same time, it seems obvious that the concept of 'architecture' has been partly constructed to coincide with what architects can in fact do, given the legal and social ramifications and the methods of representation.

Mind   Talking about his fictional city of words in the *Republic*, Plato admits that no such city exists but suggests that in heaven, there is laid up a *paradeigma* of it, 'which he who desires may behold, and beholding, may set his own house in order.'[48] This view of architectural design survived until the Middle Ages and beyond. Several medieval writers think of design as the copying of a prototype: either historical, as in the case of the Holy Sepulchre for example, or conceptual, such as an idea in the mind of the designer. Around 1200 A.D., Geoffrey de Vinsauf explained that 'if a man has to lay the foundations of a house, he does not set a rash hand to the work; the inward line of the heart measures forth the work in advance and the inner man prescribes a definite order of action; the hand of imagination designs the whole before that of the body does so. The pattern is: first the prototype, then the tangible.'[49]

45 cf. Max Black; 'How Do Pictures Represent,' in E.H. Gombrich, Julian Hochberg and Max Black; **Art, Perception, and Reality** (Baltimore: Johns Hopkins University Press, 1972).
46 Walton, Kendall; 'Categories of Art', **The Philosophical Review**, 79, 1970, pp. 334–366.
47 Goodman, **op. cit.**, 21, 27.
48 Plato; **Republic**, 592a8.
49 Harvey, John; **The Medieval Architect** (London: Wayland Publishers, 1972) p. 22.

Thomas Aquinas and Robert Grosseteste also conceived of design as the imitation of a form which pre-exists in the artist's mind. Robert wrote: 'so imagine ... in the mind of the architect the design and likeness [*forma*] of the house to be built; to this pattern and model [*exemplar*] he looks only that he may make the house an imitation of it.' Meister Eckhart arrived at the same distinction from different considerations: 'firstly, after the image in the soul I paint a rose on a material surface, and for this reason the form of the rose is an image in a soul. Secondly, I ... recognize the external rose in its inner image, even if I never want to draw it, just as I carry within me the figure of the house which I do not want to build.'[50]

Why Meister Eckhart does not want to build his dream house cannot be examined now, but the notion of a pre-existing image in the soul may be linked to the Platonic notion of *anamnesis*, according to which knowing is really just recollection, a stirring up of something that exists in our soul 'as in a dream'.[51] This doctrine was revived by the symbolists at the end of the nineteenth century, and it animates the art theories of Paul Gauguin, Wassily Kandinsky and Piet Mondrian, as Mark A. Cheetham has demonstrated.[52] A more rationalist follower of Platonic aesthetics, Le Corbusier also embraces the notion that architecture is made in the mind and only contingently represented through the drawing. In *Precisions*, he states: 'I should like to give you the hatred of drawing... Architecture is made inside one's head. The sheet of paper is useful only to fix the design, to transmit it to one's client and one's contractor.'[53]

Inspiration This view of artistic or architectural creativity is not the one endorsed by Plato. He claims that poets do not write by their wisdom or knowledge but by genius/instinct/nature (*phusis*) and divine inspiration (*enthousiasmos*). As a result, poets cannot explain their own works; they 'say many fine things but do not understand what it is they are saying.'[54] This is the reason why poetry was for him not *techne* or craft; he did not call anything a *techne* which is *alogon*, unable to give a reason or explanation of its application.[55] He assured us that 'all good poets, epic as well as lyric, compose their beautiful poems not by *techne*, but because they are inspired or possessed.'[56] Indeed, a rational approach to poetry is inferior to inspiration: 'the sane compositions will always be eclipsed by those of the inspired madman.'[57] Hence, it is precisely when they are not in control of their senses that the lyric poets compose their fine (*kala*) lyric poems.[58] This critique did not apply to poetry alone. Unlike many later philosophers, Plato found that music is an inferior form of craft, because it involves guesswork and *mimesis*; the same will be found to hold good of medicine, husbandry, piloting, generalship, and cookery.[59]

Many architects find that their art likewise transcends a rational *logos* and depends on intuition and inspiration, acting 'in a divine and unapprehended manner, beyond and above consciousness', to quote Shelley.[60] Alvar Aalto described the creative process in the following terms:

> I forget the whole maze of problems for a while, as soon as the feel of the assignment and the innumerable demands it involves have sunk into my subconscious. I then move on to a method of working that is very much like abstract art. I simply draw by instinct, not architectural syntheses, but what are sometimes quite childish compositions, and in this way, on an abstract basis, the main idea gradually takes shape, a kind of universal substance that helps me bring the numerous contradictory components into harmony.[61]

The irrationality or supra-rationality of the subconscious mind, instinct or intuition is the prerequisite for a successful architectural solution: 'In architecture, a formal solution can be attained by means of imagination and intuition; to a certain extent, you can dream up the principal

---

50 Panofsky, Erwin; **Idea. A Concept in Art Theory**. Translated by Joseph J. S. Peake (New York: Harper & Row, 1968) pp. 40–43. In the original: '**Diu eine sache ist, daz ich nach der gestalt der sele bilde eine rosen an eine lipliche materie, von der sache ist der rosen forme ein bilde in einer sele. Diu ander sache ist, daz ich in dem inneren bilde der rosen die uzern rosen einvaltecliche bekenne, ob ich si joch niemer entwerfen wil, als ich die gestalt des huses in mir trage, das ich doch niht würken wil...**'
51 Plato; **Menon**, 81d–85d, **Phaedrus** 249e–250c. In the **Statesman**, the Eleatic Stranger concludes: 'it would seem that each of us knows everything that he knows as if in a dream and then again, when he is as it were awake, knows nothing of it all.' Plato; **Statesman**, 277d. In the **Phaedrus**, Plato explains that in their earthly copies, the ideas are 'seen through a glass dimly'. Imprisoned in the body like an oyster in its shell, many people are no longer able to see beauty shining in radiance. The theories of memory as a wax block or an aviary in the **Theaetetus** are radically different: **cf. Theaetetus** 191c–200c. Perhaps the reason for Meister Eckhart's unwillingness to realize his dream is related to Aristotle's observation to the effect that 'great men are distinguished from ordinary men in the same way as beautiful people from plain ones, or as an artfully painted object from a real one; namely, that which is dispersed has been gathered into one' (**Politica** 1281b). In this sense, the actualization of the form always implies a compromise because of the imperfections of matter. During the Baroque period, this idea reached an apex when Francesco Algarotti declared that 'the truth is not as beautiful as the lie' (as quoted in Kaufmann, Emil; **Architecture in the Age of Reason** (N.Y.: Dover, 1955) p. 98).
52 Cheetham, Mark A.; **Rhetoric of Purity** (Cambridge, Mass.: MIT Press, 1988). **cf.** Gombrich, Ernst H., 'Plato in Modern Dress: Two Eyewitness Accounts of the Origins of Cubism', **Topics of Our Time** (London: Phaidon Press, 1991).
53 Le Corbusier; **Precisions**. Tr. E. S. Aujame (Cambridge, MA: MIT Press, 1991) p. 230. As quoted in Forty, Adrian; **Words and Buildings. A Vocabulary of Modern Architecture** (London: Thames & Hudson, 2000) p. 31. In the **Ville Radieuse**, Le Corbusier puts forward a different theory, insisting that the reality of architectural or urban plans is in the drawing (see below).
54 Plato; **Apologia** 22c. **(Kai gar houtoi legousi men polla kai kala, isasin de ouden hôn legousi.)**
55 Plato; **Gorgias** 465a.
56 Plato; **Ion** 533e.
57 Plato; **Phaedrus**, 245a.
58 Plato; **Ion** 533e–534a, as translated by Janaway, p. 31.
59 Plato; **Philebus** 56a. On medicine, Plato also seems to have had a different opinion. See **Protagoras** 354a where Plato explains that medicine strives for the good even if it is painful; then, medicine is not only producing pleasant effects.
60 Janaway, **op. cit.** p. 35.
61 Aalto, Alvar; **Sketches**. Ed. Göran Schildt, tr. Stuart Wrede (Cambridge, Mass.: MIT Press, 1978) p. 97.

motif… Imagination and intuition are also essential in order to bring the often conflicting elements (whether material, social, or economic) which contribute to architecture into harmony.'[62] As an example, Aalto explains that the design for the Viipuri Library came to him unconsciously, as in a dream. 'I drew all kinds of fantastic mountain landscapes, with slopes lit by many suns in different positions.'[63]

What is then the source of the dream images? While the ancients understood inspiration as divine possession, the modernists often felt that their unconscious mind was a medium for the spirit of the age. In 1939, René Huyghe wrote that 'art is for the story of the human societies what the dreams of an individual are for the psychiatrist… The soul of an age as here revealed no longer wears a mask…'[64] In this sense, then, art is the dream of society.

In Aalto's particular conception, it is the drawing that reveals the dream to those who are awake: 'it is only when you put pencil to paper that the idea becomes reality.'[65] For Aalto, then, the drawing is the real thing. Occasionally also, Le Corbusier also claimed that the reality of architecture is to be found in the drawing. In 1934, he declared that 'the Radiant City already exists on paper. And when once a technological product has been designed on paper (calculations and working drawings), it does exist. It is only for spectators, for gaping bystanders, for the impotent, that the certainty of its existence lies in the execution.'[66] In this view, architecture is not about real things at all but about plans and drawings with their dream-like status.

Technique Plato would have rejected Aalto's and Le Corbusier's notions of architecture as involving intuitive and ineffable secrets. In contrast to poetic inspiration that depends on irrationality and guesswork, he claims:

> … the art of building, which uses a number of measures and instruments, attains by their help to a greater degree of accuracy than the other arts… In ship-building and house-building, and in other branches of the art of carpentering, the builder has his rule, lathe, compass, line, and a most ingenious machine for straightening wood.[67]

Certainly, these mechanical devices may contribute to an exactness which surpasses that of the human hand, but we have to ask why such precision should be desirable. Here, Plato's notion of beauty is relevant to consider. In *Philebus*, he explains:

> I do not mean by beauty of form such beauty as that of animals or pictures … understand me to mean straight lines and circles, and the plane or solid figures which are formed out of them by turning-lathes and rulers and meas-

urers of angles; for these I affirm to be not only relatively beautiful, like other things, but they are eternally and absolutely beautiful, and they have peculiar pleasures, quite unlike the pleasures of scratching.[68]

He is not saying that good scratching would not afford even a greater pleasure than the viewing of Platonic solids, only that the latter pleasure is absolute, independent of the person and the occasion, and hence true. Moreover, the pleasure afforded by these techniques – the beauty of straight lines – is peculiar to them; only these machines can produce these values. In a similar way, the techniques used by architects produce particular values, including the view of a city as a totality, comparable to a work of art, to give just one example of what Le Corbusier dismissed as 'the illusion of plans'.[69] Likewise, the limits of the representational system often become codified as the proper domain of the profession – what is recorded in architectural drawings gets conflated with what is the essence of architecture.

In the *Biographia Literaria*, Samuel Taylor Coleridge remarks that 'language itself does, as it were, think for us – (like the sliding rule which is the mechanic's safe substitute for arithmetical knowledge).'[70] The standard methods of architectural representations also do some thinking for the architects. Michael Baxandall remarks that:

> One result of perspective was a great simplification of the physical ambiance the artist cared to tackle. There are many more right angles, many more straight angles and many more regular solids in *Quattrocento* paintings than there are in nature or had been in earlier painting. ... Systematic perspective noticeably and naturally brings systematic proportion with it: the first enables the painter to sustain the second.[71]

62 Aalto, Alvar; **Alvar Aalto in His Own Words**. Ed. Göran Schildt (New York: Rizzoli, 1998) p. 266.
63 **ibid.** p. 108.
64 Sedlmayr, Hans; **Art in Crisis. The Lost Centre**. Tr. Brian Battershaw (London: Hollis & Carter, 1957) pp. 2–3. Huyghe elaborates: 'Many think of art as a mere diversion, a thing that is purely marginal to the real business of life, they do not see that it contains the most honest confessions, confessions that have within them the least element of calculation and must therefore be accounted exceptionally sincere.'
65 Aalto **op. cit.** (1998) p. 266.
66 Le Corbusier; **La Ville Radieuse** (Paris: Editions Vincent Freal et Cie, 1964), pp. 93–94. He went so far as to declare, somewhat implausibly, that 'human happiness already exists in the plan.' **cf.** for example, Le Corbusier; **Quand les Cathédrales étaient blanches** (Paris: Plon, 1937) pp. 83–84.
67 Plato; **Philebus** 55e. Consequently, Plato praised Egyptian painting because it was bound by timeless rules. 'And you will find that their works which they had ten thousand years ago; – this is literally true and no exaggeration – their ancient paintings and sculptures are not a whit better or worse than the work of to-day, but are made with just the same skill.' Plato; **Laws**, 656e–657a.
68 Plato; **Philebus**, 51c.
69 Le Corbusier; **Vers une architecture** (Paris: Les Éditions Crès et Cie, 1924).
70 Dufrenne, Mikel; **Language and Philosophy**, tr. Henry B. Veatch (Bloomington: Indiana University Press, 1963) p. 17. See Jackson, H.J. (ed.); **Samuel Taylor Coleridge** (Oxford: Oxford University Press, 1985) p. 205n.
71 Baxandall, Michael; **Painting and Experience in Fifteenth Century Italy** (Oxford: Oxford University Press, 1974) pp. 124–127.

Comparing late Medieval and early Renaissance architecture, one can also see a similar tendency towards simplification and totalizing regularization of form. The different representations of architecture (be they buildings, models, drawings, or verbal descriptions) may be equal but certainly neither neutral nor transparent nor identical. It seems possible to construe drawings as a particular non-verbal theory of architecture in the sense that some aspects of the physical building are selected as essential while others are suppressed.[72] Moreover, drawing directs design thinking by setting up categories, problems, and solutions. These are largely based on paradigms valorized in education and reproduced in everyday practice. Among more abstract principles, the paradigms include canonic buildings, such as the Parthenon, the Pantheon, the Barcelona Pavilion, the Villa Savoye, etc. Such examples come readily to mind: they define relevant kinds of similarity and difference; they offer 'natural' solutions to problems that they help to frame; they set ideals of good composition. The crucial role of non-verbal paradigms well-illustrated in the works of Eugéne-Emmanuel Viollet-le-Duc, who continued to design symmetrical buildings even though in his verbal theory he dismissed symmetry as 'a kind of communism, enervating art and debasing those who observe it'.[73] Perhaps Plato can even be read as having recognized the importance of paradigms and techniques in determining image-making. In the *Republic*, he argued that a painter does not imitate that which originally exists in nature, but only the earlier creations of other artists.[74]

Becoming Architectural drawings imply a particular challenge for Platonic thought because they are concerned with things-in-the-making. Most Greek philosophers use the verb *einai*, 'to be', not in the sense of 'to exist' but in the veridical sense of 'to be so', 'to be the case' or 'to be true'. In general, Greek philosophy is less concerned with questions of existence than questions of 'what-ness' or essence.[75] Those things are truths that are true at all times, and they are apprehended by the mind, not the senses. Typically, Plato opposes the intelligible realm of being to the visible realm of becoming.

The Presocratic philosopher Parmenides denied the possibility of change because either a thing 'is' what it is, and hence it does not change, or it is not identical with itself, so therefore does not exist. In this sense, it was often argued that becoming (motion) is the opposite of being (rest), and as 'non-being' it cannot be real. In the *Sophist*, however, a different theory is put forward. The Stranger proposes seeing being and becoming as parallel divisions within the form of 'power' (*dunamis*).[76] Within this account, it is possible to argue that:

> Not-being necessarily exists in the case of motion and of every class; for the nature of the other entering into them all, makes each of them other than being, and so non-existent; and therefore of all of them, in like manner, we

may truly say that they are not; and again, inasmuch as they partake of being, that they are and are existent ... every class, then, has plurality of being and infinity of non-being.[77]

In the scheme of the Stranger, not-being is not the opposite of being; it may perhaps be characterized as *difference*. The varieties of non-being, including dreams and drawings of things that do not (as yet) exist, merit the same characterization. Thus, we may agree with Alberto Pérez-Gómez who argues that the substance of the human crafts is *chora*; this is the name that Plato uses in the *Timaeus* for the third condition beside being and becoming.[78]

To what extent does the Stranger's metaphysical doctrine represent Plato's own views? By calling the eristic protagonist of the *Sophist* 'the Stranger' Plato seems to underline the difference between his ideas and those of Socrates. While Theaetetus is the likeness of Socrates, the Stranger is no likeness, but the mere semblance or appearance of a philosopher, of Socrates.[79] Certainly, Socrates and Plato did not count themselves as among the sophists but, then again, Socrates is silent in the *Sophist*, as well as in the *Timaeus*, the *Statesman*, and the *Parmenides*. Is it possible that Plato is leaving Athens for Elea or Megara?

At the very beginning of the *Sophist*, Theodorus arrives with his guest, whom he introduces as a stranger from Elea. Perhaps in response to the name Theodorus, Socrates ironically asks: 'Is he not rather a god, Theodorus, who comes to us in the disguise of a stranger?'[80] A god who disguises his true being as the origin of the Forms, of identity, of Being, would be nothing more than a false image, a phantasm, a dream. But if Socrates stands for the real and the Stranger for the dream, where does that leave Plato himself? A partial answer is given by Diogenes Laertius who relates the following anecdote: 'it is stated that Socrates in a dream saw a swan on his knees,

72 Nietzsche's observation about philosophy can also be applied to representations: 'Every philosophy is a foreground philosophy – it is a one-sided judgment: 'there is something arbitrary in that he stopped **here**, looked back, looked around, that he did not dig deeper **here** but put the spade aside, – there is also something suspicious about it.' 'Every philosophy also **conceals** a philosophy; every opinion is also a hideout, every word also a mask.' Nietzsche, Friedrich; **Jenseits Gut und Böse** (Stuttgart: Alfred Kröner Verlag, 1964) §289, p. 227. In fact, the same could be perhaps said of perceptions as well. Already Goethe insisted in his **Preface** to the **Farbenlehre** that **'jedes Ansehen geht über in ein Betrachten, jedes Betrachten in ein Sinnen, jedes Sinnen in ein Verknüpfen, und so kann man sagen, das wir schon bei jedem aufmerksamen Blick in die Welt theoretisieren.'** Goethe, J. W. W.; 'Farbenlehre.' **Werke** (Munich, Hamburger Ausgabe, 1982) Vol. 13, p. 317.
73 Viollet-le-Duc, Eugène-Emmanuel; **Lectures on Architecture**, Vol. I, tr. by Benjamin Bucknall (New York: Dover, 1987) p. 474.
74 Plato; **Republic**, 598a.
75 **Einai** means 'to be', **to on** is 'that which is', an **on** is 'something which is'; **to mê on** is 'that which is-not', a **mê on** is 'something which is-not'.
76 Plato; **Sophist**, 247d–e. For a discussion, see Janaway, pp. 177–178.
77 Plato; **Sophist**, 240b–d.
78 Pérez-Gómez, Alberto; 'The Space of Architecture: Meaning as Presence and Representation', **in Questions of Perception, Phenomenology of Architecture, Architecture and Urbanism, A+U** July 1994, Special Issue, p. 13; Plato; **Timaeus** 49b.
79 In the **Statesman**, the Eleatic Stranger is juxtaposed with another Socratic stand-in, a youth called Young Socrates.
80 Plato; **Sophist**, 216a.

which all at once put forth plumage, and flew away after uttering a loud sweet note. And the next day Plato was introduced as a pupil, and thereupon he recognized in him the swan of his dreams.'[81]

---

[81] Diogenes Laertius; **Leben und Meinungen berühmter Philosophen.** Üb. Otto Apelt. (Hamburg: Verlag von Felix Meiner, 1967) III, 5.

1 Thomas Cole, **Architect's Dream**, 1840. Toledo Museum of Art, Ohio.
2 Thomas Gainsborough, **An Artist using a Claude Glass**, c. 1750-55. (b/w).
3 Adelbert Ames Jr., Ames Chair demonstration. From E. H. Gombrich, **Art and Illusion,** London: Thames and Hudson, 1960. (b/w).

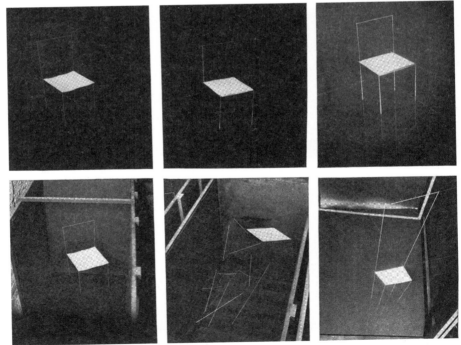

[4]

[5]

[6]

[8]

4 Cern, **Dance**. From E. H. Gombrich, 'The Visual Image,' in **The Image and the Eye.** London: Phaidon, 1982. Originally Cern's cartoon in his essay 'The Visual Image'. It was originally published in **Scientific American**, 227, pp. 82-96. (b/w)
5 Oskar Kokoschka, **Adolf Loos**, 1909. Nationalgalerie, Berlin.
6 Trude Fleischmann, **Adolf Loos**. 1930. From Heinrich Kulka, **Adolf Loos.** Vienna: Verlag Anton Schroll & Co., 1931. (b/w)
7 Alvar Aalto, **Sketch for the Viipuri Library**, 1929. Museum of Finnish Architecture, Helsinki. (b/w)
8 Apse, Notre Dame Cathedral, Chartres,

[7]

[9]

[10]

1194-1260. Photo: K. Jormakka
9 Filippo Brunelleschi, Pazzi Chapel, Florence, 1429-61. Photo: K. Jormakka
10 Akseli Gallén-Kallela, **Symposion,** 1894. Private collection (Aivi and Janne Gallén-Kallela) http://www.kalela.net/kalela%202003.htm

Section III: Design & Engineering:

Liquid Design & Engineering in Architecture & Building Technology

One of the most recent developments in architecture is formed by liquid designs of buildings. These designs have been possible since highly sophisticated computer programs enable designers to describe geometries of artifacts numerically as non-regular mathematical surfaces. A new generation of architects wants to impress with a totally new mode of design. They want to change fashion in architecture. The number of designer books on this subject is increasing. This new mode of designing buildings, although lacking a philosophical base, seems thrilling enough at first glance to convince clients, thanks to the perfection in computer imaging and renderings. The reality of materialization after cumbersome and complex building processes, however, sometimes disappoints. A considerable part of the new liquid designs do not display the naturalness of Modernist buildings. It may be too early to evaluate the architectural quality of these fluid designed buildings, but an initiative for quality ranging will be welcome. What will be the place of liquid design architecture in history?

The industrial designer, as well as automobile, ship and aircraft designers have been working for a long time now with the same computer design programs. The means of productions in their industries are based on long established experiences with aerodynamic or hydrodynamic, in any case 'fluid' designs within the transport industries. In architecture, however, these designs have only recently been introduced. And the building industry, used to low thresholds, open competition between materials, building products and building systems in many small companies, has not had the time and the occasion to develop an adequate technical vocabulary to realize these designs on the same industrial level as the current vocabulary in building technology for orthogonal buildings. Instead we see that the new liquid designed buildings are realized with old-fashioned 'handicraft' methods. In other words, the level of industrialization sinks back half a century. New methods of production have to be transferred from allied industries, adapted to the peculiarities of the building industry and its cost level and further, adapt and incorporate indispensable CAD/CAM engineering routines into the new liquid design production techniques.

Internationally a small number of architects and engineers are working together to realize these new liquid designs while maintaining a proper or high level of industrialization and efficiency. They experiment continuously, each in their own way, to materialize their designs. In the meantime a modest collective body of knowledge is being built up with individualized industrialized building techniques. 'Customization in lots of one' and 'Engineering is the core of the process' are the central creeds in the liquid design building industry. The publications on the materialization of liquid architecture are not yet numerous. Nor is the number of industries specialized in or at least familiarizing itself with liquid design materialization in order to achieve the same level of efficiency as the contemporary building industry at only a slightly higher cost level. The faculty of Architecture focuses on 'Blob Building Technology' by research, development, and designing and is interested in the opinion of peers. A number of internationally renowned architects and engineers have been invited to give their view on the quality of liquid architecture and on the future feasibility for competent materialization.

Kas Oosterhuis belongs to the first Dutch architects designing liquid design buildings straight from his computer, with the aid to control the production and building process by a 'File to factory' integration of CAD/CAM. He recently completed a 800 m long fluid design building along the motorway A2 near Utrecht NL.

Karel Vollers did his PhD on Twisted Buildings in 2001 and is now the leader of the TU Delft 'Blob Research Group'. Mick Eekhout is a full professor on the Chair of Product Development at the Faculty of Architecture who initiated the Blob research group and practices many liquid design projects with very different materials from his Delft-based Design & Build practice of Octatube International bv. He recently completed in close collaboration with architect Moshe Safdie the glass fiber reinforced sandwich shells for the Yitzhak Rabin Center in Tel Aviv as a world novelty: the return of the shell? Michael Hensel (Ocean North, Architectural Association School of Architecture, London), Henry Bardsley (RFR) and Ian Liddell (Buro Happold) also participated in this session and presented valuable contributions.

Integration and Coordination of Complex 3D 'Blob' Design Processes and Engineering

Mick Eekhout

1 Introduction

The influence of the architect and the engineer (being the sum of all involved designers, advisors and engineers) is complementary. Where the architect considers the first phase to be his/her exclusive domain, the engineer regards the final phases as his particular field. Let us say that the dominant influence of the architect is at its maximum during the first stages and that of the engineer in the final stages of design.

Nevertheless, it is important to realize that architects like Renzo Piano, Richard Rogers and Sir Norman Foster already introduce engineers or engineering considerations in the first phases. They regard both the architect and the engineer as a harmoniously performing duo in the Yin-Yang philosophy: totally complementary and ready to support each other at any time. In this process, the first group of engineers, usually called advisors, support the architect in the first stages of the building process, being paid by the client. In the following phases, engineers prove to populate the engineering departments of specialist producers and sub-contractors. It would be clarifying to make a distinction between specialists and 'jobbers': specialists absolutely must engineer the building components, while the 'jobbers' are occupied with standard products and system products or building systems with familiar characteristics. They will only apply engineering to confirm suitability. The specialists create their products as parts of the building design in its entirety. If possible, they hold on to a holistic view: the character of the building design should be reflected in the smaller components. It is clear that specialist producers are extremely important in the conceptual stages of the building design, particularly if the building has an experimental character which has to be fulfilled by sub-contractors and specialized contractors who never before worked together on a project. This ad-hoc co-operation is quite common in the building industry. In most cases, the architect will not call in these specialists instantly from the start. For instance, if he deems it better not to commit himself to one specific company before a standard process of selection and tendering has been passed through. Or in the event when he is being forced by the client, who is seeking ultimate economy and intent on low prices, if necessary, by harsh competition. In such cases, the architect has to take on competent advisors, i.e. the Ove Arups of this world. He has one alternative: namely, to personally figure out the technology and take full responsibility himself, as Renzo Piano does. Architects, with all the bravado that comes with winning a design competition and finding their way to the client's office, must appreciate that the winning gesture or the successful implementation of their design has never been achieved before and that they therefore are up for some major decision-making.

In this age of Blob designs, architects behave like artists rather than engineers who are subservient to society. Frank Gehry provides the examples for this with his models of clay, which are subsequently worked out by his excellent office organization. The freedom, provided by the design programmes to make sculptural, flowing designs and the pleasure a designer may feel when achieving an interesting result, causes him to forget the consequences of materializing: the very complex process of engineering, producing and building which is necessary to achieve an effective realization. Moreover, the new generation of engineers, here at the Faculty of Architecture of the University of Technology in Delft, in its new Master Degree programme (introduced in September 2002), will find 'sub-optimal' Technical courses. Recent graduates, as well as the Young Masters of Science Architecture, are highly skilled in Conceptual Design but, in my opinion, lack professional knowledge, skill and understanding in materialization. According to former architect Professor Carel Weeber, they will pick up these skills in practice, but even if they do, they will face years of inefficiency.

Further on in this text, two of the many 'Blob' or Fluid Design building designs of Octatube will be discussed, in a sequence reflecting their success. The first 'Blob' projects were not successful. However, as the rules of the 'Blob' game became clear, methodology and a routine began to take shape. In the most recent 'Blob' projects big steps forward in the technical sense were made. In order to publish the methodology and make it a topic for discussion, six projects were addressed in my lecture. Two of the six projects are described in detail in this publication:
- The roof of the Galleria, Wilhelmina Pier, Rotterdam, 1995
- The glass roof of the Deutsche Genossenschaftbank, Berlin, 1998
- The Provincial Floriade Paviljoen, Haarlemmermeer, 2001
- The frameless glass façades and spaghetti strips of the Town Hall in Alphen aan den Rijn, 2001/2003
- The Municipal Floriade Paviljoen, Haarlemmermeer, 2002
- The wing-shaped roofs of the Yitzhak Rabin Center, Tel Aviv, 2003-2005

This paper is given from the viewpoint of the architectural technician. The architectural values of 'Blob'/'Liquid Design'/'Freeform' architecture are not dealt with. A contribution towards that end will be welcome.

2 Co-Design for the Provincial Floriade Paviljoen, Hoofddorp

The current trend in architecture towards the direction of 'fluid designs' brings forth a new type of building that has a remarkable influence on the thoughts about spatial structures. A self-willed example of such a 'fluid' or 'Blob' building is the design by architect Kas Oosterhuis, Rotterdam, for a pavilion at the Floriade, for the province of Noord-Holland, realized in April 2002. In the design stage, architect Kas Oosterhuis and myself as the structural designer cum producer have developed the structural design up to the point where the architectonic and structural ideas and finances of the two designers showed unbridgeable differences. This contribution is mainly concerned with the various design considerations that gave the development process a strong impulse. After all, most of the fundamental decisions are made in the earliest stages of designing. The significance of design decisions in the final stages of design & engineering is of much less influence.

2/1 The first design rounds

The first draft of the architect was a kind of collapsed Gouda cheese, a round building with curved walls and a somewhat dented roof. Estimated size in the ground plan was approximately 24 m, height of the edges 7.5 m and a central height of 5.5 m. This was indeed a self-willed concept with regard to design, but due to the rotation symmetry, still very much related to the generations-long experiences with building dome structures. The negative curvature in the roof would call for a double-layered realization, while the rest of the dome was essentially single layered and tri-directional. The second version of the architect's draft had a rounded-off triangular shape, much like a wedge of Brie cheese. Other shape associations speak of a 'cobble' or a 'potato' and, more respectfully, a 'spacecraft'. Size in the ground plan: 27 × 20 m with a height of 5.4 m as the structural height and 6.3 m being the total height. From the dialogue between Oosterhuis and Eekhout, a structural outline came about for a single layered space frame with universal spherical joints, connected by bars in three directions. This outline would be capable of taking on the irregularity of the geometry, but was also based upon established industrial experience. Covering cladding and space frame would be aligned to one another. The second draft did not have any rotation symmetry, its shape being entirely arbitrary, and thus had instantly for each component a production series of one. The industrialization factor between the two described models seemed lowered by two centuries.

The familiar dome models of the recent history of three-dimensional metal dome structures consisted of single layered domes for smaller spans and double layered domes for larger spans. Furthermore, the most popular and well known nineteenth-century geometry is orthogonal with radial ribs and horizontal rings: type 'Schwedler'. Twentieth-century domes are all based upon various triangulations. Fuller's tri-way geodetic grid domes are familiar network models with icosahedral subdivisions or with horizontal rings; there are others with parallel lamella grids and even with delta truss models. The dome models that are composed of triangles all have a similar material efficiency, much larger than that of the orthogonal models. All these varying dome models usually have a half convex shape or less: their height is less than half of the diameter. On rare occasions, domes are made with a ¾ convex shape, as occurs with radar domes. In the 80's, as Eekhout did research for a 60 m. sphere in Rotterdam, the structural analysis and the soldered (1:100) model established that the strongest forces would occur in the lower bars of the dome. Three quartered domes readily buckle in their knee areas. This phenomenon would also become very clear in the Floriade design.

In the history of metal dome structures, the overall shape of the single-layered domes has always been fully dome-shaped. The Multihalle of Frei Otto in Mannheim (1973) is one of the few exceptions. Some collapses of single-layered metal domes have been scientifically analysed in the international conferences on space structures at Surrey University by Profs. Makowsky and Nooshin.

A small dent in the synclastic surface, a local breakdown of bars and joints could have a catastrophic consequence for the stability of the entire dome, with a total collapse as a final result. The Floriade design was concave at the beginning, it had a dent in the upper part of the design to which the architect, with regard to design considerations, was much attached. Therefore, the solution could actually be found in two directions:

- Local extension of the single-layered system to a double-layered system;
- Removal of the dent and making the surface overall synclastic again.

All in all, the above mentioned considerations illustrate how the architectonic design started off and was pushed forward by the possibilities for sculptural designing on the computer, as opposed to experiences and regulations for designing and building domes in the previous decades. During the first conversations between the architect and the structural designer, these considerations received ample attention. For every experiment it holds that the challenge to the designer is an incentive motivation of great importance. In this case, the structural challenge to make the improbable possible and feasible was very much present. They quickly agreed on the following starting points:

- considering the object to be a shell;
- stiffen the shell by the triangulation of bars and joints;
- stiffen the shell to loads perpendicular to the surface;
- by means of moment rigidity connections;
- alternatively, by means of shape rigid spatial angles at the connections.

The necessity to consider the object at a scale and the triangulation were agreed upon already in the first telephone conversation. The size of the triangles was a balancing act between long bars, wide angles and high shape rigidity on the one hand and the restriction of the covering 'Hylite' aluminium panels of 1.5 mm thickness on the other. These panels would be put at the project's disposal for free by Corus, with its main office in IJmuiden. The size of the standard plates was 3.0 × 1.5 m. The number of sponsored panels was limited. Each triangulation results in lots of waste (up to 50%), so the engineering had to come to an optimal use of the material. The structural rule for the space frame was: the wider the joint angles perpendicularly to the shell, the greater the resistance to external loads. Small angles to the joints often cause local snap-throughs. Therefore, a wide triangular modulation was preferred over a fine modulation.

The architect had projected an icosahedral model on the top part of the envelope of the object and the way in which the primary axes of the icosahedron

(5 meridians) were determined was typical for the nature of the triangulation. As a result of this optimizing, a foursome of triangulation alternatives occurred: from a five-sided to an eight-sided. With regard to the shell rigidity (via moment or via shape), it was decided to take a chance with these relatively small spans and to apply a single-layered hinging space frame with a maximum size of the triangles. Only if the structural analysis would show that the shell would not be rigid enough, additional measures could be taken, i.e. a moment rigid connection instead of a fully hinged joint, internal hinged posts to regulate tension on the inner skin, that could be functionally useful in the ground plan, five short stiffening frameworks diagonal to the outer walls or of under-tensioned rigidities.

The traditional scheme of deep I-profiled steel ribs, perpendicular to the surface and in an orthogonal system, was deliberately not chosen because of the expected great consumption of material, the relative simplicity of such a system and because of the fact that the development of an internally ribbed structure is common business in the aircraft and shipbuilding industries. What the structural designer wanted at all events for a new genre of design in architecture was the development of a new and appropriate lightweight structure. The architect, however, was interested in building the object; in whatever way it would be realized. Yet, the starting point of the client was to develop the design as being 'economically buildable'.

Obviously, in this phase, the industrialization factor played its part as well. The first design of the 'round cheese' was in principle rotation symmetrical. The architect, however, had already developed a systematization based on the icosahedral projection from within against the envelope of the object, thereby decreasing the repetition factor. The remnant of these considerations is the pentagonal roof of the object. Unfortunately, the roof is out of the sight of visitors moving around the Floriade. It was designed as a five-sided icosahedral roof: the five constituting icosahedral triangles were all equal and so offered a small serial benefit for the production of both the skeleton and the skin of the object. The object's structure has lost some consistency in the scientific sense of the word, but won economically.

The next consideration was the covering cladding. This would be made of 1.5 mm thin aluminium Hylite panels. The initial considerations were:
- triangulation in the shape of the panels;
- loss of material by cutting the panels from rectangular standard plates;
- individualization of the panels and edges;
- maximum size of the panels and the insufficient thickness of the material;
- the relatively non-rigidity or flexibility of the panels;

- the necessary water tightness for the entire skin of the object.

The skin of the domed panels would probably be difficult to make watertight, due to the high level of individualization. The main task would be to make an envelope of fitting panels. The watertightness as a side task would be achieved by a watertight membrane, mounted under the panels and made of PVC coated polyester fabric.

From this concept onwards, the thoughts of architect and structural designer went in two different directions. The architect thought: 'if there is a watertight membrane, then the seams between the panels do not necessarily have to be watertight, or be connected with precision. It would even be possible to apply artificial light to the space between the aluminium panels and the watertight skin, so that in the evening the object would light up in a manifold of lines'. The architect thought along the lines of building a folly. To the structural designer, the aim of the development was to make the panels as accurate as possible in the sidelong connections, so that a sealed watertighting could be applied. He regarded the watertight membrane as a secondary dam, not uncommon in buildings. It should provide an opportunity to practice the advanced insight in follow-up commissions. After all, 'Blob' designs have an increasing popularity and are worth a thorough development, in order to bridge the increasing gap between design on the one side, and engineering and production on the other.

In the meantime, the architect went on with shaping the object. He discovered that, due to the mainly visual character of the panels, the skin design should be semi-independent of the skeleton design. In phrases borrowed from the world of the car industry: the architect proposed to have the bodywork structured independently of the chassis behind. This made individualization of the panels even more difficult. Where first the panels were thought to be in line with the space frame, there now suddenly occurred spherical and concave shapes at the intersections of the panels over the space frame bars, which were the result of sculptural interventions in the design model on the architect's computer. The image of American cars from the sixties emerged. This new desire of the architect, which gradually became a demand, would eventually lead to an unbridgeable difference of opinion between architect and structural designer.

2/2 Computer Aided Design

Architect Kas Oosterhuis is well known for his digital designs. He has worked for over ten years in this field and did not realize more than one building per year in that period, but has published various colourful and inspiring books.[1] In 2000, he was appointed as a part-time professor at the University of Technology in Delft for a period of three years. In the meantime, this term has been prolonged. Thanks to pioneers like Kas Oosterhuis, new 'Blob' designs are published all over the world. In general, architects explore the boundaries of the possible by means of their CAD skills and increasingly advancing computer programs which enable designing in 3D. These programs, however, are not yet all that compatible with the usual engineering programmes. To give an example: the numbers of bars, joints and panels had to be derived from a generation list. The drawings were not sufficient as such. The via DXF file transferred computer file that the architect had certified for the use of co-designers, proved to have many shortcomings. It was only after many weeks of labour, that the geometrical data in the engineering commonly used AutoCAD 2000/14 became clearly legible, at least for the computer operator of Octatube. Yet, by the time of obtaining legible geometrical data, the structural data were still not suitable for a cost calculation. Because of the experimental character of the rough design drawings and the uncertainty with regard to feasibility, the structural designer set out to work based on a commission for co-design & pre-engineering.

The gap between the accelerating architect whose focus was on design and the carefully operating structural designer whose focus was on feasibility, proved to increase. This gap was prolonged because the computer programs as used by the architect were not compatible with those of the engineers and producers. Here, a classical dilemma occurs: to be a soloist (do everything yourself with one hand, the very basis of Octatube 20 years earlier) or to be collaborative, communicating with appropriate means. Would the architect not have been capable of realizing the entire computer generation, so that the sum of the architectonic design, structural analysis and the element/component breakdown could have been developed in one hand, while the structural designer only would have advised and the producer would have made his costs estimate as well as he possibly could?

In the course of the three months (between the end of February and the end of May 2001) of collaboration, the choice was made for the parallel method of working: 'competitive', sometimes 'collaborative'. The architect worked out the design and did his best to make proposals for its materialization. The structural designer and his engineers (Karel Vollers, Sieb Wiechers and Freek Bos), already tried in an early stage and parallel to the architect's work, to get the essence of the design, the rigidity of the structure, the composition of the elements and components and the entire water-tightness issue under control. In the light of their possible future responsibilities and liabilities with regard to structure and watertighting, they came to different thoughts than the architect's. The result was a continuous dialogue and discussion.

A strongly determining influence was the total budget for the project. Octatube's first estimate varied in two alternative realizations of EUR 430,000 and EUR 630,000. Only when the commission was granted one month later, the actual available budget for the engineering, production &

---

1 Kas Oosterhuis, **Architecture goes wild**, Rotterdam, 010 Publishers, 2002.
Kas Oosterhuis, **Kas Oosterhuis: Programmable Architecture**, L'Arcaedizioni, 2002.

assembly of the (main) support structure, the covering, the membrane, the floor and two doors became known: EUR 240,000. In reference to a shell surface area of 600 m², this meant a price of EUR 400/m², including two complex doors. This m² price hardly offered a realistic budget for an experimental project. The commission for co-design and pre-engineering was put at an approximate 5% of the real budget, but was only accepted as an obligation to make an effort, not an obligation to achieve a result!

The estimates were made as being the sum of assessed individualized element and component prices and the way in which the parts would be realized. What became clear from the large difference between the first estimate and the available budget, was the need for modesty of the level of experiment. Indeed, there is nothing wrong with a low budget, as long as it is realistic. However, in the mentioned case, the estimate could only be substantiated after three months of design development, structural engineering and the development of the skin. As soon as the layout of the space frame was determined and computer analysis had made clear that the structure was sufficiently strong, rigid and stable, the cost calculation of the space frame gained a better insight. This was determined by means of the unit prices and the number of elements and components. Basically, there was no discussion about the economy of the space frame (56% of the budget), but all the more so about the economy of the covering cladding.

2/3 Structural analysis

The second half of April 2001 was used to make the structural analysis prepared by Freek Bos, Blob graduate and building technology student and currently PhD student. It turned out 'splines' are not correct when transferred from Maya to AutoCAD. Therefore, much work was done on both sides to obtain a proper computer communication. By means of the pre-processing in FemGEN, the structural analysis in DIANA and a post-processing in FemView, insight was obtained in the behaviour of the space frame shell. The dent in the roof (with the possible breakdown as a result) had already been removed in the space frame generation. The bars had the following diameters: 82.5 × 5.0 mm, 101.6 × 7.1 mm and 203.0 × 8.0 mm for the heavy rods on the five main axes, the meridians.

Deadweight, snow loads and wind loads were applied. As a result of the deadweight and snow loads, the tail in the longitudinal intersection would lift off (vertically upward). Furthermore, the roof sagged through because of the transformation of the lower half of the rods, the knees. The material model that third-year students made later, confirmed this. Remedies were:
- to increase the structural height of the shell in all cases with a maximum of 1.000 mm, from 5.4 m to 6.4 m;
- to make the lower rods of the space frame shell considerably heavier;

- to interconnect a number of bars with moment rigidity to the five meridians across the joints by means of welded-on flanges through which tubular beam rigidity would occur with a diameter of 203.0 × 8.0 mm.
- to place five slender (structurally loaded) hinging glass rods internally on the meridians with diameters of 100.0 × 7.0 mm, which would cause the entire shell to be subdivided into a flat roof and a round 'donut' wall;
- to introduce five internal meridian frameworks that would provide the shell with a high rigidity and which were acceptable to the architect;
- to introduce a number of shell rings that would be able to control the horizontal lateral thrust. However, the two doors in the lower wall would cut right through! This alternative was not elaborated on any further;
- to place five external outriggers at the side of the meridians to reinforce the lower sidewall. This was not appreciated by the architect, although it looked very similar to the landing gear of a spacecraft.

The conclusion at this stage of the structural analysis was that in raising the height from 5.4 m to 6.4 m and with the internal enforcement by means of welded beams with a diameter of 203 mm on the meridians, the result was a reasonably rigid space frame structure with a maximum vertical deflection of 20 mm in the centre over the shortest span of 20 m, which means 1/1,000 of the span. In short, with a number of rough calculations still to come, the space frame shell would not cause any unexpected and insoluble problems. The total of the estimated self-weight of the framework shell with enforcements over the shell surface area of 600 m² would be approximately 6,000 kg (which is only 10kg/m²).

2/4 Development of the covering cladding

Problems proved to be concentrated around the covering cladding: the strongly reduced budget of the client on the one hand and the architect who required a metal skin that in its shape would be independent of the space frame on the other. Already, from the first internal estimate, it could be concluded that the budget was too low for extended engineering of the skin, now that the skeleton was, more or less, roughly determined. The quest for panels with a spatial curving started, with in the back of our minds the thought that it would always be possible to arrive at a level of flat panels, although these would not be desired by the architect. After the architect had determined by means of 'splines' and 'nurbs curves' the required curves in the skin with regard to the centres of the nodes and the axes of the framework bars, it became clear that very complex 3D components had to be made, in order to secure the covering to the space frame in its required position in free space. Furthermore, a number of rather extreme covering components had to be developed and realized. The required 3D character of the triangular covering was restricted by the flat plates, which would only be folded or curved into one direction. A triangle consists of a, more or less, flat centre and three angular points which can be bent downwards or upwards, independent of each other. The principles of bending 2D panels into star-shaped 2.5D panels have to be explored further. Based on findings so far, the definitive quotation was made with extremely deformed panelling; a few flat panel alternatives; a stretched membrane alternative. On the basis of cost-price calculation, an agreement to continue could not be reached and Octatube's activities for this project were stopped. Half a year later, the structural designer developed and realized aluminium 3D panels for the Municipal Pavilion of the Dutch Floriade. The costs for exploformed panels for that project proved to be even higher than the estimate above.

2/5 Prototypes by students

In May and June 2001, a group of ten Building Technology students in their third year were engaged in two sub-groups as part of a compulsory study module, 'the prototype'. Half of the students were to develop a regular covering component system, as mentioned above, based upon a box-shaped component, of which the body would be semi-independent of the chassis. The other half's assignment was to design a workable door that would fit into the system; it had to have a minimum size of 2.2 m width and would open by rotating, swivelling, or with the possibility for torsion or swing. They derived their principles from the double hinging doors of civil airplanes. For insight purposes, a wire model of 3 mm coppered welding wire was built, scale 1:20. At last, there was a material model, after all the virtual models on the computer, which was a relief to see and it gave its own spatial insight. The students learned much in these eight weeks, but the results of their work remained, technically speaking, disappointing. Their work confirmed the assumption that, within the given preconditions of the required design for the covering cladding, the necessary engineering and production, as well as the available budgets, a satisfactory compromise did not seem to be possible. As a result of this research the quotation for a watertight prototype cladding proved to be too expensive.

2/6 The final realization

After the parting of the ways of architect Oosterhuis and structural designer Eekhout/producer Octatube, the architect had to turn into another direction. The architect held the view that a double curved façade surface would not be realizable within the limits of the budget, as he had indicated from the beginning. Structural engineering office D3BN worked out a supporting structure on the basis of set steel strips, 20–30 mm thick and 100–400 mm in width, in the by now familiar triangulated geometry. The strips functioned simultaneously as both main support structures and covering fasteners and, to this purpose, were buckled over the full length of the members. The architect applied for a patent on this system. The covering remained consisting of slightly curved Hylite panels in a flat shape that, by means of steel brackets, were fastened to the steel strips, not being watertight. Watertighting the internal space was established by making the projected screen watertight. Thus, the object had indeed become a structure rather than a building, a folly. Still, the object was realized. The steel structure's deadweight, manufactured and assembled on site by steel structure producer Meijers Staalbouw, Serooskerke, amounts to over a 100 tonnes of steel. The Hylite skin has a negligible weight. The eventual costs were EUR 250,000. The pavilion was disassembled at the end of the Floriade Exhibition, with the intention to reassemble it with a new watertight skin as a (Blob) research laboratory at the Faculty of Architecture of the University of Technology, Delft, a task which has not been performed yet.

- The lessons learned from preliminary studies are fundamental and essential enough to present them to a national or international forum.
- The dramatic breakdown in the development of systemized lightweight space structures by unusual architectonic 'Blob' structures.
- 'Blob' structures bring along a loss of systemizing and repetition in the material supporting structure.
- The high level of individualization in engineering and producing the individual space frame components asks for further examination.
- The shifting in critical attention from spatial structure to spatial cladding asks for much design energy.
- Computer Aided Design makes it possible for architects to create 'Blob' designs and Computer Aided Engineering is vital in determining the spatial complexity of the design in its entirety, as well as the individual determination of elements and components.
- The digital link between Design and Engineering will determine, to a major degree, the feasibility and affordability of 'Blob' designs in the near future.

3 **Municipal Floriade Paviljoen, Hoofddorp**

At the beginning of 2001, after a design contest with only four competitors, the Municipal Council of Haarlemmermeer selected the building design of Asymptote Architects in New York (by the young architects Hani Rashid & Lise-Anne Couture) for the realization of the Municipal Haarlemmermeer Paviljoen at the Floriade, to be opened on 6 April 2002. The pavilion acted as the information pavilion of the municipality for six months, after which it was used as a permanent café in the park.

Together with the North Holland Paviljoen on the Floriade, it is one of the first Dutch Blob buildings. A Blob building has a non-orthogonal mathematical shape not easy to describe and a complex geometry. It is also called 'Fluid Design' or 'Free-form Architecture'. Many young architects regard 'Blob' as an honorary nickname. Opportunities and skills of the new generation can be expressed by 'Blob'. The world of the building industry, in general rather traditional, has a very different view. It is startled by the complexity of the building design and its components, hence it prefers the term 'Fluid Design Nightmares'. A good illustration of the gap between the wishes of young architects and the expertise and know-how of the traditional building industry. This gap has to be bridged.

The site for the contest was at the bank of the lake, but the realized design was built on an artificial peninsula in the lake called 'Haarlemmermeer' ('meer' is Dutch for 'lake'), upon which a building volume and two oblique roofs were designed. The large roof over the building volume houses the reception and exhibition areas, as well as the service areas. The small roof is freestanding against the embankment and forms the entrance. Over the two oblique roofs, water is continually sprayed from the top; it flows over both roofs, fills the glass pond and flows out in two gutters on both sides of the entrance, from where the water visibly and tangibly streams over the two inner glass walls to the level of the lake, allowing visitors to enter the pavilion much the same as Moses walked on the bottom of the Red Sea to the promised land. In various pavilions at the World Exhibition of Seville 1992, flowing water over façades was used as a climatological cooling. Here in the Netherlands, it is used to symbolize the wetness of the land. The theme of the pavilion was 'Nederland Waterland', not illogical when seen through the eyes of the Egyptian American Rashid with his desert background. In the pavilion, there are no specific references to the Floriade to be found, but the reference to the Dutch water landscape is true to life. The pavilion is Asymptote's first realized building, as well as a premiere in architecture. The many worldwide-published Blob designs of Asymptote had been platonic for over 15 years until then.

3/1 **First steps towards realization**

Architektenbureau Bronsvoort, Amerongen acted as the local executive architect and supervisor on behalf of the client. Infocus Bouwmanagement, Nieuwegein acted as project management on behalf of the client. The main contractor was Nijhuis bv, Houten. The project engineer was Smit/Westerman, Waddinxveen. Facing these usually traditional cooperating building team parties were the co-manufacturers, who proved to be very essential in this Blob design: for steel structures Smulders Duscon bv, Bladel, for aluminium ceiling and façade cladding Van Dam, Ridderkerk (who, unfortunately, went bankrupt only a few weeks before completion and before having finished much of the cladding) and for frameless glazing and roof panels Octatube, Delft. In addition, some thirty subcontractors acted in the classical sense, directly answerable to the main contractor.

Initially, the 'virtual office' of Asymptote was highly regarded. The idea was that from a site on the Internet project participants with a password could pick up the most up-to-the-minute drawings. Three years earlier, we had tested this for the first time at the Faculty of Architecture in a second-year unit 'Production & Realization' and now it had already become reality! However, due to communication shortcomings in the transfer of the four different computer programs of the leading players in question, it did not work. The architect worked with Microstation/Bentley, the engineer used x-steel, Ocatatube and Van Dam both used variants of AutoCAD. During the process, it became clear that there was no party in the design & engineering process that checked, co-ordinated or integrated drawings or could have control over the measuring. As it happened, all parties helplessly threw up their hands. In the short time of a real building project, there was no time to set up a good, generally working communication system between parties, to work on the basis of compatible computer programs with the various engineering co-manufacturers and at the same time to realize the actual preparations, productions and realization on the building site of the project. In this 'first' in Dutch Blob realizations, the shortcomings of the traditional infrastructure of preparation and realization of classical orthogonal designs became very clear and through publications like these (but also through lectures) this project teaches how following Blob projects can be managed in a better way.

Architect Asymptote made the images in a non-communicable program for the engineers and saw no possibilities to transfer to the AutoCAD system, which is the most popular program amongst the co-manufacturers and subcontractors. During the handing over of instructions (architect) to realization (co-manufacturers), much of the data in the DFX files went missing in transmission. The accuracy of drawing on the computer was thereby lost and it was up to the co-manufacturers to guess their appointed components' measurements based upon these data. Subsequently, no coordination or integration of the measuring of components was done. AutoCAD only allows drawing, it is not possible to perform structural analysis within this program. The steel structure companies in the Netherlands have some five different analysis and engineering programs to work with. In each of these programs, the geometry can be established, as well as the initial dimensions of steel structures and finally the influences of internal and external loads can be analyzed. These programs have been optimized in their structural performance and their production of drawings for the resulting elements. Uniformizing of the engineering programs is an absolute necessity to come to a workable **collaborative engineering**. In this case, there was only **concurrent engineering**. (The difference between the two is that the one case is about co-operation, while in the other case there is just simultaneous working in time.) Collaboration in this project was mostly done by oral explanations and by fountain pen. But, being Blob's first outing, the Floriade Paviljoen also has a guiding and infectious function. A remarkable performance has been achieved within the indicated restrictions of ICT, the partly very experimental high tech character of the building and the mostly very traditionally working partners in the building team.

As mentioned above, the design consists of two oblique roofs, one of which cuts through the building volume and the other containing a glass pond. Over both roofs, water is sprayed continuously. The design is a Blob in full splendour. The volume of the building was not determined mathematically, but geometrically, also the shape of the glass pond could not be described by means of regular mathematical functions. By parallel working and exchanging data in DXF format, the co-manufacturers have tried to approach the geometry of the 'Micro Station' drawings of Asymptote as faithfully as possible. Yet, between the glass panels of co-manufacturer Octatube and the aluminium cladding of co-manufacturer Van Dam, variations occurred of over 125 mm! It is an absolute must to have the architect draw the tendering plans in 3D AutoCAD/Inventor the next time, to have all co-manufacturers work in that 3D model and to have this 3D model certified and safeguarded by the architect. The idea is the same as in the petrochemical industry: to have data added to the 3D model by the various co-manufacturers in turn, at agreed slot times, periods in which only one co-manufacturer works on the 3D model at a time.

The computer program Catia (developed by the French aircraft builder Dassault) seems to be more extensive in this respect, because some modules can draw as well as perform structural analysis, but it is expensive to purchase (U$ 100,000 per station) and nobody in the Netherlands will make such an investment! The building industry is an easily accessible industry with much ordinary competition and low volumes and low square metre prices. Therefore, techniques can be borrowed from the aircraft building industry or the shipbuilding industry, but they have to be adapted to suit the less expensive component prices of the building industry. Nevertheless, a transfer of techniques between various divisions of industry is obvious and in that field, Delft University of Technology with its open faculties offers many good possibilities, which are rewardingly used. From 2004 onwards the open collaborations at the TU Delft between the different designing and constructing faculties would be called 'Delft Science in Design' (with an initial congress on May 26th 2005).

The process of Design & Engineering management for Blobs has to be developed as an issue in its own right. Over 30% of the total cost price is invested in Blobs for design & engineering by the parties and therefore, a suitable management for these processes is an absolute must to accomplish efficiency in the process of preparation & realization of Blob designs. Octatube took the aluminium roof under its wings, as well as the glass pond and the glazing of the façade. Mick

Eekhout, head designer of Octatube, has acted as main designer in the concerned component process and he also took care of it that one of the three very experimental aspects came to his chair of Product Development and was realized in its experimental status by a restricted flow of funds fitting into the running project.

3/2 3D Aluminium roof panels

The first experiment and innovation are the two 3D corners of the aluminium panel roof that were further developed as a research project, by the main engineering department of Octatube and the Blob Research Group, lead by Dr. Karel Vollers of the Chair of Product Development of the University of Technology, Delft, who had received his doctorate in 2001 with his thesis *Twist & Build: Creating non-orthogonal architecture*[2] and who is a specialist in the field of the twisted glazing and cladding of buildings. He formed a task force for the 3D panels with engineer Dominique Timmerman, Octatube, for the overall geometry, Ernst Janssen Groesbeek of the chair TO&I (Technisch Ontwerp & Informatica – Technical Design & Information Science) for the CAD drawing of the components, Haiko Dragtstra's computer-controlled milled foam moulds, Hugo Groeneveld's Exploform for the aluminium plate deformations and Octatube again for all the measuring, sawing, planning, welding, filling, squirting, installation and watertighting. All in all, a special purpose team for one small (18 m²) but experimental part of the building, that resulted in an innovation in the state of the art of production technique for buildings. The usual friction in the building industry between time & finances on the one hand and experiments, trial & error and uncertainty about meeting the set goal on the other, was also tangible in this particular project. Nevertheless, the task force stayed together and achieved a technique that in the world of the building industry had not been realized before. Computer models were worked out and subsequently used to activate a 3D milling machine that milled polystyrene foam blocks. The spatially curved surfaces of these moulds were smoothed and hardened with epoxy resin, after which a counter mould of integrated concrete was poured upon it (with short coarse-fibred reinforcement), which subsequently served as a lower mould during the explosion process. Hugo Groeneveld started a small business in the TNO Delft grounds by the name of Exploform, in which he develops various production methods to deform metal panels into 3D. Initially, the idea was to cast the positive moulds, because of their excellent recycling possibilities, on the polystyrene moulds in water and freeze them, and then detonate the explosive charge onto that. But high pressure turns ice into water again, which made the mould disappear too quickly. It was a great idea, but nor practical for the time being, also because Exploform could not build a large refrigerator in a short time. As it was, the aluminium panels were laid upon the concrete mould, vacuum sucked to the mould and by means of an additional water basin with a TNT ring an explosion was brought about that with its radial but equal pressure plastically deformed the 5 mm thick pure aluminium panel in the shape of the

[2] Karel Vollers, **Twist & Build: Creating non-orthogonal architecture**, Rotterdam, 010 Publishers, 2001.

concrete mould. During the explosions, water and plastic rims were blown 20 m high into the air. As a method of production this was still very experimental and far from industrial, but it proved that the accuracy that was aimed for could be achieved. The panels had imprints of the epoxy skin's glass fibre from the negative moulds and therefore had to be filled, also because of the sometimes occurring wrinkles, after which they could be coated. At Octatube's plant, a wooden 1:1 mould was made in which all panels were fitted and supplied with aluminium rims, sawed of, ground and welded into complete panel components. Karel Vollers has published extensively on the achieved production technique and on the future paths by which radical improvements can be achieved.

The same technique will currently be used for the building of deck parts of an aluminium yacht. We will go on flashing back and forth outside the sector. Even in the shipbuilding industry, 'splines' is familiar jargon for arbitrary curved lines in the intersections of ship's hulls.

In the meantime, super inventor Haiko Dragtstra has developed a professional way of making cylindrical façade panels: CAD/CAM milled moulds, as described above, aluminium panels directly vacuum applied upon them and at the rear side enforced with a thin layer of foam and epoxy resin, so as to prevent the cylindrical shape from getting straight again and to remain rigid. In this way, he has made all corner panels of the Van Dam façades in the Floriade building, and this after the experiences with the production of the 3D roof panels. The direction of development would be aluminium sandwiches and even GRP sandwiches.

3/3 Glass pond

The second innovation in the Floriade Paviljoen is the suspended glass pond, calculated on the self-weight of 1.4 m water: 1400 kg/m², which is twenty times as much as an average roof load or wind load! As far as we know, this has never been realized before in the world, not even in a James Bond film. The original tendering design had a maximum water depth of 300 mm, which little by little became 600 mm, 840 mm and finally 1410 mm in the final drawings. In the meantime, the load on the glass at the bottom had increased up to the definitive height. That load could be absorbed; it was a matter of calculation. The sizes of the frameless glazing panels, normally 2 × 2 m, have been decreased in this case to 1 × 1.4 m, resulting in a quarter of the moment. The heavy load is the reason that a point-shaped suspension, the architect's wish, was further developed into a kind of line support in the structural sense: a few node-shaped suspensions at a distance of 300 mm next to each other in the transverse direction of the pond and the larger direction of the glass panels, so that the maximum span would be 1 m.

The shape of the pond would ideally have been designed by Octatube as a composition of fluent 3D formed glass panels, 2 × 12 mm gauge, as the diagram indicates. The state of affairs has progressed in such a way, also after the boosting of the results of the above described 3D aluminium panels, that it must be possible to mill foam blocks in 3D model from CAD/CAM files and pour these, after a surface enforcement by means of epoxy and glass fibre matting, in a reinforced concrete positive mould. After sufficient cooling down, this mould will be covered with an insulating glass fibre weave, which will be used to nestle into the heated glass panels in duos of 2 × 12 mm at a temperature of 600°C and 700°C.

After cooling, the nestled 3D transformed glass panels, which have been exactly cut to measure and provided with ground rims beforehand, will be transported in wooden crates to the plant where the chemical bath is set up. There are two of these baths in Europe: one is owned by Glaverbel and one is in Switzerland. With this chemical pre-stressing, in the exterior surfaces through chemical processes, in each of the panels separately the sodium ions are replaced by potassium ions, which are larger and for that reason cause a compression zone in the exterior surface. This technique is highly developed for laboratory glass and specific industrial purposes and is therefore familiar with guaranteed high practical tensions, usually up to twice as high as with thermical pre-stressed glass. To this purpose, the chemical baths have to be enlarged, because chemically pre-stressed laboratory glass is usually much smaller. After the chemical pre-stressing, the panels are nestled again and brought back to the Netherlands where they, at a distance of 1 to 2 mm, are provided with a transparent tape all around and filled with epoxy or acrylate resin, which results in a rigid laminated connection between the two panels. Subsequently, the laminated panels are transported to the building site. This production sequence as described here is currently not in use, because it is very expensive, very labour-intensive and transport-intensive. The danger of interim breakage always lies in wait. In this particular case, the client deemed that possible replacement in the case of breakage could take up to a couple of months. This would be catastrophic, considering that the Floriade is open for only six months. The simpler solution of the polygonal variant was chosen and has been realized in the meantime. The fluently transformed and chemically pre-stressed laminated structural frameless glass was one bridge too far for the client, but remained a challenge for experimenting product developers.

One year later, in 2003 Octatube realized a 3 m dome made of 9 insulated 3D glass components on top of a 30 m glass saucer at the Bancopolis in Madrid (architect: Kevin Roche, New York).

We used 2 × 12 mm entirely pre-stressed glass, without holes, but remaining frameless by elliptic suspenders. The 20 mm bending through under the permanent weight of the water was the first foreseeable problem, the watertighting of the pond was the second and the third was the deterioration by the water to the laminate in the glass panels in a damp environment. Eventually, a liquid acetic silicone sealant (by Tremco) was used at the top, which seals the upper sheets of the laminates. For the bottom, a prefabricated silicone profile was used, grey-coloured, the colour of the rest of the sealing at the building site, which has an aerification quality. This is important, because the laminate needs aerating. If the laminate, under continuous water pressure and knowing the moist density of silicones, being under water all the time, could be capillary sucked up in the laminate this would result in a white discoloration of the laminate layer. The joints are wide because at the top a glass opening of 15 mm occurs, in the shape around the basin. Therefore, the exterior joints at the bottom are much wider: 22 to 25 mm.

3/4 Cold bend glass

The third innovation was realized in the exterior planes of the south façade. This consists of three glass façades of approximately 6 × 6 m, each of them divided into 3 × 3 = 9 panels with a maximum of 2 × 2 m. The central glass surface is flat and built up of nine flat panels of monolithic 12 mm fully pre-stressed glass. In the two side planes, the panels are 2.5D curved. The original design had a conical part and a cylindrical part. Although the making of the conical moulds was possible, it nevertheless caused the producing parties in England and Spain many problems. The architect could be convinced to change the conical shape into a narrower cylindrical shape. The three corner panels, placed on top of each other at the south side, were manufactured as monolithic thermally pre-stressed glass panels with a 12 mm gauge, thermally bent by Interglass, United Kingdom. Furthermore, the six remaining panels on each side, each of them with a maximum of 2 × 2 m, were cold-bent on site in laminated completely pre-stressed glass of 2 × 6 mm. These glass plates were offered as polygonal from the beginning. However, the architect reconsidered his decision and required bent glass. Thereupon, an alternative of on-site cold bent plexiglass was analysed. Because the elasticity modulus of glass and acrylate differs by a factor of 30, the distance of the supporting structures became 1 m, instead of 2 m and the number of nodes became four times as frequent. Furthermore, also the corner elements would have to be bent and the acrylate sizes increased considerably, up to 20 mm. All this in relation to the maximum wind load from the open field. Acrylate, an impulse during a dinner meeting with Lise-Anne Couture, offered no proper solution. After this, the separation of hot-bend corner panels and the cold-bend remaining panels came about. The cold bending of glass panels is hardly ever done. The cold-bending was carried out here on site after fixation of the four corner points by pushing outwards two points in between the two corner points, resulting in the centre of the horizontal sides in an arch of 80 mm over 2 m! The bending tension was analysed as a maximum of 50% of the total application tension of $50N/mm^2$. The remaining acceptable work tension (50%) is reserved for the absorption of the wind loads. The choice of the laminated structure of these cold-bend panels was made because of the danger of breakage during assembly. The glass installers were walking on the inside and on the outside of the cold bend panels. Obviously, the cold-bending brought along the phenomenon of too long window panes that only stressing gave them the right length (in the horizontal direction). Furthermore, cold-bending brings along an unwillingness of the vertical seams. These want to curl in the opposite direction and therefore, a number of circular button-nodes had to be placed upon the vertical seams. In principle, it must be possible to cold-bend a laminated, fully pre-stressed flat panel as a cylindrical panel with a sufficiently low bending tension to absorb the occurring wind loads. In that case, it was worthwhile to have a short span, so that the wind moments will not be big, as well as to have more supporting points, so that the glass that must be bent will be thinner and therefore easier to bend. The bending radii came to 5 m.

3/5 The ICT era

For the development of the design of the three above-mentioned building elements of the Floriade Paviljoen, the engineering and production of the components and the assembly on site, it became clear that 50% of all man-hours had to be spent on co-design & engineering phase. But, thanks to the perfect engineering, the failure costs of production were nil. That was completely different in the days before the computer era. Only ten years ago, the failure costs of production were as much as 10%! These failure costs were muster stations of all the errors from the overall and detail engineering, static analysis, shop drawings, element productions, component assemblies and installation errors. Although a non-compatible liaison between main design and co-engineering were described in this paper, in the co-engineering of Octatube no errors were made. Even the most complex components within Octatube's building parts were of a perfect fit.

The conclusion, therefore, must be that the introduction of the computer initially made repetition work easier, but very soon the complexity of the challenging designing of architects changed up into the higher gear and currently, 3D drawings must be made of components still to be produced that are more complex than ever before. Benefits are, therefore, not so much in the efficiency of traditional building anymore, but rather in the transfer to more complex (and hopefully surprising) technical compositions. 2D drawings will become 3D drawings. Right from the start of the design, elements and components have to be defined. Gone are the days when good old '6B' pencil drawings were subsequently materialized in the shop drawings phase. The design may well be drawn by an architect in Maya or 3D Studio Max, but already in the next phase, drawing must be converted to 3D AutoCAD/Inventor, or at least in a program that is compatible and communicable to the programs of co-manufacturers.

4 Conclusions for the Management of Complex Designs and Engineering Processes

After involvement of the author and Octatube in six projects with 'Liquid Design' buildings, there are several lessons to be learned. They are worthwhile discussing in a broader professional audience and involve different aspects of the 'Liquid Design' Architecture processes. I have composed the lessons for these special design & engineering processes, involving the introduction and development of innovations in Blob architecture as follows in 18 hypotheses:

1 Realizing a Blob design is an extremely experimental process of design, engineering, productions and building on site. The experimental character of the process has to be recognized and dealt with.

2 In our time of prefabrications the engineering of 'Blobs' is the absolute basis of all productions. Productions are organized on the basis of theoretical drawings of a perfect engineering perfectly coordinated and integrated with the other sub-contractors on the site. 'Measuring on the building site in order to check dimensions' as sometimes mentioned to allow architects to blame the contractors for errors in drawings or co-ordinations, has become an anachronism.

3 Coordination and integration of the engineering of all concerned co-manufacturers and sub-contractors in the total engineering of the building is essential.

4 Detailing of elements and components will have to allow for accurate 3D-measuring. Click points to be positioned accurately as the reference points both in the engineering as well as in the site surveys.

5 The architect has a choice between two forms of collaboration, either: *Hierarchic*: develop the design with the advisors, tender and have the design further developed by the engineers of the contractors, or *Building team*: by composing a team of advisors and engineering co-manufacturers that develops the design and complete engineering of the building, after which the final tendering and realization takes place.

6 The architect has three choices for the engineering:
- Only to make the design concept and the presentation drawings
- The conceptual design and presentation drawings and the initial 3D-model
- The conceptual design, the presentation drawings, the 3D CAD mother model and the integration and co-ordination of all engineering contributions from the co-manufacturers.

7 The co-ordination of the engineering of the 3D CAD mother model has to be rewarded by the client by allowing a higher fee. The Blob architect should underline this item at the presentation of the design itself. If a higher fee is not agreed, these costs are likely to be fixed as co-ordination costs in the building process and disappear out of sight, but have to be paid in the investment costs of the building.

8 The total costs of design & engineering of all parties of a Blob design will amount to 20 to 40% of the total building costs (incl. fee of architects, advisers and co-engineers). Liquid design buildings are more expensive in their engineering than orthogonal buildings.

9 Ever sophisticated computer hardware and software has not resulted in more standardized and more economical preparation and building processes, but on the contrary in more complex and surprising buildings.

10 Tender packages should be clustered in lots for each sub contractor/co-manufacturer instead of the current arrangement of collections of identical elements and components.

11 Co-engineers need to incorporate excellent engineering departments able to dimension, detail and communicate in full experience with buildings in complex geometries.

12 The different co-producers ought to get slot times to integrate their prefabricated engineering into the 3D CAD mother model subsequently, only one engineering input at a time. Between each engineering input the architect has to check and certify the additions.

13 The ISO 9001 quality system will be applicable in future to all designing and engineering parties, to all producing and all building parties, none excepted.

14 Trust between the different parties will have to be the base. If not concurrent engineering and collaborative engineering will end in contra-engineering.

15 3D-site surveys will have to be continuously connected by computer on the 3D mother model, so that frequent checks of theoretical and practical click points can be compared. The site surveyor will become an indispensable service of the main contractor.

16 Only after a few years of experience a new brand of Blob 'cluster' contractors will arise, taking over the co-ordination and integration of complete building parts under the umbrella of a main contractor.

17 The choices of elements and components of the building including their detailing will have to be able to neutralize tolerances of production and positioning.

18 If Blob designs are underestimated in their complexity, there could be different victims:
- The client, who is ill-advised and has to pay the excessive costs of the realization compared to the estimations of the architect/quantity surveyor;
- The architect who sees his design aborted if it turns out to be too expensive;
- The main contractor loses money and time;
- The building manager does not understand this technology driven design and engineering process and will be confronted with increasing costs and a longer planning;
- The sub-contractors come a cropper on their naïve cost estimates and do not want a second adventure or go bankrupt;
- Participants in the process chase each other and cause each other physical complaints.

'Liquid Design' buildings have been possible since the last decade because of the increase accuracies and 3D geometries of computer hardware and software. The design & engineering is the core of the operation and within this process the design decisions are most important. Complex issues can be dealt with by an analytical engineering approach. There is not a problem that cannot be solved. The most advanced technology has to be developed further in order to meet the new geometrical demands, which place buildings at the same level of complexity as yachts, but at a lower economical level. Cold bending and twisting of glass panels, even laminated and insulated panels mix low cost prices with complex form results.

The new generation of 'Liquid Design' buildings with their computer designed arbitrary and non-rectilinear form, are mainly generated out of sculptural considerations by architects. All lessons from the past decades, where systemized spatial structures and economical building industrialization with the most sophisticated products were developed with their salutary regularities, do's and don'ts, evidently need shifting into a higher gear. The structural glass components of these buildings require an enormous effort into collaborative design and engineering. One would recommend that at least in the design phase the concept of the building technical composition would be developed simultaneously with the architectural concept. Both in the design & engineering phase as well as in the productions & realization phase an extremely high degree of collaboration between all able building parties concerned, is an absolute necessity to reach the goal of successful 'Fluid Design' Architecture, that is, successful for all parties. Frameless glass structures contribute to that higher level of technology in Modern Architecture.

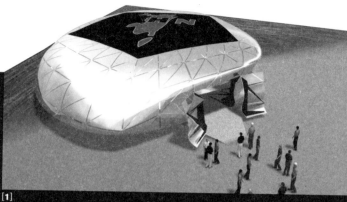
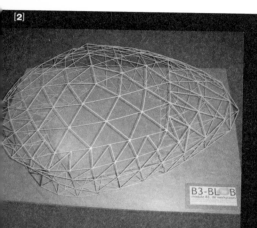

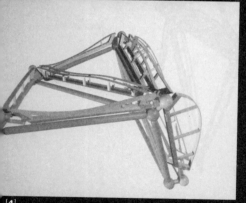

1 First model by architect Kas Oosterhuis (image courtesy: oosterhuis.nl).
2 Second model by architect Kas Oosterhuis (image courtesy: oosterhuis.nl).
3 Wire-frame model made by TU Delft students in their 3rd year.
4 Principal set-up of skeleton + cladding by students.
5 Mick Eekhout and students.
6-7 Prototype of the cladding, largely made by students.

[8] [9]

10 Rendering of the engineering set-up of the design
11 Explosion in a water basin on the grounds of Exploform
12 Redundant concrete moulds after the Exploform process
13 A single exploformed aluminium panel with welded aluminium edges
14 Test assembly of the aluminium panels before coating on a wooden jig in the factory of Octatube, Delft
15-16 The finished roof panels after installation on the roof of the Hydra Pier
17 The canopy at the entrance with the glass water pond
18 Close-up of the glass water pond
19 The cold-bend laminated glass panels in the south façade
20 View of nine laminated glass panels of 2 × 2 m, consisting of of 3 cylindrical and 6 cold-bend panels
21 Interior view of nine laminated glass panels of 2 × 2 m, consisting out of 3 cylindrical and 6 cold-bend panels
22 The cold-bend laminated glass panels in the south façade

A New Kind of Building

Kas Oosterhuis

Mass-customization Traditional vernacular building is accomplished by executing the process. There are no intermediate phases like a set of drawings, working drawings, drawings of details. The communication is direct from person to person. In modern computing lingo: through a peer-to-peer wireless sensor network. That is to say, peer-to-peer since people connect directly to their own kind; wireless since they are not physically connected; sensor network since they immediately absorb, process, and propagate information. People put their minds together, discuss and take action. Exact measurements and other relevant numeric details are decided during the process of building. The end result is unpredictable in detail, but is performed according to an agreed set of simple rules.

At the beginning of the 21st century, machines have taken the place of humans in the production and actual execution of the building elements. And now, based on digital techniques, we are able to establish a very similar peer-to-peer network of machines communicating with each other to produce an endless variety of different building elements, visually rich and complex, but still based on a set of simple rules. Humans connect to the machine-to-machine communication through conceptual interventions and through a variety of input devices. This process is called mass-customization, based on file to factory (F2F) production methods. Everything is different in absolute size and position, not because of human non-accuracy, but thanks to computational processing of diversity.

Building, as the public knows, is based on the industrial mass-production of building components. The elements are produced as generic material which will be customized later in another phase of the life of the product. The semi-products are produced in a limited range of sizes and measurements, then stored and catalogued, waiting to be taken up by the next party, eventually ending up in an assembly in the factory or on-site as part of a building. The mass-produced elements are categorized and have specialized into discrete classes: doors, beams, windows, columns, tiles, bricks, hinges, wire, piping, etc. Production according to the principle of mass-customization follows a completely different path. There is no catalogue; the products are produced starting from raw material (which in most cases is still mass-produced) for a specific purpose, to become a unique part in a unique setting in a specific building. That mass-produced part would not fit anywhere else: it is truly unique.

Architecture based on this new paradigm of mass-customization will be essentially different from the art of designing buildings that we have seen until now. Completely new tools for creating diversity and complexity are being developed to produce visual and constructive richness and diversity, yet based on simple rules being applied on conceptual procedures to generate behavioral relations between all constituting building elements. The driving forces to organize the behavior of the control points of the geometry come from both external and internal forces communicating with the evolution of the 3D model.

Looking at the world from within the paradigm of mass-customization (MC), we see that it includes all possible products along the production lines of mass-production (MP). By setting all parameters to the same value we can easily step one level down from MC to MP. The other way round is impossible. MC does include MP, while MP definitely does not include MC. Think of the inhabitants of Flatland, they are not able to experience – let alone conceive – Space. But Space inhabitants do have a notion of Flatland, as a section sliced out of Space.

A true understanding of the peer-to-peer network of machines communicating to machines connected by a flow of information leads to a completely new awareness of the architect/designer. We must go up one level, and start designing the rules for the behavior of all possible control points and the constraints of their behavior, instead of thinking of the rich and complex as exceptions to a given standard. The swarm of control points will be referred to as the **Point Cloud** in the context of this paper. All possible positions of the control points are no longer seen as exceptional states but as implicit possible states in the flocking relations between the points. The Point Cloud may be seen as a sort of Quantum State of geometry. There are no exceptions to a given standard; non-standard computation rules the control points: the exception has become the rule. Stepping up one level can be understood as stepping out of a world of plans and sections into a truly 3-dimensional space. Now we step out of mass-production and repetition into the realm of mass-customization and complexity, made possible by computational programming. We will step up one level and look at the world from there. As we will see later, I will propose to step up another level to enter the world of swarming behavioral space, leaving frozen 3D space as an experienced time-traveler would, or leaving Flatland like an inhabitant of Space.

Programming the Point Cloud The recent ONL (Oosterhuis-Lénárd) projects like the *WEB of North-Holland*, the *Acoustic Barrier* and the *Cockpit* building are based on the new building paradigm of mass-customization and the new design paradigm of programming soft design machines. Simple rules put into the machines are designed so as to create a visually complex geometry. Through a peer-to-peer communication, the data is transferred from the 3D model to the executing machines. Cutting, bending, drilling, and welding machines are operated by numbers and sequences, which are produced by scripts, routines and procedures written by ONL and executed on the points of the Point Cloud. ONL organizes the points of the Point Cloud through a variety of design strategies, using diverse programming tools. Each project has followed a slightly different path, but shares the principle of programming the Point Cloud.

To fully understand the nature of the Point Cloud, I must place it in the context of recent developments outside of the working field of architecture. There are three concepts I want to discuss here, all of them having to do with what you see when you are looking at the world from the level on complexity: Smart Dust, Utility Fog and Flocking Behavior. Subsequently, I want to dive deeper into the New Kind of Science as proposed by Stephen Wolfram, and draw conclusions on the implications it has for the architectural programming of the Point Cloud. After that, I want to take you up one more level, and discuss the Real Time Behavior of the recent ONL projects *Trans-Ports*, *Handdrawspace* and the *MUSCLE*. The behavior of the control points has in these projects become a running process, which continues running when it has been built. These constructs keep reconfiguring themselves, and produce complexity and unpredictability in real time. These projects are executables.

Building Relations between the Nodes The concepts of Smart Dust, Utility Fog and Flocks are basically all based on the concept of building local relations. One node looks at the neighboring node, but has no awareness of the whole Swarm of nodes. Intelligence is not something which can be programmed from the top down in a manner of reverse engineering, but is an awareness that emerges from the bottom up through a process of evolution by building relations between the nodes of the system. Intelligence is not necessarily aware of itself as being intelligent. Intelligence can very well emerge from swarming relatively stupid components. Together they perform as something complex, which humans may interpret as intelligent. 'Intelligence', as I use it here, is not seen as human intelligence. It is regarded as emergent behavior coming up from the complex interactions between less complex actuators. It seems to be possible to apply the same definition of intelligence to the functioning of our brains, traffic systems, people gathering, and to the growth and the shrinking of cities. Specifically, intelligence can also be applied to the relations that are built (both in the design process and in the actual operation of the construct) between all actuators/components assembled into a building.

Building relations in the concept of Smart Dust[1] is done through a peer-to-peer wireless sensor network. The concept of Smart Dust is developed by Kristofer Pister at Berkeley University and working prototypes are put together. Each micro-electromechanical mote sends and receives signals from and to other micro-sensors. They have a sensor in their backpack, all of it not bigger

---

[1] Smart Dust: Warneke, B.A. and Pister, K.S.J.; 'An Ultra-Low Energy Microcontroller for Smart Dust Wireless Sensor Networks,' Int'l Solid-State Circuits Conf. 2004, (ISSCC 2004), San Francisco, Feb. 16–18, 2004. 'The goal of the Smart Dust project is to build a self-contained, millimeter-scale sensing and communication platform for a massively distributed sensor network. This device will be around the size of a grain of sand and will contain sensors, computational ability, bi-directional wireless communications, and a power supply, while being inexpensive enough to deploy by the hundreds. The science and engineering goal of the project is to build a complete, complex system in a tiny volume using state-of-the art technologies (as opposed to futuristic technologies), which will require evolutionary and revolutionary advances in integration, miniaturization, and energy management.' Website: http://robotics.eecs.berkeley.edu/~pister/SmartDust/

than a grain of sand. The sensor is designed to pick up signals, smells, chemical substances, molecules according to the purpose of the Smart Dust particle. There is no PCU governing the swarm of Smart Dust particles. They basically sense, send and receive, propagating data and information like a rumor propagates through people in society. In the end, people are also sensors, senders, and receivers. It is my hunch that – after having taken the step to see the world from one or two levels up – that we must start designing from the awareness that buildings and everything constituting building elements are sensors, senders and receivers in the end, locally communicating with other specimens of their own and other species. Smart Dust is an operational system, be it that production costs of one mote is still something like $100 instead of the intended $1 in order to make it commercially applicable.

The concept of Utility Fog[2] by John Storrs Hall is based on the speculative assumption that we could build programmable molecules. If so, we could program these Foglets to configure into any shape or substance we might desire. The description of the possibilities goes beyond any SciFi movie you have seen. Since the Utility Fog particles are not visible – you can even breathe them freely in and out – they can spontaneously appear and disappear. They can swap from visible and tangible to non-visible and ephemeral. Utility Fog builds the ultimate bridge between the gaseous and the solid state of stuff. It can transform itself from one state into another based on its programming. Utility Fog is seen by their author as an array of molecular robots looking at each other and eventually connecting to each other to form solid material. No one could predict what it would feel or look like, but in principle it should work. The question here is whether we can learn from the concept of Utility Fog when thinking of complex structures for buildings. The way ONL has developed its latest projects shows that this is indeed the case. ONL basically regards each node as an intelligent point which is 'peer-to-peer' looking to neighboring points, and acting according to a simple set of programmed rules to form a complex consistent structure.

The constructive concept of points looking actively to each other immediately brings us to the concept of Flocks, Flocking Behavior, and Boids[3]. Boids as developed by Craig Reynolds, are active members of a flock calculating in real time their positions in relation to each other. Simple rules are underpinning their behavior. Each Boid computes a limited number of simple rules: Do not come too close to your neighbors, match your velocity with that of your neighbor, try to move towards the center of the mass of Boids. None of these rules says: 'form a flock'. The rules are entirely local, referring to what a local Boid can see and perform in its immediate vicinity. And yet the flock forms, and is recognizable as a complex whole. The importance for the procedure of architectural design here is that one does not need to define the exact overall shape beforehand in order to group the individual members together into a consistent whole. Boids can be interpreted as the flocking nodes of a constructive mesh. The designer could work with simple rules starting from the related positions of the nodes to generate the relevant data for mass-customized production. Also the behavior of the nodes might be used to form the shape of the building. Placing a bouncing box around the flock to limit their room to move remains a valid possibility since each building has to take into account the presence of other objects in their urban context.

A New Kind of Building

Building on the existing machines called Cellular Automata, Stephen Wolfram[4] recently declared his research in this field to form the foundations for a new kind of science, which he has also chosen as the title of his book weighing one kilo. Running a cellular automaton is building generation after (line after line) generation following some simple rules. By performing years of run time on thousands of possible rules, Wolfram found out that some rules lead to visually complex and unpredictable beings. Other rules tend to die out or would lead to uniform and predictable generations. And yet the rules leading to complexity are no more complicated than the other rules. Wolfram expects that these rules form the driving force behind all evolution, be it natural organisms or products induced by the interventions of humans, including scientific theories and mathematics. In theory, everything that is complex and behaves unpredictably must be based on simple rules generating this complexity. If this is indeed the case, then the development of cellular automata will outrun traditional science as the basis for further progress in all scientific fields, and, relevant in the framework of this paper, it

2 **Utility Fog: The Stuff that Dreams Are Made Of** by J. Storrs Hall, Research Fellow of the Institute for Molecular Manufacturing. 'Imagine a microscopic robot. It has a body about the size of a human cell and 12 arms sticking out in all directions. A bucket-full of such robots might form a "robot crystal" by linking their arms up into a lattice structure. Now take a room, with people, furniture, and other objects in it – it's still mostly empty air. Fill the air completely full of robots. The robots are called **Foglets** and the substance they form is **Utility Fog**, which may have many useful medical applications. And when a number of utility foglets hold hands with their neighbors, they form a reconfigurable array of "smart matter."' Website: www.imm.org

3 Boids: Reynolds, C. W.; (1987) 'Flocks, Herds, and Schools: A Distributed Behavioral Model', in **Computer Graphics**, 21(4) (SIGGRAPH '87 Conference Proceedings) pages 25–34. 'The aggregate motion of a flock of birds, a herd of land animals, or a school of fish is a beautiful and familiar part of the natural world. But this type of complex motion is rarely seen in computer animation. This paper explores an approach based on simulation as an alternative to scripting the paths of each bird individually. The simulated flock is an elaboration of a particle system, with the simulated birds being the particles. The aggregate motion of the simulated flock is created by a distributed behavioral model much like that at work in a natural flock; the birds choose their own course. Each simulated bird is implemented as an independent actor that navigates according to its local perception of the dynamic environment, the laws of simulated physics that rule its motion, and a set of behaviors programmed into it by the "animator." The aggregate motion of the simulated flock is the result of the dense interaction of the relatively simple behaviors of the individual simulated birds.' Website: www.red3d.com/cwr/boids/

4 Wolfram, S.; **A New Kind of Science** (Wolfram Media, Inc., 2002) ISBN 1–57955–008–8 'But my discovery that many very simple programs produce great complexity immediately suggests a rather different explanation. For all it takes is that systems in nature operate like typical programs and then it follows that their behavior will often be complex. And the reason that such complexity is not usually seen in human artifacts is just that in building these we tend in effect to use programs that are specially chosen to give only behavior simple enough for us to be able to see that it will achieve the purposes we want.' Website: www.wolframscience.com/

will turn out to cause a paradigm shift in the way buildings are conceived, the way geometry is generated and the methods by which the constituting parts are produced.

In essence, all points – comparable to the cells in a cellular automaton – are looking to their previous generation to decide what the next step will be, following some simple rules. Only by running the system one can find out to what class of result the simple rules will lead. Designing becomes running the computation, generation after generation, checking it, making changes, and running it again. Designing becomes to a much larger extent than it ever was an iterative process. In a traditional design process one iterates a limited number of times. When setting up a set of simple rules in a computation machine, one iterates in real time, that is, many times per second. In turbo lingo, this is designing with the speed of light, this is designing like a Formula I driver. Designing with rules, algorithms and with running the process builds the foundations for a new kind of building. These buildings are based on the behavior of an intelligent flock of swarming points, each of them executing a relatively simple rule, each of them acting according to local awareness of their immediate environment.

Specialization of the Building Detail Local rules executed by the nodes do not only create their behavior, but also the complexity of their configurations. The nodes evolve through running substitution systems, following simple rules such as: 'substitute this node by 3 nodes with small distances between the 3 new nodes'. This leads to a local specialization of the node. Or in architectural terms: to the building detail. Building details need more points, and those new points may be generated by a script describing some simple rules executed on the nodes. In the case of the Acoustic Barrier[5] each node of the Point Cloud has been multiplied to hundreds of new points in order to describe the geometry and to produce the data needed for the production of all the thousands of unique elements. It may be obvious that some of the data received by the script comes from the behavior of the points of the overall Point Cloud, and that other data used in the script comes from the top-down styling interventions of the designer, from the characteristics of the applied materials, from structural calculations, and from a variety of environmental constraints. Thus the complex swarm of flocking particles is evolving until a decision has been made to produce them.

Reading the *Scientific American* as regularly as my favorite architecture magazine (I do not read traditional architectural magazines since it is my strong belief that you have to experience the built reality architecture of your fellow architects in order to understand the essence of it, and read their theoretical texts), I stumbled upon an article on the specialization of skin into hair.[6] This idea seemed to resonate well with my attitude towards the specialization of the node into the detail as ONL has developed and built the last few years.

Hair and skin seem to be two completely different discrete elements, eventually assembled and cooperating as two separate families of elements, similar to embedding the headlights of a car in the car body. But where did the hair come from, when did it start to be a hair? The theory as described in the *Scientific American*, speculates on the concept of the specialization of the skin into a folded rim. This folded rim proved to have qualities which remained in the process of evolution. In the deepest caves of the rim a new micro-climate arose, where certain cells would become harder, yet keep growing and evolving into something hard sticking out of the skin. It soon became clear that a hair had advantages for protecting the skin against environmental conditions, and on its evolutionary path skin folded into hair on many parts of the body.

Replace now the cells by the nodes of a construct, and replace hair by the building detail – this process is exactly what happens during the evolution of the 3D model of ONL projects like the *WEB*, the *Acoustic Barrier* and the *Cockpit*. Just as hair covers the body in principle in most places, the specialized node in the form of the building detail is in principle present where it is useful. Basically in all places the specialization from node to detail is everywhere the same, but circumstantial differences in orientation create the variety of appearances of the specialized detail. Technically speaking, the detail is fully parametric; its parameters change with the changes in orientation. The end result is that of a visually rich complexity. Not a single detail out of hundreds (*WEB*) or thousands (*Acoustic Barrier*) is the same. All are different, and that illustrates the way we look at the world from one level up.

5 Acoustic Barrier, architect ONL [Oosterhuis_Lénárd], date of completion December 2004, client: Projectbureau Leidsche Rijn, product manufacturer: Meijers Staalbouw. 'The rules of the game. The brief is to combine the 1.5 km long acoustic barrier with an industrial building of 5000 m². The concept of the acoustic barrier including the Cockpit building is to design with the speed of passing traffic since the building is seen from the perspective of the driver. Cars, powerboats and planes are streamlined to diminish the drag. Along the A2 highway the Acoustic Barrier and the Cockpit do not move themselves, but they are placed in a continuous flow of cars passing by. The swarm of cars streams with a speed of 120 km/h along the acoustic barrier. The length of the built volume of the Cockpit emerging from the acoustic dike is 10 times more than the height. The concept of the Cockpit building is inspired by a cockpit as integral part of the smooth body of a Starfighter. The Cockpit building functions as a 3d logo for the commercial area hidden behind the acoustic barrier.' Website: www.oosterhuis.nl/quickstart/index.php?id=302

6 'Which Came First, the Feather or the Bird?', Richard O. Prum and Alan H. Brush, **Scientific American**, March 2003, pp. 60–69.
'Hair, scales, fur, feathers. Of all the body coverings nature has designed, feathers are the most various and the most mysterious. How did these incredibly strong, wonderfully lightweight, amazingly intricate appendages evolve? Where did they come from? Only in the past six years have we begun to answer this question. Several lines of research have recently converged on a remarkable conclusion: the feather evolved in dinosaurs before the appearance of birds. The origin of feathers is a specific instance of the much more general question of the origin of evolutionary novelties – structures that have no clear antecedents in ancestral animals and no clear related structures (homologues) in contemporary relatives. Although evolutionary theory provides a robust explanation for the appearance of minor variations in the size and shape of creatures and their component parts, it does not yet give as much guidance for understanding the emergence of entirely new structures, including digits, limbs, eyes and feathers.' Website: www.sciam.com (and type in the title in the search engine).

The detail of the *WEB*[7] is directly derived from the Point Cloud organized according to an icosahedron mesh mapped on the double curved NURBS surface. Just like needles stuck into a needle cushion, ONL generated 'normals' perpendicular to the surface pointing inward. This action doubled the number of points and generated a new Point Cloud. The points are instructed to look at their immediate neighbor and construct flat planes between the double set of points. These planes are given a thickness, and that leads to another doubling of points. From there the bolted joints are developed, leading to another multiplication of the total number of points needed to describe the geometry and hence to send those data to the cutting machines. By receiving data from interventions by the designer, in the manner of cloning and adding points according to a simple local procedure, the detail evolves from the node.

Since the doubling of the nodes is not executed along parallel lines, the connecting planes are placed at an angle in relation to each other. This leads to an evolutionary constructive advantage since the fold increases the strength of the folded plates.
It turns out that with this constructive parametric principle, ONL can virtually construct the support structure of any complex double curved surface, no matter if the curvature is round and smooth or sharply folded, no matter if the surface is convex or concave. The parametric detail of the *WEB* counts for a major invention in the construction technique for double curved surfaces. Moreover, it immediately connects the styling of the surface to the construction and the manufacturing of it. Architecture, construction and manufacturing are one, in much the same way as body, skin and hair are one.

The Point Cloud of the *Acoustic Barrier* is generated through a different procedure than was used for the *WEB*. A long-stretched NURBS surface on both sides of the barrier is bombarded with 10,000 parallel lines. The 20,000 intersection points form the nodes of the Point Cloud. Executed on the nodes, a number of scripts are evolved to develop the detail, and to generate the data needed for the production of the 40,000 unique structural members and the 10,000 unique triangular glass plates. By no means could this have been performed by traditional drawing techniques or by traditional production methods.

The Point Cloud of the *Cockpit* is directly related to the Point Cloud of the *Acoustic Barrier*. The stretched volume of the barrier pumps up so as to give space to over 5,000 m² floor surface for the Rolls Royce garage and showroom. The points are controlled along supple curves, which in their turn are controlled by a single reference curve, built in parametric *ProEngineer* software. Inside *ProEngineer*, ONL has applied a 'pattern' for the parametric detail using the points on a surface.

The architectural, structural and production concept of the *Acoustic Barrier* means another major innovation. ONL has proved in close cooperation with the steel manufacturer Meijers Staalbouw that within a regular budget, large complex structures can be built and managed without the interference of a general contractor. Thanks to the direct link between the well-evolved 3D model and the manufacturing, thanks to connecting the design machines to the production machines through scripting based on simple rules, ONL has proved that a complex building can be developed as an intelligently engineered product.

Nature and Products are Computations Based on my experiences with building the *WEB*, the *Acoustic Barrier* and the *Cockpit*, I now strongly believe that all of nature and all evolution of products are the result of a complex set of simple computations. Computations can be seen as building relations between nodes applying simple rules. The relation can vary from tracing a line (shortest connection) to exchanging data in real time (Smart Dust).

The making of architecture is the setting-up of a set of computations. ONL has a definite preference for working with raw products like sheet metal. The *WEB* is completely made out of sheet metal – steel for the construction and Hylite aluminum for the cladding panels. The *TT Monument*[8] is made exclusively from very pure cast aluminum. The more ONL can penetrate the F2F process into raw material, the simpler the rules can be to generate the outcome of the design and manufacturing process. Then the outcome of the process can be based on simple rules generating visual complexity, which is highly appreciated by the public since it feels rich and communicates the feeling of freedom.

7 Web of North-Holland, architect ONL [Oosterhuis_Lénárd], completed 2002, client: Province of North-Holland, product manufacturer: Meijers Staalbouw.
'One building one detail. The architecture of ONL has a history of minimizing the amount of different joints for constructive elements. Fifteen years ago this attitude led to minimalist buildings like the Zwolsche Algemeene and BRN Catering. At the beginning of the nineties Kas Oosterhuis realized that extreme minimalizing of the architectural language in the end will be a dead-end street. Hence in the office a new approach towards detailing was developed: parametric design for the construction details and for the cladding details. Basically this means that there is one principal detail, and that detail appears in a multitude of different angles, dimensions and thicknesses. The parametric detail is scripted like a formula, while the parameters change from one position to the other. No detail has similar parameters, but they build upon the same formula. It is fair to say that the WEB is one building with one detail. This detail is designed to suit all different faces of the building. Roof, floor and facade are treated the same. Front and back, left and right are treated equal. There is no behind, all sides are up front. In this sense parametrically based architecture displays a huge correspondence to the design of industrial objects. Parametric architecture shares a similar kind of integrity.' Website: www.oosterhuis.nl/quickstart/index.php?id=117

8 TT Monument, artist ONL [Oosterhuis_Lénárd], completed 2002, client: TT Circuit Assen, product manufacturer: Aluminiumgieterij Oldenzaal. 'We wanted to fuse the motorbike and the driver. The speed of the bike blurs the boundaries between the constituting elements. Each part of the fusion is in transition to become the other. Each mechanical part is transformed to become the mental part. The wind reshapes the wheels, the human body fuses into the new men-machine body. The fusion creates a sensual landscape of hills and depressions, sharp rims and surprising torsions. The fused body performs a wheelie, celebrating the victory and pride like a horse. The TT Monument is the ultimate horse: strong and fast, agile and smooth, proud and stubborn.' Website: http://www.oosterhuis.nl/quickstart/index.php?id=169

While everything we see around us in every room, in every car, on every street, in every city is based on simple computations creating complex behavior, it is virtually impossible to trace back the rules. The only way to find out is to run the system, to design a system which is based on simple rules generating complexity. This awareness potentially turns designers into researchers. Designers must set up systems and run the systems in order to perform. Performative architecture brings the architect and the artist back in the genetic center from where everything we see around us is generated.

Buildings are Complex Adaptive Systems This means that building relations between the nodes represent only one class of relations among many other possible and necessary relations. To evolve something as complex as a building involves many truly different actors. It is not just one system that runs in real time. It must be seen and designed as a complex set of many interrelated systems, all of them performing simple rules. In something as complex as a building, the nodes do not only communicate to other nodes, but even more to other product species. They will receive information from other systems as well, and include those data in the processing of the information, and in their behavior. In other words, a Boid is not moving in an empty world, a Cellular Automaton cannot live as an isolated machine, Smart Dust particles do have contact with other systems. All machines feed on information, and all machines produce information of some sort. Each machine is a small player in a complex structure of many interacting machines. But the necessity remains that in order to see the world from the next level, designers must start from simple rules placed in a complex environment rather than starting from a superficially complex structure without a clear concept of how to generate the data needed for customized production.

In the end, we must think of building and evolving networks relating all the different players in the dynamic process of the evolution of the 3D model. Each player in the process can be seen as having its own specific view on the data. The different constituting elements of the building have different views on the evolving 3D model. Each of them sends signals to the model which receives the signal, processes it and acts accordingly. From other disciplines the model would receive another class of signals leading to adjustment of the model for completely different reasons. In essence, this awareness leads to a process of Collaborative Design and Engineering. All players in this process – people, materials, forces, algorithms, money and energy alike – are in their own way connected to the evolutionary 3D model. Each of them is performing some simple set of rules, without complete awareness of what the other parties are doing or are capable of. They all contribute from their own systems to the complex set of related systems as a whole. In this sense, even a traditional building process behaves like a swarm. Learning from the new kind of science, we must build design processes on swarming intelligent particles in the Point Cloud communicating with each other. As humans we must learn to relate to the dynamics of super-fast real time computational processes. We must build the computational tools for Collaborative Design and Engineering in order to meet the rich expectations created by looking at the world from one or two levels up.

Based on my work with the Hyperbody Research Group at Delft University of Technology, I have started the **Protospace Lab for Collaborative Design and Engineering**. We are now entering our second operational year, Protospace 1.2. Next year we hope to continue with Protospace 2.0[9] in the resurrected *WEB* which is intended to be placed right in front of the Faculty of Architecture.

One of the issues we are dealing with is how to develop the design in collaboration with other disciplines (construction, ecology, and economy) and with the client from the Point Cloud. The Point Cloud is the raw design material, comparable to the Foglets of the Utility Fog, comparable to the Smart Dust particles and comparable to Neumann's Cellular Automata. Starting from this universe of particles we can start building rules and watch the worlds develop.

From the Point Cloud to the Soap Bubble Construct

Wolfram's *New Kind of Science* includes studies of substitution systems for the evolution of networks. The building of networks is a very appropriate tool for organizing the points of the Point Cloud. The notion of the network can almost immediately be translated into the constructive system of a building. The rule starts as: replace the one point of the T-crossing with the 4 points of a tetrahedron. Make sure that the distance between the 4 new nodes is substantially smaller than the distance between the primary nodes of the constructive system. Repeat this process with slightly adapted rules to organize the number, the direction and the positions of the new generations of the node. In this way, the new generations are nested-in or patterned-on the 3D array of primary nodes.

Repeating this procedure along the same substitution rule generates a 3D model resembling a soap bubble structure with smooth rounded transitions from floor to wall and from wall to roof. In fact, the connection between floor and roof becomes completely equivalent to the connection between wall and roof, between wall and another wall. The complete structure of a multistory building can thus be developed from one universal Point Cloud of structural nodes, each of them specialized into the building detail via a limited number of simple rules.

---

9 Protospace is a Laboratory for Collaborative Design and Engineering in Real Time, directed by Prof. Ir. Kas Oosterhuis, at Delft University of Technology. 'The transaction space for collaborative design is an augmented transaction space. Through sensors and actuators the senses of the designers are connected to the virtual prototype. The active view on the prototype is projected on a 360° panoramic screen. Active worlds are virtual environments running in real time. The active world is (re)calculating itself in real time. It exists. It represents a semi-autonomous identity developing a character. The active worlds are built according to a game structure. A game is a rule-based complex adaptive system that runs in real time. The rules of the game are subject to design. The collaborative design game is played by the players. Eventually the structure of the design game will co-evolve while playing the game.' Website: http://130.161.126.123/index.php?id=5

Point Clouds Running in Real Time For the *Architecture Biennale*, ONL created the *Handdrawspace* interactive painting[10], one of the worlds running in the installation *Trans-Ports*. This work shows with what material ONL is redefining art and architecture. ONL uses game development software (*Nemo* then, *Virtools* now) to run the system. Games are by definition running in real time; the game unfolds; the game is played by the rules. Game software is also capable of setting up multi-player worlds, which promise to be very appropriate for the process of Collaborative Design and Engineering.

In *Handdrawspace*, particles are continuously emitted from invisible 3D sketches. The number of particles, the size of the particles, their position in the universe and the colors are input values set through infrared sensors by the visitors walking around in the central space of the installation. The people connect to the Point Cloud universe. The always-changing values for the particles make sure that the same configuration will never be repeated. Each time one visits the *Handdrawspace Universe* one experiences a fresh unique world. The outcome of the real time computation is rich and complex, and never predictable in detail. The people walking around step by step learn how to cooperate with the running system: they teach themselves how to play by the rules, without changing them. Some people watch the running environment as if it were an instant movie; others involve themselves actively and change the course of the universe.

Now extrapolate this concept to the realm of architecture. When we can involve the very movements of people in the running process of architecture itself, we are really changing the static foundations that architecture has been built upon. And when we can involve the changing circumstantial conditions of the weather and other contextual data into the running process of the building itself, we can start looking at the world from yet another level. Then we are at least two levels up from where we are now. Extrapolating *Handdrawspace* into architecture leads to a major paradigm shift in the collaborative evolution of the 3D model, and it leads in the same manner to a major paradigm shift in the way we connect to buildings as running processes.

Looking at the world from there means looking at the Point Cloud as a swarm of intelligent beings communicating with each other in real time and all the time, as long as it takes them to live their process. For example, the installation *Trans-Ports* in self-explanatory mode[11] gives us another clue to build the relations between the points themselves, between the people among themselves, and between the people and the points. People and points are two different Point Clouds interacting with each other. *Trans-Ports* self-explanatory mode introduces a third active Point Cloud in the form of the pixels mapped as information on the interior skin. These pixels can be seen as a Point Cloud which can be programmed to communicate many visual complexities ranging from letters and language through signs and images, to movies and real time web-cams connected to other active environments.

Walking around in *Trans-Ports* changes the values of the positions of the nodes in the construction. The nodes are inter-connected by a building block called Cool Cloth bought by ONL via the Internet from an Australian gamer. The algorithm of Cool Cloth organizes their nodes in a 7 × 7 frequent mesh in such a way as to simulate the movements of a waving flag. ONL connected the active flag mesh to a shape which recalls that of the Saltwater pavilion, a pumped up tunnel body with open ends.

While the nodes of *Trans-Ports* communicate through Cool Cloth, the interaction with the users is built by ONL through a MIDI building block especially developed for *Trans-Ports*. Triggering the sensors is translated into MIDI numbers (between 0 and 128) which are linked to certain actions of the connected node-structure. ONL has programmed the actions in such a way that all actions can take place simultaneously, leading to complex behavior which never repeats itself.

Looking at the *Trans-Ports* machine in operation, one gets the feeling that it displays free-will, a will of its own. Since the free-will of people is in the end the result of a complex set of, in itself, simple rules being executed by the human brain in close cooperation with the human body, it seems perfectly acceptable to postulate that it is indeed a simple form of free-will. It is unpredictable for the people who have scripted it, and unpredictable for the people playing with the running system. If they are not the ones predicting what *Trans-Ports* will do exactly, it can only be the running system called *Trans-Ports* itself that decides in real time. The *Trans-Ports* machine digests the randomness of the people navigating in the installation arena.

For ONL, *Trans-Ports* has become an anchor point for Programmable Architecture. From then on, ONL was ready to lift the conceptual designers' mind up to the next level, to the level of all

10 Handdrawspace, artist ONL [Oosterhuis_Lénárd], Architecture Biennale Venice in Italian Pavilion, 2000, interactive painting. 'Handdrawspace is based on 7 intuitive 3d sketches which continuously change position and shape. The trajectories of the sketches are restlessly emitting dynamic particles. The particles are appearing and disappearing in a smooth dialogue between the 3d Handdrawspace world and the visitors at the Biennale installation Trans-Ports · When you step into the cave and go right to the centerpoint, a new colour for the background of the Handdrawspace world is launched. The inner circle of sensors triggers the geometries of the sketches to come closer, and thus to attract the particles. They become huge and fill the entire projection. Stepping into the outer ring of sensors the particles are driven away from you, and you experience the vastness of the space in which the particles are flocking.' Website: http://www.oosterhuis.nl/quickstart/index.php?id=197

11 Trans-Ports, architect ONL [Oosterhuis_Lénárd], Architecture Biennale Venice, 2000, interactive installation. 'The active structure Trans-Ports digests fresh data in real time. It is nothing like the traditional static architecture which is calculated to resist the biggest possible forces. On the contrary, the Trans-Ports structure is a lean device which relaxes when external or internal forces are modest, and tightens when the forces are fierce. It acts like a muscle. In the Trans-Ports concept the data representing external forces come from the Internet and the physical visitors who produce the data which act as the parameters for changes in the physical shape of the active structures.' Website: http://www.oosterhuis.nl/quickstart/index.php?id=346

possible interactions between all players in the game of building and architecture. Looking at the world from the position of no building being seen as static, they all move, albeit that most of them are extremely slow and extremely stupid. Since 2000, ONL has embarked on an architecture where all players (including all building materials) are seen as potential senders, processors and receivers of information, and where all players communicate with members of their own flock and with other flocks in real time.

MUSCLE at Non-Standard Architectures

Built especially for the NSA show in Paris for a budget of EUR 70,000 – ONL has applied the knowledge of the theoretical vehicle *Trans-Ports* into a working prototype called the *MUSCLE*[12]. The *MUSCLE* consists of 72 pneumatic muscles connected to each other, forming a consistent mesh wrapped around a blue inflated bubble. In this prototype for a programmable structure, it is not the nodes which are informed to move but the connecting muscles. Variable air pressure is sent in an endless stream of millisecond pulses to each individual muscle. When air pressure is pumped into the muscles they become thicker and shorter (muscles are a product of FESTO). When air pressure is let out of the muscles again they relax and regain their original maximum length. By varying the air pressure in real time (which in our physical world means: many times per second, and *per se* not absolutely continuous) for each individual muscle, the Point Cloud of nodes starts moving like the birds in a swarm.

The real time *Virtools* game as developed by ONL together with student assistants of the HRG sends out signals to the I/O boards, which are connected to the 72 valves opening or closing the airlocks. The *MUSCLE* game graph will also receive input in real time from 24 sensors on 8 sensor boards attached to 8 nodes of the constructive muscular mesh. The public can touch the sensors (infrared sensors, touch sensors and proximity sensors) so as to interfere with the running system of the *MUSCLE*.

The flock of muscles is programmed in such a way that all muscular actuators cooperate to perform a change. It is impossible for one muscle to change place without cooperating with the other connected muscles. Programmed by assembling the graphs in the *Virtools* software, the nodes are set to look at each other when changing position. The change is communicated to the neighboring nodes. From there the desired length of the connecting muscles is calculated in order to accurately perform the displacement of the nodes. The calculation is based on experimental values found by testing the system with the chosen air pressure, the chosen sizes of the air pressure tubes, and the chosen capacity of the valves.

The nodes are looking to each other all the time. While the muscles are changing their lengths, the *MUSCLE* is hopping, twisting, bending and rotating constantly. As long as the program runs and the air pressure holds, it is alive. The *MUSCLE* is ONL's first materialized construct as a running system acting out of its own free-will and at the same time interacting with the public. The process of interaction can only take place when there are at least two active parties involved, when there are at least two running systems communicating with each other. The *MUSCLE* is one running system, the human person another, both with a will of their own.

The *MUSCLE* is a 'quick and dirty' built prototype for the New Kind of Building as introduced in the title of this paper. This new kind of building is not only designed through computation, it *is* a computation. The New Kind of Building never stops calculating the positions of its thousands of primary and its millions of secondary nodes, based on input values from both the users of the building and from environmental forces acting upon the structure. The New Kind of Building is a *Hyperbody*.

911 Hypercube

Asked by Max Protetch Gallery to contribute to the Ground Zero exhibition showing the architect's response to the events of 911, ONL proposed a large fully programmable cubic volume, a *Hypercube*. ONL proposes here an Open Source Building approach, in contrast to the defensive Pavlov reaction the US took as their policy. Only by setting up an open political system based on mutual respect, can one build a society which is not based on threat, hate, or fear. To this open global society belongs an open global architecture: an architecture which is a running process and which feeds on streaming information from all sides of the globe. ONL came up with a 3D-lattice structure where all structural members are data-driven, programmable, hydraulic cylinders. The pistons act as actuators for the data-driven building. If all the pistons are at their extreme position, the building can shrink 50% of its size in all three axes. As a net result, the building can shrink or expand to eight times its original volume.

The *911 Hypercube* Building[13] responds to changes triggered by its users, but also proposes changes by itself according to a set of simple rules generating a complexity of possible configur-

---

12 MUSCLE, architect ONL [Oosterhuis_Lénárd], interactive installation in Forum des Halles Centre Pompidou, Paris, 2004. 'For the exhibition Non-Standard Architecture ONL realizes a working prototype of the Trans-Ports project, called the MUSCLE. Programmable buildings can reconfigure themselves mentally and physically, probably without considering to completely displace themselves like the Walking City as proposed by Archigram in 1964. Programmable buildings change shape by contracting and relaxing industrial muscles. The MUSCLE is a pressurized soft volume wrapped in a mesh of tensile Festo muscles, which can change their own length. Orchestrated motions of the individual muscles change the length, the height, the width and thus the overall shape of the MUSCLE prototype by varying the pressure pumped into the 72 swarming muscles. The balanced pressure-tension combination bends and tapers in all directions.' Website: http://www.oosterhuis.nl/quickstart/index.php?id=347

13 911 Hypercube, Ground Zero exhibition, Max Protetch Gallery, New York, 2002. 'The war in Afghanistan took more lives than the attack on the WTC. Why do most people feel different about the death toll in Afghanistan than about the sudden death of the WTC and 3000 users? Are some killings more just than others? Are the winners always those who kill the most people? If you examine crime movies you will find out that the "good" ones are always licensed to kill many "bad" ones. Is that why the US had to kill more Afghans and Saudis than there were

ations. The *Hyperbody* would also respond to changing weather conditions, to the behavior of people in the street, and to signals and patterns received from other buildings and other information processing vehicles from all over the world. The *911 Hypercube* is designed to be a giant interface between many different behavioral swarms, ranging from people from any culture to other built structures, both ephemeral (programs, organizations, the Internet) and tangible (buildings, cars, microwaves, air conditioning, cell phones) information processing machines. The presentation of the *911 Hypercube* comes in 12 modes, corresponding to the 12 months of the year, 12 exemplary types of weather and 12 typical New York events.

Peer-to-peer architecture means communicating between equivalent computing machines. Just as in *Smart Dust*, we look at the nodes of the *911 Hypercube* as small computing devices. Some form of intelligence has been built into the node. The nodes do at least perform some form of sensing, processing, and propagating of signals. They send signals to the actuators, the hydraulic cylinders. Thus the construction of the *911 Hypercube* is a peer-to-peer network. People can be peers; spaces can be peers; they all connect in similar peer-to-peer networks. A simple conversation between people establishes a peer-to-peer communication. It is actually this basic level of communication I am considering when thinking of programmable, pro-active hyperbuildings.

Protospace 1.1 Demo   After explicating the nature of the New Kind of Building, and looking at the world from there, how might different disciplines work together? At the Delft University of Technology, my Hyperbody Research Group has built a first rough concept for the *Protospace 1.1 Demo*.[14] Just as is the case in a complex set of peer-to-peer networks working inside *Protospace*, various disciplines want to communicate in their own way with their own kin. In a process of Collaborative Design and Engineering, one wants to express oneself to the highest level of knowledge and intuition of one's discipline. One expert in a specific field does not want to limit him/herself to constraints set by other disciplines which are either 'not obviously' or 'obviously not' relevant to one's own discipline.

The HRG has built a simple demo where the different players in the evolution of the 3D model each have their own view on the 3D model. For that I have chosen the role of the stylist, the construction engineer, the ecologist, the economist and the tourist; each of them actually sees the 3D model differently. The stylist sees a surface model which can be shaped, the construction engineer sees nodes and connecting members, the ecologist sees the surfaces separating different microclimates, the economist sees numbers and spreadsheets, and the tourist navigates through the model as it will appear visually.

Each of the players sees something different, but is still looking at the same thing. It is important that (s)he sees the essence of his/her own discipline since that effectively shows the working space where (s)he is authorized to propose changes. Each discipline has another view on the same thing, just as every single person looks differently at the same scene. Ask two people to describe what they have seen, and you end up with two different stories. Yet still they are watching the same scene.

Similar to birds in a flock, similar to the behavior of cars on the highway, similar to people in a meeting around the table, the experts in *Protospace* are looking at each other in order to adjust their positions in real time, and at the same time they are actively participating in the developing scene. In *Protospace* one is looking at the 3D model through his/her own pair of disciplinary eyes, while the other players may have a different look at things. The central theme of building tools for Collaborative Design and Engineering (CD&E) is to develop the 3D model by focused disciplinary input, synchronous with the input of the other disciplines. The ultimate goal of *Protospace* is to improve the speed and quality of the design process based on parallel processing of the knowledge of all disciplines involved from the very first stages of the design.

The players will have immediate insight in the nature of the changes that the other party is putting through. And it is then up to the flock of players to decide whether these changes are improving or deteriorating the 3D model. To facilitate this, the HRG is working on intuitive validation systems in order to validate the changes that occur in the CD&E process. None of the disciplines takes the absolute lead. Just as in a peloton of bicyclists, the players alternately lead in order to proceed as fast as possible as a swarm, as a whole. And to be perfectly honest, just as in a real tournament, someone's contributions will turn out to be advantageous and respected, and this person will eventually connect his/her name to the project.

It is very well justified to compare the process of CD&E to a game which enfolds. The rules of the game are set from the beginning. The players play by the rules. Good players make an interesting

citizens killed on 911? Come on America, wake up and find a way to take revenge in a more intelligent way. Do not waste our precious time on the easy killing of poorly armed people. Let's face it. Everybody was fascinated by the 911 event. Everyone was thrilled to watch the movie, over and over again. Only extremely disciplined individuals could resist to watch. Quickly destroying things is naturally much more appealing than slowly synthesizing things. How can we as architects appeal to people's fascinations by building new stuff?' Website: http://www.oosterhuis.nl/quickstart/index.php?id=155

14 Protospace 1.1 Demo, directed by Prof. Ir. Kas Oosterhuis, built by the Hyperbody Research Group, Delft University of Technology, 2004. 'How do the stakeholders collaborate in real time? Imagine the following scene. The game leader opens a file, the active world. Each file has a specific set of rules how to propose changes in the file. However, there will be developed a detailed Protospace protocol how to play by the rules. The referee explains to the players how to play the game. Each stakeholder chooses a view on the file. One player may choose different roles at the same time. The players come in action according to the rules of the game when it is their turn to propose a change. When playing the role of a specific stakeholder only that particular view on the database is displayed. While delivering the input through sensors and numpads the players are free to walk and chat in the group design room. The group design room is an open design studio, a social transaction space. The other players watch the active player and respond immediately like in a normal conversation.' Website: http://130.161.126.123/index.php?id=5

game. Inexperienced players make a boring game. The question which arises here, is: 'Who makes the rules?' The architecture of any outcome of the game resides inside the rules. Simple, strong rules create a higher form of complexity than shabby rules. Good architecture builds upon the strength of the set of rules. The true game of architecture in a CD&E setting creates situations where the rules are verified, tested, and eventually improved. Only then can one speak of a true evolution of the 3D model – as opposed to enrolling and developing. The one who improves the project rules can be any player at any time in the process of CD&E.

Conclusion   Architecture has become a rule-based game, played by active members of a flock, communicating with other swarms. As proven above, this is true for the F2F process of mass-customization; it is true for the New Kind of Building based on Real Time Behavior (RTB) of programmable pro-active structures; and it is true for the interactive process of CD&E. To be able to develop the F2F process of mass-customization, one must step one level up and look at the world from there: not looking from the top down, but from within into the new dimension of complexity. To be able to deal with the RTB of programmable constructs, one must step up another level and look at the world from the point of view that all nodes are executing their systems in real time and communicate in real time to their own kin and other species. In order to be able to get there – two levels up – one needs to beam oneself up into the running process of CD&E and look at the world from *within* the process. The information architect works inside evolution.

To summarize the attitude of ONL in the design and production process of the New Kind of Building:

A   One level up to Mass-Customization (MC):
 MC does not mean a single repetitive component in the built structure
 MC includes traditional mass-produced (MP) building, while traditional building excludes MC
ONL achieves MC by:
 Developing the generic parametric detail
 Establishing the File to Factory [F2F] process
MC and F2F are based on:
 Point Cloud
 Scripts, routines, and procedures to instruct the control points

B   Two levels up to Real Time Behavior (RTB):
 Constructs are developed as running processes
 The building reconfigures itself constantly

RTB includes traditional static architecture, while traditional architecture excludes dynamic RTB

ONL achieves RTB by:
- Defining building components as actuators
- Feeding the actuators with data in real time
- Relating the actuators to the game program

RTB is based on:
- Swarm behavior
- Game Theory
- Collaborative Design and Engineering (CD&E)

Multifunctional Mote

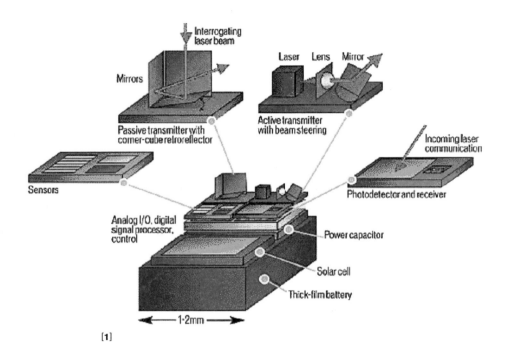

[1]

1 Smart Dust, Kristopher Pister et al., 2004, Multifunctional Micro-Mote.
2 Utility Fog, John Storrs Hall, nano-scale Foglets shaking hands.
3 Boids, Craig Reynolds 1987, Flocking Behavior.

[2]

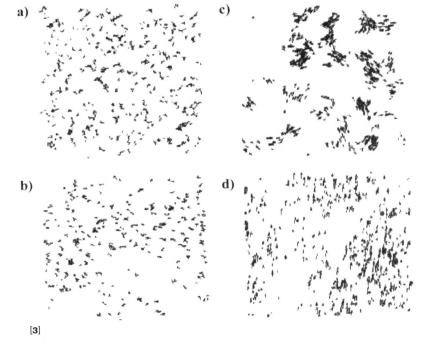

[3]

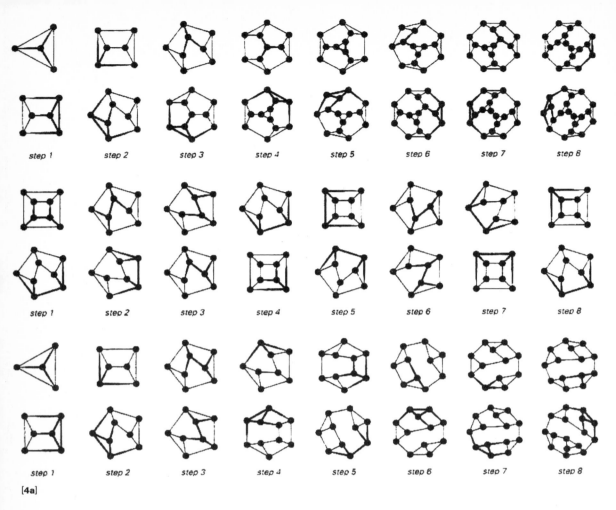

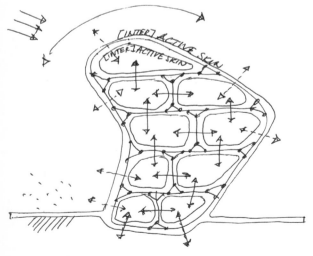

[4a]

[4b]

4a A new kind of Science, Stephen Wolfram 2002, Substitution system, Simple rules generate complex results.
4b TORS, ONL [Oosterhuis_Lénárd] 1995–2004, Specialization of the detail.
5a Acoustic Barrier, ONL [Oosterhuis_Lénárd] 2004, Point Cloud and generic script.
5b Acoustic Barrier, ONL [Oosterhuis_Lénárd] 2004, File to Factory process of Mass-Customization generates 10,000 different nodes.
5c Acoustic Barrier, ONL [Oosterhuis_Lénárd] 2004, Building site progress, 15 September 2004, the Cockpit will connect to the left end.
5d Cockpit Hessing ONL [Oosterhuis_Lénárd] 2004, Specialization of group of points to form the Cockpit.

[5a] [5b]

[5c] [5d]

6 Specialization from skin into hair, Scientific American, March 2003.
7a Web of North-Holland, ONL [Oosterhuis_Lénárd] 2002, Autoslip routine for F2F process.
7b Web of North-Holland, ONL [Oosterhuis_Lénárd] 2002, Generic parametric detail.
7c Web of North-Holland, ONL [Oosterhuis_Lénárd] 2002, Floriade World Flower Exhibition, Precision landing of spaceship.

[6]

[7a]

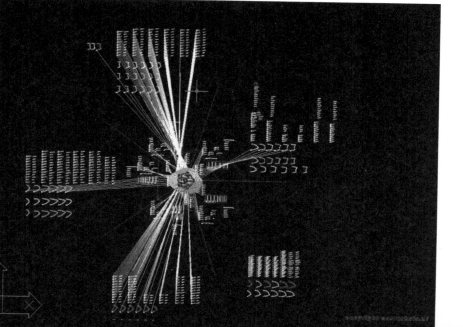

[7b]

[7c]

8 TT Monument, ONL [Oosterhuis_Lénárd] 2002, Simple rules for generating the complex surface.
9 Protospace 2.0 in the web, ONL [Oosterhuis_Lénárd] 2004, Delft University of Technology, Laboratory for Collaborative Design and Engineering [CD&E].
10 Handdrawspace, Architecture Biennale Venice 2000, Interactive painting.
11 Trans-Ports, ONL [Oosterhuis_Lénárd] 2000, Architecture Biennale 2000, Programmable architecture.
12 Muscle, ONL [Oosterhuis_Lénárd] 2004, Non-Standard-Architectures, Centre Pompidou Paris, Interactive Installation with 72 actuators.

[8]

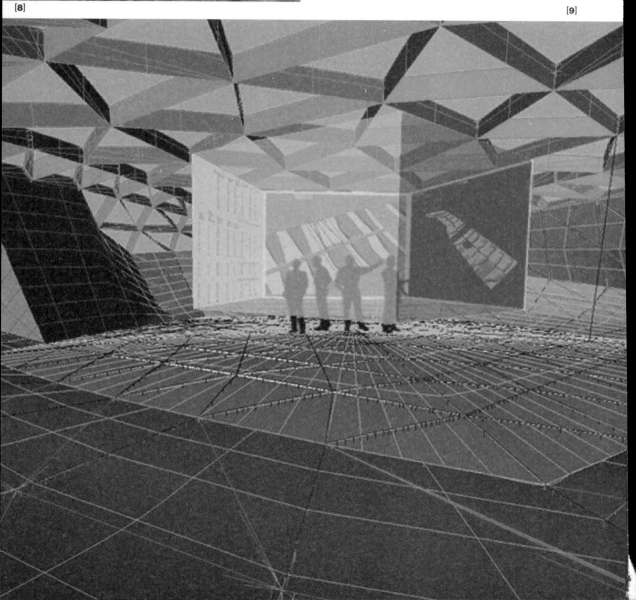

[9]

[10]　　　　　　　　　　　　　　　　　　　　　　　　　　　　　　　　　　[11]

[12]

13a 911 HYPERCUBE, ONL [Oosterhuis_Lénárd] 2002, Max Protetch Gallery New York, Open Source architecture, March mode.

13b 911 HYPERCUBE, ONL [Oosterhuis_Lénárd] 2002, Max Protetch Gallery New York, Open Source architecture, August configuration.

Bending and Folding Glass

Karel Vollers

Architects create market pull by designing buildings with complex curved façades. Simultaneously, producers create technology push by providing components for surfaces of complex geometry. This article focuses on the initial stage of a technology push in which the transforming of glass by technology transfer is crucial to develop and market a new façade system.

Technology Push and Market Pull

Many architectural designs feature freely double-curved surfaces. Though glass would have been preferred for reason of transparency, durability and appearance, the surfaces were executed in polystyrene, polycarbonate or metals like aluminium, copper and zinc. In recent years exclusive projects initiated the market for complex curved glass, for example the curtain-like panes in the restaurant of the Conde Nast Publishing Company in New York (2000) and the roof over the entrance of the TGV station Saint Lazare in Paris (2003). Their glass panes are transformed by annealing the panes on milled moulds. The panes sag on the moulds during heating in the furnace and then are slowly cooled. Milling and material costs with the additional glass processing result in high prices per panel. An adjustable mould will offer an alternative to a milled mould, saving on material and milling costs, but as yet no such mould is available.

The engineering problems of an adjustable mould can be overcome, but that implies high initial costs, whereas market demand is small. Mould manufacturers await a bigger demand that guarantees quick return on investments. Large-scale use of transformed panes requires acceptable prices and this implies in Western countries industrialised production. With milled moulds double-curved panes can be made without high investments. They enable starting a market for double-curved glass, and developing adjoining products like framing systems, sunscreens, etc. Before a big scale market has developed and adjustable moulds have become affordable, the market is initiated by processes that imply technology transfer, a high degree of manual labour as well as extensive use of materials.

Annealing Glass on Milled Moulds

Milled moulds offer a great degree in freedom of transforming with great accuracy. The annealing temperature can be high because the softened glass pane touches the mould over its whole surface. This is a great advantage over an adjustable mould that the author had designed before and that in 1998 was used to produce annealed twisted glass. With glass processing company Van Tetterode Glasatelier that mould was elaborated into a suspending surface built of high temperature resistant stainless steel bars. Because of the distance between bars, the annealing temperature was lower than would be acceptable with an over the complete surface suspending mould. The glass during annealing stayed relatively stiff and the pane's sides during the bending had to be forced down and be straightened out by heavy steel profiles resting on them. The straight bars limited the transforming to surfaces built of straight lines (ruled surfaces). Four twisted panes subsequently were fitted in a prototype of an industrially produced twisted aluminium façade, developed with Reynolds Architectural Systems (now Alcoa). Measuring the twisted panels with their varying curvature and varying corner angles was done by measuring in relation to a specially made box. It was tedious. In 2004 an improved manufacturing process was tested with milled moulds for annealing and for measuring.

Milling Moulds from Cellular Concrete

To meet the project's low budget, moulds were milled from standard cellular concrete blocks. These blocks are cheap and because of their light weight are easy to handle. Their heat capacity is low. After testing milling characteristics and material behaviour of a block in a furnace, a mould of 1.2 × 1.8 m was milled. The glass processing firm Glasbuigerij Bruining did the annealing. They have experience in bending architectural glass (especially cylindrical and spherical curved on moulds made of straight bars) and free double-curved glass (for antique car windows, mainly by suspending the panes along the contours on linear metal frames). The moulds meet the architectural demand of making relatively cheap moulds for small series. They are not suitable as yet for series bigger than three, as the moulds start to fall apart by cracking and letting loose of the cement, implying more repair work after every annealing session. The panes have surface and edge tolerances acceptable for architectural use in industrialised framing systems. Further study of characteristics of the used materials and optimising of chosen materials will improve considerably the process.

Assembling the Panes

Cutting the complex shaped panes to size is an additional technical problem. Development of a CNC cutting machine is relatively easy as compared to the making of an adjustable mould surface with low positioning tolerances that endures annealing temperatures of approx. 670°C. Such cutting machines will have, like adjustable moulds, a good return on investments when the market has grown. But the market is still extremely small. For this prototype, the annealed panes were cut to size on moulds milled from polystyrene. Assemblage to insulating panes was similar to making cylindrical and spherical glass.

First Alu-system for Freely Curved Façades

Like the first prototype (1999), the second façade prototype (2004) consists of four interconnecting glass panes, but this time freely double-curved. Again all panes differ, implying varying transforming characteristics and complicated measuring. The bottom transom is straight, while the top transom and the mullions are curved with a radius of 4 m. The prototype additionally features an opening window, which implies increased precision in the transforming of profiles. It is the first framing system in the world for double-curved façades. The optional opening window with the special designed hinges and closing mechanism therefore is the first in the world also.

A conventional aluminium framing profile must be single-curved or double-curved and be additionally twisted to connect with a parallel surface to a freely double-curved glass panel. Curving and additionally twisting an aluminium profile of simple section is complicated and labour-intensive. The complex sections of most framing profiles don't allow complex bending. The AA100Q-*Twist* system designed in collaboration with Alcoa is designed to connect panes of varying inclination. It consists of two profiles. The first, a straight or single-curved backing profile, is positioned (by preference) parallel or perpendicular to the building's superstructure. Thus the connection of the façade to the wall, floor or column is standardised. The second profile is the glazing profile which has a connecting surface parallel to the glass. The two profiles meet along an essentially cylindrical surface. Whereas the backing profile provides structural strength, the glazing profile attached to it is torsion weak. The glazing profile of complex section meets the demands of water, sound and fireproofing. It can be twisted by hand to about 10° per metre, which suffices for most façades. This industrialised system consists of standardised profiles and makes double-curved façades economically feasible.

Folding 3D Glass Panes

Further architectural freedom is offered by introducing folds within flat panes. Prior to making the freely curved prototypes' panes, we tested annealing on a milled mould, to make a folded freely curved pane for the Fashion in Architecture exhibition at the Deluxe Gallery in London in 2004. This 3D curving fold increased the complexity of transforming.

Computer animations of the blue mirroring glass pane illustrate the spectacular mirroring effects we aimed at. Fashion designers and architects have similar interests in transparency, interacting grids (moiré effects in fabrics) and tactile connotations.

Panes with bowl shaped segments that are used as washing basins, appeared on the market from 1990 onward. They are annealed on flat moulds with round holes through which the glass sags by gravity during annealing. This annealing process is repeatable with acceptable size tolerances. Similar larger panes were applied in the Prada shop in Tokyo designed by Herzog & de Meuron Architects. They were annealed by the firm Cricursa in Barcelona. The panes everywhere meet the framing profiles as flat surfaces. Costs were reduced by repeating forms.

However, the bending of such bowl shapes in a panel, is mainly a result of gravitation. The transforming control of such surfaces is limited and fluent connections between adjoining freely curved panes are impossible. Here the surface control by using milled moulds offers feasible solutions too. Folds within curved surfaces, and not only along the rim of a sagging part, can be made on moulds milled from ceramic materials. Just a few of such panes in a façade will suffice to make an eyecatcher, upgrading the aesthetic value of the whole.

A mould was made to test the making of a double-curved fold in a freely curved pane. However, its suspension in the furnace was found to require improvement. The folding occurred along a wrong line, the pane didn't sag everywhere onto the mould, and the pane with its underlying glass pane broke. This was an experiment to test extremes. It was foreseen that folding along a single-curved line would be easier, especially with the glass resting on the fold line prior to the bending. The improvised suspending technique gave insight into glass behaviour, but was a failure in so far that the desired transforming failed. The project of making folded double-curved glass has been delayed until a budget becomes available.

Final Comments　Iterative design processes allow for improvisation. When working with a limited budget on new fields of study for the researchers, better forecasting of the consequences of chosen procedures, i.e. behaviour of the mould materials and the glass, is advisable. This can be achieved by better study of material behaviour and its processing and by consulting more with specialists.

Acknowledgements

Alcoa Architectural Systems (Harderwijk)

Van Campen Aluminium (Lelystad)

Glaverned (Tiel)

Eijkelkamp (Goor)

Glasbuigerij Bruining (Dordrecht)

Tetterode Glasatelier (Voorthuizen)

TO&I at Faculty of Architecture of TU Delft

Hellevoort Visuals (Amsterdam)

Remco Wilcke Visuals (Delft)

[1]

[2]

[3]

[4]

1 TGV station in Saint Lazare, engineered by Henry Bardsley.
2 Detail of curved pane glued to metal strip, which in turn is bolted to a frame.
3 Cracking of thin layers and letting loose of cement paste.
4 Foaming of not completely dried cement paste in later filled parts.
5 Thermally deformed annealed panes awaiting assemblage into insulating glass.
6 Blue mirroring panes were cut to size on polystyrene moulds.

[5]

[6]

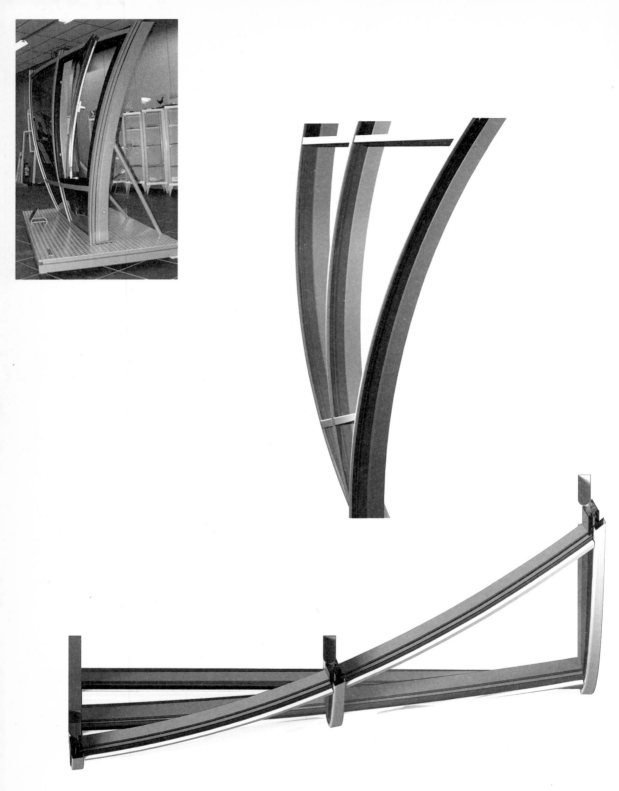

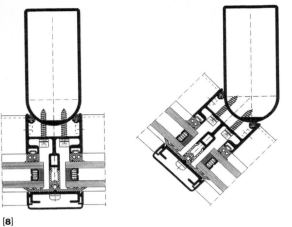

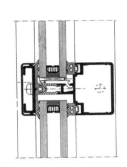

[8] [9]

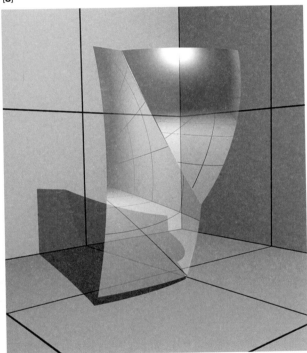

▲▼ [10]

[11]

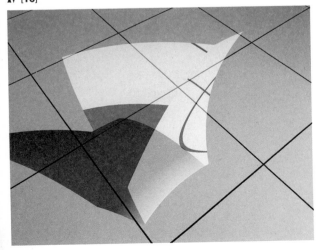

7 Side perspective and top view drawing and photograph of façade prototype.
8 The Twist-profile is split in two parts that connect along a cylinder.
9 Curving twisted aluminium transom profile.
10 Computer drawings of a blue mirroring free double-curved folded pane (2003).
11 Annealed pane lies broken on milled mould.

Section IV:

The Urban Question

Since the urban crisis of the 1960's, many scholars have been trying to capture the new developments and urban transformations that have occurred since that period. All the papers in this section have contributed to analyzing contemporary urban developments, in both a critical and historical way. From Boyer's *City of Collective Memory* to her more recent *CyberCities*, a strong critical writing guides us through the history of the city and its postmodern contemporary dispersed form. One of the major effects of urban transformation was that the geography and composition of the global economy changed so as to produce a complex duality: a spatially dispersed, yet globally integrated organization of economic activity, a theme Saskia Sassen has developed recently. Lampugnani's recent writings concentrate on contemporary phenomena like telematics and its effects on the city. He introduces new critical concepts to understand and improve our understandings of current urban place and situation. Soja's notion of *synekism* involves the formation of nucleated and hierarchically nested regional networks of settlement capable of generating innovation, growth, and societal (as well as individual) development from within defined territorial domains.

We are making a transition from a world where the urban represents a condition of social life, to one where it represents *the* condition of social life. But at the same time, the properties of urban places are assuming all manner of hybrid and newly emergent forms. The apparent disappearance of a strong local place condition in many situations can be contrasted with the fact of a simultaneous emergence of places whose extraordinary power and presence is precisely a consequence of their alignments within global networks. One of our particular tasks is to understand the mechanics of the city in terms of scale relationships and the networks that mediate them. When considered in this way, it's not difficult to see that many interesting questions of the contemporary city revolve equally around issues of adjacency and of the interpenetration of networked lives brought together in hierarchies of central places in the same geographical region, as they do around issues of the simple penetration of the global into all lives and all places. Such a perspective is likely to take us beyond the conventional wisdoms of 'community', 'neighborhood', 'urban village' and so forth, which see the local as refuge against the global, and as the psychological home of the modern urbanite, to be made as a protective buffer or fortress against the contaminating and disturbing influence of the city.

An approach which seeks to support processes which empower people – and especially those people whose power is limited within existing social and economic orders – is likely to find value in spatial processes which move towards the global – to 'de-localize' and to offer opportunity and power to people with respect to a wider world around them. People need to participate in dynamics of change, and they need to participate from places (both virtual and real) strategically situated in relation to, and energized by, a non-local space which is open, connected to streams of urban, regional and global power while not being controlled by them.

We suggest that research needs to focus not just on the procedures of city building but also on the conceptual apparatus that we bring to bear, and we will need to measure this apparatus not only against values of justice and equity, but also against sets of values that incorporate an idea of an urban future which makes, as Richard Sennett writes, 'provision for the fact of history'.

The Artful Reinvention of the Suburban Landscape

Lars Lerup

The state of the landscape is radically arrested in attenuated suburban cities like Houston. Merely representational is an arsenal of foreign species from trees to flowers that is simply grafted on to a suffocated original landscape. Detached and distant, the image of Nature (in itself a defunct concept) is simply plastered on top of the little housy-houses to produce the utterly fake dream of the happy life. How does one crack open this sinister collusion between representation and the branding of suburban living? Through a certain artfulness, beginning at the rhetorical end of the new enterprise.

We must literally reinvent Nature, and possibly first as a form of spectacle. In the 1830's, Charles Durande of Louisiana celebrating his daughter's wedding, imports a million spiders from China to spin their webs in an *Allée* of trees on his Pine Alley estate, while his servants dust the spider webs with gold and silver imported from California. It is hard to think of a more artful, yet gentle enhancement of nature, particularly in light of the current descendants of the Chinese spiders spinning their webs as we speak. Human intention joins nature in a process with measures of cooperation and uncertainty – will the Chinese spiders overpower their Louisiana cohorts? How far-fetched is it to think of such a seductive and artful process for the rebuilding of the *fieldroom*[1], and more specifically the bayous?

Let us speculate. In 1997, Olafur Eliasson, a Danish artist, runs five gallons of water per hour from a faucet in the outer wall of a museum in Johannesburg, South Africa into the street and across yards of adjacent houses. Municipal water is surreptitiously returned to Nature to begin the formation of a new creek. Likewise back in the *fieldroom* (now under the auspices of the thought-experiment) Eliasson selects a leftover field – a leap-frogged lacuna – close to a bayou domain, to dig scores of tiny shallow pools, or potholes. He opens his faucet to let water run through the lacuna. A tiny rivulet forms by allowing the small pools to connect only to find its way to the nearby bayou. Is the initial purposelessness of the action (for art's sake) eventually made purposeful by surreptitiously (the Uncanny) transforming into nature – whether a boon for the horticulturist or an environmental disgrace? From nature to culture, can the process be reversed to make of the culturally 'created' water a natural creek, or as in our case, a Houston *bayou*[2]? Can we by liquefying an act of art return it to nature? Didactically, tiny (artful) bayous bring metropolitans back to the beginning. Children will lead, and parents will follow reluctantly, while an ancient bayou is rediscovered. The *process* of creating a bayou brings it out of the background – out of its taken-for-granted objectivity. A dozen bayous and their tributaries become the literal *sea change* itself. Swept away, the metropolitans will float along.

Geographically, Eliasson is lucky. Born in Denmark but making frequent visits to Iceland, before

---

1 The **fieldroom** refers to the space formed between the endless canopy of trees and the flat ground that characterizes Houston and many other suburban cities such as Sydney, Australia.
2 The Houston system of **bayous** is the delta of run-off creeks that help drain the flat plane of the city.

moving to Berlin, he has lived at the edge of culture and nature, which poses for anyone observant an immediate question: what happens when those edges bleed together? Iceland, lying along an imaginary line between habitat and wilderness, provides no answer. It can only offer opportunities to see in the flesh what is less than easily seen in the middle of the European continent. This edge has unsettled many an artist and driven some crazy. Whether lost in the forest or tossed, seasick and bewildered, by a sudden fall storm, you find the culture residing in compass or tiny sailboat fading fast. Yet the wanderer sees also in the black lava and in the hot geysers the lineaments of human intention – because his eyes have chosen these phenomena over a myriad of others compelled by terror, on the one hand; on the other, by the idea of the aching body seeking respite in nature's warmth. Once you are immersed, all questions fade; so may the riders' on the bayous.

Eliasson's travels within the zone may have had these same disconcerting and comforting dimensions, and judging from his work, he has learned much. Along this zone, hunters and gatherers have eked out a steady living for centuries. The modern artist, confronting the spare winter light of this dazzling landscape, may be no different. The shocking difference lies between an *opening* at a museum far from the zone and the life of the zone itself. After the experience is brought to safety – maybe at the point when something slips back into the dark sea – the gatherer may want to follow whatever has slipped away to see what happens when culture (i.e., selected and found nature) returns from whence it came. It may well be that in its return alloys appear, as a form of bionic ectoplasm emerging in the future once we have left the mechanical behind.

My hunch is that it is in the actual *reversal* of the common flow from nature to culture that one of the keys to a *sea change* in public opinion lies. Eliasson hints at many such provocative events: in his 'pollution' machines, such as the one spilling a green tint into *Strömmen*, the central waterway in downtown Stockholm. Concerned citizens were calmed when the Swedish food and health administration declared the tint safe – closer to nature than to culture.

Coupled thus with the tiny bayou project, a shocking event may follow: an immense pollution project *à la Eliasson* in which each exposed bayou starts foaming in blue (the true color of 'real' water). The wakeup call and a call to action. The step between the rhetorical aspect of the grand project and its pragmatics is immense because shadowed by obscurity. Can the deliberately constructed concrete flues now passing for bayous be removed and replaced by 'slurred' versions that allow both runoff and a slowing down leading to a more effective retention of huge amounts of water? A site of *biomimicry* in which the enhanced bayou begins to tell the complex story of its history and resurrection to become the testing ground for a highly evolved artifice driven by natural principles rather than crude engineering motives. And simultaneously can the bayou domain become new linear parks, thus joining human action with Nature's? These metropolitan ambitions must remain here as thoughts and aspirations.

Cool Room In 1998, Eliasson did his most didactic pollution-machine to date. Simultaneously, a same-size block of ice is left to melt in the Musée d'Art Moderne de la Ville de Paris and outside on a sidewalk in a distant Parisian neighborhood. The first block melted in three days, while the block left to the people on the street took a month to melt. The simultaneity and the individual peculiarities of the 'same' ice blocks within their respective environments (museum and sidewalk) point to a larger cultural landscape.

In the same year in São Paulo, Eliasson installed what he called 'the very large ice floor'. Although it may still be remembered, we can assume that the floor since then has both melted and disappeared. Twenty-two meters long, the ice floor in São Paulo must have seemed as idiosyncratic and strange as the disappearance of the ice cap at the North Pole some two years later. Ice in these two cases brings us back to the *fieldroom* and its binary condition of hot and cold.

Before air-conditioning in Houston, cooling was done with ice. The relationship between ice, refrigerators, and cooled air in buildings and cars is intimate; the exposure of the two may begin the rethinking of *zoned cooling* that would turn the current binary condition into a smooth curve from cold and dry to hot and humid. These in-between technologies are slurred with natural devices: an 'improved tree canopy'; the contemporary return to giant screened public porches; semi-cooled space with fans as *sensational* features of the *fieldroom*; ice houses interspersed with mini-parks with enhanced natural cooling. This subtle shift from either/or to a range of climatic conditions enhanced by technology and nature begins to shift the project of cooling from *objects of cooling* to the *process of cooling*.[3]

Further west and north of Eliasson's Iceland, native people employ at least ten words for snow. In Swedish there are fewer words, but more than in Portuguese. C.S. Orwin, in *The Open Fields* of 1954, enumerates the words describing the complex striations of the ancient English agricultural landscape. *Assart, balk, bovate, butt, carucate, croft, fallow, flat, furlong, gore, headland, ing, merestone, oxgang, rigg, seilon, shot, stich, toft* and *wong* – now long forgotten in the aftermath of the agricultural grain combines where operators are carefully suspended above the field in air-conditioned cabs with the Muzak going. In zones where nature looms large (and close), human

---

3 A case in point in the shifting of Eliasson's works from 'strange machines' to processes is as subtle as it is telling. The ice on that floor in São Paulo was made just like the ice cubes in your refrigerator, plain and simple, so the question is: what does the floor mean? Was it a lesson in envy from the North ('we have ice, and you don't, at least not this large') or was it an act of generosity ('I know, you have ice in your drink, but here is twenty-two meters of the cold stuff, just for you')? Or is the idea much wider? Was the meaning the melting, rather than the ice? The process rather than the object, the micro-climatic change taking place simultaneously with the melting – only understood by staying very close.

senses seem to focus on its own habits, to devise language and ways to cope. Eliasson clearly uses an arsenal (menu) of ice, water, light, and steam. Critics call him a phenomenologist whose chosen task is to produce effects that, out of their normal context, startle and move you. This may be so, but for me Eliasson's work is more problematic, indeed more daring. It seems to be in search of something more consequential than the mere buzz of phenomena, by repeatedly pointing at natural cycles, at origins, at entropy, at bio-logics, and at effects. If we link the effects of all his machines into an array or geography, assembling them on an imaginary field, a map appears.[4] The geography of such a map suggests a much wider territory of possibilities than any one project. This is the nature of maps of fields, intimate and familiar.

If a map of the climatic conditions of the *fieldroom* is produced, it too may open possibilities. For the metropolitan, the field room is binary: hot and cold – at least in everyday conversation. Now add the real climate, the juxtaposition of the fronts from the cold Canadian air and the equally persistent hot and humid fronts from the Bay of Mexico that battle it out high above the canopy often resulting in torrential rains causing enormous property damage below. This new map is one of contradiction and intense tension between the natural and the artificial, especially if it was possible to map the clumsy attempts by the artificial to adjust to the natural. Temperatures rise and fall precipitously, sometimes thirty degrees in one hour, while the indoor thermostats attempt desperately to stay at 72 degrees. Hopelessly out of step, the climatic conditions in the *fieldroom* remain schizophrenic. The map tells us how difficult it is to calibrate human occupation and actual climatic conditions.

Alfred Wegener, the German inventor of the theory of continental drift, hovered intently for days on end over globes and geological maps.[5] The drifting apart of Wegener's continents is similar to Eliasson's escaping water project; it would take a concerted understanding of the movement of continents and water to reestablish the origins. The earth's plate tectonics is of course more spectacular than a minor inundation in a South African museum or a major one in our *fieldroom*, but in both cases phenomena are running away from their origins. Similarly, in viewing Eliasson's larger project as a map that may not know its origin or its future, Wegener's audacity would come in handy. It seems clear that only audacity will make us stop and take another path. The running away from an original condition (which in the case of continents and oceans is only relative since their history is long and obscure) is antithetical to most objects. Although other works by Eliasson resemble common art objects in that they are built as original pieces that sit isolated in museums, most have a *renegade dimension*, as if the objects are in some form of denial or in the process of self-annihilation. Indeed, Eliasson's machines seem to be processes first and objects second.

Objects are usually opaque and point inward toward their own authenticity, while processes point away from themselves toward a largely unpredictable future. *Processes are radical, while objects are conservative.* Shifting the attention away from peculiar incidents in the *fieldroom* to the process that underlies its climate, health and future is a radical intention.

This process is also radical in the sense that it leads away from the individual to seeing all of us *in* Nature. The first time I sensed that art may have devised an escape route from the hegemony of complete control, from authorship, was in my encounter with Joseph Beuys' *Olive Stones* in a museum in Turin, Italy. The large blocks of stone sat in an old castle. In the next room was a gold vessel by James Lee Byars. Both works overlooked a park. (At least that is how I remember it.) At first, I thought the floor around Beuys' stones is simply stained, but then I realized that the stones were emitting oil that slowly trickled down. (It reminded me of a childhood story about a boy who stood up to a giant by squeezing the rennet out of a cheese that he said was a stone.) I never understood the functional aspects of these stones, but I could imagine the castle in some distant future inundated with olive oil. With these asides, the climatic map of the field room tells us that we must go with the flow, abandon notions of the static, the binary, the fixed, in favor of the dynamic and in terms of temperature, the relative. Either/or exchanged for both/and. Radical differences are made smooth by an assortment of in-betweens: cool rooms, porches, shades, increased wind speed, fans and huge propellers, that like trees are driven by sunlight.

The concern for all collective enterprises, ranging from 'the life of the street' to 'the ecology of the entire public domain' of a city (as in the case of the *fieldroom* in Houston) must increasingly stem from the public – the 'City' being increasingly left to entrepreneurs. However, suburban dispersion and the resulting isolation and enclavism do not bode well for such collective awakening. The employment of artful spectacles may change this complacency. The progressive decline of the value and use of public space is probably best reversed by recalling the ancient rites of the City, such as the Papal parade, *Faschnacht* in Basle, *carnevale* in Rio, or modern versions

4 Such a map, or field, of eighteen projects would look like this: While the fog in [1] 'your natural denudation inverted' of 1999/2000 clouds my view, I fumble through the thicket of [2] 'Untitled [branches and saplings]' of 1998. At this point, I have reached the mouth, the 'door' in the billowing blue plastic wall of [3] 'Your inverted ego' of 1998. The light from [4] 'The inventive velocity versus your inverted ego' of 1998 blinds me. Once it is turned off, the yellow light and the smoke of [5] the Hanukkah project of 1998/99 direct my eyes to the sky. Briefly two additional works – [6] 'Untitled [Infinity]' and [7] 'Untitled [Expectations]' – draw lines of light across my face. A rainbow arches above in [8] 'Beauty' of 1994, while a dripping machine of 1994 wets my feet. Suddenly I stumble on a Buckminster Fuller igloo called [9] '8900054,' while [10] 'By means of a sudden intuitive realize' of 1996 appears now to form a habitat. Inside [11] 'Die organische und die kristalline Beschreibung' shows with its light projector, wave machine, convex mirror, and foiled windows how the fixed can be made to move. A sudden chill sends shudders through me as I encounter [12] 'Ice Pavilion' of 1998. As I put my face against [13] a moss-wall of 1994, my feet give way under me and I slide across [14] 'The very large ice floor' of 1998. All the while [15] the faucet in Johannesburg splashes, [16] 'green' with the Stockholm tint, the two ice blocks in [7] 'Nuit blanche' of 1998 drip away, and [18] 'The double sunset' of 1999 stutters on the horizon.
5 Alfred Wegener, **Origins of Continents and Oceans** (London: Dover, 1966).

such as the work of Uwe Köhler's *Titanick* in Münster and Leipzig. But it is also clear that semi-public institutions such as the Art Museum ought to bring its wealth to the street and, as has been suggested, to the *fieldroom* itself. The 'liberation' of art from politics has propelled itself deep into the *effete* backrooms of the museum; by new associations between 'public interest' and art, our common world can again come into much needed focus.

The Zoohemic Canopy

The Fieldroom

Green Walls

Olafur Eliasson, *Untitled*, 1994

Water Falls

Olafur Eliasson, *your strange certainty still kept*, 1996

Olafur Eliasson, *Ice Pavilion*, 1998

Olafur Eliasson, *your natural denudation inverted*, 1999/2000

The City of Tolerant Normality

Vittorio Magnago Lampugnani

The telematic revolution was the last of those that shook the 20th century, but by no means the only one. At the turn of the 19th to the 20th centuries, new public transport systems such as urban railways and subways radically reduced the distance in time, as well as in expense, between the outskirts of cities and their centres, promoting the development of open-plan suburbs. In addition, from the 1920's onwards, millions of automobiles arrived that made people's daily migration from their place of residence to their workplace easier – while at the same time also blocking it. Urban society developed from a very small aristocratic or upper middle-class group that held the city's destiny firmly in its hands to a widely mixed and largely anonymous mass that treats its architectural environment comparatively casually, and certainly also in contradictory ways.

The cultural assumptions that today's urban society brings to this way of dealing with architecture are no less ambivalent. The clear functional and iconographic expectations that were associated with city architecture up to the beginning of the 19th century have been replaced by an almost incomprehensible variety of ideas that cannot be reduced to a single common denominator, and are also applied in quite diffuse ways. The city is perceived in ways that are various, sometimes contradictory, and – as Walter Benjamin already noted – inattentive.

Within a comparatively short time, city life has also changed in other ways as well. New types of family and family-like communities are using houses and apartments in new ways. Working processes operate in different ways, training and education are not conducted in the way they used to be, shopping habits have changed, and leisure-time rituals have practically been revolutionised. In the face of conditions that have been transformed to this extent, are we still talking about the same type of city that we had a hundred or even a thousand years ago? Does the city not have to adapt to the new conditions as well, and will it therefore have to change completely?

Ever since the polemic conducted by the avant-garde Modernism of the 20th century against the old, traditional city, architectural culture has constantly been inventing utopias of genuinely 'new' urban structures that have been as audacious as they were unsuccessful – from Le Corbusier's Cartesian *Ville Contemporaine* of 1922 to the high-tech plug-in city proposed by the British architectural group Archigram in 1964. All of these utopias failed. They failed because although people's lifestyles have changed, their basic needs and habits have not. The revolutions and revaluations that appeared to be suggested by a regimen of technical devices did not happen. Fundamental changes in city life also took place to a much more modest extent than was suggested by the dramatic urgency with which they were announced. Above all, a comprehensive replacement of the old by the new did not take place: the widest variety of different types of social life now exist with virtually equal status alongside each other in our homes and cities – new, conventional, subversive and orthodox.

Clearly, the new lifestyles not only require a different way of using the old city, but also are stimulating the creation of new parts of the city, with different streets, different squares, different buildings, and different city parks than those we are familiar with from the historical city. Yet alongside these new urban elements, modern human beings also need traditional city spaces, and they need them in order to be able to do what they have been doing for centuries and millennia: walking, lingering, looking, showing-off, meeting other people, talking to them, exchanging views with them. These activities are as old as the streets and squares in which they are practised.

In addition, another element confirms and reinforces the insistence on historical urban spaces: the European cities themselves. They are present and existent here and now, and there is no need to tear them down and to rebuild them, nor is it even feasible. At best, they can be prudently and carefully renewed, modified, and expanded. The demand for useable space will not increase in the foreseeable future, and almost all the buildings that we will need in the coming decades are already in existence. Many other elements are superfluous, and we cannot really afford them, in view of the running and maintenance costs involved. It is not new building, but conversion and alteration that needs to be done. Today's city planners do not need to design a completely new city in the way that the Marquês de Pombal did in Lisbon following the earthquake 250 years ago. Instead, their task – only superficially more modest, but in fact tremendously complex, difficult, and demanding – is to maintain the existing city in a creative fashion.

Against Innovation Mania In fact, one of the most disastrous misunderstandings from which contemporary city planning suffers is that there is a need for inventiveness at any price. Every architect who is about to plan and build part of a city thinks he has a certified right to new creativity and that he is therefore able and even obliged to depart from everything that preceded him and everything around him. Even in historical settings, the greatest value is placed on making new intrusions recognizable as such – they absolutely have to be different from their existing surroundings. This attitude is a legacy of the period of the avant-gardes that promoted a radical renewal of art at the beginning of the last century, and who in their own ranks carried this out. Yet this tendency can be traced back even further, to the beginnings of the modern period that dawned in the mid-18th century when all the great certainties collapsed and the Romantic cult of genius began. However, the consciously cultivated obsession with innovation was not uncontroversial even then; and indeed, still does not correspond to the requirements of our own time. Today, we depend on managing our resources intelligently and economically – and not only our material resources, but also our cultural ones. We are therefore no longer able to place all our hopes on constant self-renewal – which in any case on closer inspection usually proves to be more or less virtuoso plagiarism from one -*ism* or another from the past. Instead, we need to develop and to intensify everything we have recognised as being appropriate and good and worthy of improvement. It is not happy and reckless experimenting away that we need today, nor an agitated leafing through the history books (usually these are the same thing, except that in the first case the sources are not stated), but a gentle and respectful approach to dealing with subjects and techniques.

Particularly in the art of building cities, despite all its power of visual persuasion, the primary function of this very special form of art is to serve, and it has a magnificent tradition to fall back on. If it does not do so, it runs the risk of failing in its most essential task. Because during its long development, it has produced solutions that have proven their value and usefulness outstandingly over time, and it would be rash to abandon these without good reason. A city costs a great deal, measured in terms of both investment and work; during its planning and construction, one can hardly afford to make any mistakes. And the city needs to offer large numbers of people a home for a long period of time – which equally implies that if it has any shortcomings, entire generations will have to suffer from them.

To the unprejudiced eye, it is therefore clear that most of us at the beginning of the new millennium would not be unwilling to live once again where our grandparents and great-grandparents lived. They were content with their streets and squares, with the façades lining the streets and squares, with the avenues and river embankments, with the public buildings, markets, and coffee-houses. Admittedly, their cities were not as museum-like as the historic centres are in some cities today – dressed up, laid out with pretentious cafés and high-class boutiques, equipped with stylishly designed furniture and transformed into a caricature of themselves. On the other hand, cities then were also not bursting with unnecessary inventions, and they were strongly connected with the culture of the people who lived in them and to whose life they added extra brilliance.

An unprejudiced and slightly disenchanted eye would also note that much of what has been carried out in city planning in recent decades has been at best inappropriate, and at worst evinces what Sherlock Holmes might have called 'several interesting traits'. Too much effort has been put into getting away from a past from which one wanted to distance oneself – but without having anything better and perhaps not even anything equivalent with which to replace it. Too much effort has been put into mounting dramatic scenes of novelty, originality, and sensationalism, even in places where there was no need for spectacles and dramatisations of this type.

In an article about contemporary Milan, Gadda made fun of the forced oddities of supposedly modern city architects as long ago as 1936. The city he conjures up in his satire is created at the wish of Queen RELUCTANCE and King BAD TASTE; its buildings are produced by architects whose mothers were shown camels, kangaroos and giraffes when they were pregnant, and the rules that governed the building work were:

First: any trees more than five years old will be chopped to pieces; second: none of the squares must be either square or round, they will all be irregular; third: all angles of sixty degrees, ninety degrees, or halves thereof will be banned from the plans, along with any sequential rows or straight lines; fourth: no building will be the same height as the neighbouring one, especially not on the new squares and new streets; fifth: windowless, whitewashed side walls will be paraded everywhere, so as to give the urbanised town its 'architectural physiognomy'; sixth: the roofs should be made of any material at all, at random – with pots and chimney-tops, pieces of cake, fishing rods, rusty lightning-conductors, all laid out from a scientific point of view. [Against this, the writer set as a corrective measure] the truths of nature, the simple and unchanging needs of human existence: in spite of everything, these are what give the city of RELUCTANCE the magical sound of life, these are what keep the wheel of its atmosphere and sense of life turning...[1]

The illustration accompanying the article had been published more than ten years previously by Werner Hegemann and Elbert Peets in 1922 in their *American Vitruvius,* better known by its subtitle *An Architect's Handbook of Civic Art,* and was based on a cartoon published in the journal *The Architectural Review* in 1904.[2] The picture, laconically captioned 'Chaos', shows a random sequence of volumetrically and stylistically extremely different buildings. The authors use this as a warning against the contemporary decline of city planning; as the future for the 'civic art'; they invoke the principles of 'grouping buildings into harmonious ensembles' and 'a monumental unity'.
Only two years later, in 1924, the architect and architectural theoretician A. Trystan Edwards – rather cranky, but deeply-blessed with British common sense – declared in his popular book *Good Manners and Bad Manners in Architecture*[3] that good manners were a precondition for urbanity, and he ascribed these manners to buildings. Buildings that do not belong to an 'arbitrary assemblage ... each conceived in isolation and expressing nothing but their own immediate purposes' behave towards their neighbours courteously and obligingly. They present themselves as sociable, open-minded and unselfish. They are both urban and urbane. To achieve this, they have to concern themselves with their own tradition and above all with their environment and setting, of which they regard themselves as being a respectful continuation. Every rough breach is a social affront and a failure of city planning. For this thoughtful writer, city planning does of course require innovation – just as every craft and every other form of art does. However, innovation has to be used in a targeted fashion where it is needed, and it needs to build on the past.

Simplicity as an Expression of Social Pluralism A city is not constituted by the large, impressive monuments that are seen on advertising brochures and postcards. It is determined to a far greater extent by all its individual districts, with their extending continuum of homes that are anything but remarkable – houses that play out every possible facet of that special normality on which a city's viability and quality of life are based to a far greater extent than on the highlights of its tourist itinerary. It is the normality of what is simple, but not banal; of what is ordinary, but not routine; of what is original, but not novel.

The new urban society in particular appears to be opposed to this type of normality. Not least in the aspect of being a media society; today's urban society has become accustomed to having things constantly happening, and it incessantly demands that the city, too, should provide a succession of events. In the process, it overlooks the fact that life is usually most enjoyable when nothing is happening – and, if there is any time left, in going for a walk, thinking, reading, loving. In much the same way, a city is beautiful (and worth living in) when nothing happens in

1 Gadda, Carlo Emilio; '**Pianta di Milano – decoro dei palazzi**', in: **Le meraviglie d'Italia**, (Florence: 1939), cited in: **Opere di Carlo Emilio Gadda**, published by Dante Isella, Vol. III: 'Saggi Giornali Favole I' (Milan: 1991) pp. 57–59.
'E per meglio radicare la loro podestà, già grandissima, deliberarono altresì di mutuo consenso, tanto il Cattivo Gusto re che l'Uggia sua consorte e regina: primo: tutti gli alberi maggiori di cinque anni venissero adibiti a far legna; secondo: non una piazza fosse quadrata o rotonda, ma tutte bislacche; terzo: l'angolo di sessanta gradi, quello di novanta e i loro mezzi fossero banditi dai piani, e così pure ogni allineata o rettilineo; quarto: non una casa fosse pari in altezza alla casa contigua, specie nei nuovi fori e vie nuove; quinto: i muri scialbati e senza finestra delle fiancate si levassero ovunque, conferenda alla città urbanizzata la sua "fisionomia architettonica"; sesto: i tetti fossero combinati alla meglio, e con ogni aggeggio: pentoloni, caminacci, fette di panettone, canne da pesca, parafulmini arrugginiti, disposti scientemente in visuale ed in fuga.'
'Chi è il nemico? Quali le sue milizie? Direi che sono la verità di natura, le semplici e continuate necessità degli umani: esse tuttavia conferiscono alla città dell'Uggia il tono malioso della vita, vi promuovono il circolo degli umori di vita: questo fiotto di buon sangue e di buon senso che pur seguita a correre le grigie contrade, non ostante ogni intoppo frappostovi dai razionali edili.'
2 Hegemann, Werner and Peets, Elbert; **The American Vitruvius: An Architect's Handbook of Civic Art** (New York: The Architectural Book Publishing Company, 1922) p. 1. 'Only under rare circumstances will a fine piece of work be seen to advantage if thrown into a chaos, and dignity, charm and unassuming manner are preposterous when the neighbors are wantonly different or even obnoxious. The hope that good work will show off the better for being different from its surroundings, which are to act as a foil, is an illusion. The noise produced at county fairs by many orchestras simultaneously playing different tunes is a true symbol for the architectural appearance of the typical modern city street. The fact that one of the orchestras may play Beethoven will not resolve the chaos.' This condition is detrimental to the advancement of the arts and it must be changed. One of the foremost aims of this book on civic art is to bring out the necessity of extending the architect's sphere of influence, to emphasize the essential relation between a building and its setting, the necessity of protecting the aspect of the approaches, the desirability of grouping buildings into harmonious ensembles, of securing dominance of some buildings over others, so that by the willing submission of the less to the greater there may be created a larger, more monumental unity; a unity comprising at least a group of buildings with their surroundings, if possible entire districts and finally even, it may be hoped, entire cities. Against chaos and anarchy in architecture, emphasis must be placed upon the ideal of civic art and the civilized city.
3 Edwards, Trystan A.; **Good and Bad Manners in Architecture** (London: Philip Allan & Co., 1924) p. 1. 'Can a haphazard assemblage of buildings, each conceived in isolation, and expressing nothing but its own immediate purpose, really be described as a city? What attribute is it which makes a building urban? My answer to this latter question may seem simple and tautological, but I am venturing to give it nevertheless. In order that a building may become urban it must have urbanity. I propose to analyze the precise nature of this urbanity. Now, urbanity, as everybody knows, is nothing more ore less than good manners, and the lack of it is bad manners. I think I shall have little difficulty in showing that there can be both good and bad manners in architecture....'

its architecture. And when life can unfold against its quiet, discreet background all the more unrestrictedly.

There is another, more important argument claiming to demonstrate that normal city architecture is apparently unsuitable for modern urban society. Modern society's pluralism – to mention one of the favorite slogans of contemporary city planning – is thought to demand variety. In a democratic community, each citizen, each city-dweller has an equal right to self-realisation and self-presentation. And since each citizen and city-dweller is different from his or her neighbour, the city (it is claimed) has to reflect the variety of lifestyles and cultures that it has absorbed by showing off an exuberant range of different architectural forms. The result of this attitude is a confusion of forms and associations that have lost all comprehensibility and all capacity for dialogue. In exactly the same way in which different people from different cultures can only really live together when they enter into a productive dialogue with each other, the modern city can only be a city of tolerant coexistence when it is able to offer such an existence usable locations and to give it architectural expression. Locations and expressions of this sort can never represent a mere arithmetical addition of the differences, however; nor can they consist merely of the lowest common denominator, which would function at an abstract level at best, but not at an aesthetic one. Instead, they need to symbolise the space and opportunity that are granted to individuals because of their differences, enabling them to develop themselves individually and to combine to act collectively.

In addition, tolerance is neither infinite, nor is it the first law of nature as Voltaire celebrated it. On the contrary, it is a purely cultural achievement, the limits of which need to be set by every society very precisely in order to regulate its social relations and prevent unfair advantages and even aberrations from arising. It is sometimes only a very narrow line that lies between tolerantly accepting opinions and attitudes that are contrary to one's own and standing by without helping when crimes are committed. It is not always easy to discern where this line lies, particularly since it is not a constant one, either temporally or culturally. Nevertheless, it is indispensable that it should be constantly called into question and redefined. By analogy, not every provocation against the city can be or should be permitted. And by analogy, one must constantly ask what it is that helps constitute a lively urban normality and what it is that irretrievably damages the peaceful coexistence between more or less restrained buildings, or as Edwards would say 'obliging buildings'.

Recently, the new information technologies have been opening up fresh space for individual development to an unexpected extent – from the mobile phone to the Internet chat room. By contrast, the collective, social dimension continues to be the responsibility of the city – perhaps even more so than before. The city must create the opportunities and incentives so that the nomads of the telematic era – who are already networked together in such a wide variety of ways – can raise the entertaining adventure of uncomplicated human contact to a different and more significant level, to test and to live out various forms of community. And the city needs to be able to provide a visual justification for such solidarity.

The basic motif of this justification is precisely normality, simplicity, restraint, uniformity, anonymity. Monotony and repetition of the rhythmically recurring, plain and memorable architectural elements are what characterise the large urban ensembles in the ancient Western world – both the Hellenistic *agorai* and the Roman imperial forums, as well as medieval squares (the Campo in Siena has mainly been praised for the uniformity of the buildings surrounding it), Renaissance buildings, and the dramatic urban scenes mounted in the Baroque period – from St. Peter's Square in Rome, with Gianlorenzo Bernini's gigantic colonnades, to André Le Nôtre's implacably geometric Versailles. With the breakthrough of Classicism in the mid-18th century, the principle of uniformity was largely equated with that of modernity: by Marc-Antoine Laugier[4], for example, who in his scandal-creating bestseller *Essai sur l'architecture* in 1755 and also in his *Observations sur l'architecture,* published a few years later, demanded *régularité* for the contemporary city, but at the same time also *bizarrerie*: 'extreme order in the details; confusion, noise, tumult in the whole'. In his *Cours d'architecture civile* of 1771, Jacques-François Blondel[5], the leading French architectural theorist of his time and the teacher of the so-called 'Revolution architects', also maintained the symbiotic dialectic of *unité* and *variété*. Although in an explicit polemic against the (Baroque) 'artists of the new Italy', he praised *simplicité* as constituting architecture's fundamental dimension. His pupil Jean-Nicolas-Louis Durand[6] broke more radically with the Vitruvian view of architecture as 'decorated science'. His planning system, on the basis of which he did not hesitate to redraw the Basilica of St. Peter in Rome and even improve on it, is based on simple basic geometric forms stripped of all decoration. Economy, far from appearing as an obstacle, proves to be the 'most fruitful source' for the beauty of this system. Carlo Lodoli[7] presented similar arguments, putting the emphasis on the function of buildings and preferring the reliable, essential quality of *solidità* to the incalculable, arbitrary quality of *ornamento*. The 19th century broadly agreed with these positions, which it refined and developed further. The motto of the epoch became uniformity, but an intentional application of

---

4 Laugier, Marc-Antoine; **Essai sur l'Architecture**, 2nd edition (Paris: Duchesne, 1755) p. 224. 'Ce n'est donc pas une petite affaire que de dessiner les plans d'une ville, de maniere que la magnificence du total se subdivise en une infinité de beautés de détail toutes différentes, qu'on n'y rencontre presque jamais les mêmes objets (…), qu'il y ait de l'ordre, & pourtant une sorte de confusion, que tout y soit en alignement, mais sans monotonie, & que d'une multitude de parties régulieres, il en résulte en total une certaine idée d'irrégularité & de chaos qui sied si bien aux grandes villes.'
And from Laugier, Marc-Antoine; **Observations sur l'Architecture** (1765). 'Quiconque sçait bien dessiner un parc, tracera sans peine le plan en conformité duquel une Ville doit être bâtie (…). Il faut des places, des carrefours, des rues. Il faut de la régularité & de la bizarrerie, des rapports & des oppositions, des acciden[t?]s qui varient le tableau, un grand ordre dans les détails, de la confusion, du fracas, du tumulte dans l'ensemble.'
5 Blondel, Jacques-François; **Cours d'Architecture Civile** (Paris: 1771), vol. 1, p. 396 ff. Blondel excludes monotony expressly: 'La symétrie, telle que nous l'entendons, n'est point monotone…' (p. 408).
6 Durand, Jean-Nicolas-Louis; **Précis des Leçons d'Architecture données à l'Ecole Polytechnique** (Paris: 1801) ff. quotation from: première partie, p. 21.
7 Memmo, Andrea; **Elementi d'Architettura Lodoliana, ossia l'Arte del Fabbricare con solidità scientifica e con eleganza non capricciosa** (Zara, 1834).

monotony was left to the city architecture of the 20th century – or, to be more precise, of Modernism.

Even before 1900, Hendrik Petrus Berlage[8], the key figure in modern architecture in the Netherlands, had announced that the city architecture of the new epoch would show an 'impressionistic' restraint, as this alone would correspond to the inattentive and inexpert perceptions of the new democratic masses: 'Away with all the time-consuming details, which can never be executed in the desired way! Away with everything that disturbs the grand impression of the whole! Just look for a few characteristic large surfaces and boundary lines! Today's architect must be an Impressionist'. In Berlage's peculiar, but logically consistent interpretation, Impressionism in the field of architecture implies an ever-faster smudging of impressions until all that is left is an unmoving, unbroken surface. A few years later, Karl Scheffler, an art and architectural critic, demanded typologically and also aesthetically identical city housing, so that the urban nomads who moved from metropolis to metropolis or from district to district would be able to find their way around straight away:

> 'Today's city dweller changes his apartment on average in periods of every three to five years. He expects to find a similar ground plan, or even the same one, everywhere, so that his living habits will not be disturbed. This demand is a natural consequence of modern living conditions.' From this point, it was only one small logical step further to the standardisation of exteriors. 'The uniform ground-plan results in a similarly arranged façade[9] ... The idea of interrupting this similarity arbitrarily, merely in order to appear independent, just to be different, is uncultured.'[10]

Scheffler was not arguing in favour of a banal standard, but in favour of cultivated commonness. The apartment ought to correspond as precisely as possible to the requirements of modern life, so that it would turn into a type. On the basis of this type, the new human being would be able, with a few deft movements, to make the universal apartment into his own apartment, into a home, his own little territory, without forfeiting any of his mobility and flexibility. The representatives of the modern movement in architecture were later to use similar arguments – from Heinrich Tessenow to Ernst May, from Le Corbusier to Ludwig Mies van der Rohe. Playing on the idea of the modern nomad and his yearning for anonymity, Hannes Meyer went so far as to place fictional advertisements captioned 'Furnished cities for rent' in the avant-garde journal *ABC*. Furthermore, even after these principles had already led to the perversion of the ruthlessly minimised, standardised, typologised, and industrialised dormitory estates of the 1960's, entirely lacking in anything either urban or urbane, J.D. Salinger in his short story 'Raise High the Roof Beam, Carpenters'[11] was still able to attempt a literary rescue of the reputation of the principle of uniformity and anonymity in the city. As a small boy, Zooey wishes that all the buildings and apartments in his city could be exactly the same, so that he would be able to go into a strange apartment by mistake and hug and caress the total strangers in it in exactly the same way as his own family.

Against this (but only apparently, as we will see below) stands the modern city conjured up by Robert Musil in *The Man without Qualities*:

> Like all big cities, it was made up of all sorts of irregularities, varieties, things running ahead, things not keeping in step, collisions between matters and affairs, islands of silence hovering in between, of pathways and pathlessness, of an immense rhythmic beat and the eternal friction and shifting of rhythms against each other, and the whole was like a boiling bubble suspended in a vessel made up of the enduring substance of buildings, laws, regulations and historical traditions.[12]

The modern city, at least in Europe, is the historical city – as such, it consists of layers and extensions of built-up areas that have followed each other in a sequence lasting for decades and centuries. The city therefore has a variety of architectural characteristics; but these characteristics are not only consistent with each other, but usually also anonymous – generated by reiterating what has been thought good and recommendable in various epochs. *Collage City*[13], the widespread ideal of the 1970's and 1980's that was the title of an influential book by Colin Rowe and Fred Koetter in 1978, was never the goal of the traditional city, but at best the result of rough processes of growth. If urban collages arose, they were tolerated for reasons of pragmatism and did not represent an artistic strategy.

But this is precisely what they are intended to be today. Both the ability to create uniformity and harmony and the will to do so appear to have been lost in the contemporary city. Everyone who builds anything wants to be noticed, either as client or architect. And everyone who is noticed is immediately rewarded in the media, which prefer to show, report and comment (usually favourably) on anything outside of the norm. In this way, the concrete, real, actually built city itself thus becomes (assuming it has not long since been consigned to speculators and profit-seekers, with their increasingly vulgar standard architecture) an agitated conglomeration of arrogant individual gestures, a fun-park of borrowed or even imposed emotions.

8 Berlage, Hendrik Petrus; 'Bouwkunst en impressionisme', in: **Architectura** (1894) p. 106. 'Neen, de architect die den tegenwoordigen Tijd begrijpt, en iets artistieks wil maken, gooit van zelf alle ballast overboord; weg! met al die tijdroovende details, die toch niet naar wensch kunnen worden uitgevoerd, weg! met al datgene wat tot den grooten indruk van het geheel niets afdoet; alléén gezocht naar enkele karakteristieke groote vlakken, begrenzende lijnen! de architect van heden worde impressionist!'
9 Scheffler, Karl; 'Die Psychologie des Grundrisses', in: **Deutsche Bauhütte** (1904) pp. 69/70 and 78/79, here p. 78.
10 Scheffler, Karl; 'Das Originelle in der Baukunst', in: **Deutsche Bauhütte**, issue no. 28 (1903) p. 177.
11 Salinger, Jerome David; **Raise High the Roof Beam, Carpenters, and Seymour: an Introduction** (1955 and 1959) Toronto et al. 1963, p. 68. 'Zooey was in dreamy top form. The announcer had them off on the subject of housing developments, and the little Burke girl said she hated houses that all look alike – meaning a low row of identical "development" houses. Zooey said they were "nice". He said it would be very nice to come home and be in the wrong house. To eat dinner with the wrong people by mistake, sleep in the wrong bed by mistake, and kiss everybody goodbye in the morning thinking they were your own family.'
12 Musil, Robert; **Der Mann ohne Eigenschaften** (Berlin 1930) pp. 10–11.
13 Rowe, Colin and Koetter, Fred; **Collage City** (Cambridge: MIT Press, 1978).

Even Non-Places should be Designed – Discreetly

Neutrality, viewed in a tasteful way, has nothing to do with unimportance or detachment; in fact, it is the opposite. The residual urban spaces, the waste land wedged between different districts, the embankments along urban railways, the supports for ramps on expressway viaducts, are in no way neutral; they are merely undesigned. And the buildings and areas emerging from the process of accelerated movement within the city and out of it, for which the French anthropologist Marc Augé[14] coined the term *non-lieux,* non-places, in his 1992 book of the same title, are also anything but unobtrusive – on the contrary, they are usually mounted as dramatic scenes in a quite overbearing way.

The mere fact that these dramatisations are interchangeable, makes this type of non-place into the architectural emblem of globalisation. There is no necessity for this, and it was not always so. The way in which one can deal with 'benign' residual urban spaces – for example, the space left between the carriageways on highways – was demonstrated by the landscape architect Adolphe Alphand in the thoroughly urban gardens he designed for the Paris of Louis-Napoléon and Georges-Eugène Haussmann, which was also simply his own Paris. His aesthetically and technically brilliant solutions are collected almost in catalogue form in his book *Promenades de Paris.*[15] In 1867, he produced a masterpiece in the form of the *Parc des Buttes-Chaumont,* where a quarry, a refuse disposal site, and a cemetery were converted into a park landscape with a high recreational value that did not conceal its artificiality. 'Malignant' residual spaces, by contrast, are probably hopeless cases, and should not be allowed to arise in a city. In places where it is torn up by viaducts and underpasses, the urban structure is practically beyond saving, and anything provided to correct it is merely a hopeless cosmetic exercise.

The same applies to the non-places. These also have 'malignant' forms: the large shopping centres on the outskirts of cities, for example, which drain urban energy away from the city and are perfidiously camouflaged as the very thing they are stifling – the urban structure *en miniature.* But they are merely the result of erroneous land and tax policies, and they cannot be overcome using aesthetic strategies. They are the wrong answer to a problem that needs to be solved in a completely different way.

The same does not apply to the 'benign' non-places. These are almost never completely new urban types, but rather enlargements, duplications and exaggerations of types dating from the 19th century. High-speed railway stations, for example, are not something fundamentally different from the stations we are familiar with; they just have special technical characteristics (the trains need to be able to go straight through, so that terminal stations are not possible) and different visiting qualities (journeys less and less often represent an adventure you are getting into the mood for, and more and more form part of an everyday routine that you want to get through as casually, quickly, and comfortably as possible). Several contemporary architects have already addressed themselves to this new/old task – Rafael Moneo in the elegant Atocha Station in Madrid; Santiago Calatrava in the filigree Oriente Station in Lisbon; Antonio Cruz and Antonio Ortiz in the dynamic Santa Justa Station in Seville; Nicholas Grimshaw in the elegantly curved International Terminal at Waterloo Station in London; Terry Farrell and Norman Foster in the high-tech stations on the new airport line in Hong Kong. In New York, Pennsylvania Station by McKim, Mead & White, barbarically demolished nearly 40 years ago so that the departure areas and tracks could be plunged into the stuffy cellar of a shopping centre, is being triumphantly resurrected in glass in the conversion of the Farlay Post Office by Charles and Taylor of Skidmore, Owings & Merrill. In Arnhem, Ben van Berkel's firm UN Studio is modelling the new station as an intermodal traffic junction whose complex, interlocking structure of different traffic levels is calculated from the frequency and fluctuation of passengers. Whether this will lead to a unique space or merely an inhospitable departure machine will be seen in a couple of years' time, when the complex is opened to the public. Airports are in principle nothing but outsized train stations in which it is not only high-speed trains, regional trains and urban trains that arrive and depart, but also planes as well. This makes them technically and functionally very much more complex, but it does little to alter their social and cultural purpose. Their mediating position between globalisation and local identity remains largely the same. Because even the large train stations of the 19th century were part of a worldwide network, and as such related to each other – for constructional reasons as well as programmatic ones, the traveller who journeyed from Leipzig to Milan encountered the same type of building at both his departure and arrival points and was thus aware that he was already, or still, in the midst of the journey and was able to find his way around in the large halls of iron and glass. At the same time, there could be no confusion for him about which city he was in – the stations in Leipzig and in Milan were both (and still are) unmistakable places that are closely and indissolubly linked to their own city and its culture. If the traveller had arrived by mistake in Cologne instead of Milan, he would have noticed it at once without the help of the signs – just by looking at the station's architecture, which is different again.

There is therefore nothing to suggest that contemporary airports have to be non-places in the sense that they look identical all over the world. If they do, it is only because they are in many cases built by the same architects, and above all by the same engineering companies and industrialists, who do not make the effort or do not have the talent to incorporate local characteristics and cultures into their buildings. It is therefore not merely a matter of inevitable fate, but a condition that is deplorable and can therefore be comparatively easily changed. This change is particularly expedient since in modern airports there is no question of having to process passengers as quickly as possible. On the contrary, the intention is that passengers should stay on as the airport's 'guests' for quite a long time – because only a fifth of the airport's income is derived

14 Augé, Marc; **Non-Lieux, introduction à une anthropologie de la surmodernité** (Paris: Seuil, 1992).
15 Alphand, Adolphe; **Les promenades de Paris: histoire, description des embellissements, dépenses de création et d'entretien des bois de Boulogne et de Vincennes, Champs-élysées, parcs, squares, boulevards, places plantées; étude sur l'art des jardins et arboretum** (Paris: Rothschild, 1867–1873).

from take-off and landing fees; the remaining four-fifths are earned from rentals to all sorts of service providers, from the restaurant to the drugstore and the fashion boutique. Airport terminals are therefore becoming small cities that invite people to linger and consequently try to provide spaces with attractive qualities.

Examples of architecturally ambitious airports were already being designed in the early days of flight, including André Lurçat's fantastic 'Aéroparis' of 1932 on an island in the Seine near the Eiffel Tower, and they have also been built – from Berlin's Tempelhof airport by Ernst Sagebiel, to the TWA Terminal at Kennedy International Airport in New York by Eero Saarinen; from Kansai Airport on an artificial island in Osaka Bay by Renzo Piano, to Stansted Airport near London, by Norman Foster, the airport at Bilbao by Santiago Calatrava, the one in Seville by Rafael Moneo, or the one in Kuala Lumpur by Kisho Kurokawa. All of these examples show that belonging to a cosmopolitan world in no way means that you cannot produce a specific, recognisable place that creates a sense of identity. In exactly the same way as the train station belongs to Leipzig, its airport should as well (but does not), while the airport in Milan should belong to Milan (which is even less the case – even the touching, pseudo-cosy mask that Ettore Sottsass provided for the faceless product of an engineering firm that was as efficient as it was tasteless has changed nothing in the sad fact that the capital city of Italian design – of all places – has one of the world's ugliest airports).

Similar arguments could be applied to motorway service areas, to motorways themselves, and to stops on urban railways, bus routes, and tramways. These also represent a genre of non-places, which apart from its countless debacles has also produced a few pleasant examples – for example, Alfred Grenander's subway stations in Berlin (Wittenbergplatz station in Schöneberg and Kottbusser Tor station in Kreuzberg), or the Socialist Realist ones in Moscow. In constantly varying forms, successful architectures of this and other types are able to achieve, even underground, what the aim of all tasteful design should be – to establish a high quality of life in an unmistakable location.

Yet these types should be neutral to the extent that even here, travellers should still be allowed to be human beings in spite of all their speed, hurry and blasé attitudes, and should not be showered with the stimuli with which they will be quite sufficiently pestered during their journeys or flights. Particularly in train stations, airports, and motorway service areas, it is not entertainment you are looking for, and especially not cheap entertainment. You are looking for places of rest that can give the journey its rhythm and create fresh space for ideas. Particularly in a world of display boards, monitors, public-address system messages and announcements, you want to be spared any unnecessary screens and soporific background music. Particularly in a world of densely compressed information, you would prefer forms of architecture which, although they do not hide their public and functional character, nevertheless remain discreetly silent.

Restraint and Freedom If it is true that the city is a stage – although it is one that is *sui generis*, featuring audience participation and interaction – then life on this stage should not be steered along predetermined courses, but should be able to develop itself in freedom. If it is true that the city is an expression of the highest form of social co-existence, it should not be turned into an architectural emblem of untrammelled individualism, but instead should represent precisely that which provides individuals with the social cohesion that they need no less than they need scope and freedom to move. In other words, for people to be able to live life to the full, the city should restrain itself in modesty and generosity.

All the more so in a period overwhelmed by a mass of images and stimuli. The onset of the media age indicates that the city's great task today should be to provide an antidote to precisely this type of mediatisation. This is not an aesthetic preference, but a vital necessity. People who are confronted with countless images in their everyday routine do not want an urban environment that pesters them with images that are equally innumerable. People who spend a large part of their time gazing at flickering screens to not want to see them on the walls of buildings as well. And people who find themselves being entertained whether they want to or not, whenever they happen not to be working or sleeping, might perhaps also want to immerse themselves in an architectural universe of silence, neutrality and even of leisure.

The city in the telematic age should not be the city *of* the telematic age, but rather – as it always has been and hopefully will be even more in the future – a city of human beings.

[1]   [2]

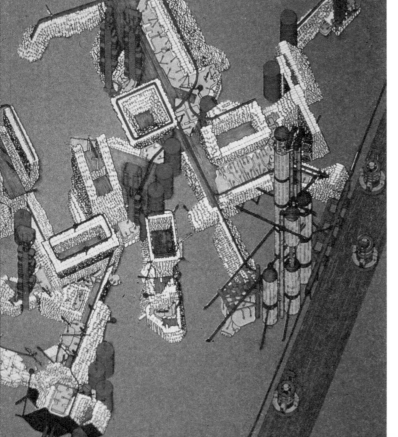

1 'Cyber-Nomaden' in: **Spiegel** 11 (1999).
2 Peter Cook (Archigram), Plug-in-City, 1964.
3 Werner Hegemann, Elbert Peets, Das Chaos der amerikanischen Geschäftsstadt 1922, in: Werner Hegemann, Elbert Peets, **American Vitruvius**, New York 1922 (first in: **The Architectural Review**, 1904).
4 A. Trystan Edwards, **Good and bad manners in architecture**, London 1924, S. 94, Fig. 26, s.o.
5 A. Trystan Edwards, **Good and bad manners in architecture**, London 1924, S. 98, Fig. 27, s.o.

[3]

FIGURE 26A

FIGURE 26B

FIGURE 26C

FIGURE 26D

[4]

FIGURE 27A

FIGURE 27B

FIGURE 27C

FIGURE 27D

[5]

6 H.T. Wijdeveld, Amsterdam Hoofdweg, 1925–27.

7 André Lurçat, Aéroparis, Vue d'Ensemble, Photomontage 1932, in: Jean-Louis Cohen, André Lurçat, 1894–1970, Liège 1995, S. 139.

8 Robert Doisneau, Jardin du Luxembourg, 1991, in: Brigitte Ollier, **Robert Doisneau**, Paris 1996.

The Urban Question in the 21st Century: Epistemological and Spatial Traumas

M. Christine Boyer

In 1970, Manuel Castells wrote his now famous book *The Urban Question: A Marxist Approach* where he examined the word 'urban'. He professed it was a concept that could be understood only through ideology. Whoever studies the 'city', he argued, must also study capital, production, distribution, consumption, politics, everyday life, grass roots politics, and so much more.[1] His specific problematic – that is the urban question – boiled down to whether there existed any such concept as the 'urban unity' or 'urban area' because the dichotomy between town and country no longer held any meaning under the monopoly phase of capitalism. It was easy to see that the countryside had been urbanized, and the town ruralized, that the compact boundary line between city and country had been erased, albeit according to the dictates of uneven development. Following Henri Lefebvre, Castells understood that each society, each mode of production, creates its own specific space. Thus Castells concluded the problem is how to define a new urban form or specify the organization of transfigured space in advanced capitalist societies.

More recently Castells has been studying the shift from industrial to postindustrial society – and the creation of the network society where production depends on the ability to generate, process and apply knowledge-based information systems. This mode of production again effects spatial relationships, power relations, and everyday experiences and sets up a new problematic: how to understand and describe the space of flows along a set of interconnected nodes, which results in the network society.[2] Castells positions this space of flow in opposition to the space of places, arguing that a placeless culture has evolved out of the increasing interconnectedness between local, regional and national communities. Organizations, groups and individuals are now connected through flows of information via telecommunication networks whose logic lies beyond the control of any given society or subgroup, yet whose impact intimately shapes the lives of these subcultures and local cultures.[3]

In an effort to couple these two problematics – the urban question and the network society –

---

[1] Castells, Manuel; **The Urban Question: A Marxist Approach** (Cambridge: MIT, 1977).
[2] Castells, Manuel; **The Information Age: Economy, Society and Culture. Volume 1: The Rise of the Network Society** (Oxford: Blackwell, 1996). Note that Jean Gottmann in his 1961 work **Megalopolis: The Urbanized Northeastern Seaboard of the United States** (New York: Twentieth Century Fund, 1961) suggests the notion of a network society.
[3] Arjun Appadurai has examined this space of flows, subdividing it into five disjunctive global flows, which he terms ethnoscapes, mediascapes, technoscapes, financescapes and ideoscapes. By appending the suffix '-scape' to each of these terms, Appadurai indicates that each flow is constructed from many different perspectives influenced by the historical, linguistic and political context in which various actors are situated – be they nation-states or communities, transnational corporations or individuals. These landscapes become the building blocks for 'imagined worlds' of consciousness, keeping in mind that these various flows are fundamentally disjunctive in relation to each other. In these different 'scapes', imaginary constructs of the mind become more important than reality. In order to study the complex configuration of cultural forms in the global society, Appadurai argues for a human version of chaos theory, which depicts the dynamics of complex, overlapping and unstable flows of persons, technologies, finance, information, and ideology. He maintains that the flows that characterize globalization processes can no longer be studied in terms of world-systems theory or any theory that contains a unitary development model spreading outwards from the metropolitan centers to the undeveloped peripheries.
Appadurai, Arjun; **Modernity at Large: Cultural Dimensions of Globalization** (Minneapolis: University of Minnesota Press, 1996) pp. 27–47; See also Li, Victor; 'What's in a Name? Questioning "Globalization"', in **Cultural Critique** 45 (Spring, 2000) pp. 1–39.

I want to explore a missing component of Castells' analysis. He offers neither an epistemological framework of knowledge in the network society nor ways of visualizing material practices and their spatial effects.[4] Since Castells does not consider these issues, I turn to another source for analysis: Stefano Boeri and the research network called *Multiplicity*, studying *The Uncertain States of Europe and Assembling Eclectic Atlases*. This group of architects, artists and researchers has recently entered the wider debate about the production of space, displaying as they arrive, their special talent for visualization that fully exploits all the advantages of computer generated graphical displays. These mapmakers are challenging traditional languages of vision, reformulating the concept of atlas and territory, and revealing fundamental changes of direction and turbulence in spatial terrains. Yet here is also where the urban question returns to haunt us with its unanswered problematic: what is the appropriate unit in the production of urbanized space in the network society and how can the geographer utilize a nuanced discourse of fluidity, contingency and multiplicity while describing complicated spatial patterns that mark the surface of the earth yet simultaneously address geopolitical questions that the superabundance of imagery often ignores? In other words, can we rub political and spatial issues, texts and images together to create an interpretative spark?

Perhaps we should keep Svetlana Alpers' message from her book *The Art of Describing* (1983) clearly in mind. She claims that geographical maps and microscopic anatomical drawings, common representations in Dutch art of the 17th century, were constructed as representation of things visible in the world. The intent was to disentangle things seen from the interpretations read into them. Artists were using these images as correct representations in order to challenge texts as rhetorical objects of subjective construction.[5] Once again, in the attempt to develop Eclectic Atlases, there may be a disjunction at play between images and texts.

The Problem of European Identity: So let us begin with the problem of European identity. Europe can be described as a superimposition of borders, hence a superimposition of heterogeneous relations to other histories and cultures of the world.[6] The operation of making distinctions between Europe and others is both a limitation and a potentiality. The limitation is exacerbated at the present, because there is a powerful and irreversible process of hybridization and multiculturalism taking place that is transforming the concept, the boundary, even the idea of Europe. Springing from former relationships that reciprocally linked the 'metropoles' of Europe with their colonies, and recognizing regional independence within national borders, this hybridization has now become a generalized pattern of relations between Europe and its 'exterior' whether the latter is within or outside its territorial space.[7] These distinctions produce a highly interactive and complex condition entailing a growing uncertainty with respect to European identity.

In addition, there are significant ethnic minority groups in every European state due to immigration and differential birth rates.[8] The potentiality for sub-group identification and individuals who resist assimilation yet cannot be considered to be outsiders because they have been born in Europe challenges the very concept of national identity before a sense of shared European identity has been forged. Castells and Nezar Al Sayyad argue there is a limit to institutional and economic integration within the European Union without a culture of sharing. They maintain '[i]n the long term, the emergence of a European network state and a fully integrated European economy without a European identity (overlayered on national and regional identities) seems to be an unsustainable situation.'[9] Apparently the concept of the network society requires a European identity, but how can this be construed?

In the past, European identity was formulated in opposition to others: be they barbarians, invaders, or the Americans. Thus it is not easy to assume that European absorption of others, and acceptance of its multi-ethnic, multi-cultural condition, will be achieved without disturbing many other Europeans who feel challenged and uncertain about their future. As stable group membership erodes, a crisis of identity emerges. Due to this blurring, others perceived as different may be accepted with difficulty. An effect of uncertainty, defensive actions may be aggressively stated and boundaries around inclusion and exclusion precisely drawn. In other words while there may be structural reorganization within Europe, there may also be conceptual closure.

For many Europeans, however, identity is not a field of struggle over insiders and outsiders, but a fluid relational concept, embedded in a network of change and linked not only to nationality and ethnicity, but social class, community, gender, or sexuality, even though many of these sources may conflict with each other. Still the problem persists: can one have a sense of belonging to a community, nation or European Union without forming a sense of identity located in place?[10] Are transnational or post-national identities emerging, anchored in a concept of the universalistic rights of personhood rather than human entitlements bound to being a citizen of a specific territorial unit? These transnational rights are determined by a nested set of identities

---

4 Visvanathan, Shiv; 'The Grand Sociology of Manuel Castells', in Johan Muller, Nico Cloete, Shireen Badat (eds) **Challenges of Globalisation** (Cape Town: Maskew Miller Longman Ltd., 2001) pp. 3–49.
5 Alpers, Stevlana; **The Art of Describing** (Chicago: University of Chicago Press, 1983).
6 Balibar, Etienne; 'Europe: Vanishing Mediator?' The First George L. Mosse Lecture, Humboldt-Universitate Berlin (November 21st, 2002) p. 22.
7 **ibid.** pp. 26–7.
8 Al Sayyad, Nezar and Castells, Manuel; 'Introduction: Islam and the Changing Identity of Europe,' in Al Sayyad, Nezar and Castells, Manuel (eds) **Muslim Europe or Euro-Islam: Politics, Culture, and Citizenship in the Age of Globalization** (Lanham, Md.: Lexington Books, 2002) pp. 1–8.
9 **ibid.** p. 4.
10 Al Sayyad, Nezar; 'Muslim Europe or Euro-Islam: On the Discourses of Identity and Culture', in Al Sayyad, Nezar and Castells, Manuel (eds) **Muslim Europe or Euro-Islam** pp. 9–29.

that challenge any homogeneous formula of who is a European and who is not, who is German and who is not, etc..[11] They upset both the space-based territorial state and the supra-state European Union at once.

In this scenario, nationality is being deconstructed from below, while simultaneously reconstructed from above. As the nation state declines in importance, the European Union imposes uniform economic and monetary policies and policing procedures on all of its member states, creating a European citizenry and European passports, eradicating and expanding boundaries as it does. The result is unification at one level, the creation of a placeless culture. But simultaneously, there is an increase in interconnections between local and national communities because the nation-state still retains control over educational, public health and welfare functions, and still is the distributor of benefits and aid. How then are identities being formulated territorially?

In this uncertainty over national identity, society is construed to be a system that differentiates between self-reference and other-reference, a differentiation that establishes the system's operative closure yet its openness to what it excludes, which forms its environment.[12] This system is operationally closed in order to maintain its boundaries and conserve its identity, otherwise Europe as a conceptual entity would cease to exist. Yet the system is energetically and materially open to its environment, undergoing constant structural change among the components within the system in order to maintain overall organizational integrity. It is a system determined by communications between itself and that which it differentiates and hence its structural arrangement is momentary and configurational. There are no limits to how it might reconfigure its structure in response to irritations from its environment. The result is constant interactions with the environment and constant self-maintenance of the whole. In other words, as we explore below, the concept of Europe is construed to be an autopoietic system; it is self-referential in that it constantly manufactures itself and all of its subsystems in order to maintain its integrity as an autonomous unit.

The Problem is Geography and Uncertain States of Europe:

Now let us link this uncertainty over national identity, this mutable and irritable system of European states, to spatial analysis. Stefano Boeri, writing about the *Uncertain States of Europe* and embarking on the creation of *Eclectic Atlases*, probably agrees with Castells' claim that the space of flows of the network society is opposed to the space of places of the geographers. Look at any map of Europe, the space is divided into states, each clearly delineated by linear boundary lines and blocked out from adjacent areas by separate colors. In such a representation, the state is a unified territorial concept, a spatial container of people and political authority. A network, on the other hand, configures

Europe as a flat field of circulatory movement: lines drawn across the land or through the air, flows of messages, people, or ideas across points in space. As we shall explore, this network lends itself to biological analogies and organic metaphors, a transfer of procedures and norms from science to geography and space.¹³

Since networks and processes have replaced older cartographic conventions of location analysis and fixity in place, there is need for new instruments to articulate the relationship individuals hold with space, new interpretative schemata and new procedures of mapping to visualize these forms; in short, the creation of *Eclectic Atlases*. But why apply the terms 'eclectic' and 'atlases' to define both the content and the ordering device behind this visual and verbal depiction?

Jean-François Lyotard wrote it in *The Postmodern Condition: A Report on Knowledge*: Eclecticism is the degree zero of contemporary general culture: one listens to reggae, watches a western, eats McDonalds food for lunch and local cuisine for dinner, wears Paris perfume in Tokyo and 'retro' clothing in Hong Kong; knowledge is a matter for TV games. It is easy to find a public for eclectic works.¹⁴

It is also easy to overlook in this witty parody of Marx, the other side of the argument. The more globalization of communication and information enable the consumption of an eclectic culture the more local cultures stubbornly refuse to disappear, feeding increasing spirals of eclecticism. The gesture is both unity and dispersion at once; there are centrifugal as well as centripetal forces at work. In other words, without local cultures the possibility of an eclectic culture disappears, indeed eclectic homogenization may even lead to the exaggeration of regional and national differences. One depends reciprocally on the other.¹⁵

11 Kumar, Krisham; 'The Nation-State, The European Union, and Transnational Identities', in Al Sayyad, Nezar and Castells, Manuel (eds) **Muslim Europe or Euro-Islam** pp. 53–68. Balibar notes that Europe has a long tradition of the secularization of politics and society, no doubt the result of its past history of religious wars. This allows recognition of religious membership as part of the constitution of 'civil society' making sure that individuals are the true bearers of rights. But this secularization can be construed as a shield against any form of religious universalism (especially Islam) and resistant to real multiculturalism if a 'culture' is deemed to be too 'religious'. **cf.** Balibar, Etienne; 'Europe: Vanishing Mediator?' pp. 28–30.
12 This analysis of systems theory is taken from Niklaus Luhmann, who in turn was borrowing conceptual structures from the biologists Humberto Maturana and Francisco Varela. For translations from Luhmann see Wolfram Malte Fues, 'Critical Theory and Systems Theory,' in Peter Uwe Hohendahl & Jaimey Fisher (eds) **Critical Theory: Current State and Future Prospects** (New York: Berghahn Books, 2001) pp. 229–47. See also Lloyd Fell and David Russell, 'Seized by Agreement, Swamped by Understanding', www.pnc.com.an/~1fell/matsbio.html. Accessed 05/31/04.
13 Mattelart, Armand; **The Invention of Communication**, translated by Susan Emanuel (Minnesota: University of Minnesota Press, 1996) pp. 3–25.
14 Jean-François Lyotard writes in **The Postmodern Condition: a Report on Knowledge** (Minneapolis: University of Minnesota Press, 1986) p. 76.
15 Hadfield, Andrew; 'National and International Knowledge: The Limits of the Histories of Nations', in Rhodes, Neil and Sawday, Jonathan (eds) **The Renaissance Computer: Knowledge Technology in the First Age of Print** (London and New York: Routledge, 2000) pp. 106–119.

Because the foundation of the self, or even an eclectic European identity, in the past has most often been linked to a strong geographical consciousness, a sense of self embedded in a sense of place, perhaps this is the reason Boeri resorts to the construction of atlases, or an archive of maps, photographs, data and words. But it probably has more to do with 'uncertainty' since the structure of this atlas with its mixture of cartographic and discursive processes betrays a transitional condition. It displays a search for both method and design that cannot be explicitly formulated in diagrammatic terms, images or illustrations – and thus requires the act of naming, of verbal description and texts. It is a project that gathers heterogeneous information on the states of Europe, and explores different mixes of textual description and graphic display. In the end, it is discourse that provides the ideological framework within which the new images of Europe are located.[16] Boeri explains that the atlas seeks new correlations between spatial elements, the words we use to name them, and mental images we project upon them, and it is eclectic because the basic correlations are often multidimensional, new and experimental. We need to explore the structural device that underlies this exploration of disparate things collected into atlases and analyze connections posed between the verbalized and visualized set of things as a framework of knowledge concerning the uncertain states of Europe.

For Boeri, the uncertainty over European national identity coincides with great uncertainty over the geographic representation of Europe. Uncertainty is erupting because the territory of Europe can neither be read topographically or in the aggregate, nor can one find a European identity or subjectivity through one's work or one's community. Mature capitalism and the network society, he claims, have produced this state of uncertainty where once the production of space and identity were joined.[17] Essentially Europe is a diffused city with differences erupting here and there – it displays unity and difference, as well as chaos and organization. Diffusion suppresses national borders, giving witness to a territory crossed with analogous processes and problems of expansion. Yet simultaneously there are too many implicit territorial boundaries carving up European space through various procedures, but not one of them helpful in forging identity.[18]

In addition, the European Union's admissions policy with its allegiance to outmoded theories of development imposes its own set of boundary lines. It defines regions to be either advanced or fragile: targeting territories of economic growth and stable investment, or weaker spots, where localities are out of synch with the Union's economic development policies. It names and locates on the map of Europe such dynamic investment zones as the Blue Banana, Sun Belt, or Alpine Redoubt. Or it names and spatially defines regions of highly interactive networks of transportation patterns, monetary transactions, economic interdependencies, and dense settlement patterns such as the Po valley network, the Paris region, or Benelux Reticular system. These top-down investment strategies deploying simplistic schemes of development overlook an array of attributes on the ground and ignore the negative effect their policies may have on local economies and the welfare of local people.

Boeri has a problem with this geographical act of naming and locating places. In general, the older conventions of geography do more harm than good – they do not allow the cartographer to assemble unlikes together, to add one layer of information upon another, to trace dynamic processes across a terrain, or to envision potentials as they begin to emerge. Nor do they stress the arbitrary construction of geographical entities or allow for the expression of uncertainty, indeterminacy, yet stability and organization. Boeri believes if the concept of European space is to acquire visibility and shape, although its boundaries may remain blurry, its space must be envisioned as a field shot throughout with subsystems having their own metabolisms of material and immaterial energies. The uncertain nature of European states, he claims, is the direct effect of action or interaction of these subsystems and hence the image of 'European space is seen as an open, available context: a surface composed of heterogeneous, continually changing geographical environments, acted on by multiple energies.'[19] Surface sprawl, or the diffused city, is the direct result of transformations by a multiplicity of autonomous subsystems reacting to changes in their environment in order to maintain organizational integrity.[20]

Consequently, Boeri proposes an alternative model to describe European spatial identity. He calls this a European '*dispositif*' – borrowing the word from Foucault as translated by Deleuze. *Dispositif* loosely means a social apparatus or device. Alternatively, it defines a tangled spatio-temporal multiplicity or multilinear ensemble composed of lines that follow certain directions, break, or bifurcate, before changing direction, becoming self-reflective or drifting about. There are four dimensions to Foucault's *dispositif*: lines of visibility or representation, enunciation or discourse, force or power and subjectification, or the process of individuation of groups and peoples. The different lines divide into lines of sedimentation (or the past) and lines of creativity (or the future). Untangling the multiplicity of lines within a social apparatus, Deleuze claims, is like drawing a map, doing cartography, or surveying an unknown landscape – in other words the '*dispositif*' has spatial effects that give form to social relationships, or map transactions and discourses in spatial terms.[21]

16 Conley, Tom; **The Self-Made Map** (Minneapolis: University of Minnesota Press, 1996) p. 205.
17 Boeri, Stefano; **Multiplicity: USE Uncertain States of Europe: Trip through a changing Europe** (Skira: 2003) p. 23.
18 Boeri, Stefano; 'Notes for a research program', in Koolhaas et al., **Mutations** (Barcelona: ACTAR, 2000) pp. 356–377.
19 **ibid.** p. 360.
20 Boeri, **Multiplicity** p. 22.
21 Deleuze, Gilles; 'What is a dispositif?', in **Michel Foucault Philosopher** (New York: Routledge, 1992) pp. 159–168. Perhaps it is better to turn to Deleuze and Guattari's own definition of a map. It is a correlation that is '…open and connectable in all of its dimensions; it is detachable, reversible, susceptible to constant modification. It can be torn, reversed, adapted to any kind of mounting, reworked by an individual, group, or social formation. It can be drawn on a wall, conceived of as a work of art, constructed as a political action or as a meditation….it always

As deployed by Boeri, such a European device may act locally but refer to long-term structural changes and processes of self-organization of inhabited space. It does not refer to operators of change such as those embedded within a State's development policies;[22] instead, this alternative model focuses on long-term interactive influences in order to observe 'certain modes of territorial mutation' or 'certain recurring mechanisms of change' within the field of operations called the *Uncertain States of Europe*.

In search of new explanatory schemata that capture these emerging effects, Boeri slips from Foucault's *'dispositif'* into the vocabulary and conceptual structure of self-organizing biological systems theory – or autopoietic systems – without acknowledging the slide. This raises a series of questions. Do we need to know the 'forgotten' science, or the full implication of the theories at work in these texts in order to understand and interpret their effect, or are the analogical borrowings sufficient unto themselves because they remain on the surface? Have these explanatory schemata – either dispositif or autopoietic – become common currency to the degree they need no further explication? And might the interpretative strategy deployed below give too much coherence to what would otherwise appear as categorical confusions, logical meanderings or incommensurable juxtapositions?

The problem is how to grasp order emerging out of complexity, entanglements, or multilinear ensembles. This is the same quandary Heinz Von Foerster wrestled with in the 1960's when he first applied information theory to biological systems. He was trying to answer why living things have a tendency to become more complex and quite distinct from the generalized chaos from which they emerged.[23] Foerster added the idea of self-organizing systems noting that not only did these systems feed upon order, but they found noise on their menu as well. In other words, noise could be interpreted as having meaning and not simply be rejected as extraneous. In fact, what might be noise at one level of the system, could be transmitted as information at a higher level.[24] Living systems were conceptualized as autonomous organisms, which rather than having inputs and outputs as mechanized systems did, responded to perturbations or irritations in their environment through structural transformations in such a way as to preserve their organizational identity. Mutations and structural transformations of the system conformed to the system's own set of laws.[25] Furthermore, living systems were composed of a hierarchical series of levels in which higher sub-systems were not reducible to lower level ones. This network of interconnected levels or subsystems communicated constantly with each other. Thus indeterminacy or noise at one level might be resolved by constraints imposed at higher levels, constraints that ensured the stability and self-maintenance of the whole.[26] So it was declared over time 'every system composed of a very large number of interacting elements will inevitably and spontaneously develop some sort of stable behavior.'[27]

It is easy to make the transference of self-organizing systems theory to the spatial terrain of cities for they too, like biological systems, are highly complex entities, not just chaotic or 'wild'.[28] They too respond to perturbations in their environment and seek stability and order according to their own set of laws. In this interpretative paradigm of autopoietic systems, order appears to emerge out of chaos. Boeri argues that 'Eclectic Atlases' are combinatorial forms that trace out various urban facts in the course of their evolution. Atlases reveal through their collection of various materials how different types of interaction persist in space, organizing the evolution of territory and reflecting levels of self-organization. 'How, behind the apparent chaos, there is in fact an excess of organization, of regularity, an excess of evolutionary patterns.'[29] He is probing the space of Europe looking for 'autopoietic' innovations in inhabited space and mutations that reflect individual acts within major waves of change. The question to be addressed is whether the paradigm of a closed biological system can adequately represent the identity problems of Europe.

Linguistic Paradigm as a Means of Exchanging Information:

Boeri's problematic, however, is not merely how to model and interpret the surface mappings of dynamic processes that pulse across the face of Europe, it is also – as any geographer must note – a matter of language and the codes and concepts which enable subjects and spaces to form their identity. Boeri proceeds as a mapmaker by exploiting a linguistic paradigm that says more about the construction of space than how subjectivities are formed in the process. He notes:

> European space, which is a palimpsest of projects sedimented in time, is also today the field of action for an indeterminate and changing number of subjects, many of whom maintain a temporary relationship with the territory. A battle of codes and interpretations ceaselessly unfolds upon this field, which is continually being rewritten, where almost nothing is ever erased, where the long-term structures are temporarily hidden by others which are less powerful and enduring, but currently more visible.[30]

has multiple entryways…' Deleuze, Gilles and Guattari, Felix; **A Thousand Plateaus** translated by Brian Massumi (Minneapolis: University of Minnesota Press, 1987) p. 12.
22 Boeri, **Multiplicity** p. 22.
23 A complex system being one in which the observer does not have complete knowledge or sufficient information to make a complete structural and operational description of the system, or describe the causal connections between the parts, or explain how its behavior is produced.
24 Paulson, William R.; **The Noise of Culture: Literary Texts in a World of Information** (Ithaca: Cornell University Press, 1988) pp. 60–72.
25 ibid. 122.
26 Dupuy, Jean-Pierre; **The Mechanization of the Mind: On the Origins of Cognitive Science** translated by M. B. DeBevoise (Princeton: Princeton University Press, 2000) pp. 132–161.
27 C. H. Waddington (1969). Quoted by Dupuy, **The Mechanization of the Mind**, p. 153.
28 Johnson, Steven; **Emergence: The Connected Lives of Ants, Brains, Cities, and Software** (New York: Scribner, 2001).
29 Boeri, 'Notes for a research program' p. 368.
30 Boeri, 'Notes for a research program' p. 375.

In order to observe these mutations and transformations, Boeri deploys two research strategies. The first involves understanding how European space constituted itself in the past forming 'sentences' or 'paragraphs'. Boeri assumes, as Aldo Rossi did before him, that spatial entities (such as urban courtyard housing, or nineteenth century cluster housing) constitute the minimal units of this spatial discourse.[31] A set of rules, or urban syntax organizes the conjunction of parts, and consequently articulates and assembles these units into a recognizable code. Like natural language, this code, syntax, or unit enables any individual to reinvent space allowing new sentences and paragraphs to emerge over time.

Boeri's claim that language articulates the geographical terrain is not restricted merely to the absorption of older forms and new translations. His second research strategy analyzes a hierarchy of levels within the language system of European space. He states that between language and discourse there resides an intermediate level where an a-centered assemblage of utterances forms the units to be studied. At this level, a logic of assemblage operates experimentally transforming words into an utterance, or a 'European sentence' that reflects the heterogeneity and diversity yet respects the historicity and traditions of its spatial language. Unholy mixtures, variations and anarchic statements proliferate, conjoined at a higher level without any perceptible plan. Focus is drawn to the awesome materiality and pragmatics of this language – its redundancies (noise), variations, and subversions grouped together by some force of attraction dictated by the codes and syntax that may control the assemblage at some higher level and that may at the same time be flouted and transformed. The urban question then focuses on how these elements and their groupings are produced, how they work as utterances, in what assemblage they are inserted. Boeri remarks:

> Urban space in Europe today means, maybe more than anything else, this intermediate sphere that, like a real 'phrase' between words and a discourse, absorbs the unpredictable variations of the world of life, and identifies them according to a code inscribed in the materiality of the urban condition. The Forum, the block, the courtyard, the suburb on the public periphery, are inventions – but we should say reinventions – of this transformational device. The point is to ask ourselves if, how and where this device is still operating.[32]

Such a research strategy recognizes that a 'European sentence' resides at an intermediary level between the word of the construction process and a larger urban discourse. Hence each sentence acts as a link between diverse parts and the urban system as a whole. Boeri claims:

> In the same way that a discursive form – an exclamation, an exhortation, an imperative – can be adapted to any part of a discourse, so the European

urban 'sentence' is the characteristic form of an unlimited spatial dimension, extensible through the action of a local *dispositif*, but nonetheless deeply rooted in and consubstantial to the experiences of European culture over time.[33]

Epistemological Trauma or the Problem of Observation:

Through reconnaissance missions, researchers move through and examine the uncertain states of the European system. With the autopoietic schema more or less consciously in their minds, they observe the metabolization of both past and innovative spatial forms, how they absorb and reinvent new forms and spatialities through the reinterpretation and colonization of older architectural styles and forms and new materials and needs. By playing with an open-ended system of rules that govern the variation of subsystems and rules that determine the conjoining of parts, these observations reveal the paradigm of European space as one of constant change both conservative and innovative.

How then to map, to envision this European space of permanence and change? First, there is a basic problem with techniques inherited from geography because the zenith view obtained from an airplane, or more recently from the synthetic images of satellite photographs, results in two-dimensional cartographic images and is inadequate for the task that Eclectic Atlases supports. This zenith view with its panoramic sweep of the eye extending in all directions may depict the 'diffused city' or regions of 'dissolving ties' and it can offer real time images of traffic flows and crowd densities, but these transfigured images of the city have produced an 'epistemological trauma' in their wake. Such images reveal isolated built objects muddled together in uncoordinated fashion, sometimes coagulating unharmoniously along communication corridors, other times strewn across open space in an irregular web of constructed sites. As Castells argued in *The Urban Question*, such chaotic entities are no longer decipherable by the outmoded vocabulary of the 'megalopolis', 'urban conurbations', 'diffuse city', even 'the city of dispersal'. Boeri notes that innovations in visual technology produce images that can be seen in real time evolution but not explained. The observer lacks the appropriate vocabulary to analyze what is spread out before his/her eyes. Thus an 'epistemological trauma' results.[34]

Even more damaging for the project of *Eclectic Atlases*, this 'zenith morphology' is misleading, for it requires distancing, and impersonal objectivity not an embodied production of knowledge. And while it may allow for the projection of thematic maps or a sequence of maps as slices of time, it cannot represent superimposed layers, nor trace out dynamic processes coursing across the terrain. Moreover it interprets what it sees as chaos, inevitable chaos, and shields the viewer from responsibility by assuming that underneath visible space lie other determining forces of

---

31 For an analogous argument with respect to Aldo Rossi's use of language, see Boyer, M. Christine; **The City of Collective Memory** (Cambridge: The MIT Press, 1983).
32 Boeri; 'Frequently Asked Questions', in **Multiplicity** p. 24.
33 Boeri; 'Notes for a research program' p. 363.
34 Stephen Walddell, J A Z, Amin Linke, Bas Princen, Alessandro Cimmino, Danilo Donzelli, 'Field Works', in **Multiplicity:** 432; Stefano Boeri, 'Eclectic Atlases: Four Possible ways of seeing the City', **Daidalos** 69/70 (December, 1998/January 1999) pp. 102–113.

economics, politics, social relations, impacting space. 'In conclusion, [Boeri notes] we can say that a paradigm that is strong in its visual tools but weak in its interpretative ability is insufficient when attempting to explain this chaos. We cannot ask it to resolve the problem it has created.'[35]

Nor can the objective aerial view, reliant on mechanical devices of observation, allow for more subjective experiences giving evidence of the 'multitudes of capillary processes', small unsynchronized leaps, deprived of hierarchies or rules of settlement, that are pulverizing and fragmenting the formation of space in every urbanized region. In this capillary action single-family homes emerge next to huge shopping centers, small factories alongside entertainment facilities, each one cut out from the ground and without any sense of interrelating, while their overall dynamics produce unpredictable trajectories.[36]

Mutations in space require new strategies of embodied observation and new vocabularies of decipherment. Although they were banished long ago, once more the polarities between center/periphery, public space/private space, or emergent areas/established parts of the city are rejected as if they were still being deployed as useful distinctions. Substituted in their place are 'Eclectic Atlases' that offer a multitude of visual thinking and lateral modes of representation combining research reports, photographic surveys, geographic descriptions, qualitative analyses, literary probes, collections of plans and projects. These atlases seek new logical connections between spatial elements, words that name the elements we see, and the mental image we project on space. They reveal that 'behind the apparent chaos, there is in fact an excess of organization, of regularity, an excess of evolutionary patterns.'[37] They are eclectic because they seek to represent the dynamics of inhabited space that are multidimensional, spurious and experimental.[38]

Metabolism or the Problem of European Identity:

There are problems with the series of analogical borrowings from self-organizing biological systems theory, to Foucault's *dispositif*, and the linguistic paradigm. Reflecting an epistemological crisis that extends to traditional methods of visualization and mapping techniques, this amalgam of theories may lead to erroneous conclusions, especially with respect to the problem of national identity and geopolitical concerns.[39] Conceptualizing the territorial space of Europe as a biological organism, in which mutation and metabolism are major concepts in the organism's search for stability and self-maintenance, Boeri and Multiplicity in an unrecognized conflation of spatial entities and subject identities assert that Europe 'is accustomed to including exogenous traditions and lifestyles: one space, therefore, built up of foreign parts.'[40] They maintain in this respect a multiculturalist perspective. Yet they also proclaim that Europe metabolizes exogenous and exotic cultures – absorbs and reinvents other forms and spaces taking an assimilationist stance.[41] This unquestioned position claims that Europe as an isolated unit of human associations has been able to deal with ethnic-religious, or social and poverty conflicts since WWII, that it is able to metabolize exogenous cultures:

> … perhaps because its ancient constitution enabled it to pit itself against other cultures from a foundation of a strong, recognizable, though constantly reworked, identity. This capacity for inclusion also originates from its extraordinary density – of people, structures, cities – which has favored and in a certain sense made opportune, diverse forms of tolerance, which has developed resources and cultures of closeness and comparison with the other, which has understood how to strongly integrate the diversity, almost to the point of dissolving it and making it difficult to rediscover the roots or retrace the evolution to its origins: Europe is a place where cultures of different spaces have found ways of relating to one another and establishing themselves in the density of the urban and territorial fabric. In Europe, the other, the stranger, the foreigner, although constantly generating conflict, has often entered in harmony with the social transformation dynamic.[42]

Contrary to this view of mutation and absorption, a position over-determined by self-organizing systems theory, there are instead unstated conflicts between assimilation theories that assume cultural differences dissolve or are metabolized as a subgroup becomes more and more like other Europeans, and multicultural theories that profess that ethnic heritages and cultural diversities must maintain their distinctiveness. No doubt both processes operate simultaneously in Uncertain States of Europe in spite of the fact they engender incompatible policies. Assimilationists, for example, see education as the route to transform foreigners into citizens, while multiculturalists regard education as the way to help new citizens value their ethnical and cultural differences.[43] The struggle between identity and unity, the maintenance or absorption of difference

---

35 Walddell, Linke, Princen, Cimmino, Donzelli, 'Field Works' p. 432.
36 Boeri, 'Notes for a research program' p. 366.
37 Boeri, 'Notes for a research program' p. 368.
38 'Most Frequently Asked Questions', in **Multiplicity** p. 19; and Boeri, 'Eclectic Atlases' p. 104.
39 The biological metaphor assumes that an organism always achieves stability or a strong sense of identity. It views reality to be the result of a play between chance and necessity; reality is path-dependent and its dynamic is non-linear and recursive. Its epistemology construes the world as an entangled web and its geography considers the evolution of such a web as the effect of interactions between the multitude of lines in the web. Emergent behavior, or order out of chaos, becomes the norm. While the system itself is neither progressing towards nor regressing from a state of perfection, it achieves states of sub-optimal equilibrium within a constantly evolving, unpredictable process. It is the behavior of such a system that becomes the focus of research. Daniel Z. Sui, 'The E-merging Geography of the Information Society: From Accessibility to Adaptability', in Donald G. Janelle and David C. Hodge (eds); **Information, Place, and Cyberspace: Issues in Accessibility** (Berlin and New York: Springer, 2000) pp. 107–129.
40 Boeri, **Multiplicity** pp. 22–3.
41 Boeri, **Multiplicity** pp. 22–3.
42 Boeri, **Multiplicity** p. 24.
43 Al Sayyad, Nezar; 'Muslim Europe or Euro-Islam: On the discourses of Identity and Culture', in Al Sayyad, Nezar and Castells, Manuel (eds) **Muslim Europe or Euro-Islam** pp. 11–12.

has yet to be addressed in the discourse of *Eclectic Atlases* as it slips in and out between spatial evolution and issues of identity as if they were the same.

The missing discourse on identity within *Uncertain States of Europe* should account for the contentious presence of Muslims in Europe and whether these subgroups, for there are a variety of different types of Muslim communities, will be assimilated into secular European societies or treated as 'protected minorities' to remain forever 'alien types'.[44] What appears dramatically missing, in the creation of *Eclectic Atlases*, is a geopolitical view of spatial determinations that deconstructs the exaggerated importance of Europe and reformulates East-West distinctions. A geopolitical view would reconsider the Euro-Mediterranean ensemble hindered heretofore by obstacles stemming back over centuries of religious and colonial conflicts. The inclusion of the other within itself, and against Samuel Huntington's theory of the clash of civilizations which erects a sharp borderline between the East and the West, could turn the Mediterranean Sea into a common space of interdependence.[45]

Forgotten Geopolitics and the Legacy of Globalization:

In the application of organic metaphors to geopolitical issues, it is important to take a historical perspective, to be aware of how these metaphorical imaginings have been deployed in the past to achieve very conservative ends. Going beyond the conventional categories of geography requires understanding how spatial practices of national identity and visualization of geographic information have been deployed in the past. That enterprise takes us back to the end of the 19th century, when geographical methods were first applied to the understanding of political and international considerations. Geographers understood that the great powers were unable to contain destructive forces being unleashed in the Balkans or dampen international competition over colonial spheres of influence, and they worried over national territorial security. They also understood that the world map for the first time in history appeared to be relatively occupied, sharpening the problem of territorial expansion in a world closed to exploration and discovery. The struggle between nations for strategic position and international power had reached a new phase.[46]

Rudolf Kjellén, a Swedish political scientist, coined the term 'geopolitics' in 1899, defined as 'the theory of the state as a geographical organism or a phenomenon in space'. He saw European powers drifting inevitably towards war and he worried over the vulnerability of Sweden in the turbulence that might follow. He considered the state, the basic unit of territorial organization, as the only source of order and protection from inevitable chaos. Yet conventional political science paid no heed to the spatial dimension of national interests. Heretofore, the state had been conceptualized as a legal and constitutional entity with its legitimacy and authority derived from treaties, dynasties, or alliances. But now, Kjellén argued, there was a higher form of legitimation grounded in geographical space.[47]

Although Kjellén has been given credit for coining the term 'geopolitics', he was merely interpreting the German natural scientist Friedrich Ratzel's ideas on how geographical factors influenced the nature and behavior of the state.[48] Ratzel imagined the state to be a biological organism existing in geographical space, with specific requirements for survival and success. Political phenomena, in Ratzel's mind, were natural and organic, and they underwent the same processes and were subjected to the same laws as any other phenomena of the natural world. As a living organism, in competitive struggle for survival, with an increasing population, a state needed space in order to sustain and nourish its civilization. A dynamic state was not a territorial space fixed for all time, but an organism endowed with the right to expand; the more living space it achieved, the greater its power would be. Ratzel predicted a bright future for those states that had large territorial space such as China, Russia and America or states with vibrant colonizing missions. In landlocked Germany, these geopolitical ideas fell on fertile soil. To be a world power, Ratzel argued, Germany must join the search for living space (*Lebensraum*) through territorial acquisitions.[49]

To ignore the legacy of organic metaphors as applied to geopolitics leads to questions of how the Uncertain States of Europe as a self-organizing system seeks stability and self-maintenance in the present. How does this organism expand and react to perturbations or obstacles in its environment if these are conceptualized as alien irritants it wants to reject or metabolize? The whole purpose of the organic analogy is to build into the political, social, economic, military, or whatever system is being imagined, the same purposive behavior witnessed in biological functions – self-maintenance being the primary objective.[50] It reduces the European system to an immaterial entity, a universe of spatial forms and communicative flows that sets up boundary maintenance as its purposive behavior.[51] If certain cultural groups are seen as a threat to this self-organizing system, the only response may be exclusion, isolation, rejection, or finally destruction.

44 ibid. p. 19.
45 Balibar, Etienne; 'Europe: Vanishing Mediator?' pp. 37–44; and Tibit, Bassam; 'Muslim Migrants in Europe: between Euro-Islam and Ghettoization', 31–52 in **Muslim Europe or Euro-Islam.**
46 Ó Tuathail, Gearóid; **Critical Geopolitics: The Politics of Writing Global Space** (Minneapolis: University of Minnesota Press, 1996) p. 25.
47 ibid. pp. 13–19, 44–45. See also Parker, Geoffrey; **Geopolitics, Past, Present and Future** (London: Pinter, 1998) p. 17.
48 Parker, Geoffrey; **Geopolitics** pp. 13–15.
49 ibid. pp. 15–18; and Ó Tuathail, **Critical Geopolitics** pp. 36–38.
50 In 1950, The US Air Defense System was described as an organic system possessing 'sensory components, communication facilities, data analyzing devices, centers of judgement, directors of action and effectors, or executing agencies.' All organisms, this report maintained, have the power to develop and grow; they are supplied with material; they can sense their outside environment and their own activities; and their main function is to interact with and alter the activity of other organisms. 'Progress Report of the Air Defense Systems Engineering Committee,' (1950). Quoted by Fox Keller, Evelyn; **Refiguring Life** (New York: Columbia University Press, 1995) pp. 90–91.
51 In the 1980's, the global market and transnational corporations were conceptualized as

There are further questions to be asked concerning the visualization of geographic space and its strategic mentality left unexamined in *Uncertain States of Europe*. Creating a geographic gaze or visualizing the entire space of the world was a pedagogical project of Halford Mackinder, a British geographer, also involved in geopolitical undertakings at the turn of the 20th century. Strongly committed to a visual and aesthetic organization of space, Mackinder argued that proper training of the faculty of sight would enable the creation of a 'mental map' of the world as a single unified whole. Addressing the Manchester Geographical Society in 1890 he asserted:

> This is geographical capacity – the mind which flits easily over the globe, which thinks in terms of the map, which quickly clothes the map in meaning, which correctly and intuitively places the commercial, historical or political drama on its stage.[52]

There is a hidden epistemology of power and abstraction within this geographic gaze, which Boeri rightly criticizes when he attacks geographical conventions. This detached pictorialization of the world, grasped by an elevated, disembodied, and scanning eye, visualizes maps as white pieces of paper on which to leave its imaginary marks. It is a willful gesture that rearranges the world and history in turn.[53] In 1895, Mackinder defined an ideal geographer in the following manner:

> He is a man of trained imagination, more especially with the power of visualizing forms and movements in space of three dimensions… He has an artistic appreciation of land forms, obtained … by pencil study in the field; he is able to depict such forms on the map, and to read them when depicted by others… He can visualize the play and the conflict of the fluids over and around solid forms; he can analyze an environment, the local resultant of world-wide systems… It may or may not be that we can think without words, but certain it is that maps can save the mind an infinitude of words.[54]

Mackinder considered world history to be the outcome of struggles for supremacy over land or sea. He believed protagonists achieved over time a balance of power so that neither land nor sea powers attained permanent dominance. But in the early years of the 20th century, when the two powers in confrontation were the British and Russian empires, he apprehensively gave the geopolitical advantage to land.[55] In a 1904 address to the Royal Geographical Society, 'The Geographical Pivot of History', Mackinder predicted the end of what he called the 'Columbian Age' – a period which began in the 15th century and gave rise to European expansion and world domination but was based on what he now saw as outmoded maritime power:

From the present time forth, in the post-Columbian age, [he wrote] we shall again have to deal with a closed political system, and none the less that it will be one of world-wide scope. Every explosion of social forces, instead of being dissipated in a surrounding circuit of unknown space and barbaric chaos, will be sharply re-echoed from the far side of the globe, and weak elements in the political and economic organism of the world will be shattered in consequence.[56]

Besides this description of a Darwinian survival of the fittest nations, Mackinder defined 'the geographical pivot of history' to be that great extent of territory in Central Asia that had no direct access to the sea. It was pivotal, because it was largely invulnerable to sea power and would eventually rise to dominate.[57] Reformulating this theory after WWI, 'pivot' became the land-based 'heartland' where the future of world power lay. His accompanying map, entitled 'The Natural Seats of Power', reduced historical and political conflicts to formulaic east/west, land/

self-regulating organic entities, in particular self-regulating systems. This rhetoric legitimated all sorts of conceptual convolutions in the name of the free market and free trade, while willfully dismantling any regulations on the commodification or privatization of the public sphere. Armand Mattelart criticized these metaphorical transpositions for failing to establish their epistemological groundings and thus blurring their reference points. 'The enterprise of the 1980's [he wrote] becomes an immaterial entity, an abstract figure, a universe of forms, symbols, and communication flows, in which the problems posed by the restructuring of the world economy and the redistribution of the dependencies and hierarchies on the planet become diluted...[into] a vaporous world of flows, fluids, and communicating vessels evolving into "dissipative structures"'. Mattelart, Armand; **The Invention of Communication** translated by Susan Emanuel (Minnesota: University of Minnesota Press, 1996) p. 306. 'This new way of thinking involves not only free enterprise but also veritable cult of enterprise,... to the point where many firms have taken their desires for reality, their project of corporate development for internal democracy, their discourse about new internal communications, for the advent of employee participation and mobilizations, and new forms of corporate self-organization for new means of personal realization.' Mattelart, Armand; **Mapping World Communication** translated by Susan Emanuel and James A. Cohen (Minnesota: University of Minnesota Press, 1996,1994) pp. 208–9.
52 Ó Tuathail, Gearóid; **Critical Geopolitics** p. 87.
53 To maintain neutrality and objectivity, the young geographer, Mackinder advised, should be trained 'in the use of an almost Ruskinian, purely descriptive language, with terms drawn from the quarryman, the stonemason, the farmer, the alpine climber, and the water-engineer'. Mackinder, Halford; 'The Human Habitat', **Scottish Geographical Magazine** 47 (1939) p. 334. Quoted in Gearóid Ó Tuathail, **Critical Geopolitics** p. 101.
54 Mackinder, Halford; 'Modern Geography, German and English', **Geographical Journal** 6 (1895) p. 376. Quoted in Ó Tuathail, Gearóid; **Critical Geopolitics** p. 106.
55 Mackinder's organic concepts of geography, extended to the consideration of 'national efficiency' or concern over the physical deterioration of military recruits due to what were assumed to be hereditary and environmental factors. The nation was an organism that needed to be kept fit and healthy if it was to survive in the competitive milieu between states. Along with sea power and land power, Mackinder added manpower. 'Let empire-builders who that they value manpower at home and in the colonies more than wealth, and the masses of our countrymen will learn to value the Empire as the protection of their manhood.' Quoted in Ó Tuathail, **Critical Geopolitics** pp. 91–92. See also Coones, Paul; **Mackinder's 'Scope and Methods of Geography' after a Hundred Years** (Oxford: School of Geography, 1987) pp. 13–22; Cuddy-Keane, Melba; 'Modernism, Geopolitics, Globalization', **Modernism/Modernity** 10, 3 (September 2003) pp. 539–558.; GoGwilt, Chris; 'The Geopolitical Image: Imperialism, Anarchism, and the Hypothesis of Culture in the Formation of Geopolitics', **Modernism/Modernity** 5, 7 (September, 1988) pp. 49–69; Parker, Geoffrey; **Geopolitics, Past, Present and Future** (London: Pinter, 1998) pp. 19–22.
56 Mackinder, Halford; 'The Geographical Pivot of History' (1904). Quoted by Ó Tuathail, **Critical Geopolitics** p. 27.
57 Ó Tuathail, **Critical Geopolitics** p. 33.

sea struggles; the simple pivot around which international politics turned and history was made. In strident imperialistic tones that would re-echo throughout most of the 20th century, Mackinder proclaimed, 'who rules East Europe commands the Heartland:/ Who rules the Heartland commands the World-Island:/ Who rules the World-Island commands the World.'[58]

Representing the state as an organism in either Ratzel's theory of competitive struggle for survival, or Mackinder's East-West 'balance of powers' led to dramatic political events and two world wars. The organic analogy as an interpretative strategy suppresses awareness that space is a political and social entity whose form is the product of historical and geographical struggles over the definition of the nation and the legitimacy of state borders. It sees territorial expansionism, imperialism and militarism as natural processes and not contentious processes of social, economic and political origin. Wars and conflicts are merely viewed as inevitable outcomes. Mackinder spoke to the leaders meeting to redraw the map of Europe after WWI, as they struggled to bring new nations into existence and redistribute the former colonies of Germany. He declared:

> The great wars of history – we have had a world war about every hundred years for the last four centuries – are the outcome, direct or indirect, of the unequal growth of nations, and that unequal growth is not wholly due to the greater genius and energy of some nations as compared to others; in large measure it is the result of the uneven distribution of fertility and strategic opportunity upon the face of the globe.[59]

Another device that needs to be historicized in the Uncertain States of Europe is the aerial view. In the closing years of WWI, the airplane gave clear demonstration of the military value of aerial photography – its realism, accuracy and reliability.[60] Its use was extended after the war to map-making and to other purposes such as geology, archaeology, and agriculture. But its application to the planning of cities was minimal: aerial photographs for map revision, or bird's-eye views of the city were common but infrequent. Only slowly did aerial photography become an essential tool to be deployed in the inventory of existing conditions and for planning the metropolitan region beyond city limits. Vertical photographs exposing an entire city allowed the planner to understand the pattern of open spaces, their interrelationships with built forms, and the role man held in the creation of an entire metropolitan region. By taking photographs at regular five-year intervals, a history of a city's development over time could be obtained. So it was argued by the American city planner Melville Branch after WWII, when urban planning begins to understand the city as a complex organism of many forces, aerial photographs will be an effective means for depicting the city as a reality of many interacting parts.

The nature of aerial photographs is instrumental and distanced; Branch provided the planner with a training manual of how to understand their potential use and cognitive effects, how to interpret and manipulate the imagery. As he outlined vertical stereo photographs, exposed with the camera directed downward toward the ground; they presented an overhead or plan view. When viewed through special instruments, these images revealed three-dimensional effects of ground features and contours of relief. Oblique photographs were much less useful for urban planning. While readily comprehensible by the untrained eye, only features in the foreground tended to be distinct, those of the middle ground less easily differentiated, while the background very small in scale and largely indistinguishable. Most streets and open areas were obscured by the interposition of structures and landforms. Oblique photographs, which tend to cover larger areas of ground than vertical photographs, were useful only when the topography was not rugged and the land not crowded with structures. Thus the vertical perspective became the accepted tool of the planner, and more recently, it has been deployed effectively to photograph environmentally impacted landscapes revealing man's destructive erasure of the natural terrain.

While *Uncertain States of Europe* correctly assesses the instrumental and completely detached nature of this aerial view when applied to mapping the urban terrain, preferring the more subjective and detailed oblique perspective which supposedly enables evolutionary processes to be grasped over time, still the vertical view cannot be so summarily dismissed. Such rejection suppresses awareness that aerial photography is still a strategic device of the military. Michel de Certeau called this a 'Cartesian attitude', defined as 'an effort to delimit one's place in a world bewitched by the invisible power of the Other. It is typical of modern science, politics and military strategy.'[61] Since geopolitics presents a flat horizontal discourse about the land, it looks across rather than cuts through the landscape.[62] It locates on a coordinate grid – or flatbed – the exact location of both stationary and moving targets. Once located by aerial surveillance or Global Positioning Systems and the coordinates inscribed on a military map, these targets are marked for aerial annihilation. Such '… is the cartographic imagination inherited from the military and political spatialities of the modern state.'[63] But network systems are also horizontal schemas, and it is questionable if they are less abstract and less prone to strategic deployment.

The United States in Iraq, and Israel in Palestine, have clearly demonstrated their technical supremacy in deploying their military systems of space-based electronic surveillance and remote targeting technologies that rely on both a communication network and a horizontal coordinate

58 Earle, Edward Mead; 'Introduction' in Mackinder, Halford J.; **Democratic Ideals and Reality** (New York: Henry Holt and Company, 1942) p. xxiii.
59 Ó Tuathail, Gearóid; **Critical Geopolitics** p. 54.
60 Branch Jr., Melville C.; **Aerial Photography in Urban Planning and Research** (Cambridge: Harvard University Press, 1948).
61 de Certeau, Michel; **The Practices of Everyday Life** translated by Steven F. Rendall (Berkeley: University of Berkeley Press, 1984) p. 36.
62 Graham, Stephen; 'Vertical Geopolitics: Baghdad and After', **Antipode** 36, 1 (January 2004) pp. 12–23.
63 Weizman, Eyal; 'The Politics of Verticality' (2003). Quoted by Graham, Stephen; 'Vertical Geopolitics: Baghdad and After' in **Antipode** 36 (1), 2004 p. 12.

grid. But such 'network-centric warfare' has unleashed in tandem the enemies' incentive to locate their forces within the complex, congested and protective terrain of cities and towns. Tariq Aziz, the former foreign minister of Iraq, on hearing the association of Iraq to Vietnam proclaimed: 'let our cities be our swamps and our buildings our jungles.'[64] Cities reduce both the effectiveness of high-tech weapons of destruction, plus the ability to fight at remote distances, for they eradicate the full deployment of a horizontal view. As the US Army commentator Ralph Peters noted: 'In fully urbanized terrain, warfare becomes profoundly vertical, reaching up to towers of steel and cement, and downward into sewers, subway lines, road tunnels, communication tunnels, and the like.'[65] In the resultant verticalization of geopolitics, the city and warfare are drawn together in ever tighter and reciprocal waves of destruction. The vertical view of cities, a sectional slice through underground bunkers to above ground towers, retains great strategic importance. When cities become verticalized defensive spaces, they threaten the military's 'full spectrum of dominance' because they block the flow of information, which 'network-centric warfare' requires. Hence they become objects to be destroyed regardless of 'collateral damage'.

Conclusion: By creating a multiplicity of entangled lines that cut across questions of European identity, cartographic procedures, linguistic analogies, and theories of biological evolution, Stefano Boeri and his research network represent European space as a multilateral, multi-noded, multi-entry construction. It is just as abstract and detached from reality as any other map used as a metaphor for spatial ordering, a surface for notating field observations, or a medium of rationality. Furthermore, their borrowings from evolutionary biology are questionable. It may be supposed that the informational code within organisms produces their characteristic structure and behavior. But can this analogy be applied to the states of Europe and the diffused city? Can the invention of codes describe their properties and how they operate and interact with each other? And if an organism *just is* – neither progressing towards nor regressing from a state of perfection – then the procedure is not only an epistemological failure to describe in words the evolutionary processes that we can merely visualize, but a failure as well to design any operational procedures that might achieve an enhanced condition or better environment. Squeezed between homogenizing forces from above, and fragmenting and fracturing energies from below, deploying an organic metaphor that assumes an organism always achieves stability or a strong sense of identity, may offer an inadequate perspective on the uncertain state of European identity and the geopolitical questions that trouble its space.

64 Aziz, Tariq (Fall, 2002). Quoted by Graham, 'Vertical Geopolitics' p. 15.
65 Peters, Ralph; 'Our Soldiers, Their Cities', **Parameters** (Spring, 1996) p. 2. Quoted by Graham, 'Vertical Geopolitics' p. 14.

What is a nation? A flag, a common history, a spirit of solidarity, a language, a culture…. Or NONE of these?

Eclectic: "The emblems of American mass culture have infiltrated the remotest outposts: the Coca-Cola logo is on street corners form Kazakhstan to Bora-Bora; CNN emanates from television sets in more then 200 countries; there are more 7-Eleven stores in Japan than in the United States. Our technology --- computerized weapons systems, medical scanners, the Internet --- sets that standard to which developing countries aspire. Even our teeth, gleaming, beveled, orthodontized into orderly white rows, are the envy of the world."
["How the World Sees Us," The New York Times, June 8, 1997]

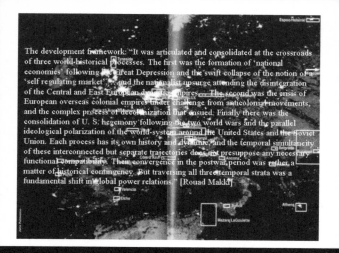

The development framework: "It was articulated and consolidated at the crossroads of three world-historical processes. The first was the formation of 'national economies' following the Great Depression and the swift collapse of the notion of a 'self regulating market'…, and the nationalist upsurge attending the disintegration of the Central and East European dynastic empires…. The second was the crisis of European overseas colonial empires under challenge from anticolonial movements, and the complex process of decolonization that ensued. Finally there was the consolidation of U. S. hegemony following the two world wars and the parallel ideological polarization of the world-system around the United States and the Soviet Union. Each process has its own history and dynamic, and the temporal simultaneity of these interconnected but separate trajectories does not presuppose any necessary functional compatibility. Their convergence in the postwar period was rather a matter of historical contingency. But traversing all three temporal strata was a fundamental shift in global power relations." [Rouad Makkl]

There are four dimensions to Foucault's dispositif: lines of visibility, enunciation, force, and subjectification, or the process of individuation of groups and people. Untangling the multiplicity of lines within a social apparatus, Deleuze claims, is like drawing a map, doing cartography, or surveying an unknown landscape….

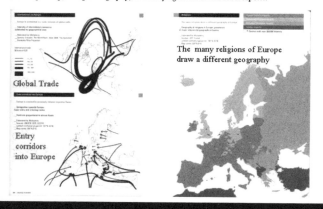

"Today more than ever before, at the end of its long search, European culture is in urgent need of a common language that might heal its linguistic fractures. Yet, at the same time, Europe needs to remain true to its historic vocation as the continent of different languages, each of which, even the most peripheral, remains the medium through which the genius of a particular ethnic group expresses itself, witness and vehicle of a millennial tradition. Is it possible to reconcile the need for a common language and the need to defend linguistic heritages?" [Umberto Eco]

"Living in the city can be seen as participating in a public experiment on our health, like when we have a little diesel particle going into our lungs... So we are all part of a laboratory in the sense that the separation between what is inside and what is outside has been blurred. ..... So I use laboratory as a metaphor for what happens to the public when it is engaged in a public experiment. In general I am trying to find bizarre terms in order to capture these new situations where the difference between science and society is so undefined." [Bruno Latour]

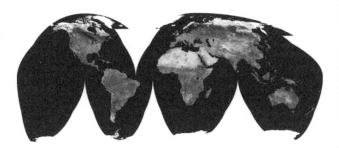

Is the aerial view always a matter of zenith arrogance?

When urban planning begins to understand the city as a complex organism of many forces, aerial photographs will be an effective means for depicting the city as a reality of many interacting parts. (Melville Branch, *Aerial Photography in City Planning*, 1948)

Terrorism and 'shadowy networks': are transnational, with a fluid mutable sense of space, adapting to external pressures, such as the destruction of gathering places by reconstituting internal cohesion in new ways. [Ettlinger and Bosco] Are all networks described as autopoietic systems?

Productive Space
Stephen Read

Everyday life is *the 'integrated by form'*
Ash Amin and Nigel Thrift[1]

Hybrid world, mobile society For a long time people concerned at the planning or design level with cities – and not just cities as logistical-functional or formal-architectural (engineered or aesthetic) objects but as more or less meaningful distributions of humans and things on the surface of an urbanised or urbanising landscape – have had to negotiate a sociology which gave scant regard to those distributions, or to the inhabited and active physical spaces of cities, which some of us believe are part of their very substance and generative principle. Thinking about cities has as a result tended to neglect the role of *cities themselves*, in their everyday materiality and specificity and in their situating power, in the constitution of our social worlds – seeing the city instead as something sitting rather passively on the receiving end of a process of social or economic or cultural 'production'. A moment of thought is all it usually takes to recognise that, spatially constituted and situated as the social, the economic and the cultural necessarily are, it is hard to imagine how dynamically inhabited physical urban spaces could *not* be productive of everyday societies, economies and cultures. But we are stuck still, it seems, as far as our ideological predispositions are concerned, in an uncomfortable situation that on the one hand regards the idea of a 'deterministic' urban space with suspicion, while on the other recognises the inadequacy of a view, in our dynamic, connected contemporary world, which honours with too much leverage social 'structures' that are founded on static and categorical orderings. These orderings into hierarchies and categories (of class for example), and their 'reflection' or 'representation' in spatially bounded entities (communities, nation states etc.) ignore completely the role dynamic processes of connection may have in driving processes of social formation in the city.

We fail to acknowledge any socially constructive effect of the material city itself. Our concern is to try to find a way to conceive the city as a product of a dynamic space, and to include an understanding of the social effects of the dynamic workings of the city itself. The question is: How do we get beyond a spatial as it is conventionally conceived as a distribution of facilities and events over a surface that is in itself neutral and already there – and approach a spatial that has to do with a modulation of fields, with tendencies to concentration and dispersion and with the orientations and movements with which these tendencies are coterminous? We, as makers and manipulators of urban form, must be looking for a spatial which can begin to indicate answers to the questions of what the city *itself* (its space and its form) is doing with respect to our social, economic, cultural and psychological lives – and with how transformations in one are connected with transformations in the other.

---

[1] Amin, Ash & Thrift, Nigel; **Cities: Reimagining the Urban** (London: Polity, 2002) p. 47.

The power of *horizontal* forces – of motion and connection tied to situation – are becoming more readily recognised today as being potentially productive (or for that matter destructive[2]) of the orders of our social, economic and cultural worlds. The mobilities and connectivities of people, information, commodities and finance at a global scale, for example, are recognised as constituting an increasingly dominant global space, which impacts on and transforms all our local ones in many profound ways. As well; the explosion of mobilities and connectivities at the metropolitan scale has shifted the centres of gravity of our cities in many profound ways we are still struggling to come to terms with, and to reincorporate into our preconceptions and understandings of the order and shape of the contemporary city.

We find also, in this increasingly horizontally mobile configuration, the until recently unproblematic division of the world into realms of 'social' and 'natural' suddenly not looking quite so straightforward. We find ourselves in the middle of a world that, while being quite clearly of our own making, is steadily increasing in dynamism and force, and galloping away from us – adding to the quotient of risk and unpredictability previously identified with a wild and untamed nature. Where does this leave the division between society and nature? – a division designed in the first place to allow us to dominate and tame the wild and 'natural' of the world. The problem has not been solved by seeing the physical as dominated and shaped by a social which somehow assembles itself before doing the same to our spatial and temporal worlds – by seeing a 'structure' of the social as something which automatically, or functionally and mechanically, reproduces and represents itself in the stuff, and the structures, we call urban. We have to somewhere, we believe, come to terms with the fact that many of the processes we encounter in the dizzying flux of our modern world have effects which are somewhat contingent and *accidental* – more or less 'natural' products of the material processes which underpin them, and *not* organically or mechanically and transparently products of the social, and directed as if by rights to our human and social ends. The great artefact of the city therefore, ostensibly socially-made, built ostensibly to our social ends, is not limited in its effects and productions to the realisations of those social ends. There is an *excess* in its productive effects that approaches the 'natural' in its properties of wildness and unplanned spontaneity. These products may nevertheless, and often do we believe, become absorbed and appropriated into our social existences in many diverse and unexpected and creative ways.

This shift of thinking raises an interesting possibility which we should also seriously consider: How many of the processes of movement and connection we encountered in the past in a less forcefully dynamic (but still much more distributed and mobile than we normally give it credit for) pre-modern and early modern world, in a similar way produced an excess of contingent, 'natural' effects; ones we have already absorbed and appropriated into the patterns of our everyday social existences? Is it at least a theoretical possibility that social and social-functional 'structures', and forms, realised and potential, of our situated social existences, emerge *out of situation*? This would place a different emphasis on our conception of social production in the first place, and would place the environment, and its built active spaces and connectivities, much more centrally into our considerations of the production and the ordering of the social.

The dialectical thinking then of many of the most influential social thinkers of the past tended to cast the problem of societies as one of thought, and neglected the role of site and of situated objects and subjects in the production of cities and of the socialities that emerged out of urban movements, connectivities and location in a situating environment. And as John Urry suggests, this does not necessarily imply falling back on the autopoeitic thinking of Luhmann or of the functionalist morpho-generationalism of Archer; the object of the city may be active and generative *without* at the same time being a completely integrated order or system the way we tend to think of these things when we think them through the metaphor of organisms.[3] Most advanced thinking about organisms has in any event left behind any assumption of purposefulness, but purposefulness remains insistently as part of a strong metaphor of organism, overspecifying processes and products that need to be seen more in terms of 'drift' or tendency, and a convergence or cogredience[4] of effects.

The fact that the world, and our cities, are becoming ever more mobile, may have forced some of this rethinking, but it is at least possible that the world has always been constituted this way. The neglect, even negation, of the city as a factor of social production may have to do with the neglect of the matter of the production of our sociality in *situation*.

The view proposed here is therefore one which emphasises action over reflection, the 'natural' and dynamic and situated over the 'structural' and assumed priority of the social – as well as tropisms, cogredience and 'perception as form'[5] over mental maps and cognitions and representa-

---

2 Transformative production is always going to appear destructive from certain perspectives. There is a substantial discourse today of 'loss'; the loss of community, of place, of a certain integrity and 'wholeness' of the world, which we should be careful not to take too quickly at face value.
3 See Chapter 1 of Urry, John; **Sociology beyond Societies** (London: Routledge, 2002).
4 Crucially for the notion of the multiplex city, it is 'cogredience' or the way in which multiple processes flow together to construct a single consistent, coherent, though multifaceted time-space event that is the key concern. The urban becomes an embedded and heterogeneous range of time-space processes. Cogredience is a relation holding between an event and a duration: an event is cogredient with a duration if it extends throughout the duration. An event cogredient with a duration is at rest with respect to the duration and to other events cogredient with the same duration. An event that is not cogredient with a given duration and its cogredient events is in motion with respect to them. **cf.** for details: Whitehead, A. N.; **An Enquiry Concerning the Principles of Natural Knowledge** (Cambridge: Cambridge University Press, 1919/1925) pp. 128–138.
5 The phrase comes from Merleau-Ponty's idea of a 'structure of behaviour'. See Merleau-Ponty, Maurice; **The Structure of Behaviour**, Pittsburgh, Penn.: Duquesne University Press, 1983).

tion.[6] Attention shifts to the active part the environment plays in the constitution of the social; the environment encountered in movements comes to constitute and generate at least a part of the social envelope in which people find themselves and act. To the extent that this envelope is a product of this movement and this encounter, the social world becomes something that is suspended within and supported by these movements rather than being pre-constituted and inserted into an environment in which movements just take place, as if against a neutral background and without regard to its dynamically relational situating substrate. The social becomes dependent on exact movement and connectivity patterns, by its very nature therefore, something radically provisional, and liable to change as configurations of mobility and connectivity change.

All this becomes much clearer in fact when we remove ourselves, and our intentions and expectations (and fears), from the exact centre of the stage. When we understand and acknowledge that many or most of the things that happen around us in fact just kind of happen – not quite the way they are intended or expected or structured to happen, but still as part of the world in which we are fully and integrally involved and immersed, and in which we deal with events as opportunities and as *openings* as much as endings and consequences. It becomes easier to see from this perspective, when we don't overdetermine the world in structures, why the world is always opening out for us, why closure eludes us – why the world is in fact much less determined and closed than we or our governors expect it to be. It becomes easier to see why the effects of a material, not entirely tamed, continually transforming city itself may be a part of this fabric of opening and part of the generation of creative potentials which affect and infect and transform our unfolding lives.

Putting cities first For a long time therefore, we have considered cities and the ways they appear and the way we interact with them to be a relatively trivial matter; a matter of style, image and 'representation'. The *social* was what we understood as being the active force behind productive processes, and the brute material of the city, what was acted upon – the more or less resistant clay, formed to more or less conform, by an active principle which sculpted it. The city was 'socially produced', and once produced, was part of the 'just there' – laid out (not always very tidily) on a horizontal surface and transparently available as a more or less intelligibly structured object to our intentionality and agency.

The question we really want to ask is: What is the *productivity* of urban space? What does urban space, of its own account, produce?

Henri Lefebvre says: 'Each network or sequence of links – and thus each space – serves exchange and use in specific ways. Each is produced – and serves a purpose;

and each wears out or is consumed, sometimes productively, sometimes unproductively.'[7] This is clearly not the end, nor even for us, really the beginning of the story. What about the creative productivity of this thing that is produced? It doesn't always wear out – it may generate its own and unexpected uses and products, regenerate in unplanned ways, it may shift spontaneously from one production or productive mode to another, and it is seldom consumed to nothing. Even in its decline we see it accommodating and becoming productive of probing, opportunistic energies of the marginal – energies which even in their marginality, perhaps because of it, identify and find creatively divergent paths to the future. It is precisely at the point where urban space *itself* becomes productive – where the urban body begins to generate its own constructive (or destructive) effects on the social body – that we can start to talk about a space that is worthy of our attention, and interesting for us as designers. The physiological processes of this urban body are the subject of our research. We act as physicians and diagnosticians on this body of our interest, calling it to account for the way it adds to or subtracts from its own potential to participate in the growth and creative transformation of the social body – as well as, of course, its potential to participate in the growth and creation of the individual and collective human body and spirit.

In order to understand how this productive capacity of urban space is activated, we need to see the city firstly as a dynamic space, capable of generation out of its own dynamism. We need to see it in the first instance as a producer of a configuration and circulation of dynamic social material, or better material in potentially social relations. We need to see it not in terms of 'representation' or 'reflection' of social or cultural 'structure', or as neutral background to individual and collective subject-centred desires, actions and identities, but rather as a gathering and patterning of connected material; a forming of lives and livelihoods within a circulation of material activated by the force of time which is the 'engine of becoming'. This is a bit like the physiological body, which is a configuration of materials which if it remained just that would be without life. A body is vitalised by dynamic process – drawn together by the force of time – which drives it and sustains it in its forms and in its vital productivity.

The city, as a material trace on the face of the earth, is in fact made twice: firstly as we pursue the everyday demands of daily lives, adding to the city as a material aggregate of work and construction and energy and exchange; and then by the aggregation of produced spaces as they intermesh and turn back on us and produce around us a dynamically integrated world of social, economic and cultural interface and form. It is made by us on the one hand as the aggregated productivity of individuals and on the other by our productivity as multitudes. As individual human subjects we enter into and inhabit a space *already created* by us as dynamically situated populations.

6 See Read (2005) 'Questions of Form'; paper presented at the 5th Space Syntax Symposium, Delft University of Technology, Delft.
7 Lefebvre, Henri; **Production of Space**, trans. D. Nicholson-Smith (Oxford: Blackwell, 1991) p. 403.

If the suburbs of the US and the rest of the 'developed' world are today the pre-eminent place of the individual, relatively poor in the sustaining and integrating fabric of the population, Asian cities and cities of the 'developing' world exist as those pre-eminently of the population. Stretched to their limits by the dynamism and momentum of numbers, they produce copiously, social and economic energies proliferating out of every crack and fissure. The response of the planning establishment, when it exists, is often one of terror, and reaction against the 'chaos' of the uncontrolled, and against the 'inhumanity' of numbers. In fact it is from these very numbers, in their situated streams which cross and overlap each other in rather precise patterns within complex and layered webs of connectivity, that interfaces of exchange and interchange are formed, supporting and sustaining existing situated social-spatial forms and producing new ones.

The aim of our research is to explore the socially creative potentials of the self-integrating and self-proliferating urban body – as well as to acknowledge and identify the real and sometimes dangerous stresses on this body. In the end we are not sure what part the city itself is playing in the lives we live, and in the changes we experience around us. The city itself and urban space may be productive of more of these effects and these changes than we customarily acknowledge, and we are particularly interested in the ways the city may be opening us to, or be constraining of change. This is a particularly important point given the speed of social and urban transformations today and the need for city and society to continually adapt to new demands and opportunities. We go further and try to see if the city itself is capable of being a space which actively *promotes* or facilitates change. The city itself may indeed be a generator of change – but not in all places, and not under all conditions. What are the conditions and parameters of an *open* and creative urban space? We need to know also what the relations are between transformations happening in the social-technological sphere and those happening in the shape of urban space and the city. We are in need of ideas about how we can profit from and intervene in the creative processes of the city without simply strangling them with our sterile and fearful visions of order and control.

But in order to throw light on these questions, we need first to know some of the technical details of how socially and economically productive interfaces are formed in the dynamic of urban space – about how the basics of social proximity and distance are established within webs of connectivity. These basics are – or should be, we believe – the most elementary building tools in an urban design practice which considers the city as a generator of social formation and transformation, and a means to creative and sustaining openings to the future.

Building blocks of urban creation: network 'shells', interface, centrality

What I hope to illustrate here is how design and research, by focussing on the specific and concrete object of the city can begin to show us where the pivotal points of urban formational (and transformational) organisation might lie. I will begin to look at key concepts like 'distance' and 'centre' and make some steps towards an understanding of cities as dynamically relational entities – not as relational entities simply strung together and expected to hold together by their own devices, but rather as relational entities integrated by a dynamical and stabilising spatial schema or diagram. These ideas revise some rather taken for granted presuppositions about the city. They privilege the topological over the metric with regard to space, and the dynamically relational and mobile and provisional over the solid, the compositional and the static.

It is part of the starting hypothesis of our research on urban form and formation, that urban-social form emerges at 'interfaces' between horizontal infrastructural 'grids' of connectivity in the overall connective fabric of the city. The first motivation for this presumption is empirical evidence of the emergence of patterns of activity and centrality in the movement grids of the Dutch city. This evidence comes from research on the form of the Dutch city[8] – and our elaboration of ideas coming out of this research is illustrated here through recent research work done in Spacelab. We have, over the last three years, been developing and deepening a vision of how the material flux of the city is implicated in establishing central locations and active areas of public space, and in locating points and areas of social and cultural vitality and creativity in the urban surface. The extension of the basic hypothesis to the role and creative potentials of larger-scaled movement and ICT networks in forming and transforming socially active space is the subject of current PhD research, but the present paper will limit itself to describing the most basic building blocks of a new urban design practice, at the level of neighbourhood centralities in urban fabric.

Our research begins to see the city itself as a organising and sorting apparatus; a coordinator, sorter and regulator, in built networks, of multiple space-times, and an organiser of the interfaces between different *speeds* of movement, reflecting different space-times, within the connective fabric of the city. It begins to see the city as a device forming, transforming or refracting of social relations by way of its sorting of space-times, and of their meeting in rather precisely defined and structured interfaces. The medium is movement, not mobility – we are not as interested in the more discussed contemporary question of mobility as in the much more fundamental idea of *movement* as an underlying principle of urban organisation.

---

8 See Read, Stephen; **Urban Life** (Amsterdam: Techne Press, 2006 forthcoming) and Read (working paper), 'The patchwork landscape and the "engendineered" web; Space and scale in the Dutch city'. Available on request.

A point about organisation needs to be made before we start, and that is that the productivity of organisations has been considered before – from the point of view of theories of systems, especially related to the organisation or self-organisation of natural and eco-systems. In general our conventional assumptions about organisation are dominated by static organisations and by organisations in equilibrium. In fact there are many systems in nature which exist on the so-called 'edge of chaos' and whose capacity for creativity is founded on the simple fact that the system as a whole topples regularly out of equilibrium, allowing the whole to reconfigure itself from the ground up in a new way. The productivity of the system in fact is founded not on equilibrium but quite the contrary on disequilibrium – or on the serial breakdown of equilibrium conditions and a rebuilding in ways which may share organisational similarities with the original system but may also differ significantly as regards the disposition and make-up of the constituent parts. Von Bertalannfy had pointed out already in the 1960s a slightly less radical version of this theme of barely organised disequilibrium, the so called 'steady-state' system, where a steady state is maintained, which is not an equilibrium, by the continual through-flow of matter or energy or information or whatever. The important point is that, in both cases, productive states are dependent not on any sort of mechanical or systematic closure, but on an *openness* of the system to invasions and disruptions, serial or continuous, from the outside.[9]

Equilibrium thinking has also played a huge role in the past in theorising about cities and continues to dominate popular presumptions about how the city hangs together and works. What I will begin to elaborate here is a story of simultaneous stability and instability – where active stable layers in the connective fabric of the city produce in the interval *between* layers, conditions which are capable of supporting activity forms which belong simultaneously to both layers, and are in a sense a new actualisation out of a convergence of different potentialities. It is these situations, *suspended in* the interval between layers, that appear, from the empirical evidence to hand, to produce precise actualisations of urban activity and what we call 'urban-social form'.

The emergence of a network 'grid'

The first point I want to address, through the research of Martine Lukkassen is one about the 'natural' – if one can use the word – contours of the city. In order to make this point I want to point out first two commonly assumed ways of spatialising and contouring the city, one related to time, the other to space, which I will oppose with another which could be called a 'psychogeographical' perspective.[10]

A very influential view on the city today reduces it to lived time. In this view, influenced by the rapid increase in speeds and reduction of travel times, the city and its locations are seen as an availability, to be taken and used in an unproblematic way, by way of time-budgets or 'space as time'. Much urban research on centrality related to travel times assumes this view. The experiential and everyday-functional shape of the city is seen as something each of us puts together ourselves out of networks of locations of everyday personal significance. This 'putting together' is something that happens in relation to a personal diary of appointments and movements. The city becomes a simple geographic mapping of a temporalised (in clock-time), and otherwise despatialised, personal existence.

Another view is connected these days to the first, though its history is older and has to do with the definition of neighbourhoods and areas by their boundaries. I'll describe it in the version in which it is connected to the 'space as time' view above: When someone lands at her destination, in the space as clock-time model described previously, on time and with no cognisance given, or needing to be given, to the interval of space between arrival time and the time of departure from the last place to appear in her diary, she finds herself in a place perfectly centred on the location where she has landed. This place has certain properties or attributes which attach to it – like architectural and historical type or period, dominant programme and use, social or ethnic composition – which then get used to define boundaries around the area where these characteristics apply.

What we get are bounded, named, spatial domains – islands in an archipelago of other islands, with attributes hanging onto each like labels – which are linked to each other in another framework entirely; that of perfectly and smoothly mobile and connected 'travellers in time'. *Connective* space becomes a smooth undifferentiated time of movement; *locational* space, an archipelago of disconnected islands, and the only remotely interesting thing we can say about contemporary urban space from the perspective of this model is that it is firstly turning into islands of the local – and then that where it is not local it is becoming 'compressed' with increasing speeds and connectivities. *Space itself* in this view is not productive – it is either 'just there' in the case of the locational and local, or it is 'overcome by time' in the case of the connective and non-local.

A view of urban space which attempts to deal better than this with its experiential properties (and as we will see with its everyday functional properties) is that of the Situationists. They proposed, as an analytical technique, the *dérive*, described by Guy Debord as 'a technique of rapid passage through varied ambiances'. Analysts

> ... let themselves be drawn by the attractions of the terrain and the encounters they find there. Chance is a less important factor in this activity than one might think: from a *dérive*

9 Von Bertalannfy; **General System Theory** (New York: George Braziller, 1969).
10 Lukkassen, Martine (2004), Research monograph: 'Transurban Situations'. Available on request.

point of view cities have psychogeographical contours, with constant currents, fixed points and vortexes that strongly discourage entry into or exit from certain zones.[11]

When Martine applied the *dérive*, in the city of Rotterdam for example, she found indeed that the city appears to have psychogeographical contours that encourage or discourage entry into or exit from certain zones.

If we take a well-known division of Rotterdam into 'islands in an archipelago' as a test case for these urban ideas[12], as soon as we start applying the *dérive* a quite different division of the city begins to suggest itself. Instead of dividing itself up by 'island' areas, with clear boundaries, what we find is that the city starts dividing itself into quite clear *horizontal* and distributed network-levels of consistent ambience. The technique of the *dérive* begins to articulate traceries of consistency – unities of speed, character and functionality in the city – which themselves map as distributed movement networks. In plan they map as webs or *grids* defined by particular characters or atmospheres – attached, I would argue, to particular *speeds* of movement. These grids tend to hold the traveller rather tightly *within* consistencies, discouraging exit from the grid itself. Entry into or exit from these horizontal grids involves a positive shift, an effort of breaking through a threshold, demarcated by a *speed* and a *time* produced in movement, from one experiential domain to another.

What we find in fact is that far from the spatial fabric of Rotterdam being overcome by time, the island-places of Rotterdam are overcome by the *dérive*, and by the experiential continuities of horizontal grids of constant ambience as they slice through island boundaries as if they were not there. The contours of the city become turned through exactly 90 degrees. From boundaries of neighbourhood areas, they become lines or *ridges* of psychogeographical continuity which when mapped in plan form a movement network grid. These ridges attest to a *continuity* and a *duration* of movement through cities and not to a hyperspace click-on-the-destination resistance-free transit. The research on Rotterdam revealed a dominant *grid of consistency* lying over the fabric of the central city – precisely tying together into a unity of movement and experience what the archipelago model proposes is untied.

We call this grid of consistency, dynamically produced by the everyday spatial practice of movement, and unifying the fabric of the central city, the *supergrid*.[13] It is a grid that, in the *experience* of motion through the fabric, floats out of the more general block grid structure of traditional central cities of a European type. We call it also sometimes the *middle-scaled* movement network because as *one* of the horizontal grids of movement affordance and performance I am talking about, it sits between two other distinct grids of consistency: that of the local neighbourhood grid of backstreets and slow movement on the one hand, and the regional or metropolitan grid of freeways and high-speed movement on the other – the one that in the Netherlands facilitates movement for the most part *between* central cities.

This alternative to the archipelago model starts taking seriously, as Bruno Latour encourages us to do, the friction of transit. It also starts articulating the actual and concrete pathways, within a continuous local, between the scales of the neighbourhood and the city and the metropole and the global, again as Latour urges us to do.[14] The distance, in this continuous local, to the scale of the city, will be the distance to the corner of the shopping street – to where we encounter the movement grid that unifies the urban centre. The distance to the metropolitan, will be the distance to the freeway or to the railway station or to those places that the metropolitan has already invaded by way of these grids. The distance to the global will in today's world be rather short, given that the global has already seeped through the global networks into the metropolitan and city networks, and by way of media networks into our very homes. The scales of the neighbourhood, of the city and of the metropolitan are represented and embodied by distributed grids – self-consistent horizontal webs of time-space – that overlap each other. These scales come via grids of consistency to us – they are not far off areas or nodes that we have to travel all the way to. Areas or locations become infected by *all* these scales, to different degrees depending on their exact relationships to the respective grids and webs. There is a certain experiential and functional 'distance' of the city then that has nothing to do with metric measurement. It has to do instead with moving 'up' or 'down' in a layering of grids embodying scales – and this is, we will see, a 'distance' we use to make 'social distance' and places for everyday social lives.

In the first place, in this research on the relation between movement space and experience in Rotterdam, it was discovered that neighbourhood centrality was linked directly to the horizontally-layered structure just outlined. Neighbourhood centres, with their high-street shops and neighbourhood identity, turned out (and this remember is within the more traditional central fabric) to be precisely the same spaces which we had noticed before as being part of a higher speed middle-scaled grid or supergrid which unified our experience of the central city. Now, because we are looking at them in a different way, we notice that they *also* centre the neighbourhood grid of local streets which surround them. What's more, the marginal in local neighbourhoods – which in this research was measured by instances of broken windows and graffiti – could be located in the number of changes of direction from these simultaneously locally centralising and

11 Debord, Guy (1958), 'Theory of the **Dérive**'; published in **Internationale Situationniste** #2. Available at http://library.nothingness.org/.
12 Palmboom, Frits; **Rotterdam, Verstedelijkt Landschap** (Rotterdam: 010 Publishers, 1987).
13 The word is used in space syntax research and reflects the origins of the idea proposed here in space syntax research I did for my PhD completed at the TU Delft in 1996. **Cf.** Read, S.A.; **Function of Urban Pattern** (Delft: Publikatiebureau Bouwkunde,1996).
14 Latour, Bruno; **We Have Never Been Modern** (Cambridge, Mass.: Harvard University Press, 1993).

wider-scaled unifying middle-scaled spaces. Most local streets more than two changes of direction from middle-scaled streets turned out in this particular study to be problematically marginal.

A great deal of research done in other neighbourhoods in other types of fabric all over the world, shows that while there is some variety in the detailed neighbourhood diagrams which work in different cases, the best space in which to inscribe these diagrams is that of layered communicative grids which firstly form horizontal, experientially coherent 'grids of consistency', and then relate to each other in a 'vertical' step-wise or topological way.

Social distance and proximity In the research of Ceren Sezer which focused on two informal settlements in Istanbul, the researcher, building on the results described above, first sought out traceries of continuous ambience in the open spaces of these informal settlements and then investigated the way these related to the life patterns of the people who lived there.[15] What became clear was that the everyday life tactics of people in these settlements involved an establishment of both 'distance' in these 'vertical' topological terms from, as well as proximity in metric and horizontal terms to, the metropolitan scale which is clearly the necessary connection in most cases for establishing a livelihood. There in fact seems to be a pattern in a lot of our research work in rapidly growing metropolitan regions, of the simultaneous holding at a distance of the metropolitan city (to which these people are often recent migrants) and a closeness in metric terms to this same metropolitan for its opportunities for securing a living. In Istanbul this was thought through in terms of what Ceren called 'everyday resistance', which involved both spatial strategies built into settlement form, and living tactics in terms of the ways this form is used for the maintenance of an economic livelihood, at the same time it is used to support patterns of culture and community that they as immigrants bring with them.

It is possible to trace the pathways to the metropolitan and to a livelihood conducted in informal trade as well as in formal and informal employment. These informal settlements may become rich and supportive living environments – they may of course also limit possibilities for self-determination and for individual growth. In the example considered here, many women in traditional marriages seldom leave the grid levels most isolated by the topological step-wise distance mentioned above from the metropolitan.

Another research, by Chintan Gohil, investigated the ways rural villages, swallowed up by the rapid expansion of the city of Ahmedabad, articulate and situate the lives of their inhabitants.[16] Again, it was found that the idea of topological 'vertical' distance is useful in understanding the processes by which people locate themselves in and inhabit the city. The movement patterns of a number of inhabitants of one particular urbanised village were traced in order to understand the role of the village itself and of the differently scaled movement grids of the surrounding city in their lives. Again, movements to the outside had to do with tactics of livelihood, but in this case the 'distance' the village establishes from the metropolitan establishes also an interior space for the production of goods like water pots which are then distributed and sold outside the village in urban and metropolitan scaled movement grids. Again livelihoods are spatial in the way they distribute themselves topologically between the metropolitan and the local. Again the village is productive of particular and 'resistant' life patterns – but here it also operates as a space of production for the outside urban and metropolitan market.

It is interesting to note that the *situations* of differently mobile people in this framework of layered communicative grids may be different – even when they occupy the same topographic place. A tourist or university researcher, who may temporarily 'drop into' the local grids of inhabitation of these parochial places, from higher scaled grids of global and metropolitan movement and travel, is differently situated to local inhabitants because location seen from this ecological perspective always refers local place to the horizons of that person's place in the world. A person's place in the world refers to the mobile and communicative 'reach' of the grids they customarily occupy or have access to.

Productive difference Having established some of the locating and social space defining ideas which will be necessary for outlining an idea of productive urban space, I want to look now at the engine-room if you like of such productivity. The important notions here are those of *difference* and of *interface* – of the infusion or intrusion of one dynamic population into another and its *interface* with another, and of the creative, productive potentials thereby generated for growth and change through an upsetting of the status quo of similarity.[17] What makes the research of Gerhard Bruyns so clear as an example of intrusion and of the creative potentials of difference is the simplicity – one could say the black and white nature – of the problem he was tackling.[18] Apartheid South Africa institutionalised racism spatially, not only through the establishment of separate areas for different races, but also, and working just as powerfully as an instrument of segregation, through the definition of racially specific movement channels. The breakdown of apartheid has removed the legislative underpinning of systematic apartheid but South Africa is *still* constructed as a space of segregation. Gerhard's research proposes a strategy for subverting the logic of this segregation machine by design – by means of a development plan for the centre of Pretoria.

15 Sezer, Ceren (2004), Research monograph: 'Resistance Spaces'. Available on request.
16 Gohil, Chintan (2004), Research monograph: 'Urban Villages'. Available on request.
17 There are two relevant discourses of **difference**. The first is on difference and the public domain which has become central also in the discussion of public space. Leading exponents are Hannah Ahrendt, through Richard Sennett to people like Iris Marion Young and Chantal Mouffe. The second is in philosophy and finds identity not in the categories partitioning similars but in a cogredience of near-chaotic and self-proliferating difference. This is the field of process and 'emergence' philosophers like Whitehead, Bergson and Deleuze.
18 Bruyns, Gerhard (2002), Research monograph: 'Ubuntu'. Available on request.

Gerhard takes the existing infrastructural connections of black and white populations respectively, from the centre to the metropolitan region, and strategically manipulates their outflows into the space of the centre towards a crossing of population trajectories. He manipulates the machine to the end of constructing a space of difference – of 'productive conflict' if you like, as an armature of 'live centrality' – in the public space of his development plan. The development plan includes a massive regeneration plan giving an important impetus to Pretoria's centre at the level of the formal economy. At the same time the movement machine of the centre promotes the logic of the informal economy, folding these two spaces of the formal and the informal through each other in a 'productively conflictual' way. Black and white space, formal and informal economies, are woven through each other, establishing a dynamic steady-state disequilibrium, and giving direction to future small-scaled development without specifying its exact form or outcome.

Generic difference machines Having proposed that the meeting of difference can be a productive force in a productive space, I need to mention a more generic dynamic urban structure and how this has generated spatial productivity as a matter of course in historically evolved urban layouts. We often think of the meeting of difference in the urban surface in terms of borders – it remains a theme for example in Richard Sennett's recent work.[19] In urban configurations however we need to think about dynamic organisation rather than static organisation, and there is a very simple and generic effect of the 'layered movement grids' urban model I mentioned previously, which facilitates the intrusion of one space into another. Grids integrating different regions – we saw earlier that the supergrid integrated the region of the urban centre of Rotterdam while the neighbourhood street and block grid integrated the neighbourhood region – bring those different regions (city and neighbourhood) into direct contact or interface with each other on the high-street. The effect generates a tension between propinquity and distantiation – out of the simple fact that things situated locally can have at the same time relations between themselves and relations with other things that are relatively distant.

How it works is that relations of propinquity demand their own grid of movement and communication for those relations to be performed and actualised. Relations of distantiation similarly require – they seem as a matter of course in real urban environments to *acquire* – their own grid of movement and communication. The two grids, each coherent and distinctive 'grids of consistency', become laid over each other, maintaining their experiential distinctiveness, while at the same time being entirely open to each other. This is the basic active pattern or diagram of the early modern European city up to the early part of the 20th century – already outlined here in Martine Lukkassen's research work on Rotterdam.

The interface of difference that we are talking about is constructed in the superposition of these two grids, as a meeting between the local and the scale above the local in the space where these two grids overlap. This is a productive order, a social 'micro-technology' maintaining a constant and relatively steady state of creative disequilibrium, and it has been enough to support the variety and diversity and small-scaled 'householder's and shopkeeper's' social productivity of the European city for three or more centuries. We can speak therefore about an *interface* between the dynamic populations of the local and the above local, rather than about a *border* as the most basic common ground of difference – and as the 'machine' of productive public space as we understand it from the great 18th and 19th century cities of Europe.

The generic interface between the local (neighbourhood) and 'middle' (city centre) scales in the fabric of the European central city produces as a 'live' space the neighbourhood shopping or high street. This can be contrasted to the relatively 'dead' streets at the local or at the middle scales where no overlap and interface occurs.

Designing by grids   A research and design project by Guillermo Vidal for the renewal of a waterfront district in Buenos Aires used interface ideas as the principle for design.[20] The renewal area was analysed first for what processes of metropolitanisation had brought in terms of changes of organisation, experience and functionality. He was particularly interested in the ways the connective grid brought the higher scales of the city differently to different places, and in the ways contingent regularities, irregularities and overlaps in grid patterns affected this.

The design became a careful rethreading of larger scale influences and flows through the site – with a view to not only restoring the openness of the urban area to influence from the outside but also to the generation of new activity patterns in line with changes in mobility and lifestyle and the changing image and function of the central city. What Guillermo attempted in fact was to establish a pattern of active potentials in the site, set up at interfaces of the local and larger scales, that could respond to changes in mobility and connectivity and accompanying social and urban formative and transformative potentials. Openness is seen in this case as an openness to the contingent meeting of difference and an openness to the contingent ways to a fundamentally open future.

Another very powerful and ambitious example of the way cities can be designed as interfaces between layered scales is that of Gonzalo Lacurcia for the redevelopment of a degenerated area

---

19 See for example: Richard Sennett (2004), 'The City as an Open System'. Paper presented at the **Leverhume International Symposium 2004**, London School of Economics, London.
20 Vidal, Guillermo (2003), Research monograph: 'Edge in Transition'. Available on request.

of Caracas into a new financial and business centre.[21] The shift in development potential in Caracas is today, as it is in countless metropolitanising cities, towards the freeway and the scale served by freeway and other regional transportation grids, and away from the fine-grained grid of the centre. One of the biggest urban design challenges many cities face today is that of the rescaling of the more traditional centre to the metropolitan scale and the creation of new centres at the metropolitan scale. The question is one of how to begin to get out of a new conflation of scales the sort of social and economic productivities that the high-street and the boulevard offered in early modern cities. The crux of the problem lies in the resistance of the freeway to being enfolded, as an urban element, into central urban fabric. Higher scaled and speeded transportation routes remain stubbornly linear and apart, their speeds establishing an almost insurmountable distance from the rest of the fabric as they resist any degree of creative re-appropriation.

Gonzalo's proposal for a new financial district for Caracas spotted the opportunity to repair a break in the central urban fabric and at the same time to overlay and interface it with the one spontaneously emerging metropolitan settlement type we know, the so-called 'edge city'. In this folding of the edge city into the central fabric the opportunities were created, in much the same way as in Gerhard Bruyns' redesign of Pretoria centre, for the creation of a public space which creatively crossed formal and informal and rich and poor into a web of mutual exchange.

A fabric of productivity The productive possibilities of urban spaces lie in their potential to be the sites of a creative interface of difference. They lie in the potentials of overlap between populations inhabiting differently scaled horizontal 'grids of consistency' and their potentials to be intruded into and contaminated. Clearly today there are serious problems with the openness of many spaces to this kind of creative productivity, and we witness this in the many lifeless contemporary spaces we refer to as 'centre' and 'neighbourhood' or 'community'. We propose the more technical and syntactical movement of *intrusion* in order to find ways of thinking directly and instrumentally about a space of *inclusion*.

We try to deal in an urbanism founded in the first instance in space-time and

---

21 Lacurcia, Gonzalo (2003), Research monograph: 'Urban Compressor'. Available on request.

movement. We take our eye off place as a given, to find it popping up again out of the flux. If we can characterise the research agenda of our group it may be here – that we regard the appearance and actualised reality of our cities as effects out of relations and movement and regard the urban connective infrastructure – seen in the most general way – as fundamentally implicated in these effects. What we begin to generate by focusing on movement – by *moving* through the city as a process of knowledge of the urban – is a practical view of issues of spatial productivity. Other views seem to have us skimming over the surface of the city as if it were no more than the plane of our mental activity. Our instruments and our research begin to point to the way the *space* of the city itself imposes its grip, engaging us in a choreography of place and situated collective existence whose effects are active and socially formative, and which we for the most part entirely misattribute to other levels of agency.

It is an urgent task, we feel, of the urban researcher and designer today to find or invent and exploit opportunities for an active cogredience of differences – for a space of co-appropriation and contamination. It is urgent that we invent strategies that avoid closing futures into the shapes constrained by the limits of our present day imaginations or current notions of what may or may not be possible. It is urgent also that we find ways of avoiding projecting our fears onto our futures and constraining and limiting future courses of inventiveness and creative assembly and reassembly of the urban. Our societies have for long subsisted on urban productivities and the everyday opportunities for livelihood, exchange and expression cities have offered. This is particularly true today of a vast urban and newly urban population which exploits generic economic opportunities offered by an urban fabric of creative margins and centres. It is important we maintain open fabrics if we are not as societies to fall victim to that curiously passive, unresponsive, brittle and unforgiving stuff that urban space would otherwise become.

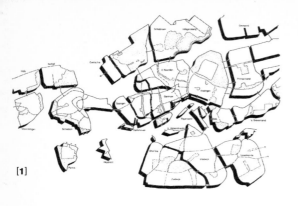

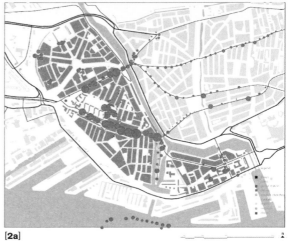

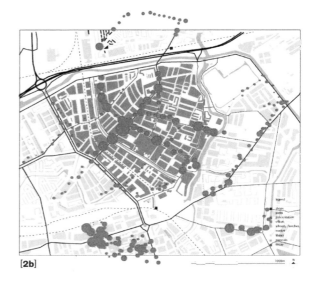

1 Frits Palmboom's division of Rotterdam into an 'archipelago' of 'island' places.
2-2c Patterns of activity in 'archipelago' place – the violation of the boundary.
3 An extra-local scaled pattern of activity 'projected' through Rotterdam centre.
4 The 'middle-scaled' infrastructural grid as an armature of this extra-local pattern of activity.

[3]

[4]

[5a]

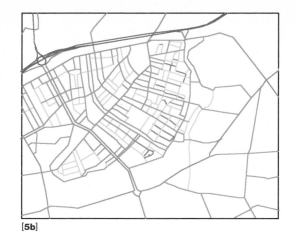

[5b]

[6a]

[6b]

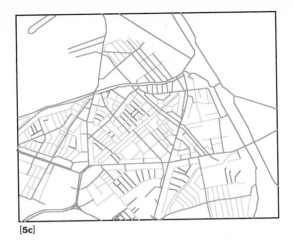

[5c]

5a-c Depth (graded from red to blue) from the 'middle-scaled' infrastructural grid — an instrument used to investigate instances of vandalism and graffiti (tagging) in the local area.
6a-b Pictures: High street in Rotterdam / Backstreet in Rotterdam.
7 The Gaziosmanpasa informal settlement in Istanbul.

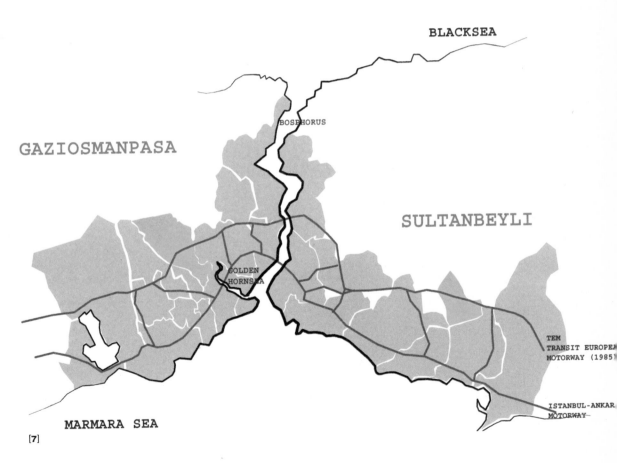

[7]

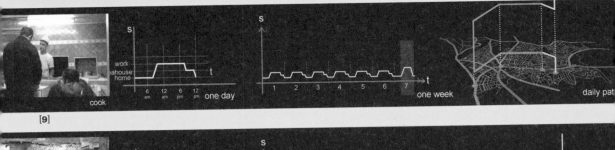

[9]

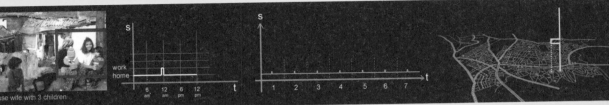

[10]

[11]

8 The distinguishing of different characters (and speeds) of public space attached to different infrastructural web layers.
9 Space-time mappings of people's movements related to the layerings – the use of these layers to establish 'social distance' and place.
10 The movements of women, structured by the layerings of infrastructural grids.
11 The overlap of layers – the creation of particularity in the overlap.

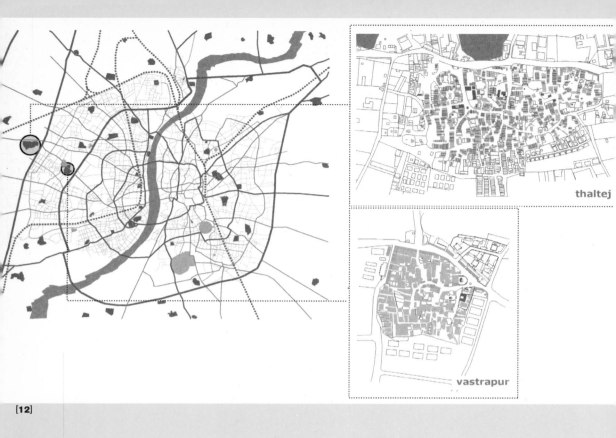

[12]

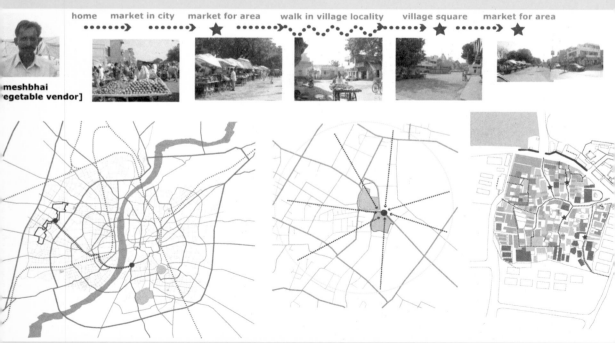

[13]

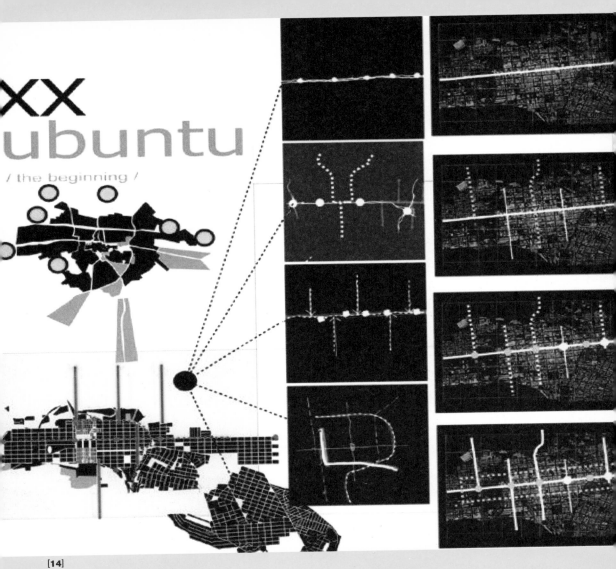

12 Two villages absorbed into the expanding fabric of Ahmedabad.
13 Tracing the movements of a vegetable vendor and situating him in relation to village and metropolis.
14 Physical armatures of 'black' and 'white' space.

15 Subverting the apartheid machine.
16 The metropolitanisation of a waterfront area.
17 'Interfacing' the metropolitan with the local area.

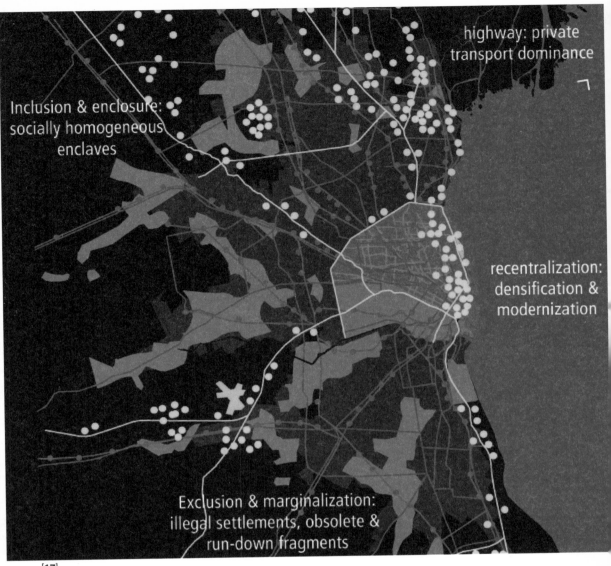

[18]

18 Repairing the grid alongside the transnational motorway.
19 A new 'edge city' business district for Caracas as a fully integrated extension of the existing centre.

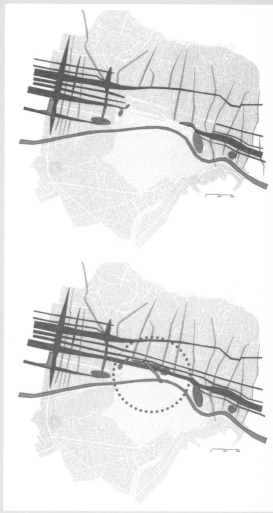

[19]

Opportunities of Clustering: IBIS Network Research and the Urban Question

Marisa Carmona

This paper is oriented towards revising the theoretical framework of the research network IBIS and establishing a thematic clustering in order to improve collaborative network research. This paper elaborates the urban question and cross links of each of the clusters within the general framework. The IBIS research group is a global network which is composed by 40 universities engaged in research on 25 cities.[1] The main research subject is the impact of globalisation in the urban form, the search for new explanations about its transformation, as well as the changing condition between social and spatial dimensions steered by the technological revolution. The aims of the network research is to search for new forms of city governance, city financing and urban forms in order to cope with city sustainability under the condition of interlinked relations between economic growth, poverty, and the environment.

1 Globalisation, Urban Form, and Governance

The analytical position of the IBIS research group is that the intensification of global trade and flows caused by changes in world production and the ICT revolution, brought about unprecedented change and expansion of urban agglomerations, the development of infrastructure networks, and of advanced global services giving a new importance to cities and bringing significant co-modification of culture. There is a condition of continuity and rupture since the weakening of the former phase of capitalist accumulation and the recommendations of multilateral agencies to adjust productive structures to achieve competitiveness and growth have underpinned the present stage of capitalist development known as globalisation. The neo-liberal model provides the ideological framework of globalisation.

The development of information technologies have indeed contributed to changing the world productive system through the mass production of individually made goods organised by transnational corporations, according to new organisational integration forms and intra-firm relationships.[2] Globalisation has intensified the relocation of industrial production to 'just-in-time' industrial production and outsourcing, thereby fostering a new-world division of labour. This new stage of capital accumulation has developed a new creative class, has intensified mobility, and has exacerbated the dispersion and extension of metropolitan areas favouring diversity and fragmentation. Globalisation has made a significant number of functions and activities associated with the former Fordism-based production obsolete, has downgraded labour and has intensified individualism and a consumerism culture. It has resulted in a large amount of urban land becoming vacant and at the same time has created a new demand for land for new functions and lifestyles. It has significantly enlarged social and economical imbalances between world regions, countries and localities and has put society and nature at risk (Beck 2002). Globalisation has also created wealth in an unprecedented way: 70% of Foreign Direct Investments flows is concentrated in OECD countries, whilst 2/3 of the world's population live on less than US $2 per day. Moreover, 2 billion of the world's population still live on US $1 per day in cities of developing countries (Cohen 2004).

Globalisation has imposed a new dynamic on cities as it is both source and generator of economic growth and to achieve city competitiveness has become an essential goal of local economic development. A new type of urban agglomeration has emerged: spatially expanded, polynuclear and highly dispersed, socially polarised and unsafe, economically imbalanced and culturally diverse and fragmented. Without having completely lost its indigenous identity, it has been invaded by a set of symbols and landmarks innate to the new stage of capitalistic modernisation. The rapid extension of cities towards metropolisation has brought profound changes both in its functionality and organisation, accelerating flows of movement and relocation of functions 'in' and 'out' of cities, changing the appearance and image of the centre-periphery, in many cases locating poor and rich sectors closely to each other and thereby making poverty more visible but also socially more articulated.

The IBIS network research takes globalisation as an important component in understanding the transformation of cities, yet sees the city as a social process of building and negotiation, in which different actors state their interests, propose solutions and generate decisions. The IBIS network research seeks to enhance innovation and creativity in the city using a historical and holistic approach that aims to overcome a partial vision of urban problems (architecturally, sociologically, technically and managerially oriented). The ability to create a socially and environmentally sustainable urban form is based on negotiation and good governance that is able to adapt local forces, cultural context and local opportunities to the new pressures presented by globalisation.

2 Theoretical Base of the IBIS Network

The work of John Friedman in the early 1980's on the global city, expanded on by the substantial work of Sassen, Castells and Borja, has been the leading theoretic and empirical underpinning of the IBIS network research framework on Globalisation. Important contributions to the framework research have been established in the book 'The Challenge of Sustainable Cities' (Burgess et al.) and the series 'Globalisation, Urban Form and Governance'. With regards to society and development the theoretical framework is based on the structural materialism approach of Althusser, Poulantzas, Bethelheim, Marcuse, Mandel and Samir Amin and more recently the work of Beck, Giddens and Habermas. In term of society and organisation of space, the approach followed is that space is socially produced, in line with the ideas of the French School (Topalov, Silegni, Mingione, Acher); the Latin American work of Pradilla, Singer, Toledo Silva; the theoretical and empirical work of the Boston-based Lincoln Institute of Land Policies (Smolka and Lungo, Clivenshi, Morales); the concepts of the regulationist school of

Aglietta, Deak, Ominami which have influenced the issues of flexible accumulation; and finally Preteiceille, Terrail and Magri on the issues on production, consumption and social needs. The regional question, which has developed substantially over the last 20 years, is based on the work by de Mattos' Ibero-Latino-American network (RII) on Globalisation and Territory, which deals with the failure of the thesis on 'polarisation reversal' of the 1980's (Richardson, Gilbert) and has developed the theory on 'Deconcentration with polarisation' which explains the new type of metropolisation-dispersion processes.

Morphology, mobility and vitality are crucial elements related to time and space compression, and express the shift from the family and home to the individual and public space (Rusk, van Cammeren, 2001) and is associated with energy consumption and sustainability in the compact city debate (Jenks and Burgess, 2001). Network city (Drewe) and networked infrastructures (Graham, 2001), together with mobility and accessibility concepts are interweaving since they are associated with vitality and the creation of centralities (Valenzuela, 2002) and thus to activity nodes and functional networks and are crucial for urban restructuring, slum upgrading, and the repopulating of the inner city. The theoretical framework of globalisation necessitates the analysis of the new role and nature of Large Urban-Regional Agglomerations, Large Infrastructural Networks and Large Urban Projects, linked to Urban Governance and the ability to enable processes and projects which are key factors for achieving the goals of city competitiveness. Good governance is underpinned by concepts of city elasticity (Rusk), decentralisation deregulation, municipal financial autonomy and the concept of vitality and mutual supporting uses (Valenzuela).

3 IBIS Research Clusters

Within the general framework of Globalisation, Urban Form and Governance, five research clusters have been emerging:

3/1 Capital and Urban Form

The first cluster focuses on the new type of metropolisation process taking place in different contexts. Over the last decades, there has been a revival in interest in the role of large metropolitan areas and mega-cities, which are no longer considered to be pathologies. Rapid urban growth and expansion have taken place. The cluster analyses the significance for the regional sustainability of the changing principles of a balanced system of cities and contained urban growth which have been abandoned in favour of new ideas on regional city networks. This shift has resulted in the emergence of several terms to describe the phenomenon of the emerging metropolis: Informational City (Castells); Global City (Friedmann and Sassen); Dual City (Castells and Sassen); Diffused City (Indovina); Unbordered Metropolis (Monclus); Network City (Dematteis); Postmetropolis and Fractal City (Soja); Regional City (van Cammeren) etc. The transformations occurring in Latin America, and some

---

1 The network theoretical framework and publications are to be found at www.bk.tudelft.nl/urbanism/alfa.
2 Sassen, 1991, 94, 99.

African and Asian metropolitan areas are compared to the transformations occurring in central countries, in a quest to determine whether the transformations are the result of global pressures or are linked to the internal dynamic of large metropolitan areas subjected to modernisation of capitalist relations under neo-liberalism, flexibility of financial capital and ITC transformations. What is clear is that the metropolisation process – especially in developing economies – is different to that of the previous period marked by Keynesian beliefs, the welfare state and industrialisation for national markets. The current stage is strongly affected by the transformation to a subsidiary and neutral state which enhances neo-liberal ideologies regarding production and consumption, involving the deregulation of urban development, the flexibilisation of labour, privatisation, FDI allocations, the operation of new technologies and the opening up of markets for goods, financial capital and services.

Research needs to be done on organisational integration needed to accommodate the transition from an economic activity controlled by the nation-state to the new situation, which has predominantly inter-enterprise and cross-border relations. It should be discovered how different regions and metropolitan areas react to this new situation, in which enterprises operate in partnership and in cross-border relations, what choices and opportunities exist for mergers and acquisition, and which sectors are working together (productive, financial, commercial but also cultural, scientific and tertiary education).

The flexibility of financial capital to move from place to place and the amount of Foreign Direct Investments entering and flowing out, display particular differences in the way in which different regions open their markets and transform. It is clear that FDI flows – which have doubled over the past 5 years and currently amount to US $1.5bn – are overwhelmingly concentrated in a small number of developed countries and global cities.

Deak and Schiffer are conducting research in São Paulo, while de Mattos within the Ibero-American research network covers Spain, Portugal and Latin America. The research of Ledo and Crespo is based in the case of Cochabamba, whilst Rocco focuses on comparative research in São Paulo (a mega city in terms of demographic and development features and a global city in terms of financial and accumulation features) and the Randstad-Holland (an institutionally multi-level large agglomeration).

The research of Rocco points to similarities and differences between the metropolitisation process in the Randstad-Holland and the São Paulo metropolitan region and seeks to understand features related to agglomeration economies and their comparative advantages to attract FDI. The similarities and differences are scrutinised in terms of economic attractiveness and advantages as well as political, cultural and social conditions. This research forms the basis for a broader identification of new built environments associated with intensive accumulation and rapid globalisation and in terms of the new forms of centrality and regional network. Similarities are present in the formation of new global centralities and large urban projects. Important in both cases is the analysis of the way in which surplus value, created by the concentration of command functions and global services activities connected with modern infrastructure, is either being privately appropriated and captured or socially distributed. The research creates possibilities to analyse the social polarisation process in relation to the structure of the labour market in both situations as well as the spatial fragmentation of the city.

3/2 Nodes, Networks and Metromorphosis

As global changes unfolded in the world economy, metropolitan areas became the economic nodes of the globalised dynamics. Several transformations in the organisation and functioning of metropolitan areas affect the economic, social and spatial morphology. The location of nodes and links of the different cross-border networks changes the metropolitan economic base. This change is associated with the growth of the service sector and the significance of urbanisation in the economy.

The historical, political, spatial, and economic context, as well as the size and hierarchy of agglomerations, differentiate metropolitan areas in a world context. Strong hierarchies are being established with regards to the global command functions. Large differences can be found between and amongst developed economies as well as in developing economies. Differences in the process of deconcentration exist between the large megacities and megalopoli developed during the Keynesian decades of industrialisation directed towards internal markets and the smaller agglomerations in Latin America, although they exhibit similar high rates of urbanisation and demographic characteristics. Despite these differences, many similarities and linkages have been identified in economic, cultural and spatial terms (Cohen, Sassen). The research of de Mattos has identified the way in which decentralisation and industrial de-concentration and metropolitan dispersion have taken place in Mexico City (Parnreiter, 2002), São Paulo (Taschner y Bogus, 2001) and Buenos Aires (Cicolella, 1999) on the one hand, and in other agglomerations such as Santiago (de Mattos, 1999), Caracas (Cariola y Lacabana, 1999) and Bogota (1999) where indigenous productive forces have reacted in different ways to the opening up of markets, structural adjustment and globalisation. The differences can be attributed to the size of the agglomeration and internal market, the type of mass consumption society, the historical form of policy making, the social polarisation which determines forms of industrial modernisation, the development of employment in the service sector, labour productivity and labour flexibility, the structure and nature of agrarian production, and in general the historic relationship between productivity gains and social distribution.

IBIS Research in this cluster is multifaceted and is associated with the change from a balanced system of cities to regional networks and the emergence of the new urban environments and centralities. It is carried out by Peresthu (Djakarta), Soto (Santiago), Gordillo and Marengo

(Cordoba), Sepulveda and Valenzuela (Santiago), Di Lullo (Tucuman) and Lungo (Central America).

Soto and Marengo analyse the impact of globalisation on spatial form in different contexts: the Santiago macro central region (6.5m inhabitants) in Chile and the second largest agglomeration in Argentina, the city of Cordoba (2.0m inhabitants). The research in the case of Santiago is based on the identification of scenarios for measuring socio-spatial sustainability of systems of centralities guided by new infrastructure. Three different scenarios have been identified: The first includes a broader definition of the central region of Chile based on cross-border relations; the second scenario covers the pressures of 'inflows' and 'outflows' from the metropolitan area of Santiago towards the Santiago region, consisting of three provinces where new urban-rural and centre-periphery relations are reshaping a vast area and a system of secondary activity nodes and radial corridors is developing around the magnet of the central node; the third scenario entails strategic planning for consolidation of the large metropolitan area, as a compact and sustainable city in which growth is contained.

In the case of Cordoba, the regional-metropolitan space is analysed from the point of view of the processes of social segregation. The Cordoba region is smaller than that of Santiago, and less subject to centrifugal forces since it is located in Argentina's pampas. The 'inflows' and 'outflows' respond to a self-contained city affected by globalisation, structural adjustments and a long period of economical crisis. The new system of centralities is different to the former polycentric city of the welfare state with inward-directed industrialisation, but with the former model social segregation based on class relations is exacerbated. Segregation is more fragmented than in the case of Santiago, which is an example of structural dualism.

Also in Santiago, Sepulveda analyses the complexity of public space fluxes and networks in a highly socially polarised city. The study covers different levels: metropolitan, urban and local (the Spanish block). The huge differences in development opportunities of the various fragments of the city are related to historical development, types of institutions, and governance issues.

Sepulveda identifies four urban processes (globalisation, gentrification, renewal, re-functioning) in three areas, that explain metropolitan/local changes and opportunities. At urban level he develops analyses of vitality and accessibility related to the various centralities to determine the potential of public space to restructure the city. Vitality is based on concepts developed by Valenzuela and is defined as the combination of mobility and mix of land uses. In order to measure centrality potential he defines three forms of accessibility: infrastructure, income (people) and economic (of the location). A model of bi-directional flows for measuring current centrality potentials is developed. Finally he superimposes these results on the centralities created by the new transport system currently under construction (Transantiago).

Valenzuela, also working in Santiago, in his research on Decay and Urban Sustainability in a Developing Metropolis accepts that Santiago's rapid metropolitan growth results from its increased participation in the global economy, which creates obvious benefits, but also inequality, unemployment and other problems, including environmental decay. He states that the decline in the quality of life threatens the sustainability of metropolitan development, as well as its economic function (Jacobs), but emphasises that focused local action can counter-balance these threats (Batley). He shows that local strategies applied in Santiago and other developing metropolises can prevent central area decay and peripheral disintegration by intervening in two key variables of specific urban areas: their accessibility; and their physical-functional attributes (layout, environment, land use and aesthetic characteristics). Valenzuela focuses on strategic planning and governance to repopulate and re-vitalise the inner city, and evaluates the effectiveness of the 1985 strategic plan.

3/3 Exclusion and Governance

Urban fragmentation and residential segregation have historical precedents. They are not only linked to location but also to exclusion, not only to racial and ethnic issues but also to socio-economic and cultural factors. Poverty is a multidimensional issue not reflected only by income and assets, but also by accessibility to shelter, to public infrastructure, safety nets, protection of rights, time cost, labour availability and political power. This places more attention on the imbalances between production and social reproduction than on hierarchical positions of cities in relation to the global cities. The theories of van Kempen, Marcuse, and Goldsmith on racial discrimination (in the case of the US and UK) highlight the role of residential segregation in the maintenance and enlarging of imbalances between blacks and whites. Residential segregation in Latin America is analysed by the group of Rodriguez and Sabatini from a socio-economic perspective and reflects the fact that although there has been a reduction of poverty in absolute terms, segregation and violence have increased because fragmentation has made poverty more evident. Fragmentation and the improvement of mobility and accessibility have brought the rich and the poor spatially closer together in most developing cities, necessitating a completely new framework for understanding cities and urban governance. An understanding of the different ways in which to finance urban development and especially the potential created by the mobilisation of income created by value creation of land and urban properties, is crucial to the analysis of poverty alleviation. The locally and spatially oriented empirical analysis of layout and usage patterns can be useful for some short term actions, but can actually hinder the struggle against the real causes of poverty over the long term, while accentuating social disparities. A more productive form of research focuses on the creation of land and property values, through the creation of place, that allows for the mobilisation of income generated through increased collaborative and negotiated actions between districts with different levels of income, resources and appeal. Faced with increasing fragmentation and asymmetry between locations, places and segments of society, negotiation and interlinked operations are becoming crucial for poverty alleviation and the protection of the environment.

With regard to exclusion the main research questions centre on the issues of complexity and diversity and with regard to governance, the main research questions focus on the issues of decentralisation and the appropriate level at which to resolve social and environmental problems within the context of neo-liberalism. It terms of complexity and diversity, increasingly complex local forces are coming to the fore in response to global pressures, to address the idiosyncrasies of different cities. These forces are expressed through different development paradigms (Cohen, de Mattos). Most governments promote strategies to increase globalisation and economic growth and improve urban productivity. However, some focus on developing endogenous forces and internal dynamics, which allow for strategies to alleviate poverty and protect the environment and have revisited the World Bank's Redistribution with Growth (R & G) strategies of the early 1980's.

For many, diversity has become a pre-condition for urbanity. From one perspective it is linked to mobility and opportunities, and as such diversity is a form of voluntarism or 'lottery where risks and rewards are on display' (Montgomery, Stren and others). From another perspective diversity is the expression of inequalities manifested in space as a result of unbalanced mobility and frustrated opportunities. Vitality and diversity cannot be purely attributed to location but also to socially determined factors. Therefore new indicators, methods and data are being developed that range from non-spatial attributes such as access to decision-making and finance to spatial attributes such as land use, mobility, density and typologies. Growing social polarisation, spatial and social segregation and uneven mobility and access to opportunities contribute to the inability of public policies and strategies to cope with the growing complexities that current development requires in different contexts.

In terms of governance, numerous models have been proposed for decentralisation and significant differences exist between them. In Developing Countries, the most commonly applied versions have been those where executive responsibilities have been transferred to municipalities but regulatory powers and fiscal resources remained at the central level. The central state retains control of management and flows of finance to local authorities and monitors the planning and execution of projects. However, local authorities have greatly expanded powers to choose, plan, prioritise and implement projects and have been empowered to generate revenues from services and local taxes.

In some decentralisation models, empowerment has been associated with the extension and consolidation of democratic rights. Local authorities have increased their role in co-ordinating community organisations, citizens groups, co-operatives, NGOs, enterprises and households. Many believe that increased participation and accountability is a prerequisite for 'bottom-up' social environmental improvements. The viability of neo-liberal decentralisation measures, on the other hand, centres on fiscal issues. Although the flow of central transfers is clearly critical for social and local environmental improvements, emphasis is placed on the ability to increase income generated from local taxes and user charges.

Some argue that the effects of municipal decentralisation on equitable social distribution and environmental improvements will be negative. Some municipalities are just too small and their population too poor to afford either the equipment or professional skills required for the provision and regulation of services. The effects could be an intensification of many of the irrationalities in the urban form that threaten its social and environmental support capacity: the uneven distribution of infrastructure and services, green and open areas, differential land use and zoning standards etc. Critics of the universal applicability of political and administrative decentralisation measures have pointed to the absence of strong local government traditions in many countries, and have doubted the ability of current policies to make much of an impact over a long period of time. In cases where regulatory competence has been transferred to the local level, many doubt that local authorities will have sufficient power to deal with greatly strengthened private sector interests.

The work of Carrion (Quito), Schoonraad (Pretoria), Perez (Costa Rica), Sugranyez (Chile), Di Paula (Montevideo) and Peinado (Bogotá) fall within this cluster.

The research of Carrion focuses on the feasibility of planning urban transformation. He states that the new scenario based on free market principles has explicitly neglected the ideology and practice of planning as a tool for strategically addressing urban, regional and national development requirements and is able to ensure the social function of land and access to goods and services. State planning agencies have been dismantled or severely reduced in most countries and local, regional and national spatial plans have been minimised in the majority of the cases. Only macro-economic plans and strategic planning oriented towards city 'competitiveness' have been supported. These are mostly short-term negotiations with private investors linked to deregulations to allow for the unregulated operation of market forces. These processes are increasing urban segregation, social exclusion, and housing and service deficiencies for increasingly more people. Carrion develops a metropolitan area information system that enables the preparation of a strategic plan for the city that can guide urban transformations on a long-term and proactive basis. The plan enables negotiation between the various actors in the formulation of strategies. Schoonraad's research centres on the contradictions between plan and policy, which are regarded as the main source of current segregation and inequality within the context of neo-liberalism. Local government planning in the City of Tshwane (Pretoria), Integrated Development Plans and other programs such as subsidized housing are analysed to explain these contradictions. The research measures the extent of spatial segregation in the city and the ways in which local government influenced the pattern of spatial segregation/integration. For Schoonraad, the fall of apartheid has not meant improved living conditions of the poor, despite several governmental programmes to address poverty.

Four findings have emerged from this research: Firstly, that the spatial segregation patterns which have changed over the past ten years are increasingly taking the form of socio-economic segregation rather than racial segregation. Secondly, that land-use decisions taken by the local government contradict the spatial plan and vision of the city to promote spatial integration and equality. Thirdly, that the attitude of planners, developers and local communities do not support the values of the spatial plan to promote spatial integration and equality. And lastly, that design guidelines and sectoral policy plans, based on presumed 'neutral' standards, contradict the vision and spatial plan of the city to promote spatial integration and equality.

The research of Sugranyez (Santiago) and Di Paula (Montevideo) aims to find explanations for the increasing segregation patterns caused by efficient free market oriented housing policies. Both make reference to the form of metropolitan institutions and the lack of effective metropolitan plans able to control market oriented profit making and the weakness of local authorities.

Because of its economic stability over the last 15 years, and the neo-liberal development model followed for the last 30 years, Chile has largely decentralised local powers. In Santiago there are 34 financially autonomous municipalities for 5 million inhabitants. This municipalisation process is an important factor in explaining the drastic polarisation together with the different public policies that have been implemented. One of the social policies that played a major role over the last year in creating large homogenous areas of either poor or rich inhabitants, has been the housing policy implemented, planned and designed by the private sector. This housing policy, by which the state subsidises 80% of the social housing stock, has produced 10 housing units per 1000 population, and has resulted in an overwhelming concentration of high density housing developments in some areas. Sugranyez has developed a method for measuring the spatial and social impacts of the policy and aims to find recommendations to improve the living conditions of the 400,000 households that live in these mono-functional residential areas.

The study of Di Paula refers to similar problems: efficient housing policies in quantitative terms but inadequate in terms of urban sustainability and 'habitability'. He also addresses the issue of contradictory competencies between metropolitan and local authorities, which stimulate segregation and inequalities. This research points to the lack of metropolitan guidelines to deal with fragmented urban form and the contradictions between the traditional mutual help housing programs of the municipality of Montevideo and the market oriented morphologies of the new policies.

3/4 Infrastructure and Large Urban Projects

Most governments advocate globalisation as the best way to escape underdevelopment. They argue that globalisation would enhance the trading opportunities of the less developed regions and would promote their internal modernisation by transferring advanced management and technology from global firms. Structural adjustment, formation of large agglomerations and infrastructure modernisation appear simultaneously as a requirement and as an outcome of the new age of globalisation.

The extensive development of highways, roads and railway infrastructure has contributed to the configuration of a new city image, characterised by the formation of new centralities and a productive system organised in networks in a rapidly growing metropolitan area. This leads to questions of: decay and de-population of the inner city; the growth of the suburbs and peripheries; the formation of multiple centres; the emergence of new urban environments between the peri-urban intersections; the proliferation of new large urban projects; and the transformation of compact mono-centric development into dispersed, radial forms. The modernisation of national infrastructure and the development of Large Urban Projects are therefore the most significant expression of these global transformations.

The need for intensification of trade and flows has necessitated the rapid development of cross-border and national road and train infrastructure linking the modernisation of seaports, airports, and regional corridors.

These developments are made possible by the privatisation and deregulation of national and urban infrastructure, the merging and acquisition of public assets by large TNCs and the deregulation of urban development and partnerships between public and private sectors, which facilitate the development of Large Urban Projects.

The cluster focuses specifically on who captures the increase of values created by large infrastructure development and large urban projects. The new methods of financing urban development underpinned by privatisation and liberalisation of markets are central to the question of public-private partnerships and negotiation.

Questions can also be raised from the point of view of sustainable development, which have challenged, from an environmental perspective, the metropolitisation and the further growth of mega cities under market and 'efficiency-oriented' investment criteria (Burgess and Camona, 1997). From this viewpoint the complexity of the relationship between the built and natural environment cannot be fully comprehended by traditional economic parameters since these are unable to recognise critical social and environmental thresholds and have failed to internalise negative environmental externalities.

Sustainability arguments challenge the notion that city size doesn't matter when dealing with urban environmental issues. Many believe that spatial decentralisation and regional policies are important for dealing with social and urban environmental problems. The environmental rationale demands the reassertion of a spatial basis for planning, focused on the integration of socio-economic and environmental parameters at various spatial scales.

A number of issues and questions can also be raised about the social and environmental impact of policies to privatise and deregulate the provision of urban services, infrastructure and land development. Neo-liberal policymakers argue that these measures are necessary because the previous Keynesian models, which recognised the 'public character' of services and infrastructure provisions, were neither efficient nor equitable, leading to the development of significant environmental externalities. They argue that privatisation will eliminate these inefficiencies, allow extended coverage of services and improvements in quality and will prove to be more equitable.

Questions can be raised, both in terms of efficiency and equity, about the ability of privatisation to achieve these effects in cities in developing countries. Firstly, models for privatisation were first developed in developed countries, and data thus covers a longer time span. Secondly, the range of the term 'privatisation' is considerable, and it is clear that only in certain sectors and activities (which vary from country to country and city to city) has there been a full conversion from the public sector to the private sector. More commonly there has been a merging of public and private capital and much remains unknown about the nature and implications of this new relationship.

A primary target for privatisation has been the publicly-owned, centralised and hierarchical monopolies that dominate the provision of technical infrastructure (water, electricity, sewerage, roads, drainage, transport and communications). The search for profit and competition has involved the transformation of profitable monopolies or segments of public monopolies into private ones; the residualisation of the less profitable segments; the privatisation of fragments of networks according to local or regional capabilities and the division of public and private responsibilities according to functional operations within the system. These networks are natural or technical monopolies in which market laws often do not apply and where regulation is essential to achieve efficiency at the level of the whole system (Toledo Silva). Privatised segmentation of these systems can jeopardize their larger rationale. On the other hand the expansion of central processing capacity without coordinated increases in distribution networks can lead to inefficiencies and high costs associated with under-utilization of capital and equipment, whilst rapid expansion of distribution systems without increases in central processing capacity can lead to system failures.

Differences between urban realities in developed and developing countries are another reason why doubt has been cast on the appropriateness of privatisation and deregulation criteria. In developed countries, urban population growth rates are low; there are high levels of consolidation of investment in the basic system; virtually complete coverage and high levels of technological homogeneity. As the costs of the basic system have been paid off, privatisation is largely concerned with adjusting and rehabilitating elements of the system and with upgrading quality standards. In developing countries, on the other hand, privatisation occurs under circumstances of rapid urban growth, a low level of consolidation of investments in the basic system and high levels of technological heterogeneity. Here the main issue for privatisation has to be the expansion of the basic system in ways that enhance productivity and hold social and environment benefits. The ability of the private sector to realise what are massive investments with slow rates of return has been questioned.

One such research project concerns the transformation of the central area of Mexico City during the Porfiriato (1900) and Globalisation (2000) periods. It aims to discover, through an analysis of urban and architectural projects during these two periods, economic and technological development similarities in the vision of the politicians, the languages of professionals' expression (architects or planners) that have been the creators of the formal ideas, and the economic attractiveness of Mexico City for transnational investors. The main objective is to determine the fundamental elements of urban transformations and to develop a method of historical analysis. To achieve this, Large Urban Projects from both periods will be systematised according to their urban and architectural context and main characteristics, trying to explain the historical and cultural context and to identify the role of the different actors.

The research of Cuenya deals with Large Urban Projects and Social Actors in the Retiro project in Buenos Aires. Cuenya states that the production of space for international business by means of large urban projects generates extraordinary benefits and social costs. Contradictory forces put into action by different actors (state, economic and community) to support or oppose a large urban project during its formulation, are strategies aimed at obtaining benefits or compensation for damages, not only in an economical sense, but also political and cultural, derived from its notable features and potential impact on a strategic area of the city. It deals with the knowledge of the process through which the built environment is produced and used. Particular attention is given to the understanding of two key issues:

- The relationship between structure (in terms of what drives the urban process and produces distinctive patterns during specific periods) and agency (in terms of the way in which the agents involved, such as public officers, landowners, investors, developers, politicians, consultants, designers and community groups, pursue their strategies);
- The manner in which this relationship is reflected in the production of large urban projects in the context of globalisation.

3/5 Managing Culture

Two studies deal with the topic of culture, that of Villegas (Peru) and Qu Lei (Beijing). Villegas analyses Local Forces and Global Pressures in the transformation of the Peruvian South Andean Urban Structures, focussing on three Andean cities: Cusco, Puyo and Cuno. He developed the following hypotheses:

- When powerful cultural forces in the South Andean Peru struggle with external pressures, extremely different urban transformations can occur in nearby locations. These can range from destruction of traditional societies and built structures to rebuilding them, from urban sprawl to urban conservation.
- In the framework of proactive national changes regarding the management of natural and cultural resources, severe and particularised divisions of socio-economic cultural links between urban and local action can evolve.
- When traditional urban structures undergo transformation as a result of new socio-economic settings, a cultural 're-appropriation' process can increase inequalities and social features.

The research includes historical, economic, social and physical analyses at metropolitan level as well as a survey of buildings and their cultural features.

The research of Qu Lei contributes to the analysis of the renewal process of the inner city of Beijing taking into consideration the interest of the various actors and experience of this field. The actors includes the city authority and the housing institutions as well as the people living in the inner city neighbourhoods, who have developed a specific culture and attachment to the place. The aim is to determine the feasibility of reorientating the inner city housing policy towards integrated revitalisation of residential areas through a social-cultural based policy for human settlement planning and through improving the city elasticity (Rusk). It is expected that an integrated revitalisation policy could improve accessibility and break up the concentrated pattern of urban poor; providing continuity of culture and urban identity.

4 Conclusions

It is becoming clear that physical space and societies are affected by globalisation in an unprecedented manner through complex global pressures and that it constitutes a new spatial and social geography with unpredictable consequences exacerbating old contradictions and creating new ones. Similarities in spatial transformation and large urban projects that are supported by global and local forces hide great and significant differences.

Globalisation is here to stay (Castells, 2001), and most governments look to increasing globalisation as the only way to tackle poverty and underdevelopment (Silva, 1995). Most countries have opened their borders, dismantled their obsolete inward-oriented industries, opened up to international services and facilitated rapid spatial changes through developing two main comparative advantages: the formation of large spatial agglomerations and the development of technical and social infrastructure to become competitive and attract FDI.

The complex relationship between globalisation, urban form and type of urban governance is as much an issue of power and negotiations amongst the different levels of authority and stakeholders as it is linked to spatial solutions addressing separate sectoral policies.

The importance of clustering urban research is to deepen the theoretical and empirical dimension of the organisation of space under globalisation and to contribute to preparing the ground for negotiations and decision making between public and private sectors from different angles. It implies the development of multisectoral, multilevel and multi-layered strategic platforms and general guidelines to deal with spatial issues in an integrated and holistic way taking into consideration the new global and local driving forces that shape and organise space. It includes the recognition of flexibility, mobility and internationalisation of capital according to the needs of international corporations and the understanding that the dimensions of space and time are rapidly being changed by the new ITC technologies.

Bibliography

Alquier F. (1971). Contribution a l'étude de la rente foncière sûr les terrains urbains. **Espace en Société** n. 2–6. Paris.

Acher, F. (1972). Quelques critiques de l' 'économie urbaine». **Espace en Société** n.4–9. Paris.

Aguilar, A. (1996). Reestructuración económica y costo social en la Ciudad de México. Una metrópoli periférica en la escena global. Ponencia presentada en el Seminario 'Economia y Urbanización: Problemas y Retos del Nuevo Siglo'. unam. Mexico-City.

Balbo, M. (1998). La planificación y gestión urbana local, en 'Ciudades Intermedias en América Latina y el Caribe: Propuestas Para la Gestión Urbana', Ministerio degli Affari Cooperazione Italiana. cepal, compiladores R. Jordan y D. Simioni.

Beck, U. (2002) **Capitalismo o Libertad**. Editorial Gili. Barcelona.

Boisier, S. (1994). Crisis y alternativas en los procesos de regionalización, **ECLA Journal** n. 52. Santiago.

Boltvinik, J. (1994). 'Los organismos multilaterales frente a la pobreza', in **Pobreza, Ajuste y Equidad**, Sarmiento, L. (ed.), Secretaria Presidencial para la Politica Social, Bogotá.

Borja, J. (1994). **Barcelona, Planificación Estratégica y Desarrollo**. Ed. Gustavo Gili. Barcelona.

Borja, J. and Castells, M. (1997). **Local y global. La Gestión de las Ciudades en la Era de la Información**. Capitulo 3. 'Planes estratégicos y proyectos metropolitanos'. Taurus, Madrid.

Burgess, R.; Carmona, M., Kolstee, T. (1999). **The Challenge of Sustainable Cities**. Zed Books.

Carmona, M. & Burgess, R. (2003). **Strategic Planning and Urban Projects. Responding to Globalisation from 15 Cities**. DUP. Delft.

Castells, M. (1972). **The Urban Question**, Edward Arnold, London.

Castells, M. (2000) **The Rise of the Network Society**. Blackwell, London.

Clichevsky, N. (2002) 'El contexto de la tierra vacante en America Latina'. In **Tierra vacante en ciudades Latinoamericanas**. Lincoln Institute of Land Policy. Cambridge, Massachusetts.

Cohen, M. (1996). 'The Hypothesis of Urban Convergence: are Cities in the North and South becoming more alike in an Age of Globalisation?', in **Preparing for the Urban Future, Global Pressures and Local Forces**, (ed.) Cohen M., Ruble B. and others, The Woodrow Wilson Centre Press, Baltimore.

DeFazio, Kimberly (2004), Urban Post-Theory, Class and the City. www.geocities.com/redtheory.

De Mattos C. et al. (1998a), **Globalisacion y Territorio. Impactos y Perspectivas**. Pontificia Universidad catolica de Chile, Instituto de Estudios Urbanos, Fondo de Cultura Economica. Santiago.

De Mattos C. et al. (1998b). 'Restructuracion, Crecimiento y Expansion Metropolitana en la economias emergentes Latinoamericanas'. In **Economia Sociedad y Territorio**, Vol i no. 4, pp. 723–754. Santiago.

Garay, A. 2002. Modalidades de gestión de grandes proyectos. Concepto y contexto de algunas grandes intervenciones de urbanismo. Lincoln Institute of Land Policy, 24 to 30 June 2002. Cambridge USA.

Graham, Stephen and Marvin, Simon (2001). **Splintering Urbanism. Networked Infrastructures, Technological Mobilities and the Urban Condition**, Routledge, London.

Lojkine, J. (1976). Contribution to a Marxist theory of capitalist urbanization, in **Urban Sociology**, Pickvance C. (ed.), Tavistock Publications, London.

Lungo, M. and Oporto, F.(2002). 'Tierra vacante en el distrito comercial central de San Salvador'. In **In Tierra vacante en ciudades Latinoamericanas**. Lincoln Institute of Land Policy. Cambridge, Massachusetts.

Parnreiter, C. (2002). Mexico City: The Making of a Global City? Internet.

Richardson H.W. (1973). **The Economics of Urban Size**. Saxon House. London.

Rusk, David (1999). **Cities without Suburbs**.

Toledo Silva, R. and Schiffer, S. (1993). 'Globalization, Infrastructure and Competitive Advantages in the Prospect of Privatization', Paper presented to the xiiith World Congress of Sociology, Biefeld.

Sassen S. (1991). **The Global City.** Princeton University Press, Princeton.

Sassen S. (1994). **Cities in a World Economy**. Pine Forge Press, London.

Sassen S. (1999). **Global Financial Centers**. Foreign Affairs 78.

Sassen S. (1999). **Globalization and the City**. Longman Harlow.

Valenzuela, J. (1997), 'Urban decay and local management strategies for the metropolitan centre: the experience of the municipality of Santiago, Chile', in **Latin American Regional Development in an Era of Transition. The Challenge of Decentralization, Privatization and Globalization**, United Centre for Regional Development. Nagoya, Japan.

World Bank (1990). **World Development Report: Poverty**, Oxford University Press, Oxford.

World Bank (1990). **World Development Report: Poverty**, Oxford University Press, Oxford.

Section V:

The Museum and the Media

If there is one building type or urban institution that embodies the changes in the economies and the social conditions of cities over the past three decades it may well be the museum. Since the 1980's the number of newly founded and built museums has increased substantially, and this development is reflected in a steady stream of publications, both in the mainstream press and the professional media. Established museums and initiatives for new ones have become essential in the strategies of politicians and planners aimed at transforming cities into touristic destinations and from centres of production into environments for new services industries. This faith in the magical capacity of museums to improve a city's place in the regional, national or global pecking orders regularly produces strange fruits. As Michael Müller and Franz Dröge have observed, 'every village has its own bread-, cork-, toy-, or beer museum, its museum of plaster casts, crochet work or flower arrangements.'[1] In many cases the assumption is that building new museums is an answer to de-industrialization and the fundamental change of the role of urban centres is reflected by the fact that the physical structures are realized before there is a collection or a curatorial concept to fill it, and sometimes not even a prospect that these might develop. The choice of a star architect then becomes essential to compensate for the absence of content and, as Wouter Davidts notes, 'what is required from the museum in the first place is that it should provide a landmark and a signature, a place and an identity.'[2] The consequences of this logic became most visible in the case of the Guggenheim Museum in Bilbao where the collection of modern art presented here, though in itself not unattractive, is rendered insignificant by the status of the building and its globally mediated image. The Guggenheim also offered an example of the converse when it opened a branch on Berlin's Unter den Linden boulevard modelled on the presentations of corporate businesses. The small gallery occupied by the Guggenheim in the centre of the German federal capital, surrounded by the showrooms of Volkswagen and Mercedes-Benz, relies neither on its architecture nor its collection, but wishes to establish the presence of the museum as global brand.

Meanwhile much of the media coverage of new museums is characterized by a celebratory tone, refraining from a critical discussion of what the role of the museum is or should be in contemporary culture and cities. Publications on the architecture of museums seldom problematize the ways in which museums present interpretations of history or contemporary culture, nor do they examine how these new public buildings might be seen as environments in which collective experience finds a physical expression. As Davidts argues, the absence of a general discourse on the role of architecture in relation to these issues, which are treated as part of a separate, professional debate among the curators of museums, is further enhanced by ignoring the fundamental differences between categories of museums of contemporary art, general history, nature, technology or objects of everyday life, to name a few categories. Any generalized debate on *the* museum as such is bound to suffer from the fact that different types of museums represent different cultural discourses and, therefore, address their audiences differently. From the perspective of the relationship of architectural design and museum practice it seems that a critical examination might usefully focus on the category of the museum of contemporary art. It is in this type of museum that changing concepts of what a museum should present and how its users are addressed as active participants or passive audience seem to be most directly reflected in the architectural designs of buildings. The development of museums of contemporary art, particularly since the late 1950's, appears as an area that shows a particularly broad range of different approaches to issues of curatorial practice, art production and architectural design. More than other fields of cultural production, contemporary art has engaged in an investigation of its own position in the culture of cities, resulting in an almost continuous process of defining and re-defining the relationship between its own specific environment and the context of cities and urban cultures. Changes in the production of art works are immediately followed by curatorial concepts, generating new requirements for the design of the buildings. By comparison museums specifically dedicated to architecture seem to be governed by relatively conventional ideas about the presentation of their material and about addressing their audiences.

Museum and city – mediated images of cities

The development of museum typologies and of curatorial concepts reflects the trajectory from the *Kunsttempel* of the nineteenth century, 'the auratic and alienated space of the bourgeois autonomy of the arts', to the contemporary museum and the attempted explosion of the hallowed space of collective memory as intended by various movements in the twentieth-century avant-gardes.[3] In the terminology of current curatorial politics museums address their audiences as 'visitors' or even 'clients'. The use of this generalizing term, as Müller and Dröge point out, suggests that the museum as an institution has lost its 'social backbone', an established urban *Bildungsbürgertum*[4] that would identify the museum as a place of national collective representation, of enlightenment and education.[5]

In the post-war years a number of influential museum directors formulated the ambition of their institutions to act as catalysts for social and cultural change. In a number of cases this agenda materialized in proposals for new buildings in which the relationship between the interior of the museum – and its content – and the surrounding city – the context from which art was to derive its significance – was radically re-defined. The opening of the Centre Pompidou in 1977 marked both the climax and the end of this search for the 'museum that is no longer a museum'. Combining the experiments with flexible galleries, performance spaces of the previous decade, Pompidou 'promulgated a vision of the museum as a civic institution providing education and even entertainment. Rather than insisting that visitors reverently behold its masterpieces, the Pompidou invites us to walk, talk, think, read, shop, and even eat within its doors. Hugely popular with the public, if not the traditional arbiters of high taste, its controversial exterior form and interior program radically reversed notions of what a museum could be or do.'[6] Almost immediately the Centre Pompidou and its architecture attracted profound criticism. The reaction against the museum as a factory producing as much as presenting cultural artefacts materialized in a series of museums realized in the Federal Republic of Germany during the subsequent decade that sought to re-establish

---

1 Michael Müller/ Franz Dröge, **Die ausgestellte Stadt – Zur Differenz von Ort und Raum** (Basle/ Boston/ Berlin: Birkhäuser, 2005) p. 119.
2 Wouter Davidts, **Museumarchitectuur van Centre Pompidou to Tate Modern**, doctoral thesis (Universiteit Gent), 2002/2003, p. 9.
3 Müller/ Dröge, p. 137.
4 **Bildungsbürgertum** might be translated as 'educated middle class', but in German clearly refers to the concept of a collective of educated individuals with a clearly defined self-image and formed by Humboldt's educational ideal, which would distinguish it from its English equivalent.
5 Op. cit., p. 134.
6 Douglas Davis, **The Museum Transformed – Design and Culture in the Post-Pompidou Age** (New York: Abbeville Press, 1990) dustjacket, quoted in Davidts, p. 225.

continuities with traditional, nineteenth-century museum typologies. While restating the special status of the museum as a space in which art is confronted rather than made, and re-establishing the formal characteristics of the nineteenth-century gallery, the museums in Mönchengladbach (Hans Hollein), Stuttgart (James Stirling) and Frankfurt (Hollein and Richard Meier) were consciously presented as part of the rediscovery of the traditional European city and the development of entire urban ensembles into museum environments.7 At the same time the museum became firmly established as a building programme that was singularly suitable for the exploration of the formal ideas of an emerging formation of star architects, prompting the painter Markus Lüpertz to write: 'The classical museum is built like this: four walls, roof lights, two doors, one for entering, the other for leaving. This simple principle had to give way to art, the art of architecture. All these new museums are often beautiful, noteworthy buildings but, like all art forms, hostile to other art practices. They do not give simple, innocent paintings, simple innocent sculptures any chance – no, they generated their own loud machinery filling the space and originating from decorative and pedagogical impulses.'8

Despite Lüpertz' complaint about the intrusive character of the architecture of the museums realized during the 1980's, the specifically personal signature of the architect was restricted. It was mostly allowed and sought where the representative nature of the museum and the effect of the building as an *urban* image was at stake. While the museum was re-established as a monument to collective, urban memory, the city itself started to invade the building in the form of a new emphasis on additional commercial programmes; since the 1980's no European museum is complete without a range of shops, bars and restaurants.

Museum as loft – the post-industrial interior

Another aspect of the relationship between museum architecture and the production of images has become increasingly visible in the interior of museums, and concerns the ways in which the presentation of objects is related to images of contemporaneity and cultural sophistication. In presenting its exhibition spaces as a sophisticated factory floor, the Centre Pompidou had exploited the image of the building as a workshop, reflecting the idea that museums of contemporary art had turned themselves into centres of production of art pieces and performances rather than places of contemplation.
The image of the museum not as 'a static place of keeping art works, but a dynamic place of production'9, appeared to allow the visitor to participate fully in the process of creating art works and to have immediate contact with the 'living artist'.

Following the experiences with 'alternative' art spaces in the late 1960's in New York, the explicit presentation of museums as buildings inherited from the industrial past has become another possible option for expressing the experimental and therefore contemporary character of the museum. Museums such as the Stichting de Pont in Tilburg (Netherlands) designed by Benthem Crouwel in 1987, the Dia Center for the Arts in New York designed by Richard Gluckman in 1987, Frank Gehry's Temporary Contemporary in Los Angeles or the Hamburger Bahnhof in Berlin adapted for use as a museum for contemporary art in 1990–96 by Josef Paul Kleihues present themselves as environments for an ongoing work in progress, evoking the image of the artist's loft as the paradigmatic concept for art spaces. Yet, however strong the reference to the industrial past of the environment in which the objects of art are shown may be, the exhibition spaces themselves often continue to conform to the model of the white box. When, in 1983, Charles Saatchi opened his gallery in a former paint factory in North London, the uneven surfaces of the building's brick walls were covered behind pristine white panels, to establish the visual neutrality thought necessary to view works of contemporary art properly.[10] The distinction between the presented image of the former industrial environment, in the context of cities where industrial production is no longer part of everyday experience, is poignantly visible in the case of Herzog & de Meuron's Tate Modern, a building which seems to contain and synthesize all current trends in the presentation of art and the museum. Presented as a building that 'will not be a place where the high-art jet set crowd goes', the Tate consciously exploits the traces of the industrial past, its dirt and patina which allows 'the art to be comfortable rather than simply on show'.[11] Originating in venues of avant-garde art practice, the image of the artist's loft has become part of the readily available imagery that has been adopted across the entire consumer environment, from the museum into the interiors of bars, restaurants and high street retail outlets.

With Wouter Davidts we might question these strategies of evoking loft or industrial environments and the implicit suggestion of the museum as a place of 'work in progress'. According to Davidts, the use of the loft aesthetics, recently taken to new extremes in the wilful stripping of the 1930's monumental Palais de Tokyo into a squatters' canteen housing contemporary art, is the expression of a three-fold misunderstanding; the idea that the museum can appear as a decorum for artistic production, that its environment therefore has to evoke the *atmosphere* that has been associated with artistic practice since the late 1960's and that, finally, this operation can be carried out by literally housing museums in vintage industrial spaces – or spaces that suggest such an origin. It is here that Davidts and Müller/Dröge seem to concur in demanding that museums should be allowed to be viewed as places of *difference*, accepting the distinct status of objects viewed and framed in a specific environment 'of aesthetic stability'. This would allow the museum to remain an institution demanding reflection and distance, showing the fragmentation of the world around us but also to experience (temporally and spatially confined) claims to a view of totality – possibly even Utopias – that cannot be constructed anywhere else. Is it not exactly the 'Entwurzelung der Werke'[12], the alienation of the objects on view, that allows us to learn, or re-learn, the ability to look and question what we see; and the 'capacity to experience distinctions and to view passively the works which themselves are condemned to passivity' on the wall and in the spaces of the museum?

In this section, the contributors were invited to elucidate their views on the current state of museums and curatorial practice. The contributions by Wouter Davidts (Universiteit Gent, Belgium) and Chris Dercon (Haus der Kunst, Munich, Germany) question some of the fashionable ideas about the contemporary museum in challenging the predominance of the visual in a culture of

7   Müller/Dröge, p. 133.
8   Markus Lüpertz, 'Kunst und Architektur – Art and Architecture, Heinrich Klotz/ Waltraud Krase', in **Neue Museumsbauten in der Bundesrepublik Deutschland** (Frankfurt am Main: Deutsches Architekturmuseum, 1985) p. 30/31.
9   Davidts, p. 25.
10 Rita Hatton/John Walker, **Supercollector – a critique of Charles Saatchi** (London: Ellipsis, 2000) p. 133.
11 Cynthia C. Davidson, 'An interview with Nicholas Serota and Richard Burdett', **ANY**, no. 13, 1996, pp. 21–58.
12 Müller/ Dröge, p. 186.

accelerated image production and the subservient role of architecture as a provider of consumable images. Michael Müller's (Universität Bremen, Germany) excursion into the history of collecting as a strategy of ordering the environment, and making sense of it, allows the author to propose the museum as 'a magic mirror of the times in which we live', places that invite us to reconsider what is familiar or alien, and to accept difference as a productive cultural force. The contribution by Christoph Grafe (TU Delft) is a study of the 'cultural centre' as a building type invented and widely realized in postwar Europe as part of efforts to give access to culture to larger audiences and the questioning of boundaries between 'high' and 'low' culture in this period. The presentation of Rosina Gómez-Baeza (ARCO, Madrid, Spain) is an account of the development of ARCO, an arts event and presentation held annually in Madrid, and its efforts of extending its role as a place of experimentation and enquiry. Hal Foster's (Princeton University, USA) contribution, finally, could be seen as an essay on the interactions and tensions between urban politics and image production, architectural stardom and academic theory, vested interests and the perceived need to propose newness that affects museum practice and contemporary cultures.

You can be a museum or you can be modern, but you can't be both

Chris Dercon

You can be a museum or you can be modern, but you can't be both
Gertrude Stein

I would like to share with you some of the disappointments and frustrations, but also the dreams and ideals that have been with me over the two years that I have been working as director of the grand old museum Boijmans van Beuningen in Rotterdam. Like all museums, this museum is in a continuous process of change. How can this process best be characterised? I would like to call it a process of a number of 're'-factors: renovation, revitalisation, refinancing, reorganisation etc. In other words, it is about the revision of all the conceptions we have of a museum. Let me name a few concrete examples. We are currently renovating and enlarging the Boijmans. Why are we making a museum that is already gigantic even bigger? We are changing the allocation of areas to offer the viewers another experience of the collection. We are not only aiming to attract more visitors – which is an economic necessity – but also would like to better meet the different expectations of the individual visitors.

In this way, the revision of our conceptions of a museum also implies rethinking the role of the curator. Should the curator in a museum like the Boijmans be a scholar in its conventional sense? My answer is unambiguous: our curators should indeed be scholars in the first place. But I would not like this to be interpreted as a case for a return to a dogmatic historical approach. What matters to me is a critical attitude on the part of the curator, which I consider to be an indispensable alternative to a purely academic or educational style of fulfilling the role of a curator. What I have in mind are curators who are capable of designing alternative models of exhibitions and including various disciplines, from sociology to literary studies.

Over the past decade, a number of things have changed in and around the museum Boijmans van Beuningen. For more than hundred and forty years, the museum used to exist in an ivory tower. Around 1990, suddenly art institutions sprang up in the vicinity that pursued different goals, like the Witte de With or the Kunsthal. The Kunsthal is known for organising more populist exhibitions that attract large numbers of visitors – significantly more than the Boijmans. The Witte de With receives only a few thousand visitors a year, but its innovative and concentrated programme of exhibitions is nonetheless unique, not only in Rotterdam. Both these institutions are valuable counterparts for the Boijmans in comparison. This influence does not need to be shown directly; but in fact, they play an important role in the process of reflection on our own institution.

The collection of the Boijmans encompasses one thousand old masters, three thousand new masters, thirty thousand objects of art and craftwork, hundreds of thousands of drawings and engravings as well as one hundred and forty thousand books. Nonetheless, it is a crucial part of the legitimacy of a collection like that of the Boijmans to continue collecting and thereby extending the collection. However, purchasing key objects is not easy nowadays, for example in the areas of design and contemporary art. For one, it's simply cheaper to purchase a Dutch painting of the 17th century than a Gerhard Richter. But there are other reasons as well. In an interview William Rubin gave in 1974, the then director of the museum of Modern Art in New York confessed, 'that the concept of a museum was not able to be extended indefinitely'. He attributed this concept to the break between the traditional categories of painting and sculpture on the one hand, and Land Art and concept art on the other – two types of art that had been *en vogue* at the time. The latter, Rubin explained, required a completely different presentation in museums, and also, he added, possibly a different type of audience. By saying 'that the concept of a museum was not able to be extended indefinitely', Rubin, in my understanding, referred to the very type of problem that is linked to the museum as a public institution.

When the Centre Pompidou opened its doors in 1977, the sociologist Pierre Bourdieu predicted that the profanation of various objects of cultural value in a mundane environment that assumed different cultural functions, was turning the museum into a prime example of a public institution. He thought that this would not only undo the traditional aesthetic categories, but also result in changing the perverse attitudes of cultural consumption for the better. The Centre Pompidou was also confronted with a 'different kind of audience'. This change was, however, not seen as a problem, but rather as a solution to a much more essential problem, the problem of the museum.

Rubin was not able to foresee in 1974 that, twenty years on, another round of plans for renovations and extensions of his museum would spark a discussion that was rather unusual for the MOMA. The video artist Bill Viola came up with a metaphor that precisely captured the future redesign and reorganisation of the MOMA: an internet website that enabled the viewer 'to move through space and time vertically and horizontally at the same time.' Not only the slogan that 'these collections tell the history of modern art' was attacked, but it was also demanded that the new designs raise the presence of 'contemporary content', by giving more space to experiments. The new MOMA was supposed to be a heterotopical museum, a new model with a lot of unprogrammed areas. By now we know, this only meant lots of space for all kinds of things that pass for visual culture – in the form of photography, videography, cinematography and all the things that are usually credited with the term info-aesthetics. In short, a mixture of practices that deny the forms and possibilities of presentations characteristic of a conventional museum.

The MOMA, which has a memorable collection from the areas of photography, cinema and video, does not seem very much at ease with this emphasis on virtual media and this turn towards a cultural diversity on a global scale, to name just a few of the characteristics of the more recent artistic production. For that reason, the 'university' MOMA founded a 'technical college', when it teamed up with the small, but highly advanced P.S.1 in Long Island City, Queens. The P.S.1 comes from a completely different conception of a museum, and also has a different audience – just as Rubin had predicted.

At the same time and similarly, around twenty years later, Bourdieu and other supporters of the Centre Pompidou had to admit that the democratisation of high culture was rather a side issue, if not a mere illusion. The diversity of activities in the Centre Pompidou had not changed the existing hierarchy of cultural manifestations in any way. The audience was the same as everywhere else – the fans of contemporary art were jostling their way in next to the users of the library. However, their numbers had increased, and to an extent that the structure of the activities themselves began to suffer. From the latest renovations and plans for extensions of the building, we can infer that the division into departments and themes will be more rigid than in the past. Moreover, in the near future the activities will be spread over various locations in the city. The glass façade, that used to invite exchanges, has turned into an opaque surface behind which art and culture are divided up.

Which place and which significance do museums like the MOMA and the Centre Pompidou, but also all the other museums of the 21st century, occupy? In our present culture, where the borderlines between what is called 'low' and what is called 'high' are becoming blurred, the realm of high culture is most of all an unoccupied territory. Nobody can claim it as his own, but at the same time everybody should aspire to get there or be enabled to do so. Is the museum a showcase of art history or a centre of visual culture? Is the museum a buffer against contemporary culture or does it take part in it?

By now it has become clear that not only the questions asked in progressive as well as more conservative circles are similar, but that both sides have also arrived at the same answers, which they emphatically put forward. They do so with such fervour that the representatives of the left, no differently from those of the right, simply dismiss the rise and fall of the avant-garde, considering it an error of history. 'Throw it out' is what they say. But let us not forget that the rise and fall of the avant-garde only reflected a fundamental change in the relationship between art and society, a change that was set in motion as the public gained access to the institutions of art. Before that era in which art became a factor of social life, art had never been public and hence there had never been a public for contemporary art. Viewed historically, the public presence of art is a factor that has grown gradually, and increased explosively since 1960. We have to live with that. Not even geniuses like Markus Lüpertz can reverse this process!

And there is another factor. The layman – and not only the layman – pays little or no attention to the differences between the various forms of art production (even if, for the time being, we ignore the poor visual and verbal competence of most of the more recent art) and the visual culture that determines our social environment. Most art and most consumer goods follow the same logic, the same patterns of construction of visual thinking.

In this situation, the museum is no longer a priori a separate environment. Or is the museum, as Boris Groys formulated, 'the only guarantee that allows us to make a distinction'? A buffer that endows objects with meanings that are otherwise meaningless?

We have understood by now, that all those things we see in the growing number of museums are merely fragments, a small selection from a far bigger whole. Every single piece in the museum has become an exemplar, a piece of evidence. As a result, the real and the imaginary space of the museum turn into a nearly virtual space that is equally open to works of art and the audience.

One could even go as far as to say that today the museum has partly become a model, a representation of itself. A direct consequence of this development is the spectacular and most of all photogenic, architecture of museums. In most museums – and this is equally true of museums of old art as of contemporary museums – the aspect of time is increasingly played down, in favour of architectural significations that convey an intense experience of the space. The museum and its objects are, more and more, driven out of history and displaced into an overly aesthetic space. Or let us think of the bombastic halls artists, curators, politicians and business people envisage when they imagine a museum. Bilbao's Guggenheim (no matter how powerful Gehry's quotidian architecture may be) is based on similar misunderstandings, for instance that the confrontation of internationalism (i.e. Americanism) and PR will create a vibrant cultural scene that is manifested by the construction of a new museum.

To summarise, the universal accessibility of museums has resulted in them no longer being 'something special'. Their accessibility has also contributed to turning the museum into a virtual space and a machine of representation. How can we then still differentiate between the concern for the works of art and the concern for the audience?

As one further factor, we have to take into account a number of completely new developments, like the rise of hypermedia and the boom of photography and cinematography. In my view, these are decisive areas affecting the relationship between the public and the museum. Given these new developments, we can only hope that these questions will be given more priority. In view of the collaboration between artists working in different fields, the continued expansion of media and new technologies, in addition to the desire on the part of many artists and curators to create products that are to some extent useful and have their place in the real world as well as in a much wider visual discourse, we should at least think about the possibility of a redesign of the museum. The function of the audience would be the central element within this kind of design. But are these factors reflected in the present museum projects or extension plans? Hardly.

The result of the competition for the new MOMA in New York, not to mention the recent merger with the P.S.1, is wholly in line with the status quo. Instead of taking up innovative proposals that would have a direct impact on museum activities themselves – proposals like the ones by Rem Koolhaas or Bernard Tschumi – the powerful board of the museum played it safe and decided in favour of the modest, package-like design by Yoshio Taniguchi. New curatorial ideas, like the ones that were so vigorously discussed by the staff of MOMA and that were also commented on by outsiders, have suddenly ceased to be of interest. The book *Imagining the Future of the Museum of Modern Art*, created by the MOMA's research and science programme and published on the occasion of the competition, will be an important source when, one future day, someone attempts to write or rewrite the history of the museum. But unfortunately it's a book and not a museum.

It could be argued that, by now, popular 'alternative models of museums', like the Dia Center for the Arts in New York or the Museum für Moderne Kunst in Frankfurt, commendable as they may be, are grappling with ideas of the past rather than those of the future. These institutions have clearly adopted an approach that is centred around the function of the artist. The architecture as well as the activities of these museums reflects the position: what is good for the artist, must also be good for the audience. This strategy does not always work out.

The interest we can currently observe in many architects and other figures of cultural life, in transparency and mobility should also be applied to the architecture and programmes of museums. There are some great examples, like Daniel Libeskind's fascinating Jewish Museum in Berlin or the great Museum for Sculpture/Public Square in São Paulo by Paulo Mendes da Roche. Both of these buildings question the forms of presentation through forms of presentation, as a slight variation of Andrew Benjamin's wording in his work *Present Hope: Philosophy,*

Architecture, Judaism. On Libeskind's building, Benjamin writes that 'it draws attention to the ability to question any kind of representation itself. It prevents the representation from taking on a final character. This self-restraint opens up a space that, on the one hand, delineates the fundamental issues, and on the other can be used as a space for presentation.' And in fact, the Berlin and São Paulo museums were built in the sense of these questions. At the same time, these buildings clearly indicated that they wanted to be museums (or, alternatively, something else: a monument or a public square). But where are the customers who would dare to entrust this kind of architect with the construction or extension of their museums? The proposal by Herzog & de Meuron for the new Tate Gallery on the London South Bank, in itself, does not go beyond the pioneering concept proposed by Renzo Piano and Richard Rogers for structuring the space above and below the ground in the Centre Pompidou. And there is the additional factor that, in developing their conception of a lively museum, including some very commendable initiatives for an increased involvement of the audience, the directors of the Tate and similar projects have primarily been led by methods that have proven successful in exemplary international exhibitions of the past decade. However, an interesting and challenging model of an exhibition is not the same as an innovative institutional strategy. And we need new, innovative models of institutions, to further develop the significance of the museum. Or are there limits to the extension of the concept of the 'museum'?

Initially in museum presentation, an encyclopaedia of images in rows on a white wall was guided by the principle of chronology and/or style. Today, the audience is aware that the museographic project encompasses more, for which, however, the museum is not a suitable environment. I am thinking of an *archive* in which the latest information technologies, image-and-text systems are available. The digitalisation, but most of all the binary codes that are the fundamental units of the operation of databases, is responsible for a more recent phenomenon, namely the fact that art is hiding behind its antithesis – behind a kind of 'anthropomorphic fetishism'.

In his excellent essay 'The Archive without Museums' the American theoretician Hal Foster presents a number of examples of this phenomenon, taken from several recent covers of the magazine *Artforum*: O.J. Simpson, Courtney Love, Broadway Boogie Woogie, Matthew Barney, Prada, the architecture of Christian de Portzamparc, Larry Clark, Hugh Grant, Georg Baselitz, Gilbert & George, Calvin Klein etc., etc.. In fact, it does not even seem absurd to state that these examples, under the guise of current interests, have made their way into contemporary art exhibitions and publications, owing their presence directly to this virtual space that the museum claims, not only metaphorically, but quite literally: film and art, architecture and art, fashion and art. In this respect, the question of expanded public access and wider interest in certain forms of cultural expression plays an important role. And I would venture to claim that this is no longer a friendly gesture towards the audience, but it's the audience demanding its 'interactive' rights and directly addressing the museum.

But be that as it may, this only goes to show that the museum no longer divides and rules, displays and preserves. The museum has become a venue among others, part of a wider museographic project that is also taking place in other places. In this respect, the MOMA's new motto is informative: 'The basic reading of the collection is interrupted, at various points, by alternative readings or by possibilities to become engrossed in the work of a particular artist, an era or a particular subject.' One of the biggest challenges in this regard is linked to the concept of 'reading' and the tension between image and text.

The audience's strong interest in photo books is just one piece of evidence for this phenomenon. The countless exhibitions by photo artists – that all look more or less similar – come across like pointers or advertising campaigns for some publication or other by these artists; their character is one of illustrations rather than reproductions. The exhibitions and the exhibits remind us of enlargements and images of the objects in the book. This is where we come full-circle: for aren't the disciplines of art history, and in some sense also the museum, children of photography?

The museum does indeed present itself as a photographic-cinematographic space. The current interest in photography and cinematography, so typical of many museums, but also of the problems they face, once again underscores the urgency of the question of whether we know what a museum is. That is why I consider the discourse on the function of photography and cinematography within museums crucial to all reflection on the future of the museum and its relationship to the public.

If we give credence to Walter Benjamin, then photography has put an end to the effect an exhibition can bring about. According to Benjamin, photos should have stayed where they came from – in books, magazines, posters and archives. Today we know that it has all turned out differently. Maybe Benjamin had missed, or underestimated, the fact that the possibility of mechanical reproduction of works of art has kept pace with the reduplication of the exhibition effect, or rather with the peculiar reduplication of the very institute of exhibitions, the museum. Photography was not only reproducible, but also able to function as a mark of honour of an exhibition. That's why it was dressed up according to the rules of the museum. There are any number of examples. We need only think of the history of the MOMA and the role of photography in this history. How was photography supposed to be shown, and more importantly, which photography? Confusion still reigns on this matter.

We don't need to go looking very far to find concrete examples. The museum Boijmans Van Beuningen held an exhibition entitled 'Aim Left' that was about investigating the artistic climate in Rotterdam in the thirties. The exhibits contained photographs by Paul Schuitema, Piet Zwart and Wally Elenbaas, along with objects and typographical designs by the same or similar artists. This exhibition effect, this reconstruction, played a crucial role in maintaining the radical aura of these photos, preserving their status as 'New Photography'. Another photo exhibition in the Boijmans museum consisted of holiday snapshots taken by the author Paul Bowles in Morocco. The photos were put under glass and shown with the sounds of Berber music in the background. The amateur photographs were not only enlarged for the exhibition, but most of them were reproductions, even photographs of Bowles' originals that were intended for publication by a commercial publisher. This practice is the exact contrary of a photo exhibition in the Witte de With. Here, elegantly framed heliogravures were shown, Muybridge's studies of motion that had long been torn out of their album context, along with contact prints of working photographs by Jan Dibbets. In this surprising confrontation, Muybridge's photographs suddenly appeared like autonomous works of art. Recently, an exhibition by the American concept artist Chris Williams in the Boijmans museum evoked memories of Renger-Patsch, the pioneer of New Photography, and of the photographic distancing that is so characteristic of the Neue Sachlichkeit. But the ontological difference between photography and painting, the difference between taking and painting pictures, was destroyed by presenting photographs, as if they were precious paintings, on big white walls. Or was this theatrical presentation merely a harbinger of the institutionalisation of photography, of its literal and metaphorical placement/misplacement in museums?

And what about exhibition rooms in which thousands of projected images and even more pixels of information, accompanied by oceanic background sounds or heavy techno beats, let the viewer forget time, leading some museums to consider issuing multiple-entry tickets for the exhibition? Aren't they somewhat reminiscent of modernistic Greenbergian white cubes? Of course, the one difference is that the white cube has been painted black. Is there anyone among the present audience who has recently been to a museum or an exhibition of contemporary art, who has not found that one or several of the white rooms were dimmed? Ironically, the architecture of the prime example of a European museum of visual culture, Le Fresnoy in Lille, designed by the deconstructivist Bernard Tschumi, did not provide enough darkness. For that reason, the whole building was wrapped in black plastic, which, in turn, was lightened up by the pigeon droppings. As the artist Jeff Wall phrased it: 'The notion of the "museum" seems to be associated with daylight, while the cinema requires a dark room. But since its beginnings, the museum has claimed to be a universal institution. And as such it has to reflect both, day and night. That is why we need more dark rooms in museums.'

Maybe we should even think about a 'sun area' and a 'moon area' in the museum. I am curious to see whether Wall will prove right. In any case, we will soon need a museum architecture that is relevant to our time and that is oriented towards an individual rather than a collective time. In view of some outstanding recent examples of information architecture, such as science museums, new libraries, or archives, it should be possible to realize this kind of new museum topology. But still, these considerations have nothing to do with the break between the traditional aesthetic categories or criteria, that Rubin feared. But what is involved is the question of the purpose of collections, of the one who exhibits and of preservation of works of art in general. *Omne bonum est diffusivum sui* – everything good finds its own place. But what happens when this piece of wisdom no longer holds?

Museums are in fact confronted with a phenomenon that has far-reaching consequences for their activity of collecting and exhibiting. Much of what is currently labelled as art is simply not suitable for acquiring, preserving and storing in a museum, at least not if we proceed by customary standards. This development has become more and more noticeable since the early sixties, so that we are increasingly reminded of Rubin's remark that 'the concept of the "museum" cannot infinitely be extended.' The consequences of this development for the degree of legitimacy of our collections have slowly begun to emerge, at least if we assume that a collection is primarily legitimised by its additions, by incorporating things that were outside of it. And in fact, certain 'things' are missing in our museum galleries and warehouses. The majority of all museum people resort to the verdict that 'it doesn't fit into our collection' as an alibi for the admission that 'we don't know how to deal with this stuff'. This kind of attitude, displayed by many museum curators and directors these days, is the main reason why much of the art produced today does not find its way into the museums.

What would it mean for the museum, but most of all for the public, if the notion of what is 'easy to store' were replaced by the notion of what is 'hard to reconstruct', such as has been common practice in theatres, opera houses and concert halls for over a hundred years? The first presentation in my time as director of the museum Boijmans Von Beuningen was the film installation *Four Rotating Walls* by Bruce Nauman. The Boijmans museum had acquired this work in 1970, but had neglected to preserve and to exhibit it. Therefore the premiere and restoration of

Nauman's pioneering work in the field of media art was not able to take place until 1996. In fact, in museum circles we hardly ever hear serious discussions about showings and repeat showings, and hence about the status of the archive and its accessibility. Restoration remains a technical notion, devoid of ideological content or context.

Many museums have understood by now that big illuminated photo cases, projections of changing slides, showings of film loops, multiple video projections that are automatically repeated or interactive computers with internet access remove the illusion of a static world from the exhibition rooms. Or should we say that we suddenly start to become aware that there are indeed non-moving images? Because of new uses of photography, cinema and video we also seem to be able to reflect on the meaning of the still image. In Jeff Wall's view, the static nature of still images has changed. Otherwise we could not really explain the immense current interest in still photography. And it was the cinema that liberated photography from the rather solemn orthodox theories of photography geared towards painting. Photography was able to reinvent itself as an art form by reflecting on, conveying, and interpreting cinematography.

But there is something else. Just as the coining of the term 'still life' coincided historically with the rise of photography (previously they were called 'kitchen' or 'banquet'), so also art history drew widely on photographic techniques to place a host of different objects within a classification scheme. But what would the consequences of a digital reorganisation be, Hal Foster asks provocatively? Art as a text of images, as information pixels? An archive without a museum? Art museums have got rid of the static world. Now they also have to let go of the world of images. While movement was the project of the 20th century, now maybe another project is ensuing, a project that has to do with growth, with the capacity of an organism to incorporate things. Cinema and other media, that are inherently temporal, could conceivably (at least that is what gurus like Rem Koolhaas and Bruce Mau think) fall by the wayside.

The moral we can draw from all this is simple: The primacy of visuality in visual culture may be mere semblance. The reorganisation of the field follows the course of a digital logic, in which the logic of the word and that of the image fuse, just as the photo camera and the film projector fused in the computer. According to Hal Foster, the developments in film and photography prevent art at the end of the 20th century from fully differentiating towards purity, whether in terms of visual appearance or of information content.

In other words, art can no longer escape from its corrupt double: mass culture. For artists like Hans Haacke, Daniel Buren, John Knight, Susan Hiller, David

Hammons or even Andrea Fraser, the museum and its public are still something to conquer and to change. For a younger generation of artists, curators and critics alike, the museum is just one of the many places where art can be shown. Its public is taken for granted. We often tend to forget that this development is only possible because those who practise 'new ways of making art' have other capital available to them, other than the capital of the museum, the gallery, or the collector. The generosity of this new capital has created a new kind of institutionalisation. Within those, so to speak, 'new institutions', conventional values of the museum or the gallery, such as authenticity and criticality, are mainly considered mere properties of the site – the choice of a particular neighbourhood, building or other architectural infrastructure – engaged by the artist. Such displacements are easily marketable – comparisons immediately suggest themselves with the worlds of fashion, design and of course architecture – as well as conveniently controlled by and through social and political bodies. The object of contestation remains the same: the institution of autonomous art and its thousands of objects or fragments of objects. This resistance makes the 'new practitioners' who are often heard publicly stating that one needs 'to connect', Robin Hoods in the eyes of many, not least of those who know little of the artists mentioned above and of their history, artists who do critically examine institutions.

There is not much new to be discovered about these new practices of art – at least nothing in the sense of questioning the power over meanings exerted by established cultural institutions. After all, the museum of (modern) art was born the moment art was no longer possible. And as Hal Foster has shown, most of the basic assumptions of the old productionist model as well as the legitimising figures of the avant-garde persist. What is really new is that art has recently begun to look for a global style and discursive form. And it is photography and cinema, above all, that can, more than other arts, claim to be a genuinely global media. Photography and cinema do not seem to be confined by differences in cultural factors, and that is exactly what art is looking for. Furthermore, it is now generally agreed that the so-called 'new art practices', such as for instance 'hyper public art' – interventions, observations, info-aesthetics, etc. in the urban realm – are thought to offer a 'relief' which the autonomous art object can not. As a result, visual art is slowly but steadily vanishing into an expanded field of culture and of other artistic disciplines as well. The linguistic and visual complexity and competence of the art object can only be diluted by such an 'enthusiastic cultural anthropology of modernity'. The difficulties that come with this new flexibility, unrestricted curiosity, and apparent selflessness which characterises so much recent contemporary art, are indeed manifold. Not only does the question of value judgement become much less important than ever before, there is also the naive belief that everything good will find its own place.

And add to these the fact that many art practitioners aspire to give their work the status of field work, drawing from the start on the basic principles of sociology and anthropology, and we will end up in the realm of knowledge which hitherto had been the province of the social sciences. There is nothing basically wrong with such rivalry, if only there is some evaluation at work. Instead we are confronted with an endless archive of highly unverifiable observations and theories which hopefully 'might find their own place'. But things go really wrong when the practice stays unnoticed or is only accessible for a select audience. Will this lead to a situation where visual art will no longer be able to shape its own cultural space, thereby invoking not only a pseudology of social sciences but a poor imitation of other art disciplines as well? Is that what we want to achieve? Of course, many artists who work in these new fields of practice have created innovative and significant works. But it does make me sceptical when artists turn their backs on what Thierry de Duve has described as an aesthetic history of art institutions. For it is not only the faculty of comparing and evaluating that belongs to this history, but also and primarily the faculty of judging. This right to cast one's vote is not only one of the basic principles of our democracies, but also of our museums. All those who practice 'new ways of making art', for instance, outside the museum, must allow for situations whereby many people should be able to put themselves in a position in which they can say: 'This is a very important step, a very important element of our culture'. That responsibility must be given back to the average viewer, not just to a specific client, a 'community' or a select audience.

And this is why I call on you – however idealistic this may sound – to reintroduce the right and the capacity to assess – as something that all cultural producers should nurture and support. We have to search for new models of evaluation, models that do justice to the changes in the production and the significance of contemporary art, but also to the changed function of art institutions. Heated debates about the future of art and of the museum often end just before asking the question of how the existing disciplines and institutions need to react to the new social and technological developments. What the participants in these debates lack is the readiness to truly take into account all relevant factors. And the readiness – if that proves necessary – to give up everything in order to pursue completely different activities. These would be activities that we no longer identify by the fields in which they take place, but by the effects they bring about.

Bibliography

Ernst van Alphen, 'Artists as Observing Scientists and Artists as Critical Observers'. In: **Chambres Séparés. Over hedendaagse kunst en macht.** Ghent (Rijksuniversiteit), 1999

Jean Baudrillard, 'The Beaubourg Effect: Implosion and Deference', In: **October**, no. 20, 1981/1982

Andrew Benjamin, **Present Hope: Philosophy, Architecture, Judaism**, London: Routledge, 1997

Stefaan Decostere, Chris Dercon, John Wyver, 'The New Museum', In: **Mediamatic**, Vol. 3, no. 4, summer 1997

Chris Dercon, 'Am I Now Getting Sentimental?', in: **Parkett**, no. 33, 1992

Chris Dercon, 'Business As Usual', in: **Archis**, no. 10, 1998

Hal Foster, 'The Artist as Ethnographer?', in: **Global Visions. Towards a New Internationalism in the Visual Arts**, London: Kala Press, 1994

Hal Foster, Dennis Holler, Silvia Kolbowski and Rosalind Krauss, 'The MoMA Expansion; a Conservation with Terence Riley', In: **October**, no. 84, 1998

Boris Groys, 'Kunst im Museum, lecture delivered at Basel Museum für Gegenwartskunst', 1998, unpublished lecture

Boris Groys, 'The Restoration of Destruction', in: **Witte de With, Cahiers**, no. 4, Rotterdam

'Imagining the Future of Modern Art', In: **Studies in Modern Art, Research and Scholarly Publications Program of MoMA**, New York, no. 7, 1997

Rosalind Kraus, 'The Cultural Logic of the Late Capitalist Museum', in: **October**, no. 54, 1990

Donald Preziosi, 'Avoiding Museocannibalism', **Katalog XXIV Bienal de São Paulo**, 1998

Donald Preziosi, 'Brain of the Earth's Body: Museums & the Fabrication of Modernity', unpublished lecture, Deutscher Kunsthistoriker Tag, Technische Universität München, 1997

Gianni Vattimos, 'Ort möglicher Welten zur Rolle des Museums in der Postmoderne', in: **Lettre International**, no. I, 1997

Why Bother (about) Architecture? Contemporary Art, Architecture and the Museum

Wouter Davidts

In November 2004, the Spanish artist Santiago Sierra intervened in the Museum Dhondt Dhaenens in Deurle, Belgium. In line with his reputation of being one of the most controversial contemporary artists, he made both a simple and a radical gesture. He took out all the artworks from the museum space and then removed all the glass from exterior doors and windows. The museum was stripped to the bone, reduced to a bare structure, where wind and rain had free reign. Sierra has a record of such drastic acts against architecture. For his contribution to the Venice Biennial in 2003, he had the main entrance of the Spanish pavilion walled up. To the visitors' indignation, Sierra re-routed the entry via a dreary back door that was guarded by a Spanish police officer who only allowed passage to those who could present a valid Spanish passport. The handful of visitors who were able to comply were confronted with nothing but empty rooms inside. In Kunsthaus Bregenz, he loaded the upper floor of the building with 300 tons of bricks. The work *300 Tonnen, 300 Tons* pushed the loading capacity of the KUB structure to the limit, to such an extent that the weight had to be dispersed by pillars on the lower floors. In all three of these cases, architecture – and by extension first and foremost the art institution that it houses – was tested in its capacity to endure artistic intrusion. Whether the injured building is laid bare, locked, or put under pressure, the institution is incapable of functioning in a regular manner, or *in extremis*, any further. Sierra's interventions fit within the fairly recent tradition of symbolic and ever more violent gestures on architecture, and on the architecture of the museum institution in particular, starting with Yves Klein's 'Le Vide' (1958), Armand's 'Le Plein' (1960), Daniel Buren's sealing of the entrance of the Galleria Apollinaire (1968), Robert Barry's 'During the exhibition the gallery will be closed' (1969), Michael Asher's removal of the windows of the Clocktower Gallery in New York (1976), Gordon Matta-Clark's 'window blow-out' in the New York Institute for Architecture and Urban Studies (1976), Chris Burden's 'Exposing the Foundations of the Museum' in the Temporary Contemporary in Los Angeles (1986) to more recent intrusions such as Ingmar & Dragset's 'Taking Place' in the Kunsthalle Zurich (2001) to Kendell Geers' blowing up of a temporary wall in the Antwerp Museum of Contemporary Art (The Devil never rests … 6 June 2004).[1] Since the 1960's, architecture is incessantly perceived and deemed as an instance to be acted *against*. Architecture is regarded as the discipline and practice that represents and enforces the system – its institutions and the social order – and needs therefore to be put on trial, pierced, cut, demolished, split, torn apart, etc. Architecture gives form and identity to institutions, and is therefore the most exquisite target to be able to attack them. By intervening on architectural elements such as doors, windows, stairs or foundations that define and make up the space of the institution, the institutional conditioning of that interior can be assailed, questioned, and ultimately discussed. But haven't such gestures had their day? After decades of all sorts of attacks on the museum and its architecture, aren't we yet convinced of the fatal role that architecture plays within the constitution of the

---

[1] For a brilliant discussion of the way these gestures use architecture to critique the institutional conditioning of exhibition spaces, see the last chapter 'The gallery as a gesture' that was added to the 1999 edition of Brian O'Doherty's 'Inside the White Cube. The Ideology of the Gallery Space' (1999, Berkeley, University of California Press).

museum; that is, delimiting, fixing and affirming the boundaries of the institution? Haven't these kind of assaults on architecture merely become pathetic and hysterical? Is architecture still the most appropriate target to critically re-evaluate the museum, and by extension, institutions for contemporary art in general? If we follow Benjamin Buchloh's statement that every artistic practice needs to develop a critical attitude towards architecture, then how are we to define the nature of that criticality?[2] Within the vital reflection on new stages for contemporary art, is architecture still an instance to bother, or rather to bother about?

In May 2003, a conference entitled 'Museum in Motion' was held at the arts centre De Balie in Amsterdam, after the seminal book *Museum in ¿Motion?* of 1979 edited by Carel Blotkamp.[3] The book of 1979 and the conference of 2003 were launched under comparable circumstances. The book was published upon the occasion of the departure of director Jean Leering from the Van Abbe museum in Eindhoven. Leering's direction of the museum was considered so influential that it merited review. The year 2003 saw three very similar cases: the directors of the most important Dutch museums of modern and contemporary art were about to leave: Rudi Fuchs from the Stedelijk Museum in Amsterdam, Jan Debbaut from the Van Abbe museum in Eindhoven and Chris Dercon from the Boijmans Van Beuningen in Rotterdam.[4] This collective exodus was experienced as both urgent and promising. After all, new directors always get things moving, they set things 'in motion'. They come up with fresh ideas, take a new direction, and reform the existing institution. Or at least, that's what they are expected to do. A new director who merely continues the policy of his predecessor is readily accused of having neither personality nor vision. However, the 'fresh wind' a new director gets to blow through the institution is often translated into building ambitions. The construction of a new wing or a brand-new building – and most of all, the fund-raising that such an enterprise requires – are increasingly considered as one of the most important achievements of a directorship. The irony of the situation in the Netherlands (at the time of the 2003 'Museum in Motion' conference) was that two of the three directors were leaving just after they finished a major, and in both cases, very strenuous and demanding building process: Debbaut at the Van Abbe and Dercon at the Boijmans. While the third, Rudi Fuchs of the Amsterdam Stedelijk, partly resigned because of the desperate dead-end situation of the planned extension project. All three directors were thus engaged in what Stephen E. Weil once aptly described as the 'edifice complex' of the contemporary museum world.[5] In recent decades, just about every museum has drastically renovated, expanded or added to the existing building, at least once.[6] After all, building plans for museums create high expectations. Although architecture is stable, fixed by nature and thus motionless, museums seem to look upon it as the most appropriate medium to break new ground. Architecture is the medium *par excellence* to redefine and re-articulate their institutional position as well as their attitude. In the countless plans for additions and extensions, museum directors are seldom satisfied with making more space available, or just renovating the existing premises. On the contrary, with every museum building enterprise – whether an extension, an additional wing or a brand-new building – they explicitly express the ambition to tackle the 'institutional' space as well. Architecture is used as a vehicle to fundamentally re-think the museum on both a micro and a macro level – not only the commissioning institution itself, but the entire concept of 'the museum' as well. Architecture is capable – or so we are made to believe – of extending the museum's boundaries in both the literal and figurative senses. Thus, while preparing the recently finished renovation of the Museum of Modern Art in New York, Glen D. Lowry claimed that the project would entail more than an expansion of the existing facilities; the museum would 'fundamentally alter its space'.[7] Whenever a museum starts to build, it pretends to do more than give itself a facelift, an implant, a correction, or an 'enlargement', to use beauty industry jargon. The phantasmagorical desire that the restyling of your body will guarantee a better and more rewarding life – epitomized by such television programmes as 'Extreme Makeover' on ABC, 'Beautiful' on VT4 or 'I Want a Famous Face' on MTV – would seem to have infected museums and their directors too.

2 Buchloh, Benjamin H. D.; 'Cargo and Cult: the Displays of Thomas Hirschhorn', in: **Artforum-International** 40; no.3, Nov. 2001, pp. 109–110.
3 Blotkamp, Carel (ed.); **Museum in ¿motion?: the modern art museum at issue / Museum in ¿beweging?: het museum voor moderne kunst ter diskussie** ('s-Gravenhage: Govt. Pub. Office, 1979). The most remarkable difference between the book and the symposium, however, was a peculiar feature of the title of the book. It may be regarded as a detail, but the questions marks that the editors of the book **Museum in Motion** put in the title in 1979, were left in the designation of the 2003 conference. The editors of the book not only put a question mark at the end of the title, but also a reversed one before the word 'motion'. Although this may be considered as a mere typographical joke, it represents the then 'disputable' state of the museum discussion. At that moment in time, there was still lot of discussion, conflict, and disagreement about whether the museum of modern and contemporary art could be set in motion, and how this had to happen. I used the title **Museum in ¿motion?** again for a conference I organized in November 2004 in Sittard and Maastricht, the Netherlands. To consult the programme, visit: http://www.museuminmotion.tk
4 In the meantime, all three museums have new directors: Sjarel Ex at the Boijmans, Charles Esche at the Van Abbe museum and Gijs van Tuyl at the Stedelijk.
5 Weil, Stephen E.; 'A Brief Meditation on Museums and the Metaphor of Institutional Growth', in Weil, Stephen E.(ed.); **A Cabinet of Curiosities: Inquiries into Museums and their Prospects** (Washington: Smithsonian Institution Press, 1995) p. 42.
6 The most recent expansion of the Boijmans van Beuningen by the architects Robbrecht & Daem was already its fourth building campaign. For a discussion of that project, see Davidts, Wouter; 'Robbrecht en Daem and the Museum Boijmans van Beuningen. Architectural Interventions so that Things May Overlap' in: **Maandberichten Museum Boijmans van Beuningen** (May 2003) pp. 2–7.
7 Lowry, Glenn D.; 'The New Museum of Modern Art Expansion: A Process of Discovery' in Elderfield (ed.); **Imagining the Future of the Museum of Modern Art,** New York, Museum of Modern Art / Harry N. Abrams, Inc., 1998, p. 21. This discourse is often enhanced by the rhetoric of the architects themselves. See for example, Rem Koolhaas O.M.A. **Charrette. M(oMA) 1997** (Rotterdam / New York: Office for Metropolitan Architecture, 1997). 'Throughout its history the Museum of Modern Art had used architecture as a vehicle of self-expression and regeneration, articulating and re-articulating its evolving understanding of modern art in built form…At no other time since its founding has the Museum had such a unique opportunity to undertake so extensive a redefinition of itself.'

But what are the results of this general quest for fundamentally new spatial concepts for the museum? From the Neue Staatsgalerie, the Groninger Museum, Guggenheim Bilbao, Milwaukee Art Museum to Tate Modern, we have been regaled with the most diverse and spectacular architectural appearances, ranging from museums that look like hospitals, prisons, jewel boxes, spacecraft, offices, and even all sorts of fish. But has this architectural extravaganza offered a similar amount of thought-provoking institutional structures in exchange? In other words, did these buildings 'imply', bring about, even provoke totally different museum policies? Did all these exquisite *bodies* generate an equivalent amount of innovative and pioneering institutional *personalities*?

Upon closer scrutiny of the kaleidoscopic collection of new museums and museum extensions of the last three decades, we must admit that, despite the euphoric, exhilarated tone of the discourse on museum architecture, very few genuinely innovative museum projects – with the same kind of combined architectural and institutional vigour as the Centre Pompidou, the eminent start of the so-called museum boom – have been completed. Few actual building projects, if any at all, have succeeded in setting the traditional museum typology – architectural as well as institutional – 'in motion'. Yoshio Taniguchi's rebuilding of the Museum of Modern Art in New York may be the largest and most expensive museum building enterprise of the last decades, but it certainly does not convey a substantial breakthrough in our thinking about contemporary museum space, let alone the fundamental spatial alteration that was envisioned and promised – unless in terms of surface and scale, of course.

The rather thin crop is due to the paradoxical position architecture is forced to occupy within a museum commission on the one hand, and to the rather elliptical discourse on museum architecture on the other. Despite all the rhetoric, architecture has rarely been permitted to intervene in the actual spatial development of the museum programme.[8] All too often, the ambition to use architecture to rethink the museum's programme and, by consequence, to develop a novel spatial framework to house that programme, is paradoxically shattered in the name of flexibility or programmatic freedom. Museums, with the museum of contemporary art as the absolute champion, simply do not allow architecture to get in the way of their ambitions.

The museum of contemporary art wants to be at the absolute service of art and artists, so it is troubled by an almost paranoid desire for an architecture that is receptive, adaptable, and adjustable, or, in other words, flexible. But here we face the first paradox. Although architecture is compelled to apply the strategy of self-effacement, it must simultaneously address itself to helping the museum overcome its problems with art. Because after all – as it was defined as the core problem of the *Museum in ¿Motion?* Book of 1979 – art causes the museum a lot of trouble. Since the 1960's, art has drastically altered its nature and strategies: it has become ever more agile, critical toward the institutional framework of the museum, and eager to operate on more specific sites. The museum of contemporary art wants to keep up pace, but is confronted with spatial, institutional, and socio-political problems and limitations. It suffers from the unhappy conscience that it is never able to occupy a true place in the artistic present, as it always 'frames' art. This identity crisis incites the museum to indulge in ongoing self-critique, institutional introspection, and ultimately, self-denial. In recent decades, we have been confronted with dozens of museums that, following the artists, contest their own space and develop an anti-museum policy, some even going so far as to pretend to stop being a museum. The nature of this crisis, however, is fundamentally spatial. A quick glance at the metaphors used by museums to question their status, reveals the architectural bias of the crisis; if the museum of contemporary art wants to transform itself from a static *repository* for the art of the past into a dynamic *workshop* for the art of the present, it has to tear down its walls, open up its space, leave the premises, push back its frontiers, etc. Both the words 'repository' and 'workshop' imply a different spatial, and hence, architectural connotation. So it seems that architecture has ended up in a quite ambiguous position; while it is obliged to refrain from intervention or mediation in the museum programme and is expected to produce so-called flexible and neutral spaces, it is nevertheless always put at stake within the critical questioning of that programme.

When Marcel Broodthaers was asked in an interview what space hid, he compared that pursuit to the children's game '*Lou es-tu là*'.[9] The relentless search for fundamental spatial alteration or the continuous drive to redefine the space of the museum amounts to nothing but a phoney game of hide-and-seek, merely a desperate attempt to deny the institutional conditioning of the museum interior. This search does nothing but obscure the essence of art: its institutional encompassment and its resulting reification. The inexorable quest for new concepts of museum space is just a misleading game in which the players – artists, museum staff, but also architects – go to great pains to evade the true answer. Continuously, architecture is asked to meet the problematic desires of museums and other art institutions. They believe that architecture will enable them to transform themselves from a motionless stock into a vibrant workplace, from a place of passive spectatorship into a locus of active and animated cultural production, into an institution that is ultimately as un-institutional as possible. This ambition was achieved, both fiercely and tragically, in the Centre Pompidou. The building tried to deliver a solution for the unpredictable

8 For a discussion of an architectural project that did engage itself with the definition of the museological program through an architectural design, see my discussion of the design of Xaveer De Geyter Architecten for the Antwerp Historical Museum or Museum Aan de Stroom (MAS): Davidts, Wouter; 'The Museum as Warehouse' in Bekaert, Geert (ed.); **Xaveer De Geyter Architects. 12 projects** (Ghent / Amsterdam: Ludion, 2001) pp. 36–39.
9 Broodthaers, Marcel; 'Dix Mille Francs de Récompense. Une interview d'Irmeline Libeer', in Gevaert, Yves (ed.); **Marcel Broodthaers. Catalogue – Catalogus** (Bruxelles, Société des Expositions du Palais des beaux arts de Bruxelles, 1974) p. 66: '**En effet, le loup dit chaque fois qu'il est est ailleurs, et cependant il est là. Et l'on sait qu'il va se retourner et attraper quelqu'un. La recherche constante d'une définition de l'espace ne servirait qu'à cacher la structure essentielle de l'Art, un processus de réification**.'

development – the spatial and exhibition requirements of the contemporary work of art – and express the image of a popular and iconoclastic art institution. And as Reyner Banham once remarked, it drove that question so far that it elliptically handed it back.[10] It is therefore not surprising that the Centre Pompidou was not only experienced as 'too flexible', but that its immense popularity also meant it was worn bare within nearly two decades.

Although many international examples of purpose-built museums of modern and contemporary art could be regarded as praiseworthy responses to the innovative manifesto of Piano & Rogers, they could never rival its – albeit extremely problematic – radicality. None of the icons of the recent museum frenzy – such as the Neue Staatsgalerie in Stuttgart, the Getty Center in Los Angeles, The San Francisco Museum of Modern Art, the Guggenheim Bilbao, or Tate Modern – are truly innovative projects. On the contrary, most investment has been done, in one way or the other, in what Alma Wittlin could still categorize in 1970 as 'peripheral functions'.[11] Whereas the core programme of the museum – the conservation, study and presentation of artefacts – used to take up about 90% of the total surface of museums, this has shrunk to a mere 50%. In the post-Pompidou era, about every museum has an elaborate gift shop, a fancy restaurant, a well-equipped concert hall or movie theatre, and in extreme cases, even a supermarket or shopping mall. The classic museum programme is seldom the key element of a building operation. Indeed, art museums are built for various reasons, few of them having to do with art. What this means for architecture is that attention is now focused chiefly on the way it gives shape to this external programme. A museum design is no longer assessed primarily in terms of its intrinsic museological qualities, but on its response to the external programming package: whether it provides the city with a landmark, how it fits into the cityscape, whether it adds value to the surrounding urban fabric, stimulates city planning, distributes the museum's different peripheral functions in an interesting manner, and so on and so forth. All these kinds of design qualities have their importance, as they situate the role of museum buildings within a broader socio-economic, urban, and political context. Nevertheless, they demand evaluation criteria that, in a sense, are entirely detached from the assumed fundamental institutional change. They do not provide any new insight into the way in which the museum can function as a stage for contemporary art, and ultimately, the form that the museum – as an arsenal of memory – could or ought to take. They may result in a building that – like the Guggenheim Bilbao – functions as the icon, sign, and logo of a city, all at once, but no longer cares about what's being shown inside, whether motorcycles, Armani costumes, or artworks. The building takes care of the spectacle. When you visit the website of the Guggenheim Bilbao, the first heading you can click on is 'the building', the second is 'the exhibitions', and the third and last is 'the permanent collection'.

But maybe the design of these gaudy sculptures – the audio guide of the Guggenheim Bilbao even wants you to believe that the building has erotic qualities, as brilliantly 'performed' by Andrea Fraser in 'Little Frank and His Carp' (2001) – is the only challenge that is left for architecture within future museum commissions. A quarter of a century after the publication of the *Museum in ¿Motion?* book, the situation has drastically changed. Museums are no longer confronted with the same problems as at the beginning of the 1970's. The critical questions – graphically represented by the double question mark – that the editors of the book in 1979 were still able to ask, and the answers that the museum officials, artists, critics, theoreticians, and academics tried to formulate, have now been completely superseded by the contemporary state and conditions of the art world. At the beginning of this new millennium, the eventual mobility and liveability of the museum is no longer a point of discussion; the critical relationship between art and museum even less.

The core of the present museum discussion is simply not occupied by art anymore. Whoever thinks that it is still art that brings the museum in an awkward position is terribly naïve. Museums no longer feel impotent or helpless towards art that critiques the institution, leaves, or even destroys, the building, or asks for help for large-scale and complicated projects.[12] Quite the contrary, the former rebels have been domesticated; they are welcomed with the greatest cordiality, and almost cuddled to death. William Rubin was quite accurate when he warned artists, as early as 1974, that they'd better be warier of the open arms than of the closed doors of museums.[13] But it's too late. The willingness of museums to go along with so-called 'transgressive' artistic adventures is limitless. They have made them merely 'part of the program' as they are estimated to enhance their credibility and guarantee their reputation of being rebellious, critical and controversial. Nowadays, it is hard to find a museum that does not function as a platform for contemporary art – even scientific or history museums nowadays invite contemporary artists to mess around in their collection and exhibition spaces. In October 2000, when The Museum of Modern Art in New York converted The Abby Aldrich Rockefeller Sculpture Garden into a staging area for the construction of the Museum's new building, the museum itself invited the artist Mark Dion to perform a series of archaeological excavations. Museums have adopted the critical strategies of

10 Banham, Reyner; 'The Pompidolium', in **Architectural Review**, no. 963 (May, 1977) pp. 277–278.
11 Wittlin, Alma Stephanie; **Museums: In Search of a Usable Future** (Cambridge, Mass.: MIT Press, 1970) p. 1.
12 These issues are still formulated in the 1979 **Museum in ¿Motion?** book by Hans Haks in the introduction as the 'major problems' that museums are facing. If we take a closer look at the questions, it becomes immediately clear that a lot of them, if not the majority, seem no longer to be a true issue of discussion. Would there be anyone to argue that the museum should deal with visual arts only, or with theater, music, literature, architecture and dance as well? Or that the museum should engage with 'high art' only, or with any cultural phenomenon? Let alone that someone would contest the idea that a museum should organize temporary exhibitions. Or, just imagine that we would question the idea that the museum's activities are delimited to its own building. And, finally, who on earth would contest the idea that a museum commissions artworks?
13 Alloway, Lawrence and Coplans, John; 'Talking with William Rubin' in **Artforum** (October 1974) as reprinted in Blotkamp, Carel (ed.); **Museum in ¿motion?: the modern art museum at issue / Museum in ¿beweging?: het museum voor moderne kunst ter diskussie** ('s-Gravenhage Netherlands, Govt. Pub. Office, 1979) pp. 311–319.

artists to such an extent that they pretend to share the same interests. As a result, the space of the museum – and by extension the role of architecture – is no longer brought up for any real discussion, unless within self-deceptive, tragi-comic and narcissistic gestures – on both an artistic and institutional level – such as those of Dion and Sierra. While the so-called exploratory intervention of the first and the violent action of the latter still pretend to question the museum space by hassling architecture, they simply profit from the institution's sadomasochistic desire to be subjected to it.

This dramatic shift in the position and attitude of the museum has serious ramifications for architecture, and especially for architectural design and practice. The institutional problem and complementary desire that lay, for example, at the basis of Centre Pompidou, and that were consequently translated into the building brief, have simply dissolved. The flexible attitude towards contemporary art that the museum aspired to and that architecture was supposed to frame, has become standard procedure. But this major shift has happened without, almost in spite of architecture. It is no longer up to architecture to develop an operational form for the institutional programme, to *design* a building that embodies it. Quite the contrary, as museums are convinced that they function as workshops, architecture is forced into a position of mere accommodation, *once more*. And thus we end up with the paradoxical call for flexibility *again*. Or, the other option – one that, since a decade or so, is considered by many as the hottest trend in museum design – is the plea for minimalist and loft-like interiors.[14] The strategy to reconvert former factories and industrial buildings into museums – with such famous examples as the Temporary Contemporary in Los Angeles, Tate Modern, and, more recently, Dia:Beacon – is the new paragon of architectural self-effacement, and is therefore often regarded as the true alternative for the architectural extravaganza. This strategy, however, often amounts to either a fetishistic glorification of raw and often large spaces, or a cosmetic persiflage of a historical spatial paradigm of artistic production. It is based on the false and too easy assumption that the museum, in order to function as a space of artistic production, needs to adapt the guise of that space that is historically considered paradigmatic for it.[15]

But does this mean that there is no critical space left for architecture in museum design? Is there no vital role and significance to discern anymore for contemporary architecture within the construction of future museums? Is the only thing architects are still allowed to do, to put their signature at the entrance, as Hans Hollein did already more than two decades ago in the Museum Abteiberg? Has architecture failed so dramatically that it is now being forced into a mere subservient and benign position? A museum may still be one of the most prestigious commissions an architect can get, but is it as challenging as it used to be, or is said to be? Isn't it quite an exaggeration to declare that museums are 'seismographs of architectural culture'?[16] They may generate architectural discourse, but is it really the discourse that represents the most avant-garde practices in architectural theory and practice?

Maybe the issue is not that much of an architectural, but rather of an institutional nature. Yet, it is time to question the assumption that contemporary art needs a new museum typology of its own, a typology that first on an institutional and then on an architectural level corresponds to its strategies of production, and facilitates them as well. Do we, as Hans-Ulrich Obrist suggested in an interview with Cedric Price, really need to invent 'a certain type of institution' *again*?[17] The museum concept is, as William Rubin stated in 1974, not infinitely expandable, let alone that it would be endlessly renewable. The ambition to rethink the museum has become so compulsory that it is on the verge of becoming preposterous. The fact that many – artists, curators as well as critics – find it necessary to batter the museum over and over again, is entirely wretched. Why do they still consider the museum as an enemy-institution? It is a sign of total idiocy to think that within the reflection on new stages for contemporary art, one first and foremost has to finish with the museum. As Thomas Keenan rightly pointed out, being critical about the museum does not imply that one needs to demolish it. What difference does it make that the critique takes place in this very place, the museum, the place they seek to contest? The challenges are far-reaching, but they do not simply proclaim that the museum is finished. The question of what museums might be 'for' testifies to a certain fidelity to it.[18] The boundaries and possibilities are always subject to precisely the renegotiation it seeks to render possible, by virtue of its publicity.

Moreover, the traditional tasks of conserving, studying, and presenting artworks haven't lost their (public) relevance at all. Contemporary cultural production is no longer static and slowly evolving as in the 19th century, but almost totally commercialized, fleeting and mediatized. Within a society that is reigned by short-term agendas, instant memories, temporary regimes, provisional programmes, ephemeral networks, and impermanent flows, the traditional – some may call it old-fashioned, even conservative – programme of preservation and memory has become even more important than before. But, as Charles Esche stated so blatantly at the *Museum in Motion* conference in 2003 in Amsterdam: the management and presentation of

14 Helen Searing, 'The Brillo Box in the Warehouse: Museums of Contemporary Art and Industrial Conversions' in Weingartner, Fannia (ed.); **The Andy Warhol Museum** (Pittsburgh / New York / Stuttgart: Andy Warhol Museum / D.A.P. / Cantz, 1994) pp. 39–65.
15 The popularity of the loft stems from the idea that it is an 'original' or 'authentic' space for the art of recent decades: it 'belongs there'. As much art, from Minimalism onwards, has been made in industrial spaces – such as lofts – it is frequently said to be fitting to exhibit these works in similar spaces. See for example: Foster, Hal; 'Illuminated Structure, Embodied Space' in Gluckman, Richard (ed.); **Space Framed: Richard Gluckman, Architect** (New York: Monacelli Press, 2000) p. 184. For an extended critique of the loft as the new typology, see Davidts, Wouter; 'Musea en de belofte van artistieke productie. Van Centre Pompidou tot Tate Modern' in: **De Witte Raaf**, no.101 (2003) pp. 7–11.
16 Lampugnani, Vittorio Magnago; 'The Architecture of Art: The Museums of the 1990s' in Lampugnani (ed.); **Museums for a New Millennium** (Munich: Prestel, 1999) p. 13.
17 Price, Cedric; **Re:CP** (Basle: Birkhauser Verlag AG, 2003) p. 63.
18 Keenan, Thomas; 'No Ends in Sight' in Borja-Villel, Manuel J., John G. Hanhardt and Thomas Keenan (eds); **The End(s) of the Museum / Els límits des museu** (Barcelona: Fundació Antonio Tàpies, 1995) pp. 19.

a collection is simply not all that exciting, while the production of exhibitions with contemporary artists is 'sexy'.[19] But isn't making sexy art exhibitions, seductive museums, and tempting architecture a rather meagre challenge? Museums can function as sites that provide the indispensable spatial and temporal enclave to study the inflationary field of culture from the sidelines and that install the necessary temporal margin to decide what in the end may be worth preserving, to decide which things we want to remember, after all.

Contemporary art runs the risk of disappearing in the visual sludge of our culture, or in what Hal Foster has rightly labelled as the 'total design culture'.[20] When every fringe, rough edge, or unnecessary remnant – whether of an interior, a body, or a company – is neatly smoothed away and subjected to an appropriate design solution, we end up in a situation of *indifference*. When everything is streamlined, polished, and, above all, stylized to perfection, there is no margin left for culture, and art in particular, to *distinguish* itself. Instead, its only task is to deliver artistic *surplus*. In the current era of the total blending of artistic disciplines, the dissolution of institutional domains, and the liquidation of critical distances, it seems all the more important to create temporal and spatial enclaves that allow for distinction, that afford room for *difference*. If art is one of the few domains where one can still work and research with some measure of freedom and independence on the meanings that constitute our contemporary culture and society, then the museum remains one of the most appropriate sites to discuss the results of that investigation, seriously, and most of all, publicly. Its public constitution guarantees that the discussion never reaches a consensus, but maintains a status of critical *dissensus*. And it is precisely at this point that one can see a valuable role for architecture to play. If the museum remains one of the pre-eminent places for bringing art up for discussion, for negotiating and disputing its public nature, then architecture can contribute to the specificity of that debate. Architecture can never participate in the debate, neither anticipate its unpredictability, nor guarantee it a smooth progress. Architecture can only create the conditions to 'ground' it, provide concrete parameters. Architecture is the medium *par excellence* for demarcating a specific space, differentiating a certain area within the ever increasing and nebulous field of cultural production, and for defining and materializing the boundaries of the framework within which the museum can deploy its institutional programme. Then again, it is up to the museums to let it interfere. But that requires *guts*, and, as Stephen Weil once

---

19 The irony being that in the meantime, Esche has become the director of the Van Abbe museum.
20 Foster, Hal; **Design and Crime and Other Diatribes** (London / New York:Verso, 2002).
21 Weil, Stephen E.; **A Cabinet of Curiosities. Inquiries into Museums and Their Prospects** (Washington: Smithsonian Institution Press, 1995).
22 Broodthaers, Marcel; **Lettre Ouverte (Departement des Aigles**, Düsseldorf, 19 September 1968), reprinted in Blotkamp (ed.); **Museum In ¿Motion?**, p. 250.

laconically remarked, 'Courage is rarely an institutional quality.'[21] It is easier to bother architecture, than to be bothered by it. It requires less bravery to molest and demolish a building than to cope with it, to think *through* it. Maybe Broodthaers was right when he stated that museums are eternal playgrounds.[22] But even that doesn't rule out architecture. Even a game needs rules and boundaries. And that makes it all the more dismal to attack architecture over and over *again*.

1 Daniel Buren, **Travail in situ**, Galerie Apollinaire, Milan, 1968.
2 James Lee Byars, performance on the facade of the Museum Fredericianum, Documenta V, Kassel, 1972.
3 Xaveer De Geyter Architects, Competition Design for Antwerp Museum Aan de Stroom (MAS), 2000.

[1] [2] [3] ►

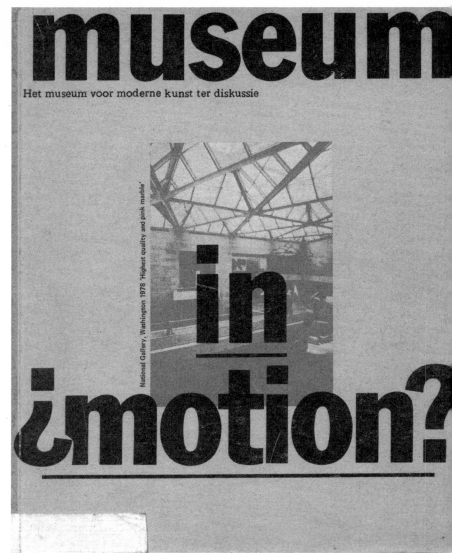

[4]

[5]   [6]

4 Carel Blotkamp (ed.), **Museum in ¿Motion?: the modern art museum at issue / Museum in ¿beweging?: het museum voor moderne kunst ter diskussie,** 1979. Cover.
5 Hans Hollein, Städtisches Museum Abteiberg, Mönchengladbach, 1986.
6 Renzo Piano & Richard Rogers, Centre Pompidou, Paris, 1977.

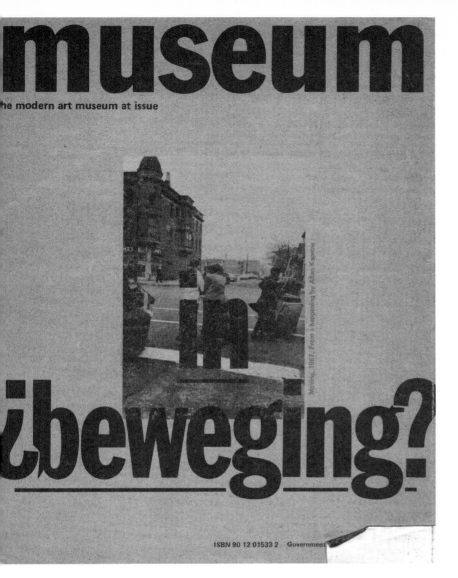

[7]   [8]

7 Santiago Sierra, **Removal of the glass windows and doors of a museum**, Museum Dhondt-Dhaenens, Deurle, 2004.

8 Santiago Sierra, **Removal of the glass windows and doors of a museum**, Museum Dhondt-Dhaenens, Deurle, 2004.

Order on Display
Michael Müller

This paper proceeds from the proposition that the European city is the place of both accumulation and collection of materials that reflect the value hierarchy of a structured world. With his magnificent fresco, 'the Good and Bad Government' (Siena 1337–1339), Lorenzetti has shown us impressively how much our collective identification with this place depends on an order being manifested in these objects, an order in which these materials are sorted and classified and whose norms transcend this place, even if this order does not yet have the rationality of planning that began to pervade the history of the European city with the first utopian visions of cities in Renaissance times. In other words, in this interconnected order of the city, the social process is seen to be meaningful and beneficent. The identity of a place is then dependent on the various grades of objectifying. Achieving this transformation by the production of objects (buildings, utensils, devices, clothing, etc.) is a precondition for spatialization, i.e. the dispersal of objects in space. Clearly such an arrangement belongs to visibility.

In further pursuing this thought, we will discover that we can only achieve this formation of spaces (whether real or in our imagination) if we are capable of viewing this variety of objects arranged in space, or these places, as coherent. This arrangement, in turn, requires a capacity of abstraction in a place like a museum, a place that only arises from a collection of objects, that allows us to practice particularly well. If, now, we follow the authors of the catalogue of 'Wunderkammer des Abendlandes'[1] – Western European curiosity cabinets, or rarity cabinets – then this fact is perfectly in keeping with what they call a 'uniquely European phenomenon', i.e. a compulsion to arrange objects in order to understand them as a whole and to join tradition with innovation.

Therefore we can only understand the European city, its configuration and culture and the arrangement of the space, as well as the arrangement of objects in the space, if we duly take into account the significance of collecting and its influence on the city. The question of whether it's the one that influenced the other or vice versa, is of no concern to me here. After all, viewed historically, it is the city which set in motion the 'museum experience'. Since at least the dawn of modernity, the specific experience of museums has become one of the constitutive elements of the experience of cities. Their mutual influence plays an important role in the practice of constituting these spaces, and indeed of constituting meanings in these spaces. This role is all the more true for residential spaces and their interior. For the 19th-century middle-class citizen, living in residential spaces and collecting objects that he derives from the commercial exchange of goods are one and the same thing. 'It says a lot about the middle classes of that time', Peter Gay writes in his excellent study *Bourgeois and Bohème*,[2] 'about their conception of their self-worth and about their search for a place in the world, to observe how their passion for collecting truly flourished. Conceivably, the fact that the bourgeoisie in the 19th-century was a nation of collec-

---

1 **Wunderkammer des Abendlandes** (1994). Catalogue of the exhibition 'Wunderkammer des Abendlandes. Museum und Sammlung im Spiegel der Zeit' (Bonn, 1994/95).
2 Gay, Peter; **Bourgeois and Bohème** (Munich, 1999).

tors can be interpreted as meaning that they regarded this activity of collecting as emblematic of the triumphant individualism.' The activity of collecting provides a 'sense of control, i.e. power over a selected area'.[3]

We shouldn't be distracted by the fact that the items these (petit-)bourgeois gathered around them were mostly no more than a better kind of 'trash'. Nor by the fact that they share this passion with the *Lumpensammler*, or 'rag-picker', who was similarly poetically heroic. The kind of man, in Baudelaire's work, who would 'have to pick up the big city's garbage of the previous day'.[4] 'Everything that', as Walter Benjamin writes, 'the big city discarded, everything it lost, everything it despised, everything it trampled on – he would list in his inventory and collect.'[5]

It is certainly no coincidence that the cities, as they lose their medieval defensive walls and thereby open themselves up as 'collections', start to erect new walls around the now precious perimeters of their collections, but close their eyes to their own unbridled growth. Consequently, the constitution of the museum coincides historically with the introduction of changes in the urban space, in which borders are continually crossed; an exchange marked by the dislocation of values, and simultaneous growth and decline.

The city, as a place of mutual interpenetration, gives rise to experiences that permeate the space of the museum and the attitude towards collection and exhibitions. The great American scholar of the city, Lewis Mumford, emphatically stressed again and again this mutual interaction of the museum and the city. This charged and changed relation of the city to the museum, becomes for him not some necessary consequence of the growth of the Metropolis, but rather, is the most typical institution of the Metropolis.[6] What the gymnasium was for the ideal life of the Greek city-state, or, the hospital for the medieval city, so the museum is for the metropolis. Indeed, by the beginning of the 19th century we will see that the museum takes on the aspect of being a universal space. Just as the metropolis serves 'the museum in an essential way, [thanks to] its inner breadth and its long history, a city, which has grown over the centuries, possesses a larger and richer collection of cultural examples than can be found anywhere else'.[7]

A Universal Space In this universality, the city finds its full match in the museum. As early as in the beginning of the 19th century, we witness the development of this aspect of the universal and universalizing of the museum space since it is at this time that the museum begins to claim for itself the title of being the exemplary place of the universal language of art, of all art. The museum, in the emphatic sense conceived of in the Age of Enlightenment, is a universal space because it gathers together all human knowledge and achievements; it brings together and unites paintings, drawings, books, objects, and specimens of natural history. The Museum of Calais founded in 1836, for example, perfects the goals of the Enlightenment, not only as an institution, but also as a vivid symbol of the success of the purpose and aim of the Enlightenment.

However, there was nothing abstract about this for contemporaries. Just how little this place was thought of as a mere place of contemplation in the 19th century, is reflected in the fact that between 1806 and 1914, in France alone, over 70 newspapers, journals and albums used the name 'museum' in their title. The notion of the 'printed museum' existed as an encyclopaedic institution intended for the education and cultivation of the whole society.

Comparable to the multitudinous museum-related publications were the 75 magazines, in roughly the same time frame, which published details about the contents of department stores and goods as their exclusive theme. In them, we find numerous references to the association of the museum with commerce and money. There is a parallel between the perceptions and movements that had to be learned in department stores, on the one hand, and in museums on the other. Thus, in the actual decoration of their interiors many museums copied the model of the *boutique*. In addition, the theoretical and conceptual vocabulary of museums is borrowed from the world of commerce. One example may suffice: the vitrines or glass display cases (*montres*, in French) were about viewing as well as showing. Guided tours were offered around the department stores, which could hardly be distinguished from those provided in museums, and vice versa. For this reason, most visitors to museums, who were not trained in the history of the arts probably perceived the curiosities on display first and foremost as expensive objects – commodities of extraordinary value, yes, but still commodities.

In the 19th century, the press, the museum, and the warehouses of commercial goods all played a complementary role within specific ideological systems. Just as the great commodity depots invited their customers to have the pleasure of privately consuming the accumulation of the process of industrialization, the press sold to the readers the pleasure of 'accumulated' information and the museum allowed the visitor a symbolic appropriation of the extraordinary which itself could not be bought, nor even understood, except by a small elite of amateurs, yet through the process one acquired the prestige of high culture. The press, the museum, and the department store play complementary roles within a single ideological system. Out of these three results of industrialization, the department store offers its customers the pleasure of privately consuming accumulated products; the press sells its readers the pleasure of accumulated infor-

3 **ibid.**, pp. 194**ff.**
4 Benjamin, Walter; **Gesammelte Schriften** Bd. I,2 (Frankfurt am Main, 1974) p. 582.
5 **ibid.**
6 Mumford, Lewis; **Die Stadt: Geschichte und Ausblick** Bd.1 (Munich, 1979) p. 656.
7 **ibid.**

mation; and the museum permits its visitors a symbolic possession of inaccessible objects – objects they can neither buy nor fully understand, with the exception of a small elite of *amateurs* – and which thus carry the prestige of high culture.

Identity Formation and the Act of Displaying

This closeness of the capitalist production and consumption of goods to the presentation and perception of collected objects in museums sheds light on an additional common aspect in our attitude towards the world of objects – be it in the space of the city, the museum, the department store or within one's own four walls.

At this point we can benefit from the results of research by various scholars, such as C.B. Macpherson, Susan Stewart, Colin Campbell, Grant McCracken or James Clifford. They all arrive at the conclusion that the acts of collecting and displaying have deeply shaped the process of Western identity formation, with consequences that continue to this day. What they too find remarkable is the fact that the act of collecting has created a world that is centered around and created by man, with its own internal system of meaning, and whose physical content, the collected objects, forms a structured environment with a temporal reference frame of its own.

Macpherson[8], in particular, analysed the possessive individualism in Western culture in the 17th century, and investigated the emergence of an ideal Ego as an owner of property. According to him, the property accumulated by the individual is a reflection of the world of commodities. The very material world, we could say, is a limit case for the structure of subjectivity, so that identity is a 'kingdom' of objects, knowledge, memories and experience. Clifford remarks on this point: 'A kind of "collecting" of objects around oneself and one's group – the accumulation of a material "world", the delineation of a subjective realm that is not "another" – can be considered universal. All these collections embody hierarchies of values, exclusions, a territory of the ego governed by rules. However, the conception that one part of this act of collecting is the accumulation of property, the thought that an identity springs from a wealth (of objects, knowledge, memories, experience) – these ideas are', according to Clifford, 'certainly not universal.'[9]

In the catalogue for the exhibition 'Wunderkammer des Abendlandes' of 1994, the 'accumulative drive' is discussed, i.e. a drive, that has been characteristic of the European centers of growth since Renaissance times, a drive 'to collect, to preserve, to catalogue and to analyse the expressions of other cultures.'[10] Colin Campbell[11] adds force to this observation and traces for the era of Romanticism, a constitutive connection between the determination of the subject and the newly arising patterns of consumption. Contrary to the thinking of Enlightenment and Idealism, in Romanticism the subject is conceived of as an individual. The transcendental characterization of the subject is transformed into an empirical claim. Self-*fulfillment* and self-*realization* become more important than the earlier aspects of self-*preservation* and self-*determination*. The subject's uniqueness and autonomy go hand in hand with an insistence on self-realization through experience and creativity, both being driving factors in the Consumer Revolution. According to Campbell, the individuals are willing to accept that the Ego is dominated by consumption, and finds in it its expressive agency. Clearly, commodities possess cultural meanings. As a result, goods attain a cultural significance and are used by consumers for cultural purposes, to express cultural principles and cultivate ideals and to form their own lifestyle and self-image. Over time the system that produces and designs these goods becomes itself a form of cultural understanding. It would not be an exaggeration to subscribe to the view that culture in Western cultural spaces is consistently dependent on consumption. Without consumer goods, our society would lose its key instruments for the reproduction, representation, and manipulation of its culture. Without consumer goods, self-realization and collective differentiation would not be possible.

These commodities form the material side of the 'expressive turn', a term Charles Taylor uses to denote the constitution of the modern individual. He writes that since the end of the 18th century, the characteristic feature of human life is 'the expression of the Ego as a conscious agent'.[12] This 'expressive individualism' – with its urge to actualize its inner richness through expression – matches ideally, corresponds to, and is served by the universal dissemination of the production of goods as a projection screen. The flipside of this alliance was recognized early on by Marx as the reification to which all human efforts of expression are subject.

Crystal Palace   Perhaps we have no better proof of this than the World Exhibition of 1851 in London. The American art historian Donald Preziosi has shown how the relation between the exhibited goods and the formation of identity in the Great Exhibition of the Arts and Manufactures of All Nations, together with the fantastic architecture of the Crystal Palace, turned into an overwhelming experience. In Preziosi's words, this building became a true container of the world. The Crystal Palace seemed to arch over the whole of London.

This event was the first high point of an alliance of art and capital that now culminates in the universalization of aestheticism. Preziosi[13] speaks of the link between the enterprise of art history as a modern discipline and that aesthetically metaphorical configuration of modern technology that visually united all the images, symbols and signs of all the cultures of Europe and

---

8 Macpherson, C.B.; **The Political Theory of Possessive Individualism** (Oxford, 1969).
9 Clifford, James; **Ausgestellte Kulturen** in **Lettre Internationale** no.29 (1996) p. 28.
10 Wunderkammer des Abendlandes, op. cit., p. 17.
11 Campbell, Colin; **The Romantic Ethic and the Spirit of Modern Consumerism** (Cambridge, 1987) p. 193.
12 Taylor, Charles; **Hegel** (Frankfurt am Main, 1978) p. 33.
13 Preziosi, Donald; 'Reckoning with Things' in **Kritische Berichte** I (Marburg, 2002) pp. **9ff.**

beyond. In doing so, the differences were reduced to mere matters of style, letting the visitor to the exhibition experience the alienness, or 'otherness' of the colonized non-European cultures, as well as historically distant cultures, as something that had been disciplined and controlled impressively well. The Crystal Palace, as such became the pinnacle (*summum capax*) of all that went before in culture and architecture, an embodiment of the wonderfully ordered system of, if I may use the term, museology, of consumerism, of art, style and history.

In summary, we can draw two conclusions from this overview. Firstly, the Western European city is a place where the strategy of 'possessive individualism' of culture and of authenticity was upheld. In contrast to the eastern city, the European city is a site or place of revolution, a place where the act of 'collecting' involves a continual transformation of resources based on the simultaneous destruction and construction of accumulated objects. Consequently, 'collecting' also is implicated in the 'death of the object'. As such, the revolutionary potential becomes a precondition for the continuous museumification of the accumulated material world.

Secondly, commodity production and consumption, as a mystical glorification of material goods, are inseparably linked as components in the constitution of the subject of the 19th century, and are indeed unthinkable as separable. The World Exhibition, as a large collective experience, demonstrated for the first time that one function of the physical presence of goods – i.e. their visible surface – was to exercise control over the foreign cultures that they represent in a context of universal aesthetization. The parameters of order that determine the way in which things and objects, and the object world are viewed, seen, and understood, drives an aesthetic as spectacle, or as a collectively experienced mega-event. Up until today, art history once again is obliged to fulfill the role of disciplining, through its technique of viewing in a way that is capable of differentiation, and to attribute meanings, characteristics, races, creeds and mentalities through its medium of style and authorship. For after all, Fordism, over the course of several decades, made increasingly less use of a discipline of differentiation, the distribution of objects and subjects in an ordered space, a phase where the mere results of a disciplining way of viewing were sufficient. The cities of those years remind us of uncreatively lined up display cases of museums and department stores. Symbolization and representation seemed to have fallen into oblivion.

Today – besides the shopping paradises – museums are paragons of places where the skill of viewing objects can be practiced in exemplary fashion, where aesthetic values can be tested, and movements and encounters in the space can be experienced. They therefore influence our demands from our experience of the real urban space, demands that are already reflected in many places in the aestheticism of urban space.

Collecting   'A museum is like the lungs of a metropolis' (Georges Bataille)
A 'true collector', the sociologist Alois Hahn from Trier contended, 'is someone who has abandoned the low life of pragmatism, and instead lingers in the higher sphere of culture and knowledge'.[14] On the practical side, collecting means something like stocking up provisions, which seems to be not only reasonable, but also necessary, as it guarantees the future in the present.

In this pursuit, the city, has, since its very beginnings, been an exemplary model, as a 'warehouse, a place for storing and collecting'. Due to its 'capacity for holding objects' the city is a unique storage medium. But it is also a medium that – as we have seen – conveys the superiority of the urban way of life through the diversity of the stored objects. For the city 'not only keeps a larger number of people and institutions together than any other community, it also transmits a larger share of the lives of its inhabitants, more than the human memory is capable of transmitting through simple speech'.[15] What emerges here is the long-standing knowledge of the dependence of the skill of memorization and remembering on the possibility of experiencing a space. Even in Cicero's times, it was known that spatial order and sequence are most suitable mnemonic aids. The places and images bring forth this *Artificiosa memoria*, the artificial memory. Mnemosyne, we may remember, was not only the ancient goddess of memory and remembering – she was also the goddess of the Muses and museums. Following Lewis Mumford, we would have to say that it is also, and above all, the city that achieves this coherent memory.

According to Lorenzetti's conception of the city, the city is a wonderful example of a future in which accumulating and collecting diverse things serves to create a spatial unity. This unity is, at least, what the Council of Nine of the city of Siena wanted: to convey an image of the kind of structure that symbolically represents the continuity, that is to say, the future of the urban society. In that image, the aesthetic synthesis appears to align every part with the whole, the substance with the shape, the sensory images with a sensible order.

In returning to Lorenzetti, the image shows how the city is a gathering of different things into a spatial unity. Much later, in the world of thoughts and ideas of the Enlightenment the 'consensus' emerging from the work of art would be transmitted to the audience – as a happy consensus between individuals and society. Hence the exhibition-like character of all the aggregated virtues of human existence, inherent in Lorenzetti's visual depiction of urban life, represents one of the classical characteristics of the condensed life in European cities, the non-violent encounter with

14 Hahn, Alois; **Konstruktion des Selbst, der Welt und der Geschichte** (Frankfurt am Main, 2000) p. 441.
15 Mumford, Lewis; **op. cit.,** pp. 115ff.

strangers. We can trust *Bon Governo*, while in *Mal Governo* sheer violence rules. As we recognize the presence and arrangement of people and things in a place, or in the space as a result of deliberate action, they are *given meaning*, this attribution being the very characteristic of later museum representation.

The act of collecting in which diverse things are ordered sensibly or put into a sensible relationship in order to arrange them in a homogeneous context – the context of the city – does not yet mean, at least not in Lorenzetti's picture, extracting the individual segments of urban life and work from their original contexts. At this point, their shared context in the city still gives them meaning, outside of which no other contexts exist. With the advent of industrialization, at the latest, this position is no longer tenable – not even as an illusion.

In the museum, the strategy of classification allows the reality of the collection to dissociate itself from the specific layers of production and acquisition. In this way, the museum conveys the illusion of a relationship between the things instead of social relationships. As a result, museums are places where a specific conception of the world is learned and practiced; they accumulate possessions and thereby represent a collective identity that is tied to wealth. Museums consequently carry a big responsibility for they set an example. They act as models in the way that they represent the activity of collecting and displaying; they lay down definitions and thereby provide a sense of control over the objects. As a result, in the words of Lewis Mumford, they offer the opportunity, 'to gain access to a world whose immense size and diversity would otherwise far exceed human comprehension. In this sensible form, the museum, as an instrument of selection, is an indispensable contribution to urban culture.'[16] Yet – which urban culture?

Childish Collecting You are certainly familiar with Walter Benjamin's observations on the 'tidy child', observations that – by serving as a contrast – impress on us that the peculiar control over things expressed in the act of collecting is an activity that is socially and culturally conditioned. To an 'untidy child – each stone he finds, each flower he picks, and each butterfly he catches is already the start of a collection. In him this passion shows its true face, the stern Indian expression that lingers on, but with a dimmed and manic glow, in antiquarians, researchers, bibliomaniacs.' (Marx calls this 'individual mania'.) 'His drawers should become armoury and zoo, museum of crime and crypt. "To tidy up" would mean destroying a burrow full of thorny chest-nuts that are morning stars, a pile of aluminium foil that is a silver treasure, toy bricks that are coffins, cactuses that are totem poles and copper pennies that are shields.'[17]

This childish way of collecting creates disorder, and by tidying it up, it would inevitably be destroyed. James Clifford argues differently. For him, the childish act of collecting stones, mussels or the 'jealously guarded little bowl with the remnants left from sharpening coloured pencils', are little rituals that he perceives as the 'channelling of obsession', 'an exercise in controlling the world and collecting things around oneself in an appropriate way'.[18] Less prosaically than Benjamin, Clifford is convinced that a child, whether it 'collects dinosaur models or dolls, will sooner or later be encouraged to line up his possessions on a shelf or keep them in a special box or set up a dollhouse. Personal treasures become public.'[19]

Benjamin wants to save the passion in the childish act of collecting from being channeled according to the general cultural norms (which are, in Clifford's view, systematic rationality, gender and aestheticism). According to Clifford, channeling means the transformation of a predatory need to possess into a sensible rule-bound desire. From the point of view of the cultural anthropologist, this is a precondition for enabling the Ego, 'that needs to possess, but cannot possess everything', to learn 'to select, to order and to classify in hierarchies – i.e. to establish "good" collections'.[20]

De Certeau's 'Rhetoric of Walking'

At this point, I am finally returning to the city. What interests me is whether Benjamin's view of the childish act of collecting can be applied to the ways of dealing with urban space. For that reason, I would like to direct your attention to a certain manner of moving in the urban space, described by Michel de Certeau in his 'Rhetoric of Walking'.[21] The importance of this rhetoric lies in the fact that its cartography, similarly, has little in common with the big inventories and ordering systems. Instead, its typical demeanour resembles the childish way of collecting.

In de Certeau's description, the individual subverts the design of a planned city, by claiming the right to restructure the space through 'the play of (his) steps'. The interesting thing about these steps is that they cannot be localized in advance, as 'their' space is only created in the course of playing. If one were to 'tidy up' the impressions, observations, smells, etc. in retrospect, by imposing on their result – i.e., the space constituted by the movement – a rational intent permitting to repeat the rhetoric or the play of the steps, then this would be just as incongruous as viewing the childish collection as something ordered. As a result of the constitution of the space that is created by the rhetoric of walking, parts of the totality of the city are enlivened, against their diagrammatic order.

This play of the steps certainly does not occur in moments of absent-mindedness – a view, by the way, that is perfectly in line with similar ideas of the Situationists. Which is why the individual,

---

16 ibid., pp. 656ff.
17 Benjamin, Walter; **Gesammelte Schriften** Bd. IV, 1 (Frankfurt am Main, 1972) pp. 115 and 286.
18 Clifford, James; **op. cit.,** p. 289.
19 ibid.
20 ibid.
21 de Certeau, Michel; **Kunst des Handelns** (Berlin, 1988) pp. 192ff.

as he walks, is constantly making choices. Without these choices, however, taking on the character of ordering, hierarchically classifying, establishing 'good' collections. The conversion of the 'predatory need to possess' into a 'sensible desire'[22] occurs in the process of appropriating a space controlled by rules by the (subversive) means of producing one's own rules. Once more, this brings to light the arbitrariness or the social construction that lies at the basis of the distinction between 'good' and 'bad' collections.

We could therefore say that, in the rhetoric of walking, the experiences characterizing the childish act of collecting are subjected to a test of maturity. While the childish experience is still enchanted with disorder and obsession, in the adult it turns into one of the fundamental pillars of urban behavior. While physically complying with the design of the city, the adult frees himself from the societal conformity expressed in it and demanded from him.

Therefore, a successful assimilation of the childish approach to collecting – free of defensive attitudes – by remembering how it was for us, would be a precondition for an open, proactive way of dealing with the fears and constraints that have obstructed the discourse on the city and reduced it to a discourse of order since the earliest Utopian visions. There is certainly an urgent need for this, since rationalism is no longer an appropriate response for dealing with the disorder of cities, and ignores our emotional and subconscious reactions to every single aspect of urban life. Instead, we need to learn to acknowledge the differences that are expressed in the continuous interaction between rationality and irrationality, order and disorder, harmony and conflict, to accept them as the inherent determining factors of our existence, and to use suitable ways of raising them as topics and processing them.

Conclusion

Can any of this be applied to the displayed order of the objects in our museums? I think yes! We have said about the European city, that, as a collection, it continuously loses exhibits, as a result of the destruction caused by the on-going modernization. And that for this reason, the museum is all the more called upon to be a mirror of the past.

But wouldn't it also be possible to view collections as a magic mirror of the present times in which we live? By taking this approach seriously, museums –

22 Clifford, James; **op. cit.,** p. 28.

especially museums of ethnography and urban and cultural history – would have an opportunity to talk about our present times (and do so also and especially by reference to its history), as a world that resists the categorizing approach that one would expect there more than anywhere. It is certainly worthwhile studying, once again, the ethnographic perspective taken by the surrealists. They purposefully resume the tradition of the curiosity cabinets, to be able to give new answers to the question: 'What is a thing?' Their theme of the alienness we perceive in muted objects, could prompt us to try applying this concept to the perception of urban space. In a place where what is familiar should be made unfamiliar, and what is unfamiliar should be made familiar, we would be in the center of what is meant by urbanness and its taxing demand on us: to live with strangers, knowing full well that we are strangers ourselves.

Cultural Centres in Europe
Christoph Grafe

The cultural centre is a type of public institution and a building type that emerged, had its heyday, and almost completely disappeared again within a period of no more than two decades in the third quarter of the twentieth century. From the late 1950's until the mid-seventies, the 'cultural centre' was the default solution for the requirements of state administered provision of culture in Western Europe. By the end of this period, most smaller or medium sized towns in Sweden, France, Holland, Germany or England had obtained such a building housing a variety of cultural institutions, from performance spaces to galleries and cafeterias. In most cases a library would form a core to the building or, as the French came to know them, *mediathèques*, accumulations of culture recorded and printed, open to all.

In the neo-liberal eighties, cultural centres seriously went out of fashion. Partly this was a result of a more general rejection of everything, and especially the architecture, which the mild humanist social set-up of the late welfare state had produced and which looked so bland and suburban, against the wilder cultural explorations of so-called subcultures, the effusions of punk-baroque, gay glitz or squatters' fantasia, explored in former industrial buildings. The concrete cultural centres of the sixties and seventies now looked at best boring, but more often years of neglect had left their marks to such an extent that they started to be perceived as ruins, vestiges of an era which a new decade besotted with the endless opportunities of the free market did its best to forget.

So the cultural centres in French provincial towns, in the satellite settlements of Stockholm, Paris or the British inner cities, and even those right in the centres of these cities, crumbled away, often just kept afloat by their administrators trying to make ends meet with less and less money. Sometimes the agencies established to programme a mixture of cultural offerings for a wide audience were replaced by *quangos* whose main concern it became to develop the commercial potential of a site rather than exploring how these institutions might take a role in redefining culture and access to it. Sometimes, cultural centres just renamed themselves and now operate under the name 'theatre', as if to hide their origins in the consensus of the post-war welfare state.

While the cultural consumers of the new middle classes of the *Wirtschaftswunder* abandoned the grey cultural centres, either in favour of established and new opera houses and theatres or spectacles staged in former industrial sheds, those who could not or would not participate in the liberal reverie of the re-invented traditional European city (and its theatres, museums and grands cafés), found that the programme of opening culture to new audiences had disappeared. The end of the social experiment, uncooked as it may have been from the start, found its reflection in newly fixed cultural canons, national curricula emphasizing the value of approved traditions, and the blossoming of precisely those bourgeois cultural institutions which had been questioned radically only a few years earlier.

Meanwhile we live in a Europe celebrating its urban renaissances with anxiety, as if to remind ourselves of our cultural identity, and only reluctantly coming to terms with the demographic and cultural developments which have taken place in the past two-and-a-half decades. The assumption that the existing social and cultural fabric can somehow absorb the demographic change that has occurred seems very optimistic and the insistence on the necessarily exclusive character of culture, however justified in many respects, impedes the exploration of opportunities to reformulate the role of culture, or to quote Raymond Williams, a 'common culture' in a newly emerging civic society.

What has this to do with cultural centres? The cultural centres, with all their faults and naïve proposals, are one of the very few attempts at establishing places which combined the reality of accessibility with the opportunity to transgress social and cultural boundaries. The absence of a clear programmatic definition, the fact that there is not one established form of arts at the heart of the institution, and the building, allows them to develop into environments where one is not only able, but actively invited to engage with a variety of social and cultural practices. In doing this, the more successful examples of this institutional type prove to be surprisingly good at absorbing different audiences and constructing collective experiences of the sort not found in any other public space.

The cultural centre was not a Western European invention. In the Soviet Union workers' clubs and houses of culture had been opened after 1918 and this model was exported to countries in Eastern Europe after 1945. In France, Northern Germany and Scandinavia similar institutions had also existed before World War II, in the form of the *Maison du Peuple* and the *Folketshus*, initiated and run by trade union movements.

As a standard ingredient of the cultural infrastructure established in the postwar period, however, there is a notable concentration in developed Northern European societies of the welfare state type. Most of us growing up in the 1970's and 80's will know such a building intimately, and many of us will have seen their first art house movie, borrowed their first copy of Stendhal or Salinger or perhaps witnessed their first dance performance in some darkish space contained in a volume of, preferably, exposed concrete and surrounded by footbridges.

The cultural centre is part of the wider agenda for modernizing European societies after 1945. The effect of this agenda became visible in the programmes for housing, health care and education which resulted in the large-scale production of new buildings, ranging from residential new towns to hospitals and schools.

Within this campaign the cultural centre was – obviously – not a first priority. During the 1950's the emphasis was clearly on housing and rebuilding the industrial fabric. Yet it was also in this first decade after the war that the administrative structure, which was to bring forth buildings for culture became established. New public bodies intended for the advancement of the arts and their wider distribution were set up. In Britain this happened in almost direct continuation of the war effort, when the committees charged with bringing arts and entertainment to soldiers and the population at home were absorbed into the new Arts Council of Great Britain, set up and initially chaired by John Maynard Keynes who stated in a radio broadcast in 1946: 'State patronage of the arts has crept in. It has happened in a very English, informal, unostentatious way – half-baked if you like'.[1] In France, a minister for cultural affairs was installed only in 1957-58 with the establishment of the Fifth Republic.

The building of the cultural centres occurred as part of what might be seen as the phase of fulfilment of the social programme which proposed – or, to keep more distance, professed to propose – the creation of an egalitarian society, aiming at material redistribution, but also allowing more evenly spread access to education and culture.

This programme gained a new momentum in the reform policies of the early 1960's (just before being radically questioned in '68) and was directly responsible for the provision of a large number of buildings. The conception and construction of buildings dedicated to culture, and eventually a significantly enlarged concept of culture, which was emerging while the building campaigns were already under way, stands out as an attempt at re-creating Western-European societies by altering the conditions for access to cultural production.

Cultural Politics in Post-war Europe

The sets of values and cultural assumptions informing the cultural politics that led to the construction of cultural centres varied significantly from country to country. The definitions of what would constitute a modern democratic liberal society were formulated within each of the nation states and informed by specific historic circumstances.

The arrangement of the welfare state in post-1945 Britain, for instance, with the introduction of the National Health Service and educational reform reflects a preoccupation with overcoming class divisions; an aspect which was certainly a matter of concern elsewhere, but with different

[1] John Maynard Keynes, **The Listener**, 12 July 1945, p. 6; quoted in: Raymond Williams, **A Policy for the Arts – Comments on the 1965 White Paper**, Tribune, 5 March 1965.

tones and stresses. The debate in Germany was more focused on the democratic nature of the organization of society and on the social and cultural conditions necessary to foster liberal democracy and free public opinion. By comparison, the question of how democratic institutions could reach citizens in remote communities would have been a greater concern in Sweden than in other, more densely populated countries. In Holland it seems that a tacit coalition between Social Democrats and reform Catholics was instrumental in proposing a policy of *cultuurspreiding* (cultural dispersion) also resulting in a building programme in smaller and medium sized towns across the country.

The two most consistent and centralized policies for cultural development could be found in Sweden and France. In Sweden this may hardly surprise, given the dominance of the political social democracy since the 1930's and against the background of a fully developed welfare state by the late 1940's. Consequently, the Swedish cultural debate in the 1960's and early 1970's is marked by a large degree of consistency: first the political system had given the individual citizen a share in the material offerings of the welfare state, now it was time to redistribute non-material stakes. It was along these lines that the debate was led and eventually laid down in a new cultural policy in the early seventies. To the non-Swedish observer it is particularly interesting to register that there seems to be little difference between the public discourse in newspapers and magazines and the official political debate. The tone of the arguments is clear, even pedantic and they rely on the assumption that the outcome of a rational public argument will inevitably materialize through collective, that is, state action.

In Britain the establishment of the Arts Council and cultural policy in general was a characteristically ad hoc affair, but here, too, the debate focused on how to bring the arts to new audiences. Interestingly, it is here that the idea of cultural diversity and cultural resistance is also a factor. In his 1946 broadcast John Maynard Keynes stated this explicitly: 'We look forward to a time when the theatre and the concert-hall and the gallery will be a living element in everyone's upbringing, and regular attendance at the theatre and at concerts a part of organized education. ... How satisfactory it would be if different parts of this country would again walk their several ways as they once did and learn to develop something different from their neighbours and characteristic of themselves. Nothing can be more damaging than the excessive prestige of metropolitan standards and fashions. Let every part of Merry England be merry in its own way. Death to Hollywood!'[2]

France was an altogether different case: here de Gaulle had appointed André Malraux as Minister for Culture. This appointment, which Herman Leibovics in his study *Mona Lisa's Escort – André Malraux and the reinvention of French*

Culture[3] presents as a strategic move by the general in his attempt to pacify the intelligentsia, was instrumental. In using his area of policy to the full, Malraux launched the single largest campaign for the building of cultural centres in any Western European country. The building of Centre Pompidou, although conceived and realized after Malraux left office in 1968, may be taken as the fag end of this programme which brought *Maisons de Culture* to Amiens, Grenoble, Rennes, Rheims, Nevers, Chalon, St. Etienne, Le Havre, Caen, Bourges and Firminy. At bottom these buildings were presented as part of a policy to take metropolitan culture out of Paris and provide a cultural infrastructure to match the higher mobility of the population. More than that, however, the French initiative seems to have been informed by the idea of creating cultural strongholds across the country, places where a socially varied audience could gain access to theatre and dance performances, films and art shows of quality. The audience also was mostly young: about sixty per cent of the users were under the age of thirty. And, although the manifesto for this part of Malraux' *action culturelle*, written in 1962, formulated a claim 'to transform a privilege into a common possession', the remit of the cultural centres was also informed by the idea to infuse the audiences with an understanding of the Frenchness of the culture presented here.

As politicians adopted cultural policy as a tool for social reform or cultural consolidation within the arrangement of a liberal democracy, questions about who had access to the institutions and how the audience could be enlarged acquired a greater urgency everywhere and increasingly during the 1960's. Raymond Williams repeatedly made the point that the original 1946 charter for the Arts Council[4] specified 'the fine arts exclusively' as the working field of the institution. This was eventually changed in the 1967 charter to 'the arts',[5] which, as Williams dryly notes 'of course returns us to the problem'. In any case, the public nature of many of the older cultural institutions came under scrutiny. Theatres, museums, galleries and opera houses became criticized as places of privilege, catering for the bourgoisie instead of realizing experiences of equality. The notion of the public character, or *Öffentlichkeit*, was established at the heart of debates on how to transform established cultural institutions and conceive new ones where none had existed.

The idea of culture as something that should be accessible to all is a recurrent theme in the public debates on the nature of democratic society in post-war Europe, changing in tone but not in intensity from the late 1950's to the early seventies. Based on the experience of the collapse of liberal democracy in Germany or reflecting the experience of the collective war effort in Britain, the belief that a common culture was necessary to rebuild or develop democratic society was widely held and informed education and culture policy. This consensus, 'half-baked' as it may have been (to follow Keynes), from the outset implied the redefinition of the role of culture in general, and the accessibility of its products for a greatly enlarged public.

---

2 Eric W. White, **The Arts Council of Great Britain** (London: Davis-Poynter, 1975), p. 60.
3 Herman Leibovics, **Mona Lisa's Escort: Andre Malraux and the Reinvention of French Culture** (Ithaca: Cornell University Press, 1999).
4 Raymond Williams, **The Arts Council: Politics and Policies** (London: The Arts Council, 1981), p. 4.
5 idem, p. 4.

Indeed, it was this public access to education and culture where *equality of opportunity* could be demonstrated and experienced, or simulated, an equality which in the 'real' world of economic relations was still far away. Raymond Williams formulated this in his conclusion to his study of the idea of culture in modern Britain: 'A common culture is not, at any level, an equal culture. Yet equality of being is always necessary to it, or common experience will not be valued. A common culture can place no absolute restrictions on any of its activities: this is the reality of the claim to equality of opportunity.'[6]

The Architecture of the Cultural Centres – Typologies and Forms of Buildings

The focus on the notion of *Öffentlichkeit*, on the public character of the buildings provided by the state for culture coincides with a strong interest in this notion among architects in this period. Adrian Forty noted that, in the course of the 1950's, the distinction between the public and the private and an investigation of what these terms might mean in a developing consumer society replaced the earlier focus on ideas of community.[7] Modern architecture had tended to translate the requirement for public decorum into a display of transparency. By the end of the 1950's this transparency was acquiring an extended significance, representing the openness of the public spaces designed for the new enlarged audiences addressed by the cultural centres.

Architecture's capacity to create places in which equality could be experienced provided a valuable instrument for demonstrating the good intentions of the welfare state. A revision of the experience of those places defined as public was therefore necessary to state the claim that fundamental social change was intended and imminent. This involved nothing less than an entirely new architectural proposal for the cultural centre: not only because the variety of functions 'under one roof' could not be organized using established principles of composition, but also because the architecture needed to represent and aestheticize the collective, classless experience, shedding off all symbols of importance and propriety which were associated with the established theatres, opera houses or museums.

In many of the cultural centres the older architectural models for cultural institutions were abandoned and reworked into large-scale buildings of great complexity, which sought to engage with the surrounding city by imitating and transforming it. The availability of a new building task for architects prompted a variety of solutions ranging from the then current preoccupation with megastructures, via functionalist picturesque compositions to a minority of clearly defined urban buildings with a specifically individual handwriting of the architect.

Designed within the planning tradition of post-war reconstruction, the majority of cultural centres of the 1960's are informed by functionalist preoccupations developed in the design of housing and new towns while at the same time anticipating the demands of an affluent consumer society with an emphasis on choice and self-fulfilment. This dilemma within the architectural culture reflects the tension between the tradition of enlightened provision of fine arts for the masses and the renegotiations of what might constitute culture in a pluralistic society which played at the level of the cultural institutions.

It is almost literally tangible in, for example, the photographs of the South Bank Arts Centre in London published on its completion in 1967. Here we see a building that is outspokenly unconventional, reminiscent of film sets representing a brave new world of technologically astute and progressive individuals, populated by a decidedly traditional audience, which seems at best uneasy in this environment brought to them by radical young architects. It is a world that is still halfway between the austere, restricted realities of the 1950's and a future society of endless opportunities. The music performed here may be Brahms and Dvořák, the environment suggests something different. At the same time: while the photographers were taking their pictures of people in black suits and cocktail dresses, elsewhere in London the paving stones were loosened in search of the beach, the Beatles had launched their Sergeant Pepper Album and the Stones were about to play to thousands in Hyde Park.

South Bank Arts Centre   While the South Bank Arts Centre presents itself to the uninformed first-time visitor as an entity unified by a particular architectural language and the bridges leading to and from it, the ensemble in fact contains two fairly separate parts, each housing a specific function and institution and each with its own entrance off the deck system. The combination of music and visual arts in one complex was primarily the result of an ad hoc decision to provide spaces for music performances and an art gallery. Although the combination of various institutions under one roof was already the favoured approach to cultural centres all over Western Europe while the South Bank was planned, the Arts Centre itself is not a *maison de culture* seeking to serve a newly defined broader range of cultural activities under one roof. Consequently, there are no internal connections between the gallery and the concert halls, nor is there a shared foyer space. The two parts of the building merely co-exist and acquired separate names as soon as they were completed. The perception of the building as one structure with a complex geometry is almost entirely the result of the elaborate, apparently endless system of outdoor walkways and bridges connecting the Arts Centre with the river promenade, the adjacent Waterloo Bridge and further into the Underground railway system.

6 Raymond Williams, **Culture and Society 1780–1950** (Harmondsworth: Penguin Books, (1958)1971), pp. 304–305.
7 Adrian Forty, **Words and Buildings** (London: Thames and Hudson, 2000), p. 105.

This openness contrasts with the rather more traditional internal planning where we find both parts of the building being developed from a brief that was formulated along well-established lines and which is informed by existing models of concert halls and public galleries. None of the agencies involved in the planning of the project formulated a statement that would unify the programmes covered in the buildings. The only reference mentioned in the extensive correspondence and the public reports on the initiative users was the project for the Lincoln Center, a centre for the performing arts planned at the same time in New York and comprising separate theatre buildings for the Metropolitan Opera, the New York State Theater and a concert hall arranged around a plaza. Even this comparison with the New York project, however, was only made after the decision to combine the two concert halls and a gallery. In any case, there is no evidence that arranging spaces for different types of cultural production on one site was seen as part of a programme of diffusing boundaries between visual and performing arts or experimenting with new forms of interaction between producers and their audience.

Yet it is the comparison with the Lincoln Center that reveals how much the London building departs from the established model of a cultural compound made up of freestanding buildings and how much, despite the clearly separated internal planning of the two component parts, the Arts Centre's architectural conception is that of a unified structure. It is this architectural proposal and the fact that the building in its entirety came under one management, which was also responsible for running the adjacent Festival Hall, that allows the Arts Centre to be categorized as a cultural centre. Even though the day-to-day planning of programmes for performances and exhibitions was done by separate bodies, the County Council had the overriding responsibility for the provision of these cultural facilities and the general public was invited to view the new centre as one single building offering a variety of different cultural activities. The difference between the organizational structures of the institutions housed in the Arts Centre, in fact, exposes the contradictions in the conception and the design of the South Bank. These are contradictions which can be explained against the background of a debate on how the welfare state should provide for the cultural need of its citizens. In Britain this debate had only begun in the aftermath of World War II and was to develop into a larger, and increasingly controversial, discussion when the major decisions about the planning of the South Bank had already been taken. Seen in this light, the South Bank marks a moment of transition from one concept of state-administered provision of culture to another, and from making forms of established culture accessible to an enlarged audience to a more thorough questioning of the definition of culture in a democratic society.

The beginnings of the South Bank Arts Centre go back as far as the mid 1950's, but the definitive project was designed by a group of then very young architects within the architects' department of the LCC including Warren Chalk, Ron Herron and Dennis Crompton. After the presentation of the design in 1960 it took eight years for the building to be completed, during which the project architects remained faithful to the original design while the young designers moved on to become leading members of Archigram.

According to Warren Chalk the deck was conceived as an answer to the conflict between the pedestrian flow and car traffic. The fact that the deck takes the shape of a rail system of bridges is explained by the examples of the entries for the *Hauptstadt Berlin* competition of 1958 by the Smithsons and Arthur Korn. In this vision the walkways were to establish a new ground level offering comfortable and uninhibited access to the spaces it connected. This access scheme which later seemed inappropriately complicated and was partly demolished in the late 1990's, as a matter of fact, negotiated between the different levels surrounding the site in a rather sensible way. The interiors of the concert halls and the gallery were interpreted as conventional boxes whose functions allowed little or no relationship with their surroundings. The external deck and the foyer serving the concert halls, both meandering around the boxes and through the cavity separating the two parts of the building, took over the role of a public concourse extending far beyond the immediate surroundings.

The formal language of the cluster with its allusions to the solid concrete volumes of late Le Corbusier and the bold, rough treatment of balustrades – Chalk makes the connection with Japanese contemporaries – disguise the fact that essentially the building was conceived as a large machine.[8] Channels of movement, of people and cars but also of air or electricity, were expressed as independent from the enclosed rooms which they serve. The main feeder duct to the auditorium, for instance, appears as a heavy ribbon wrapped around the top, and air ducts materialize as cantilevered concrete bands resembling the pedestrian balconies winding around the building. A few years later this conception of the building as a conglomerate of boxes and separate elements plugged in to feed them undoubtedly would have been expressed in a lightweight steel or even inflatable structure.

The building is conceived and presented as the sum of efficient responses to particular problems. These problems are identified as technical issues: in an article written on its completion Charles Jencks admired the superb air treatment and the sophisticated arrangements for sealing the auditorium acoustically, as well as the Helmholtz resonators controlling the reverberation time inside the hall. There may be a sarcastic undertone in his detailed listing of the technical

---

[8] This was also how Banham interpreted the building to illustrate his argument about the well-tempered environment. Reyner Banham, **The Architecture of the Well-Tempered Environment** (London: Architectural Press, 1969), pp. 256–261.

achievements, implying that for all their efforts the designers had lost control against the nerdy preoccupations of the specialists – including the 'Director of Mechanical and Electrical Services of the GLC' (the Greater London Council which replaced the LCC in 1965) who receives a special credit. The general tone of the article and its praise of the 'professional ingenuity', however, appear genuinely respectful of the work. What you see is what you get: the cultural centre is a machine and appears as one, and there is no problem with this proposal.

The mixture of technological accomplishment and rejection of history, even of the Modernist examples of the previous decade, contains a statement on the cultural programme of the South Bank and its implications for the city and society at large. The question is, however, how this position can be understood in terms other than simply negative ones, beyond doing things in a way different from a preceding generation. In an article published when the complex was finished eight years after its initial design Warren Chalk stated somewhat defensively: 'At the time it was designed the architects thought they had something to say, and said it consistently.'9 Chalk implies that the building had been conceived from a set of ideas that, while appearing to be valuable at the moment of the conception around 1960, had somehow lost their validity when their product was finally available at the end of the decade. Were the considerations informing the design superseded by new ones, and if so in what way is the original design a response to a situation as it existed in the early 1960's and what had changed in the eight years between the design and the eventual realization? Was the design proposal watered down by someone else; the executing agency or the institutions using it? And if a change occurred, what was its nature: was it the architectural proposal that had become outmoded or the programmatic conception? Was the architect disappointed at what seems to have been a reasonably traditional institutional use as concert halls and a gallery, inhabited by strangely uncomfortable people wearing suits and ties? Or had the Centre's design optimistically anticipated a situation that through adjustments made on the way, in the detailed realization of the building and its programme, was not matched by the reality of the end product?

The architect's account does not give an answer to these questions. Perhaps Chalk was just taking a defensive position in order to brace himself for what was to come. In any case, the South Bank Arts Centre received a mixed reaction when it was completed towards the end of the 1960's and, rather than settling down after a phase of initial adjustment to its architectural proposal, it has had a rough ride ever since. Most commentators will explain the problematic history of the combined concert halls and art gallery with a reference to the appearance and layout of the building. Yet, the vicissitudes of the complicated existence of the Arts Centre may be more successfully explained by the dismantling and disempowerment of the institutions that were responsible for its creation in the first place: the ensemble suffered both the wilful destruction of the Greater London Council (GLC) which succeeded the London County Council (LCC), original client and managing body of the buildings, and the restructuring of the Arts Council of Great Britain, one of its main users, which entailed a shift of its activities from actively producing exhibitions to distributing money.

Appropriations Observation of the South Bank in the early 21st century allows us to see a series of buildings and open spaces, bridges and terraces occupied by an extraordinary mixture of flaneurs, tourists, amateur artists and homeless people who find what has become a rare commodity elsewhere in London: places which are freely accessible and not dedicated to commerce or consumption. These buildings, which were erected to allow access to culture to all, are doing just that, and in ways which were entirely unforeseen by their planners and clients. The most poignant illustration for the level of appropriation of the South Bank are the skateboarders who have found in the sloping floors of the caves under the Queen Elizabeth Hall an unrivalled environment for their pursuit. Against these realities the South Bank seems to have anticipated forms of urban behaviour emerging only decades later: the kinds of private behaviour in the open air that we have come to understand as public.

The South Bank Arts Centre can be discussed as a failure, a failure because it responded to a view of what a public building destined for the provision of culture should be that was current in the late nineteen fifties and it represents a response to this view rooted in the architectural culture in Britain of this period. The architects anticipated a type of use that seemed progressive when the building was designed, which focused on an emphasis of movement and change facilitated by a design approach free of sentimentality but full of nostalgic futurist imagery. The circumstances of the use of the realized building were, however, different from those of the late fifties. In a period enchanted with experiment and interaction the suggestion of a building offering technically efficient facilities had lost much of its appeal both for the producers of culture and their audience. The proposal of the cultural centre as a machine that could be experienced without the obstructive influences of the conventional typologies of auditorium and gallery, was no longer sufficient to generate the excitement which the designers had undoubtedly envisaged in the early 1960's. The expectations of the audience, especially the younger metropolitan public, were by this time influenced by the context of a cultural production ranging from the Rolling Stones appearing in Hyde Park to old railway barns used for concerts and theatre. Launched into this situation as a combination of two conventional programmes disguised as a building representing rather than enabling openness, the South Bank could only be seen as the London equivalent of the concrete campuses, schools, and community and shopping centres that were realized in the same period, too tough and unglamorous for many, too conventional for those who had embraced popular culture and artistic experiment.

But then, is failure not part of the destiny of any building designed to cater for something as fluid as culture? Are ensembles such as the New York Lincoln Center that abstained from any attempt at redefining the nature of the public building and of the experience of culture really more successful, as the chief executive of the South Bank Centre suggested in 2004, simply because they turn out to be commercially viable?[10] In his curious review of the South Bank Arts Centre published in 1967[11], Charles Jencks made a point about the building that seems valid, although its implications have never been fully realized. Jencks imagines 'deliberate acts of the burlesque' in a building that is contradictory and intellectually unfinished. The South Bank was conceived as a modern machine, and it fails when approached as such. It would be a very successful venue for experimenting with new forms of artistic expression if it were to be used like a very well made old car that allows robust yet loving tinkering, not with the desire to mend its loose ends, but as a repository of opportunities.

10 'Hayward threat as South Bank speeds up', **Building Design**, April 23 2004, p. 2.
11 Charles Jencks, 'Adhocism on the South Bank', **Architectural Review** no. 144, July 1968, pp. 27–30.

[1]

1 Cultural centre **de Meerpaal**, Dronten.
2 South Bank Arts Centre, Queen Elizabeth Hall Foyer (1967).
3a South Bank Arts Centre.
3b Fun Palace (Cedric Price, 1964).

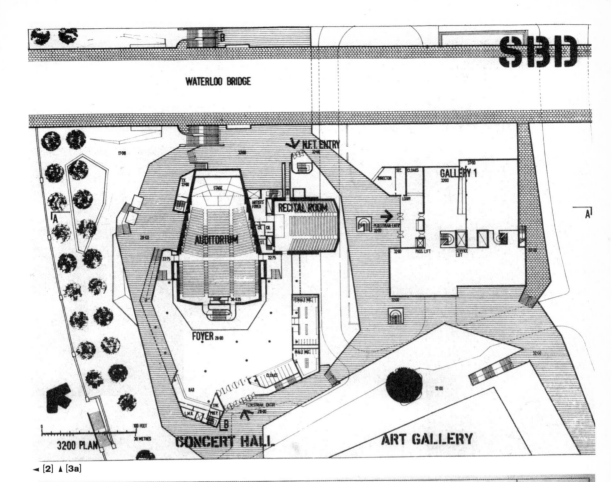

◄ [2] ▲ [3a]

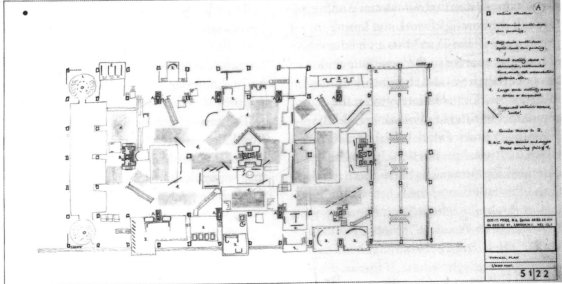

[3b]

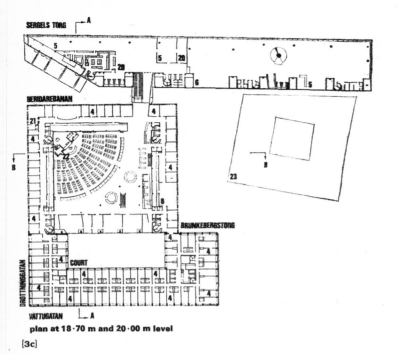

3c Kulturhus, Stockholm.
3d Centre Pompidou.
4 South Bank Arts Centre, Queen Elizabeth Hall Foyer (1967).
5 South Bank Arts Centre.

plan at 18·70 m and 20·00 m level

[3c]

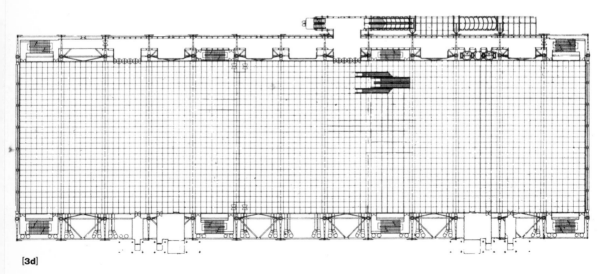

[3d]

[4]

[5]

[6]

6 South Bank Arts Centre, Hayward Gallery.
7-8 South Bank Arts Centre (2005).

[7]

[8]

ARCO Spatial Project
(On Ephemeral Architecture and Art)

Rosina Gómez-Baeza

The town exists only as a function of circulation and of circuits; it is a singular point on the circuits which create it and which it creates. It is defined by entries and exits: something must enter it and exit from it. It imposes a frequency.[1]
Gilles Deleuze and Felix Guattari

In the special edition of the *ARCO contemporary art* magazine, published to mark ARCO'05, Arie Graafland compares the spatial design of this edition of the Madrid's International Contemporary Art Fair to the street layout of New York City. The Fair's architecture can be considered as a concentrated, ephemeral city, 'a grid of shops, criss-crossed by streets and avenues, following a hierarchical order in terms of accessibility and visibility'.[2] The Fair can also be seen as a map, where urban *flâneurs* can lose themselves and stop to take a look at whatever catches their eye; or experienced as a Benjaminesque shopping arcade, a 'micro-world' in which people construct their own nonlinear experience, based on the juxtaposed images in their memory.

ARCO is all of this, and also a commercial and cultural event in which all of the elements that it comprises, along with their spatial layout, must respond to meticulously thought-out reasoning, in order to enable the spectator to enjoy an agile, thorough and professional visit. The Fair is growing, and more and more features have to be accommodated, such as the galleries' stands (belonging to the general programme, the various curated invitationals, and the special guest country's pavilion), institutional stands, publication stands, media stands, signage, restaurants and cafés, rest areas (chill-outs), catalogue sales points, and information stands.

Also, the exhibition programme itself has to be partnered by a large spatial and logistical infrastructure, with enough conference halls to house the International Art Experts' Forum (over 60 panel debates on current issues involving the contemporary art scene) and the wide-ranging Social Programme (presentations of cultural projects, cocktail parties and receptions hosted by various institutions to celebrate the event). Future editions of the Fair will incorporate the Convention Centre, where the Forum is held, into the exhibition as a whole, along with restricted-access VIP lounges.

ARCO was created with – and continues to have – a cultural, as well as commercial, goal in mind: in 1982, when democracy had only recently been reinstated in Spain, ARCO was conceived as a showcase for contemporaneity, ready to adapt to its container, content, and public. This goal was remarkable seeing that Spain did not always match the standard of its European neighbours, not to mention American or Asian countries, where cultural affairs are the responsibility of other institutions. As the historian Nacho Ruiz points out in *ARCO: Arte y mercado en la España democrática* (ARCO: Art and the Market in Democratic Spain):

---

1 Deleuze, Gilles and Guattari, Felix; **A Thousand Plateaus** (London: Continuum International Publishing Group, 2001) pp. 35–36.
2 Arie Graafland, 'ARCO's Art Space as an Aesthetic Condenser', **ARCO'05 Special** (Madrid: Amigos de ARCO Association, 2005) pp. 46–49.

> In cultural terms, the birth of ARCO implied the desire to achieve the standard of other advanced nations… Internationalism and modernity were seen as the watchwords, and so the Fair had to succeed, at times even at the expense of sacrificing financial profits in order to further the promotion and dissemination of culture.[3]

Importantly, the space at ARCO is not just a series of modular rectangles; rather, it is a part of a Fair that is committed to contemporary creative output in all of its disciplinary expressions. As the architect Fernando Quesada said in the *ARCO'04 Special* magazine:

> Every year, ARCO does its best to use the event as a venue for experimental architecture, in which the architects involved can try to free themselves from the norm, introducing droplets of diversity. … and, every year, we take part in a collective experience in which the architecture takes on gentle nuances of diversity that are always meaningful and thought-provoking.[4]

ARCO belongs to IFEMA (the Madrid Trade Fair Institution), which organises and accommodates no less than 70 trade shows and exhibitions each year. ARCO itself is held in two of IFEMA's exhibition halls (numbers 7 and 9), which between them cover 38,590 m² (of which 58% are occupied, that is, 22,500 m²). The different architectural studios that are commissioned to design the floor-plan each year have to deal with a space in which they need to house around 300 galleries, 60 magazines, and 20 institutional stands, whilst taking into account the approximately 180,000 visitors who arrive over only six days. Therefore, the final layout has to be a structured grid that combines the mobility of visitors with the static nature of more conventional architectural features, and the standardised white cubes of the display booths with more versatile features offering greater scope for experimentation.

The ARCO'04 Spatial Project: With all this in mind, the spatial design for ARCO'04, by the young Madrid-based architecture studio Equipo Tent-T, contributed some important ideas. On the one hand, the central core of galleries was regrouped and condensed into a grid layout. The institutional stands were relocated at the end of Hall 9, whilst the entire perimeter of this same hall was devoted to that year's curated invitationals (*Art Unknown*, *Project Rooms* and *Up & Coming*). Another qualitative solution was clearing the two entrances (one for each hall), in order to direct the openings towards the outside. This way, the entrances were free of anything superfluous, and were clearly visible from the exterior, thanks to large canopies shaped like three-dimensional bows.

This rational and minimalist solution enabled the Equipo Ten-T team to highlight the wonderful zig-zagging aluminium bench by Ignacio Carnicero inside the entrance which served as an information stand, as well as the bubble-lamps designed by Ten-T themselves in collaboration with Eduardo Cajal from TDI-Huesca. These lamps, which were made on-site using bubble-wrap, filtered the light at the entrances and information stands, humanising the scale and reducing the feeling of intimidation so often felt when walking into a huge exhibition centre.

Another issue, in terms of defining the layout, was that of the nexus connecting the two exhibitions halls. In these intersections, Equipo Ten-T's solution was to channel the traffic flow whilst also, on a practical level, providing accommodation for two sections: *Written Word* and *Written Image*, both devoted to cultural and contemporary art publications. The space was distributed so that it both moulded itself to the flow of visitors and held them in. Thus, the nexus between halls 7 and 9 encouraged visitors to pass straight through, whereas the one between 5 and 7 was a cul-de-sac which sent them back to their starting point. To emphasise both movements, Ten-T designed a number of objects. In the nexus between 7 and 9 they placed the *Musshops* (mini museum shops), which were aerodynamic, oval aluminium shapes that successfully controlled the transit flow. In the 5-7 nexus, they created large, bow-shaped tables made of aluminium and leather, on which that section's magazines could be displayed, serving both as reading tables and a resting place.

Finally, another of Ten-T's contributions was the ironic signage. Again in collaboration with Eduardo Caja of TDI-Huesca, they used a Spanish play on words to design map-holding easels in the shape of safety pins, or *imperdibles* in Spanish, which literally means 'unlose-able', so that visitors would not get 'lost'.

The ARCO'05 Spatial Project:
The spatial design for ARCO'05 was the work of the young team at the TABLE architecture studio, who defined their project as follows:

> TABLE sees ARCO as more than just AR*te* CO*ntemporáneo* (CO*ntemporary* AR*t*). ARCO is also an ephemeral and constantly evolving exchange, in which walking around, seeking a reflex, a glance, in fact every real-life gesture, all physically converge in every square metre of the air and in all its dimensions. This is where the encounter takes place between a person and a representation, where each person should feel free, alien and comfortable, provoked into feeling and thinking beyond artwork, fully aware of where, and above all, what, they are.[5]

The layout devised by TABLE cleared the space even more than before, reducing the number of

---

3 Nacho Ruiz, **ARCO: Arte y mercado en la España democrática** (Murcia: Región de Murcia Consejeria de Educación y Cultura, 2004), pp. 35–36.
4 Fernando Quesada, 'A Look at Space', **ARCO'04 Special** (Madrid: Amigos de ARCO Association, 2004) pp. 44–45.
5 TABLE Architecture and Design Studio, 'ARCO'05 Spatial Project', **ARCO Contemporary Art** no. 34, Winter 2004 (Madrid: Amigos de ARCO Association, 2004) pp. 84–85.

chill-out areas and eliminating other rest features, such as avenues and boulevards. The design maintained the location of the curated programmes on the outer edge of Hall 9, as well as the institutional stands at the end of it. In terms of the entrances, TABLE had clear-cut ideas about signage and markers, highlighting the routes and making it easier to visitors to see where they were at all times. The hall entrances were placed perpendicular to the façade, so that they were visible from the outside and would capture visitor flow. In other words, the idea was for the walk to the entrance of the Fair to be part of the whole sensory experience of the event.

The entrance lobbies inside the exhibition halls were devised as 'spaces where the exchange of information and meetings can take place, in an area between the outside and the inside, where the bulk of the visit is concentrated.'[6] TABLE enhanced the unique qualities of both these spaces by changing the colour of the flooring and with careful lighting, unifying the colours and shades of light. The benches designed by Ten-T the previous year were used again, as were the Musshops, which this time served as sales points for catalogues and VIP reception.

The problem of the hall links was solved by using the same idea as the previous year, which had worked very well, but adding some qualitative improvements and giving the space a more industrial look. Ten-T placed the *Written Image* section in the 7–9 link, with a different layout which made it possible to add a number of individual stands to the central area, creating paths that crossed or ran parallel to the main visitor flow. The internal structure of these extra stands, made of wood, steel plate and translucent polycarbonate, contained a small locker with lighting and a seat for the exhibitor.

TABLE gave the 5–7 nexus three contrasting, but compatible, uses: the galleries in the ARCOLatino curated programme, the chill-out SPOT, and the *Written Word* section (devoted to emerging art magazines). The galleries were placed in the corners of the quadrangle, whilst the chill-out was placed so that it highlighted the route marked out by the entrance hall. The magazines were displayed on some original and specially designed oil-drums.

Another of the great new ideas at ARCO'05 was the curated invitational *TheBlackBox@ARCO*, comprising a selection of new media art galleries, each showing work by a single artist. Architecturally, the project was housed in a large black cube which stood out on the skyline of the Fair due to its height and colour. The Black Box was divided into two floors, each of which housed in its unlit interior stands containing pieces of cutting-edge media art. An information booth and reading stand was placed at the centre, where visitors could find interactive information on the most up-to-date proposals.

ARCO'05 also featured architectural projects located outside the exhibition halls, on the central avenue of the exhibition complex. Casa México, also designed by TABLE, embodied Mexico's status as this year's special guest country. The Casa México was an open-plan, abstract, and almost transparent construction, made of polycarbonate cubes on a steel, iron and glass structure, calling into question the binary opposition of indoors and outdoors. The centre of the pavilion was made up of five spaces which represented Contemporary, Colonial and Archaeological Mexico, Sun and Beach, and Natural Mexico. Amongst other events, Casa México hosted the inaugural press conference, a brunch for collectors, and a number of DJ sessions.

Next to Casa México, the Avión (Aeroplane in Spanish) Cultural Warehouse project landed and remained for the entire Fair. The work of Eduardo Cajal, in collaboration with Equipo Ten-T, this DC-9 was found by Cajal on a scrap-heap in the northeast Spanish province of Huesca, after it had suffered a spectacular accident in which, amazingly, no-one was hurt. The plane became a mobile venue for socio-cultural exchanges, devised to accommodate artistic endeavours at ARCO'05, such as performance art, theatre or concerts.

Finally, a replica of the wall that separates Tijuana, Mexico from the US frontier was set up all along the fence outside the exhibition complex. On it, several artists created pieces protesting against the social situation that has arisen there as a result of increasing levels of illegal immigration to the United States.

In sum, ARCO has backed a number of simultaneous and independent pieces of architecture which were, simultaneously, interrelated, producing an intricate rhizome of varying intensity. As Martín Lejárraga rightly said

> We should be driven by art and architecture to go beyond the relationship to the environment that was established by the orthodox attitude of static contemplation based on a 19th-century view of nature and which, ultimately, leads to a deceptive vision. This can be seen from its fake landscapes and true illusions. For these reasons, I believe that art and artistic expression, architecture and architectural space, are actions that always occur simultaneously alongside other ones. They take place at the same time as other things and are actually no more than the backdrop to our life, the screen onto which we are projected.[7]

Chill-Outs: Chill-outs were introduced during ARCO'03, to provide visitors with open spaces along their way, where they could rest and recover their energy. These rest areas were also designed as architectural research labs, giving younger professionals the chance to create land-

---

6 TABLE Architecture and Design Studio, **op. cit.**, pp. 84–85.
7 Martin Lejárraga, 'Architecture at ARCO'04: A Space for Art', **ARCO'04 Special** (Madrid: Amigos de ARCO Association, 2004) pp. 46–48.

scapes with a clear social interrelationship. As Pablo Berzal and David Pastor commented: 'From the very start, the experience was based on the clear purpose of researching into this area, as ephemeral architecture is considered an extremely valuable test-bench for architecture.'[8]

ARCO'03 presented eighteen chill-out areas, which also acted as boundary markers between the various elements of the Fair programme. From then on, we decided to reduce the number of chill-outs and make calls for projects to cover this feature, both in the special guest country and at Spanish schools of architecture. This endeavour was co-ordinated by the art historian David Pastor and the architect Pablo Berzal.

At ARCO'04, two chill-outs were presented, and three schools of architecture competed for the projects. For the chill-out area in Hall 7, which housed the section devoted to the special guest country, Greece, proposals were accepted from the Athens School of Architecture. The winning project, by Chrysokona Mavrou and Grigoris Stavridakis, was entitled *Under the Sea… I See Light*. The design, which covered 180 m² of floor-space, explored topography as a creator of spaces, articulated in a fluid and natural manner. The project reproduced the seabed, the ideal place for rest. Pastor and Berzal highlighted 'the novel use of building techniques and materials to create a delightful abstraction of the most important Greek cultural symbol, the sea.'[9] The second winning project (*nada*), (nothing), was the work of José Antonio González Casares from the Granada School of Architecture. Measuring 236 m², the design was based on the visual impact of ARCO, and provided a rest area devoid of any visual references, in other words, 'nothing'.

ARCO'05 also contained two chill-out areas. The call for ephemeral architecture projects was made at Spain's University of Navarre, and the University of Mexico City. The winning projects worked in opposite ways, the positive and negative aspects of an analysis that had a large number of common features. The winning Mexican team (Héctor Luis Hernández Carrillo and Neyra Villamar Téllez), proposed a structure in which visitors could enter and isolate themselves from the tumult of the Fair. The winning Navarran team (Pablo Ausucua García, Iñaki Esteban Valencia and Joseba Garraza Álvarez) proposed the opposite: an open space in which people could interrelate publicly and socially, with a series of sensually appealing hemispheres.

8 Pablo Berzal and David Pastor, 'ArchLab: An Architecture Laboratory at ARCO', **ARCO Contemporary Art** no. 34, Winter 2004 (Madrid: Amigos de ARCO Association, 2004) pp. 81–84.
9 Pablo Berzal and David Pastor, **op. cit.**, pp. 81–84.
10 Martín Lejárraga, **op. cit.**, pp. 46–48.
11 Martín Lejárraga, **op. cit.**, pp. 46–48.

Conclusion: Every year, ARCO evolves and regenerates, in terms of both contents and container. Martín Lejárraga described the purpose of ARCO perfectly: ARCO 'is a space that envelopes us; it is heterogeneous, despite the regularity and orthodoxy of its map-like layout; changing with the hours, questioning at times, it is like an ambitious stage set, where layers of information in all kinds of conditions are piled up, expressing the stratigraphic status of the art world.'[10] He goes on to say that ARCO must be 'above all, a space for life'.[11]

[1]

1 Mexican galleries at ARCO'05
All photos by Lorena Sánchez Pereira
2 Aluminium bench designed by Ignacio Carnicero ARCO'05

[2]

[3]

3 Tijuana wall at ARCO'05
4 Avión Cultural Warehouse at ARCO'05
5 View of ARCO'05
6 TheBlackBox@ARCO
7 Navarrean Chill Out at ARCO'05
8 Mexican Chill Out at ARCO'05

[4]

[5]

[6]

[7] [8]

A Little Dictionary of Design Clichés
Hal Foster

A is for Architecture: What else? Clearly architecture has a new importance in the culture at large. Although this prominence stems from the initial debates about postmodernism in the 1970s, which centered on architecture, it is clinched by more recent developments, such as the great inflation of design and display in so many aspects of consumer capitalism today – art, fashion, retail in general, corporate relations, and so on. Yet the significance granted architecture today also has a compensatory dimension: in many ways the architect is our latest figure of the artistic genius, of a creator endowed with magisterial vision and a worldly agency in a way that the rest of us in mass society do not, cannot, possess. Despite the great gap between vanguard architecture and everyday building, a given person on the street is likely to come up with the names of a few architects, but not of a few artists, writers, or film directors.

The often paranoid structure of architectural discourse today – the manner in which, in all kinds of statements, visions of grandeur alternate with feelings of impotence – also points to the compensatory dimension of the status of architecture today. 'Since the early 1990s the "market" has ruled, so the only tool we seem to have left is seduction,' the Dutch architect Kamiel Klaasse remarks. 'This creates the unpleasant condition of dependency. Architects combine arrogance with impotence; we are beggars and braggarts.' 'These days architects can do everything – and, at the same time, nothing,' Maarten Kloos, another Dutch architect, adds. 'Architecture has become an amorphous, evasive concept that just hangs like a scent in the air or the last fad.'[1]

B is for Bonaventure: In his celebrated analysis of postmodern space in terms of 'the cultural logic of late capitalism', Fredric Jameson used the vast atrium of the Bonaventure Hotel in Los Angeles designed by John Portman as a symptom of a new kind of architectural sublime: a sort of hyper-space that overwhelms the human sensorium.[2] Jameson took this spatial delirium as a particular instance of our general incapacity to comprehend the advanced capitalist universe, to map it cognitively. Strangely, what Jameson offered as a *critique* of postmodern culture is what many architects (Frank Gehry foremost among them) seem to take as a *paragon*: the creation of extravagant spaces that work to seduce and/or to subsume the subject, a neo-Baroque Sublime dedicated, for the most part, to the glory of the Church of our age – the Corporation. Such architects appear to design not in any suspicion of 'the cultural logic of late capitalism' but according to its specifications. Of course, one might object, what else can they do? But doesn't 'architecture' arise, like desire, out of a lack beyond need and arrive, like love, at an effect beyond expectation?

C is for Carcassonne: Carcassonne is a tourist destination in southern France, a medieval *cité* replete with chateau, church, and fortifications. Viollet-le-Duc restored its towers and

---

1 See **Hunch** 6/7 (Amsterdam: Berlage Institute, 2003) pp. 269, 276; all quotations followed by page numbers in my text derive from this publication.
2 See Jameson, Fredric; 'Postmodernism, or The Cultural Logic of Capitalism', **New Left Review** 146 (July–August 1984).

turrets in the mid nineteenth century, and the site retains an unreal sheen: a historical town turned into a theme park, with its walls whitened and capped like TV-star teeth. At least Americans make their Disneylands from scratch, or at least they once did. More and more this Carcassonnization – the canonization of the urban carcass – is at work in American cities as well. For example, the cast-iron buildings of my SoHo neighborhood in Manhattan now gleam with the shine of artifacts-become-commodities. Like Viollet-le-Duc, developers undertake these face-lifts in the name of historical preservation, but of course the purpose is financial aggrandizement. And like victims of cosmetic surgery, these façades may mask historical age only to advance mnemonic decay.

D is for Design: Again, what else? Today everything – from architecture and art to jeans and genes – is treated as so much design (this is the main claim of my little book *Design and Crime* [2002]). Those old heroes of industrial modernism, the artist-as-engineer and the author-as-producer, are long gone, and the postindustrial designer now rules supreme. Today you don't have to be rich to be cast as designer and designed in one – whether the product in question is your home or business, your sagging face (designer surgery) or lagging personality (designer drugs), your historical memory (designer museum) or DNA future (designer children). Might this 'designed subject' of consumerism be the unintended offspring of the 'constructed subject' of postmodernism? One thing seems clear: today design abets a near-perfect circuit of production and consumption.

Where are architects left in this situation? Again, they seem able to do everything – 'and, at the same time, nothing'. It is the architect redefined as designer who occupies this paradoxical position: as architecture expands into total design, it also runs the risk of dissolution there. 'It seems that it will soon be necessary for architects to offer a complete array of services,' Hani Rashid (of the design firm Asymptote) comments, 'and to do so with a seamless integration of all aspects of the process' (Hunch 391). (Today, it sometimes seems, if you don't have an acronymed design-consultancy – an AMO to complement your OMA – you might not count as a major player.) Some architects celebrate this condition of total design. 'The architect is going to be the fashion designer of the future,' Ben van Berkel and Caroline Bos (of UN Studio) state. 'The new architectural network studio is a hybrid mixture of club, atelier, laboratory, and automobile plant, encouraging plug-in professionalism' (Hunch 90). Others decry it. 'Architects have transformed themselves from creators into coordinators and managers,' the Swiss architect Mario Botta remarks. 'They have become directors of sorts, shackled by consultants and specialists from multiple sectors; their duties have become limited to mediating diverse technical, economic, juridical, and functional requirements' (Hunch 112). Here, however, one should differentiate, dialectically, between the progressive aspects of the tendency toward total design (for example, the way in which architecture, landscape design, and urban planning are no longer so divided in practice or in pedagogy) and the problematic aspects of this tendency (for which this comment by Dutch designer Kas Oosterhuis might stand as an instance: 'we architects must focus now on emotive styling ... we give shape to the flow of data; we sculpt information' [Hunch 371]).

E is for Environment: The world of total design is an old dream of early modernism, but it only comes true, in perverse form, in our pan-capitalist present. With post-Fordist production, commodities can be tweaked and markets niched in such a way that a product can be mass in quantity yet appear personal in address. In an economic world retooled around digitizing and computing, commodities are no longer objects to be produced so much as data to be manipulated – designed and redesigned, consumed and reconsumed. In large part it is this perpetual profiling of the commodity that drives the contemporary inflation of design. Yet what will happen when this political economy of design breaks down, as markets crash, sweatshop workers resist, or environments give out?

F is for Finitude: An early version of total design was advanced 100 years ago in Art Nouveau with its will to ornament in all things. This *Style 1900* found its great nemesis in Adolf Loos, who attacked it in several texts. One of his attacks was an allegorical skit about 'a poor little rich man' who commissioned a designer to put 'Art in each and every thing': 'The architect has forgotten nothing, absolutely nothing. Cigar ashtrays, cutlery, light switches – everything, everything was made by him.' This *Gesamtkunstwerk* did more than combine art, architecture, and craft; it commingled human subject and inorganic object: 'the individuality of the owner was expressed in every ornament, every form, every nail.' For the Art Nouveau designer the result was perfection: 'You are complete!' he exults to the owner. But the owner is not so sure; rather than a sanctuary from modern stress, he sees his Art Nouveau interior as a deepening of its malaise. 'The happy man suddenly felt deeply, deeply unhappy... He was precluded from all future living and striving, developing and desiring. He thought, this is what it means to learn to go about life with one's own corpse. Yes indeed. He is finished. *He is complete!*' For the Art Nouveau designer such completion reunited art and life, with all signs of death banished. For Loos this triumphant overcoming of limits was a catastrophic loss of the same – the loss of the objective constraints required to define any 'future living and striving, developing and desiring'. Far from a transcendence of death, this loss of finitude was a death-in-life, living 'with one's own corpse'.[3] Sometimes, in the neo-Art Nouveau, total-design world of today, I know what he means.

---

3 See Loos, Adolf; 'The Poor Little Rich Man,' in **Spoken into the Void: Collected Essays 1897–1900**, trans. Jane O. Newman and John H. Smith (Cambridge MA: MIT Press, 1982) p. 125.

G is for Gesamtkunstwerk: The notion of the Gesamtkunstwerk emerged in the 19th century as an imaginary way to undo the separation of art from life; this was its role in Art Nouveau as well. However, in our time the culture industry has solved this separation of art from life, perversely, for its own purposes. In our situation the Gesamtkunstwerk is a given condition, a default program. Consider, as but one example, how shows of art and architecture are judged according less to the work displayed than to the display as such: exhibition-value trumps art-value, indeed all other values, and every exhibition seems to be an installation piece, every museum a Gesamtkunstwerk (or a Guggenheimwerk). Even the apparently restrained Museum of Modern Art is its own master effect.

The Gesamtkunstwerk has a nasty side today. Recall the horrendous remarks of the avant-garde composer Karlheinz Stockhausen on the 9/11 attacks (as reported in *The New York Times*): 'What happened there – they all have to rearrange their brains now – is the greatest work of art ever: that characters can bring about in one act what we in music cannot dream of, that people practice madly for ten years, completely, fanatically, for a concert and then die. That is the greatest work of art for the whole of the cosmos. I could not do that. Against that we composers are nothing.' The category-mistake in this statement is profound: Stockhausen conflates avant-garde transgression with mass murder. And yet such crypto-fascist feelings of sublimity were in play, very soon after 9/11, in Afghanistan and Iraq, and more in this vein is sure to come. In *Mao II* (Hunch 199) Don Delillo suggested that the terrorist has outpaced the novelist as imaginative shaper of reality. In this sense, too, the famous architect now complements the infamous terrorist as master builder to spectacular destroyer (I will come back to this point at the end.)

H is for High-Rise: In *Delirious New York* (1978), a 'retroactive manifesto for Manhattan', Rem Koolhaas published an old tinted postcard of the city skyline of the early 1930's. It presents the Empire State, Chrysler, and other landmark buildings of the time with a visionary twist – a dirigible set to dock at the spire of the Empire State. It is an image of the 20th-century city as a spectacle of new tourism, to be sure, but also as a utopia of new spaces – of people free to circulate from the street, through the tower, to the sky, and back down again. (The image is not strictly capitalist: the utopian conjunction of skyscraper and airship appears in revolutionary Russian designs of the 1920s as well.) The attack on the World Trade Center – of the two jets flown into the two towers – was a dystopian perversion of this modernist dream of free movement through cosmopolitan space. Much damage was done to this great vision of the skyscraper city – and to New York as the capital of this old dream.

I is for Indiscipline: Several of these notes circle around a single thesis: contemporary design is part of a greater revenge of advanced capitalism on postmodernist culture – a recouping of its crossings of arts and disciplines, a routinization of its transgressions. We know that autonomy, even semi-autonomy, is a fiction, but periodically this fiction is useful, even necessary, as it was at the modernist moment of Loos and company 100 years ago. Periodically, too, it can become repressive, even deadening, as it was a few decades ago when late modernism had petrified into medium-specificity and postmodernism promised an interdisciplinary opening. But this is no longer our situation. It is time to recapture a sense of the political situatedness of both autonomy and its transgression, a sense of the historical dialectic of disciplinarity and its contestation.

This is not to turn against critical theory and interdisciplinary work; instead it is to set them in historical perspective in order that they might be practiced anew. Today one often hears that we have too much theory and interdisciplinarity; on the contrary, we have never had enough. In my experience at the School of Architecture at Princeton, which is known for its emphasis on these approaches, the critiques are not often well informed in relevant philosophy or other pertinent fields (including art). We are not sufficiently theoretical, and we have not yet been critical.

J is for Jewel Box: No term is more important to modern architecture than 'transparency'. For Sigfried Giedion this transparency was predicated on technologies such as steel-and-glass and ferro-concrete that allowed a thorough exposition of architectural space. For Lazlo Moholy-Nagy it allowed architecture, potentially, to integrate the various transparencies of other mediums like photography and film. Less concerned with space than light, Moholy saw this integration as fundamental to the 'new vision' of modernist culture in general. Yet this vision did not fare well after the war. In 'Transparency: Literal and Phenomenal' (1963) Colin Rowe and Robert Slutzky devalued literal transparency in favor of phenomenal transparency, in which 'Cubist' surfaces 'interpenetrate without optical destruction of each other'.[4] This revaluation marked the moment when, once more, articulation of surface became as important as that of space, and understanding of skin as important as that of structure. That is, it marked the discursive advent of postmodern architecture in its two principal versions: first, architecture as a scenographic surface of symbols (as in pastiche postmodernism from Robert Venturi on) and, later, architecture as an autonomous transformation of forms (as in deconstructivist postmodernism from Peter Eisenman on). Today many prominent architects, such as Koolhaas, Herzog & de Meuron, Kazuyo Sejima, and Richard Gluckman, do not fit neatly into either camp: they hold on to literal transparency even as they elaborate phenomenal transparency with projective skins and luminous scrims. Sometimes, however, these skins and scrims only dazzle or confuse, and the architecture becomes an illuminated sculpture, a radiant jewel. It can be beautiful, but it can

---

4 Rowe, Colin and Slutzky, Robert; 'Transparency: Literal and Phenomenal,' **Perspecta** 8 (1963) p. 46.

also be spectacular in the negative sense of Guy Debord – a kind of commodity-fetish on a grand scale, a mysterious object whose production is mystified. (Incidentally, this 'Lord of the Rings' is what Charles Jencks now celebrates as 'the enigmatic signifier' in architecture [Hunch 257], and declares as the proper style for public buildings today.)

K is for Kool House: 'This architecture relates to the forces of the *Groszstadt* [the metropolis] like a surfer to the waves,' Koolhaas once remarked of the old skyscrapers of Manhattan.[5] With his recent interventions in global cities, especially in China, the same might be said of his own work. What does it mean for an architect to surf the global *Groszstadt* today – to perfect its curve, to extend its trajectory? Even if an architect could be empowered enough to make the attempt, can he or she do more than crash on the beach?

L is for Liebestod: Worse than an opportunistic surfing of the present is a traumatic fixing of the past – that is, an architectural fixing of a traumatic view of history. Here the exemplar is Daniel Libeskind, especially in his design for the World Trade Center site. Such 'trauma architecture', with its allegorical use of literal ruptures, might work in his Jewish Museum in Berlin; but that is a museum, not an entire section of a metropolis, and it is distant enough from the commemorated event to serve in a memorial manner. However horrific, the World Trade Center atrocity can hardly be compared with the Holocaust. That 'our lives were changed forever by the events of 9/11' seems true to many Americans, even ones who are not as history-challenged as our leaders, but this statement is also deeply ideological (it was originally credited to former Attorney General John Ashcroft), and Libeskind has adapted its rhetoric to support his trauma architecture. 'From now on architecture will never be the same,' he proclaimed in *The New York Times*. And the implication is that his architecture will best register this supposed fact.

M is for Media: Once more, what else? And yet, in architectural discourse today, every call to follow the forces of media dematerialization seems to be met with an equal and opposite demand to recover the experience of traditional materiality. Thus, for example, Mark C. Taylor insists on the one side, 'All architecture must become network architecture,' while Kazuyo Sejima states on the other side, 'In an age in which people communicate through various media in nonphysical spaces, it is the architect's responsibility to make actual space for physical and direct communication between people' (Hunch 445, 407). Perhaps, like all antinomies, these two positions – the one for the virtual, the other for the material – belong to one another, even partake of one another. What, then, is the stake that lies beneath the opposition, and how can we grasp it and move beyond? (One possibility: in very different ways both positions insist on 'communication', which suggests that a felt lack of community is held in common – as well as, perhaps, a felt predicament of what 'architecture in the public sphere' might be today.)

N is for Nobrow: One aspect of our mediated world is a merging of culture and marketing. For some commentators this merging has produced a 'nobrow' culture in which the old hierarchy of highbrow, middlebrow, and lowbrow cultures no longer applies.[6] For fans of this development 'nobrow' is not a dumbing down of intellectual culture so much as a wising up to commercial culture, which becomes a form of sophistication and a source of status in its own right. Today, this argument runs, we are all in the same 'megastore', only in different aisles, and that is a good thing – that is democracy. This, of course, confuses democracy with consumption, and this conflation in turn underwrites the principal commodity on sale in this marketplace: the fantasy that class divisions are thereby resolved. This fantasy is the contemporary complement to a foundational myth of the United States – that such divisions never existed here in the first place – a delusion that allows millions of Americans to vote against their class interests every four years at least. Much celebrated architecture – from the faux-populist postmodernism of Venturi et alia to the spectacle designs of Gehry and company – has supported this delusion.

O is for Occupation: While some of us live in a megastore democracy, others inhabit a security state, and still others exist in both at once. Here too, in matters of security, architecture figures centrally, for architects are still involved in the disposition of bodies in space and time (even, perhaps especially, when they have gone 'network'). That is, they are still involved, crucially, in questions of law, order, and surveillance – and not only with structures that might be deemed somehow panoptical. With every new project, every new proposal, architects must decide how to intervene in this political terrain.

One such intervention is especially noteworthy here: *A Civilian Occupation: The Politics of Israeli Architecture* by Rafi Segal and Eyal Weizman. This ensemble of plans, maps, photographs, and texts examines the settlements in the West Bank in many dimensions – architectural, urbanistic, environmental, historical, military, political. The conclusions of these two Israeli architects, now in exile as a result of this work, are damning: 'The mundane elements of planning and architecture have been conscripted as tactical tools in the Israeli state strategy, seeking national and geopolitical objectives in the organization of space… Space becomes the physical embodiment of a matrix of forces, manifested across the landscape in the construction of roads, hilltop settlements, development towns and garden-suburbs.' For Segal and Weizman the settlements amount to war continued by other means – architecture and planning in the service of conquest and colonization (this is an old story, of course, but it is made new here). For all the specificity of the West

5   Koolhaas, Rem; **S, M, L, XL** (New York: Monacelli Press, 1995) p. 937.
6   See Seabrook, John; **Nobrow: The Culture of Marketing, the Marketing of Culture** (New York: Alfred A. Knopf, 2000).

Bank, they also suggest that the settlements might not be entirely unique, that they might serve as prototypes for militarized societies of the near future, 'a worst-case scenario of capitalist globalization and its spatial fall-out'.7

P is for Post-Fordism: The object world of modern cities was born of a Fordist economy that was relatively fixed: factories and warehouses, skyscrapers and bridges, railways and highways. However, as our economy has become more Post-Fordist, capital has flowed ever more rapidly in search of cheap labor, dispersed manufacture, financial deregulation, and new markets, and the life expectancy of many buildings has fallen dramatically. (Many cities are now hybrids of the two economies, with Fordist structures often retro-fitted to Post-Fordist needs.) This process is pronounced in the United States, of course, but it is rapacious where development is even less restricted. How can architectural design adapt to such a rapid rate of physical turnover? It seems a matter less of intermittent plug-in than of continuous retro-fit.
Forty years ago advocates of Pop architecture like Reyner Banham, John McHale, and Archigram responded with proposals of 'throwaway architecture', plastic design, and 'instant cities'. But throwaways don't necessarily go away, and plastic has a half-life that rivals that of the Parthenon; in large part Pop architecture became another recipe for Junkspace. The same is true of the decorated shed of Venturi postmodernism, which was explicitly offered as an accommodation to 'the ordinary and the ugly'; today the periodic adjusting of the sign out front and the structure behind is standard practice everywhere. Recently architectural historian Sylvia Lavin has updated this Pop-postmodern position with a call for a 'reorientation of architecture toward a field of [special visual] effects' (Hunch 296). If this is what 'Learning from Las Vegas' or Los Angeles means today, perhaps we should look elsewhere for models.

Q is for Quarantine: For Koolhaas the skyscraper is the crux of the 'culture of congestion' of the old Manhattan, and he sees it as a mating of two emblematic forms – 'the needle' and 'the globe'. The needle grabs 'attention', while the globe promises 'receptivity', and 'the history of Manhattanism is a dialectic between these two forms.'8 Since 9/11 the discursive frame of this Manhattanism has shifted. New fears cling to the skyscraper as a terrorist target, and the values of 'attention' and 'receptivity' are rendered suspect. The same holds for the values of congestion and 'delirious space'; they are overshadowed by calls for surveillance and 'defensible space'. In short, the 'urbanistic ego' and cultural diversity that Koolhaas celebrates in *Delirious New York* are under enormous pressure, and not only in New York. They need advocates like never before, for, to paraphrase the Surrealists, cosmopolitan beauty will be delirious or will not be.

R is for Running-Room: As much as interdisciplinarity is crucial to cultural practice, so too are distinctions, as the great Viennese critic Karl Kraus insisted in 1912: 'Adolf Loos and I – he literally and I linguistically – have done nothing more than show that there is a distinction between an urn and a chamber pot and that it is this distinction above all that provides culture with running-room [*Spielraum*]. The others, the positive ones [i.e., those who fail to make this distinction], are divided into those who use the urn as a chamber pot and those who use the chamber pot as an urn.'9 'Those who use the urn as a chamber pot' were Art Nouveau designers who wanted to infuse art (the urn) into the utilitarian object (the chamber pot). Those who did the reverse were functionalist modernists who wanted to elevate the utilitarian object into art. For Kraus the two mistakes were symmetrical – both confused use-value and art-value – and both risked a regressive indistinction: they failed to safeguard 'the running-room' necessary to liberal subjectivity and culture. Note that nothing is said about a natural 'essence' of art or architecture, or an absolute 'autonomy' of culture; the stake is simply one of 'distinctions' and 'running-room', of proposed differences and provisional spaces.

S is for Spectacle: Far more than 'trauma architecture' 'spectacle architecture' is a dominant tendency in contemporary design. (The two are not mutually exclusive, as the World Trade Center design attests. Here a pessimist might glimpse a Trauma Theme Park in the making, the transformation of historical tragedy into urban spectacle.) Certainly, to make a big splash in the global pond of spectacle culture today, one has to have a big rock to drop, maybe as big as the Guggenheim Bilbao, and here architects like Gehry have an obvious advantage over artists in other media. In *The Society of the Spectacle* (1967) Debord defined spectacle as 'capital accumulated to the point where it becomes an image'. Of course spectacle has become only more intensive in the four decades since then, to the point where media-communications-and-entertainment conglomerates are the dominant ideological apparatuses in our society, powerful enough to refashion other institutions (such as architecture, not to mention art) in their guise. Today, then, the corollary of the Debordian definition appears true as well: spectacle is 'an image accumulated to the point where it becomes capital'. Such is the logic of many cultural centers today, as they are designed, alongside theme parks and sports complexes, to assist in the corporate 'revival' of the city – that is, in its being made safe for shopping, spectating, and spacing out. This is the true 'Bilbao-Effect', and it is indicative of our political condition that 'liberal' organs like *The New York Times* continue to promote this spectacle logic as a democratic architecture of a public sphere. After *The Lord of the Rings* movies, then, the Bilbao-Effect might be rethought as the Bilbo-Effect: the designer as hobbit who seizes a ring of power that he cannot control.

T is for Tectonics: For all the futurism of the computer-assisted designs of architects like Gehry, his structures often recall the Statue of Liberty, with a separate skin hung over

---

7 Segal, Rafi and Weizman, Eyal; **A Civilian Occupation: The Politics of Israeli Architecture** (London: Verso Press, 2003).
8 See Koolhaas, Rem; **Delirious New York** (New York: Oxford University Press, 1978).
9 Kraus, Karl; **Werke**, vol. 3 (Munich: Kösel Verlag, 1952–66) p. 341.

a hidden armature, and with exterior surfaces that rarely match up with interior spaces. With the putative passing of the industrial age, the structural transparency of modern architecture was declared outmoded, and now the Pop aesthetic of postmodern architecture looks dated as well. The search for the architecture of the computer age is on; ironically, however, it has led Gehry and his followers to 19th-century sculpture as a model, at least in part. The disconnection between skin and structure represented by this academic model has two problematic effects. First, again, it can lead to strained spaces that are mistaken for a new kind of architectural sublime. Second, it can abet a further disconnection between building and site. This is not a plea for a return to structural transparency; it is simply a caution against a new Potemkin architecture of conjured surfaces (all those blobs and flows out there) driven by computer programs.

U is for Utopia: Often the utopian impulse is an expression of a new class in society and/or a new mode of production on the rise. Think of the projects of Ledoux or Boullée in light of the French Revolution, the bourgeois breakthrough, or the Crystal Palace (1851) at the height of the Industrial Revolution, with the bourgeoisie confident of its control. It emerged again, as the expression of other political forces, in the Russian Revolution, and here the emblematic project was the Tatlin *Monument to the Third International* (1920). The utopian impulse was evident in the International Style too, but in our own time it was misread as totalitarian, and the International Style was condemned for its own corporate abuse. The utopian in architecture became taboo.
There is, however, a small revival of the utopian impulse today – in architecture as well as in art. One instance comes by way of a young Japanese architect named Yusuke Obuchi. His *Wave Garden* project is a 480-acre field designed to float, like a Suprematist rectangle, off the coast of California. Made up of 1800 Piezoelectric sheets supported by 1800 buoys, it is an electrical generator during the week and a marine park on the weekend. In its first mode the sheets of the garden are bent by the sea waves in a way that generates electricity which is then transferred to the energy grid of the Golden State. In its second mode electricity is run through the sheets in a manner that shapes them into a metamorphic island of coastal leisure and maritime play. Neither entirely fantastic nor quite practical, *Wave Garden* is precisely utopian: it forces us to think why-not in a way that questions what-is.

Obuchi calls up different precedents from Gaudí to the Earthworks of the late 1960s and early '70s. However, he does not partake of the fascination with entropy so evident in Robert Smithson, say; on the contrary, *Wave Garden* works to generate alternative energy rather than to submit to its doom-day dissipation. At the same time the project is not as redemptive as it may first appear. Early on Robert Morris was sensitive to the ideological recuperation of the Earthwork idea — that despoilers of the environment might use Earthworks as so much artistic camouflage. *Wave Garden* skirts this danger: unlike many designers today, Obuchi does not seek to naturalize his architecture; rather, his project is continuous with the greater human project to acculturate nature, which it proposes to play with as well as to exploit. In the era of Enron, Obuchi conjures a vision of energy, physical and social, that is utopian in its force.

V is for Vernacular: Postmodern architecture pretended to revive vernacular forms, but for the most part it replaced them with commercial signs, and Pop images became as important as articulated spaces. In our design world this development has reached a new level: now commodity-image and space are often melded through design. Designers such as Bruce Mau strive for programs 'in which brand identity, signage systems, interiors, and architecture would be totally integrated'.[10] This integration depends on a deterritorializing of both image and space, which depends in turn on a digitizing of the photograph, its loosening from old referential ties, and on a computing of architecture, its loosening from old tectonic principles. (As Deleuze and Guattari, let alone Marx, taught us long ago, this deterritorializing is the path of capital, not the avant-garde.) One already 'experiences' this seamless mélange of sign-space in many malls, actual and cyberspatial. This is one version of 'network architecture', and it may well qualify as nobrow.

W is for Wound: Should urbanism be rethought in terms of trauma? Libeskind argued so for Berlin: 'The lost center cannot be reconnected like an artificial limb to an old body, but must generate an overall transformation of the city.' Should New York be remapped through Ground Zero? The memorial elements of the original Libeskind design — the foundation walls, the open wedge, the symbolic spire — wanted to fix lower Manhattan in monumental terms, which is alien to a city defined from its beginnings by its very embrace of change. Such traumatic monumentalism was problematic enough, but there was another motive here as well. 'Build them higher than before,' one often heard after the fall of the Towers, as if our problem were penile dysfunction — and, perhaps, imperially speaking, it is. If the hole in the ground figures the attack for Libeskind, his spire 'will let the world know that the terrorists have failed.' A large part of the popular attraction of his scheme is disclosed right there: it gives us both a pathos pit and an imperial thrust, both the traumatic and the triumphal. Or, more exactly, it gives us trauma troped as triumph, a site of civilian tragedy turned into a symbol of militaristic defiance. This is a dangerous concoction; certainly the historical precedents when the wounded have linked up with the hubristic are not savory to contemplate.

X is for Xed: As in crossed out or canceled. Such, it seems, is the status of 'critical' archi-

10 See Mau, Bruce; **Life Style** (London: Phaidon Press, 2000).

tecture today, at least according to 'post-critical' advocates. Sometimes the call for a post-critical architecture is a complaint about an instrumental use of theory, and here one can only agree. But why drown the baby in the bathwater, and pronounce all critical thinking in design dead? What is the difference, politically, between such post-critical affirmation and the dominant neo-conservativism? 'There are vested interests that want us to believe that "there is no alternative",' the critic-historian Hilde Heynen comments. 'We should not denounce this dimension but rather seek to reevaluate and resuscitate it' (Hunch 242–43). Indeed, it is the wrong time – tragically wrong, architecturally as well as politically – to give up on critique.

Often implicit in post-critical discourse is a futurist faith that new materials and new media are somehow progressive per se. Certainly these things hold great possibilities – an immense expansion in techniques for designers and a partial recovery of the means of production for architects – but they also invite pre-critical naivetés. (A small instance: no one seems to question the return of perspective in the computer images that have become standard in architectural presentations.) Also often implicit in post-critical discourse is a willful opposition between 'critique' and 'invention', as if the two were really opposed, and an equally willful narrative in which 'pragmatism' comes 'after theory', as if pragmatism were not also a theory and theory not also pragmatic. What is meant by pragmatism here? An innovative reengagement, through 'design intelligence', with the world. 'Innovation operates by an affirmative, non-linear process of continuous feedback,' the critic Michael Speaks argues, 'through which opportunities are discovered that are exploited and transformed into designs not posed or foreseen by the problem' (Hunch 417). Yet this formulation can also sound suspiciously like a new kind of design formalism or process fetishism.

Y is for Yahoos: Fill in the blank as you like.

Z is for Zero: Not Ground Zero. That has become too mobile a site: lower Manhattan, Afghanistan, Baghdad, Bali, Istanbul, Beslan… Rather, Z as in Zero Degree, an Architecture Zero Degree on the model of the Writing Degree Zero championed by Roland Barthes 50 years ago – that is, an architecture not so driven, so consumed, by ideology, an average architecture that would be inventive, an everyday architecture that would be reflexive, nonetheless.

For all the massive attention given a few projects, reflexive architecture appears ever more marginal to the actual construction of the everyday world. There are many reasons for this situation – from the old separation of architecture and engineering on the more recent domination of big construction firms – but some blame must be laid on the self-involvement of the architectural vanguard. (Self-involvement is hardly the same thing as creative research, let alone self-critique, and it should not be pinned on the usual scapegoats – academy and theory – as even some academic theorists tend to do today.) This relative lack of reflexive architecture permits the proliferation of what Koolhaas calls Junkspace and what Luis Fernández-Galiano calls the Horizontal Babel.

One can attend to 'market realism' and still not submit to this Junk Babel. How might developers and architects alike be induced to pay more attention to everyday building, to average architecture, in a not-so-throwaway manner that neither reiterates 'the ordinary and the ugly' nor opts for 'special visual effects'?

Section VI:

Urban Compositions in the 21st century

Urban Compositions
Han Meyer

During the last 25 years, urbanism has been developed at the Delft University of Technology as a discipline with its own scientific standards and methods, and as a specific type of research. The DSD-program in Urbanism tries to build upon these recent results and insights and to enrich and elaborate the existing insights with the work of the students and researchers.

Randstad Holland: a laboratory of designing and planning a networkcity in a vulnerable territory

Recent developments in urbanism at the Delft University are on the one hand based on the rich 'Fine Dutch Tradition' of urbanism and land-use planning in the Netherlands during five centuries; on the other hand, it is a reaction to the technocratic and functionalistic approach which characterized urbanism in the Netherlands in the post-war decades. During the 20th century the practice of urbanism in the Netherlands rested largely on the paradigm of the *Plan*, marked as this was by an indissoluble bond between preliminary social-scientific research and urban design, and by the strong conviction that it should be possible to steer and control the developments of society and land-use from a top-down approach.

During the 20th century, the Randstad Holland was developed as a relatively clearly organized network of cities, settled in a vulnerable territory (as a delta of large European rivers) which seemed to be 100% under control, thanks to an advanced civil engineering technology. However, this approach became outdated and unfruitful to solve present-day problems and to anticipate a new urban and regional reality. This new reality concerns the transformation of the Randstad Holland from a network of cities to a new, unprecedented and complex *networkcity* or '*Deltametropolis*'. At the same time, climate changes forced us to reconsider the idea of 100% control of the territory, and to develop a more 'elastic' type of water management.

These new realities asked for new concepts and methods to steer the spatial development of this Deltametropolis in a fruitful direction, with new relations among territorial conditions, land use, infrastructures and urban qualities. As such, the Randstad Holland can be considered as a laboratory of new concepts for urban and regional design and planning, with a broad international significance.

To be able to face these new tasks, two types of design research have been developed at the Delft University:
- Design analysis and typo-morphological research;
- Research by design as a means of exploring and discovering possibilities.

This development has ushered in two monumental books: one about the various methods of design research and research by design (De Jong et al. 2002), the other about architectonic and urban composition as a subject for research (Steenbergen et al. 2002).
Both types of this 'new urbanism' are focussed on three 'main questions':
- the question of sustainable territories and landscapes;

- the question of the transformation of a city-network to a network-city;
- the question of the urban fabric and the urban project.

The question of sustainable territories and landscapes

In the next 50 years, a doubling of the world population living in cities is expected (Unesco 2005). The largest part of this urbanization process will take place in extreme vulnerable territories: river delta areas, earthquake zones, hurricane areas. Taking into consideration the changes of the climate, attention to a careful management of the territory will be of great significance to the worldwide development of the urbanization process. The territory of the Randstad Holland, which de facto is a vulnerable river delta area, can be considered as an important laboratory, where on the one hand experiences of a centuries-long tradition of urbanizing a delta area can be evaluated, and on the other new concepts and approaches concerning urbanization of a vulnerable territory can be tested.

The Delft Department of Urbanism pays special attention to two aspects:

a the changing *technical conditions* of the territory because of climate changes, rising sea-level and increasing river-water supply. Questions of land use in the Randstad Holland can not be considered without the development of new concepts and approaches concerning water management technology. The concern with this technology has been expressed with the production of several PhD researches (Tjallingii, van Eyck). This work is strongly related to the program of the Water Research Centre of the TU-Delft (www.water.tudelft.nl).

b The second aspect concerns the *composition* of the landscape, focusing on the landscape as an architectonic composition. This consideration is especially relevant for Holland, where the landscape has been transformed from a 'wild' natural landscape into an almost completely man-made landscape during the last 1000 years. Not only merely technical but also economic, aesthetic and cultural motives have played a role in the composition of the landscape and will also play a role in the transformation of the landscape in the nearby future. This field of research is strongly rooted in the Delft Faculty of Architecture since the 1980's, expressed with several PhD works and other publications which are considered as standard works worldwide, some of them translated in many languages (Steenbergen 1990, Reh 1995, Steenbergen & Reh 1996, Tummers 1997, Steenbergen & Reh 2005).

The question of the urban project

During the last decades, from the 'paradigm of the Plan' to the present-day situation, the meaning of the urban project has changed fundamentally. Some decades ago, especially in the Netherlands, an urban project was a part of the implementation of an overall plan. Since then, approaches have been developed that consider the urban project as a strategic element to generate a new development on a larger scale. To be able to implement this idea in reality, it is necessary to develop new concepts concerning the relation between the design of short-term projects and the planning of large-scale regions on the long term. It supposes also a reconsideration of the relation between landscape, urban land-use and infrastructural networks. The urban project concerns the scale where this relation can be influenced and transformed most directly. Also from this perspective, the Randstad Holland can be considered as a laboratory where it is necessary to investigate the significance of relatively small-scale urban projects for the development of the region with a new coherence.

In order to be able to develop urban projects in an effective way, knowledge and research is supposed concerning:

a The relation between the characteristics of the landscape and the structure and fabric of the city. The morphology of the Dutch city and the typology of buildings and public space are strongly related to the development of the city as a hydraulic construction. From the 1980's this relation between territory, hydraulic engineering and city design has been a central focus in the research program of the Department of Urbanism, expressed in many publications (Louwe & vd Hoeven 1985, Palmboom 1987, Geurtsen 1988, Heeling et al. 2002, Hooimeijer et al. 2005).

b The changing relations between private and public domain, expressed in the change of typologies of residential ensembles and the infrastructure of public space (Komossa et al. 2002, Meyer et al. 2005).

c The increasing complexity concerning the relation between large scale infrastructures (seaports, highways, railways, airports) and the small scale structures of urban centres and areas. Developing a new relation between large scale infrastructure and the urban fabric is one of the most important design tasks of urban projects, and has been studied in research projects and PhD work during the last ten years (Meyer 1996, vd Hoeven 2001, vd Spek 2002, Calabrese 2004).

The question of the transformation from city-network to network-city

As it becomes clear from comparative research (van Susteren 2005), the Randstad Holland is one of the very few urban regions in the world which shows the birth of a new reality of a network-city. It is true that all over the world many urban systems can be considered as metropolized regions with a variety of centres and nodes, but most of them are multi-nodal systems where the networks of roads, railroads and airports are just connecting elements between the nodes. The new reality of the network-city is the appearance of the network itself as an urban centrality. The total area of the Randstad Holland is developing into a large urbanized area, with an infrastructure which plays more and more a role as a generator of new urban centralities. Research to get knowledge and to be able to anticipate this development is important not only for the Randstad itself but for all metropolitan areas in the world.

The research concerning the network-city at the Faculty of Architecture is focussing on the following aspects:

a Comparative analysis of metropolitan systems and the transformation of these systems from multi-nodal to network-systems (van Susteren 2005);

b Understanding networks: analysis of the conditions which play a role to transform a network from a mere connecting system towards a system which generates urban centralities (Read et al.)

c Developing new concepts and typologies: classifications of different models and types of network-cities (Frieling, Jacobs, et al.)

d Developing new design and planning strategies: which type of interventions and which means of managing the planning process should be developed to be able to develop the discipline in an effective way (Meyer & van den Burg 2005).

Urbanism, Landscape Architecture and Civil Engineering Urban design in the 21st century is not possible without a multi-disciplinary approach. Urban design concerns the spatial organization of *society*, by means of working up and parcelling the *territory*. This is impossible without a close collaboration with the disciplines which are specialized in society and territory: Concerning society we have to deal with social geography, economic geography, spatial economics, etc – these scientific areas we usually summarize with the term 'Planning Sciences'. Concerning the territory we have to deal with landscape architecture, civil engineering and geodesy.

The disappearance of the 'paradigm of the Plan', and the increasing necessity of a stronger emphasis on urban design as a technical discipline, as well as on the development of urban design as a way of research, resulted into new relationships among urbanism, landscape architecture and civil engineering. This trinity is fundamental for the investigation (by designing) of the possibilities of the territory for urban settlements and other human activities.

In the meantime, changes also took place in the field of the planning sciences, with an increasing interest in the possibilities and results of research by design at the TU-Delft. Planning science as such, is more and more moving to become a discipline which is specialised in fundamental knowledge concerning the relation between land use and social-economic processes. As such, the relation between Planning Science and urban design is moving from a linear relationship towards interaction and dialogue.

In Conclusion Present day urban design can be regarded as the culmination of a development that has seen the design discipline conquer its fears and present its research work as 'real' science. Designers, drawing on their particular expertise and domain, are making their own contribution to the discourse on the spatial development of town and country, alongside the input of, say, the spatial sciences. The significance of this research in the near future will be in strengthening and building upon that expertise, namely by continuing to explore in depth the methodological and theoretical issues of design research and research by design.

...ese, **Reweaving Urbanism, Mobility & Architecture,** ...lft 2004

...k, **Vernieuwen met water. Een participatieve strategie voor de gebouwde omgeving,** DUP Delft 2003

Dirk Frieling, 'Metropoolvorming', in W. Reh, D. Frieling, C. Weeber, **Delta Darlings,** Delft 2003

Jan Heeling, Han Meyer and John Westrik, **Het ontwerp van de stadsplattegrond,** SUN Nijmegen 2002

Rein Geurtsen, **Delft Zuidpoort Atlas, een stadsmorfologische analyse,** DUP Delft 1989

Casper van der Hoeven and Jos Louwe, **Amsterdam als Stedelijk Bouwwerk. Een morfologiese analyse**, SUN Nijmegen 1985; reissued 2003

Frank van der Hoeven, **RingRing – Ondergronds bouwen voor meervoudig ruimtegebruik boven en langs de Ring Rotterdam en de Ring Amsterdam,** 010 Publishers Rotterdam 2002

Marc Jacobs, **Multinodal Urban Structures. A comparative analysis and strategies for design,** DUP Delft 2000

T.M. de Jong and D.J.M. van der Voordt (eds), **Ways to study and research urban, architectural and technical design,** DUP Science Delft 2002

Suzanne Komossa, Han Meyer, Max Risselada, Sabien Thomaes, Nynke Jutten, **Atlas van het Hollandse Bouwblok,** Thoth Bussum 2003

Fransje Hooimeijer, Han Meyer, Arjan Nienhuis, **Atlas Dutch Watercities,** SUN/BOOM Amsterdam 2005

Han Meyer, **City and Port. Transformations of Port Cities: London Barcelona New York Rotterdam,** International Books Utrecht, 1999

Han Meyer, Leo van den Burg (eds), **In Dienst van de Stad / Working for the City. 25 years of work from the urban design departments of Amsterdam, Rotterdam, The Hague,** SUN Amsterdam 2005

Han Meyer, Frank de Josselin de Jong, MaartenJan Hoekstra, **Het Ontwerp van de Openbare Ruimte,** SUN Amsterdam 2005

Frits Palmboom, **Rotterdam Verstedelijkt Landschap**, 010 Publishers Rotterdam 1987

Stephen Read, **Function of urban pattern. Pattern of urban function,** DUP Delft 1998

Wouter Reh, **Arcadia en Metropolis. Het landschapsexperiment van de verlichting,** DUP Delft 1995

Stefan van der Spek, **Connectors. The Way beyond Transferring,** DUP Delft 2003

Arjen van Susteren, **Metropolitan World Atlas**, 010 Publishers Rotterdam 2005

Clemens Steenbergen, **De stap over de horizon. Een ontleding van het formele ontwerp in de landschapsarchitectuur,** DUP Delft 1990

Clemens Steenbergen and Wouter Reh, **Architecture and Landscape. The Design Experiment of the Great European Gardens and Landscapes**, Thoth Bussum 1996

Clemens Steenbergen, Henk Mihl, Wouter Reh and Ferry Aerts (eds), **Architectural Design and Composition**, Thoth Bussum 2002

Clemens Steenbergen, Wouter Reh, Diederik Aten, **Zee van Land. De droogmakerij als atlas van de Hollandse landschapsarchitectuur,** Architectura & Natura Press, Amsterdam 2005

Sybrand Tjallingii, **Ecological condition – strategies and structures in environmental planning,** Wageningen (IBN/DLO), 1996

L.J.M. Tummers and J.M. Tummers-Zuurmond, **Het land in de stad. De stedebouw van de grote agglomeratie**,Thoth Bussum 1997

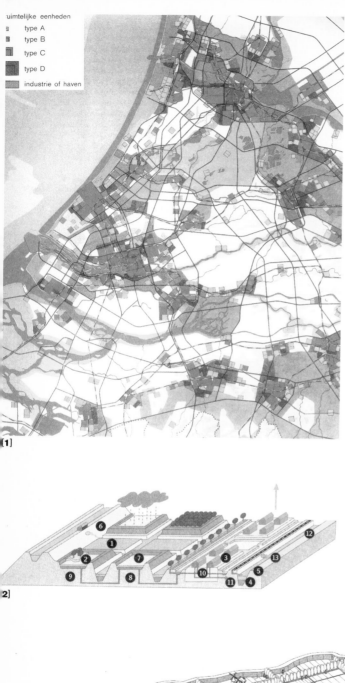

[1]

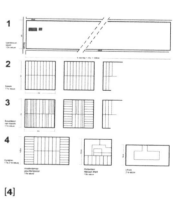

[2]

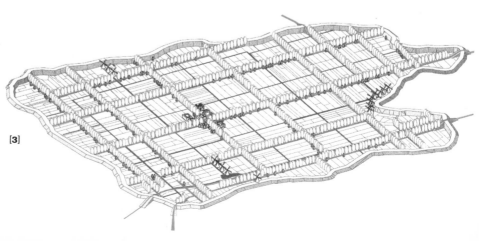

[3]

1 The paradigm of the Plan: 'Randstad Holland' in 2nd Report on Spatial Planning, 1966.
2 Integration of urban design and water management system in Haarlemmermeer, 1995.
3 The composition of the Dutch Polder: analysis of Beemster by Steenbergen and Reh.
4 The evolution of parcelling and urban blocks in the Dutch urban landscape:
1: original agrarian parcel 12th century;
2: urban block and parcels according to Simon Stevin, c. 1600;
3: transformations of the ideal urban blocks;
4: urban blocks in 17th-century Amsterdam (left), 19th-century Rotterdam (middle); 21st-century Amsterdam-IJburg (right). (From Heeling, Meyer, Westrik, op. cit.)

[4]

Urban Composition: City Design in the 21st Century

Joan Busquets

As we progress through the first years of a newly unwrapped century, we have become more aware of the acute transformations currently occurring in our cities. On the one hand, we are experiencing great satisfaction as we observe new urban phenomena emerge. On the other, we feel great frustration as we see that many urban problems are not getting resolved, and are even becoming more severe.

Some of the latest statistics help us understand this contemporary urban problem: The population in urbanized regions, in comparison to rural areas, continues to grow at an exponential rate. If the world today has 3 billion urban dwellers, there will be 4 billion in 2030. This fact evidently shows that urbanity and urbanism are increasingly important dimensions in everyday life. Furthermore, the greatest concentration of urban growth can be found in developing countries that generally have weak urban structures, and are unable to provide the required infrastructure to handle such high-paced levels of growth. If we take a glance at the ranking of world cities, we can see clearly that cities referred to commonly as global, such as New York or London, are stepping down in contrast to former minor cities, such as Bombay, Manila, or Lagos, which are rapidly surfacing.

Today, urban development is concentrated predominately in two open-ended forms of growth and/or transformation: the restructuring of the traditional city and the development of an amorphous conurbation. In the first case, the traditional city, one that had clearly defined boundaries between urban and rural areas, has ceased to exist. The urban/rural distinction might be a concept that our culture is very resistant to lose, but from a morphological perspective, it is rarely visible in our current environment. In the second case, a large percentage of urban settlements are emerging as amorphous amalgams. Such development has no apparent formal structure and stretches over vast surfaces, resulting in an extended mixture of built spaces, voids, and natural areas. This kind of urbanization has been categorized conventionally as having weak visual definition, and therefore, comes across as illegible and lacking in rationality. We must then ask ourselves, can we simply label these new forms as fallacious, or must we find other lines of thought that are able to understand and structure these new urban paradigms?

Given the new urban domain described above, cities are being confronted with severe restructuring processes. These changes are mostly linked to the centralization of inter-modal systems. Consequently, major alterations in the traditional urban fabric are necessary in order to accommodate new spatial systems that have a greater distribution logic. Even Western cities that have limited demographic growth have been subject to the introduction of revisited infrastructural forms.

Throughout the course of the twentieth century, cities and territories have undergone traumatic transformations. Many apocalyptic voices present the city today as a chaotic and untameable beast. In Europe, industrialization caused the greatest level of unplanned growth in the history of the traditional city. Despite the initial upheaval, such growth achieved greater levels of improvement in urban life. As urbanists, we cannot plan for a predictable future, but it is evident that our cities, municipalities, and its agents have a more ample palate of methods and techniques to better engage current urban challenges. The potential of our disciplines lies in the ability of urbanism to operate in a flexible manner that can respond quickly to a wide variety of ever-changing contexts and scales.

Cities are now undergoing a major shift from industrial to postindustrial forms of activity. In this new *modus operandi*, information and telecommunication technologies play a huge role in the restructuring of large tracts of land. Old systems of industrial production have adopted automated control systems that have drastically changed the way in which products and services are assembled and distributed.

In the last decades, changes in technology have had a major impact on the city's operative nature and, therefore, its form. We can see clearly a great shift from the traditional model in which the residential quarters were grouped around places of labor in order to maximize accessibility, to a more open system in which greater distances are traversed between domestic and corporate districts. This distribution has been facilitated primarily by the development and proliferation of transportation systems, both collective and private. New transport infrastructures, such as the high-speed train, and its intentional overlap with other forms of mobility, have established a new set of hierarchies at urban and territorial scales. This new organizational logic establishes new kinds of centrality that result in unprecedented patterns of urbanization.

Urbanism and the design professions, at large, participate in a constant process of re-invention. We can clearly observe new ways of engaging the urban problem, but new urban strategies have not been consolidated fully. We still perceive in our cities fragments of utopian visions rendered by the Modern Movement and their agenda for social renovation. We continue to admire their social drive, but we have to understand that their theories no longer apply. The profession must endorse and validate new procedures for the city. The role of the urbanist is not only to speculate on new forms of urbanization and the kind of city these might proffer, but also to develop methods and techniques that allow us to better understand our multiple urban realities, and help us define well-tempered strategies for the formatting and/or re-formatting of space.

For example, issues of mobility are of great importance in the definition of new urban spaces, being either of people or information, or both. The inter-modal nature of space is acquiring a greater importance in the restructuring of traditional fabrics and infrastructure, and also in the definition of new urban settlements. Public space has also acquired a new set of values. Urban culture and its setting, the city, knowingly participate as a kind of marketed product. The city's public and collective spaces have a much higher profile than ever before and have become objects of desire and intensive use for the majority of its dwellers.

It would be impossible at this point in time to define a unitary urban model that could guide urban growth. The urban realities that we deal with today are many and quite varied, and the arguments to support such a model cannot be fixed. If we take ideal or desirable density as an example, one argument might lean in favor of greater density. This approach might be supported by the idea that a greater concentration can result in more efficient and attractive urban settings, as well as ecologically friendly environments. Another line of thought might go against densification in favor of dispersion, arguing that density results only in congestion and that transport infrastructure today allows for dispersed but efficient forms of urbanization. If the traditional compact city is efficient in reducing infrastructure costs for both the public and private sectors, one must consider if it does not also reduce benefits for the city's population. Therefore, from a broader perspective, we must evaluate not only the economic costs but also the benefits associated with each of them.

As active participants within the agency of design, we must be interested in the formal properties of both the natural and the built environment. Urban form is embedded in places according to time-honored cultural beliefs, laws, and social practices that result in an urban complexity that must unfold in the hands of the designer. Only by fully understanding the operative nature of these new spatial dynamics will we be able to establish fresh urban methods that can help us reveal the identity of these emerging urban configurations. This attitude departs from the common professional desire to negate an urban form that is different from that of the traditional city.

Given the framework presented above, we must redefine urban composition; the way in which we give shape to ideas about the city must be revisited. First and foremost, we must understand that today we can find abundant lines of work within the city. The idea of a single, overriding ideology has been superceded by a multiplicity of voices that cover and/or address a wider set of issues than ever before. Urbanism and its compositional strategies must be tested by measuring the success of their ability to exist and to operate at multiple levels, starting as an intellectual construct and resulting in a built project or legal agreement for development. In this translation process, urbanism must be both strategic and tactical, acknowledging that certain decisions must be affirmative and consistent and others must be malleable and open-ended. The latter ones will obtain their definition from future decisions about program, cultural habits, and site specifications.

Urban composition can be defined as the ability to bring together coherently through an urban strategy the multiple realities of our current environment, to work with and within the complexity of our cities, and to acknowledge that any action within the field will be more a form of negotiation between different forces than the deployment of one prescriptive ideology. The lines of work presented below, perhaps the ten most salient ones operating today, generally evolve from the ideas presented in the preceding paragraphs. Let's take a look at them in detail:

1 Synthetical Gestures: An outstanding aspect of recent urban projects is the simultaneous presence of strategic, high-ambition interventions based around other, clearly delimited actions that use their spectacular nature to trigger the operation as a whole. This strategy is an extreme but potentially very effective position, and one that has its supporters. One 'key' building can give rise to a spin-off, producing a thoroughgoing transformation of which that one building is a minimum expression.

Synthetical gestures usually involve the deployment of iconographic forms in a given city or territory. They have been imposed traditionally in a top-down manner, even though in recent times a stronger sensibility has emerged toward acknowledging the pre-existing context. Such compositions tend to emerge from a highly internalized formal structure, and aim at being self-sufficient entities within their environments. Through these compositional strategies, the aim is to achieve symbolic form.

Today the tendency is to introduce iconic gestures of a lesser scale; in most cases, they appear at the scale of a building. These new gestures might still be highly internalized from an iconographic point of view, but at an operative level they are much more attuned to the reality in which they sit. Furthermore, they tend to act as the tip of the iceberg. The iconographic gesture is only the formal fragment of a much broader strategic plan. The most prominent example of such urban action can be seen in the proliferation of key buildings, which through their iconic power drive more complex urban redevelopment plans.

2 Multiplied Grounds: Artificial Surfaces: The main mechanism behind this strategy is the multiplication of ground, and the creation of artificial surfaces that act as inter-modal points between different forms of collective and private mobility. The greatest strength of this strategy is its potential for sectional integration of uses and services. This strategy is more commonly executed on the fringe of traditional cities where mono-functional train stations are being transformed into mixed-use transportation hubs. Perhaps the greatest benefit of the multiplication of grounds is the creation of surplus space that can accommodate a great variety of smaller programs and services that capitalize on high volume flows. These interventions are today perceived as new and unique spaces in the city and/or act as international reference points, spurring development well beyond its envelope.

Urban projects that tackle the transformation of emblematic parts of the city use the high profile of converted infrastructures and/or a brief for high-density reuse. This approach responds to the idea of the 'big project' – though in this case going beyond a single brief and attempting to combine in the space of the project decision-making, a series of briefs and/or actions that, because they are inter-linked, take on a surprising nature that endows them with greater influence.

3 Spinal Chords: The spinal chord is a compositional mechanism that can provide a clear structure of growth for urban development. The composition usually defines a clear spatial syntax to which less structured kinds of development can be clamped. The spinal chord's function tends to be for circulation or for the movement of infrastructural services. This strategy is most commonly used in extreme affordable-housing projects for which neither the government nor the private sector can provide a fully finalized product. Most recent examples of this kind can be seen in affordable-housing initiatives or in experimental housing projects in the developing world.

These projects, generally driven by an extreme economy of means involve reducing the intervention to the least possible dimension. Here lies the intervention's strength and, perhaps its reason for success. Above all, the intervention represents a bid to show that there is almost 'always something to be improved' in the city, and that this 'something' can be undertaken using relatively restrained, controllable instruments.

4 Thick Bi-Dimensionality: This compositional line is usually deployed in small pedestrian pockets within the city. The ground plane is mainly conceived in plan with minimal manipulations in section. Although such interventions are conceived individually and appear as scattered throughout the city, in sum they make up a broader revitalization strategy. These projects are frequently the result of a general desire for urban improvement, but it is applied in the form of discontinuous, piecemeal actions. An initiative that is initially more global in approach may not be very explicit, and only a proliferation of manifold projects, all with different briefs and conditions, can express a more general change of threshold.

Furthermore, thick bi-dimensionality serves as a substantial way to reconfigure a wide variety of under-utilized spaces. Among these are: derelict spaces created by the extensive geometry of vehicular infrastructure, existing but outdated spaces that need to be reprogrammed, and the creation of new spaces that serve as anchors for urban growth.

Traditional city centers benefit highly from this strategy, providing a new lease on life without the cost attached to larger restructuring operations. Building the city on the basis of empty spaces, and finding new ways of re-understanding urban mobility in the form of interventions in transport, albeit relatively modest ones, can give historical centers a new lease on life, and improve outlying districts more effectively than can costly restructuring operations.

5 The Urban Fragment – Piecemeal Development: The main objective behind fragmentation mechanisms is the re-scaling of obsolete infrastructural tracts of land into a more piecemeal structure in order to accommodate new programs that are more attuned to current needs. Generally driven by the reorganization of public works, the scale of infrastructure is altered to accommodate new and distinct uses. The re-stitching of the infrastructure in relation to open spaces, and the development of newly reconfigured urban blocks are the main compositional moves for such operations.

The urban fragment is usually conceived from morphological speculation in which the basic structure of the block is defined through a set of rules that delineate its massing. It is through the execution of the overall project, which generally involves many different hands, that this morphological hypothesis is tested.

The most significant issue in this compositional method is that of devising a strategy to design the overall structure of the urban fragment without designing its specific architecture. A new form of city is the result, in which the quality of the whole is much greater than the summation of its parts.

6 'Traditional' Layouts: The structure of the city block is still and will continue to be a key element in defining the nature of future urban growth. The open-versus-closed block systems are the two most significant compositional methods used to generate or re-generate a city block. The open-block, or grid system, defines the street as positive space, which denotes its interior as a void. In the closed block system, the interior is defined as a solid, usually built fabric, in which the street is conceived as a void. Both mechanisms are still commonly used and the choice between the two is generally made by considering the level of desired density, issues of scale, parcel size, zoning, and private mobility.

Perhaps the most significant difference between block structures can be seen when observing the traditional city versus newer and more extended forms of urbanization. The first favors a block structure in which the built-up defines the space of the street. The latter depends more on an open, grid system where transportation is the main element that fosters development.

In recent years, many revivalist projects emerge from the hypothesis that the urban form appreciated by most future users of a sector is comparable to that of the late-nineteenth and early-twentieth century residential city. Consequently, the old city, 'brought up-to-date' by the functional criteria of today provides the pattern of the city of the future.

7 Large Urban Formations: Within the most dominant lines of work, one can detect concern for actions outside the consolidated or metropolitan city. Evidently, new urban tensions are taking place in open territories. Traditional urban models structured around existing nuclei are being superseded by decentralized activities within an open landscape, and new forms of urban composition must be found to engage them.

This line of work formulates interventions based on the dynamic qualities of its territory and the intrinsic logic of its natural environment, resulting in the restructuring of large tracts of land in which human settlement becomes a single element that participates in a much broader ecological system. Rather than imposing pre-established patterns for growth, this mechanism allows for the calibration of the dynamic entities within a given territory in which human settlement is only one element of a much broader ecology. The resultant territorial development is less subject to a prescribed form and participates more in a process of formation, where morphology is never fixed.

8 Retrofits into Historic Fabrics: The main challenge in re-organizing traditional and historic fabrics is that of restructuring their flows and updating their infrastructures in order to guarantee their operative potential as active urban centers. Through precise modes of representation and documentation, we must find the most tactical ways in which certain infrastructures, such as vehicular circulation and provision of basic services, can be updated without altering the city's most delicate tissues. Through such re-workings more efficient ties can be made between city centers and their outer boroughs.

In general, this line of work deals with new ways of relating the historical and the modern center with the surrounding urban reality. This retrofit, therefore, involves questions of access to the center and of the new uses of the facilities and/or services that guarantee their central condition, along with the restriction of traffic, parking schemes, public transport routes, loading and unloading, refuse collection, etc. Such strategies can also involve the rehabilitation for residential purposes of urban typologies that may be obsolete or require planning reinterpretation, being also creative and respectful of the original architectural layout. Other elements that are key to this line of work are the re-working of historic typologies for new uses, the clearing of over-crowded urban fabric to introduce open pockets, and the precise introduction of new types to accommodate programs that have been foreign to the historic core.

9 Analog Compositions: Today, a question of increasing relevance is that of the role of the urban master plan or pilot plan, and its inherent need to transmute from an all-encompassing formal structure for urban growth into an analog system. An analog system structures and monitors through an indexical system the more piecemeal and fine-grain development.

If in the immediate postwar years the master plan was conceived as the initial and primary step in a process that required a series of well-structured and highly prescribed phases, today the master plan should be more the sublime result of a series of smaller-scale projects that in sum compose a comprehensive plan. The true value of a plan is in its ability to acknowledge and to group together interventions of varied objectives and scales. The master plan has the capacity to offer a future 'vision' of the city as a whole. A plan, however, must have recourse to past plans, which must be reconsidered from the point of view of new urban phenomena – of urban projects on the basis of other working scales, particularly the intermediate or specific projects that the master plan can then in turn re-evaluate, integrate or nuance.

Above all, the master plan must seek a 'projectual' direction with contents different from those of the traditional general plan. Today the city is seen as an open phenomenon. Attempts to define its boundaries with elements such as ring roads no longer apply. Zoning-by-uses also seems to be fairly outdated. Subsequently, the new principles of master planning are still up for grabs.

10 Speculative Procedures: The urbanistic project receives a major stimulus from experimental investigation into the application of concepts adapted from other theory-based disciplines. While in the 1960's, structuralism established links between the analytical categories of language and architecture and urbanism; other lines of influence – principally philosophy and thought – have enriched the concepts of urban analysis and the urbanistic project.

Research lines that found many independent followers have used extremely varied references to incorporate the most diverse analogies: of science and technology, of hydraulics, of thermodynamics, of the computer, etc., all paving the way for new ways of understanding a changing urban reality and, most of all, of formulating new planning principles. These analogies lead to experimental lines of investigation that, though they may not transform the 'mass' of the disciplinary corpus, do offer some critical references for more commercial professional practices. This group of initiatives includes the creation of some formal repertories of interpretation and representation of the city that are of great innovative value. Their main field of work is the architecture competition, and they are very frequent and widely accepted in schools of architecture for their critical and experimental capacity.

Approaches to City and Open Territory Design

In trying to bring together a coherent urban strategy in order to cope with the multiple realities of our contemporary environment, the ten strategies or lines of work have been explicated. These ten are merely the most urgent with which to work in and with our complex urban situation. In revisiting the distinct lines of work that are currently shaping our environment, it is important to note the significance of Urbanism as a form of practical knowledge. Urbanism, at large, allows us to understand the built environment in order to intervene and to facilitate distinct forms of urbanization. Throughout the twentieth century, a series of design, planning, and administrative disciplines have hinted toward a form of consolidation that seeks to blur the boundaries between the different design practices. These disciplines might be addressed under distinct names, depending on the context in which they are being used – architecture, landscape architecture, urbanism, landscape urbanism, urban planning, urban design, etc. These design domains still operate today within their loosely defined fields; but ultimately, they all have a common thread that is urban development. They all seek a form of governance within the political dimension of the city.

At this point in time, it is necessary to reflect on the state of Urbanism at large. Only through revisiting it and reshaping its values can we bypass its stagnant state, characterized by closed domains that have proven to be fairly inefficient in engaging the urban problems of our social, political, and cultural engines. New urban practices must be found – practices that envision the integration and resolution of urban questions at large rather than ones that seek just to reproduce the recipe of its own domains.

The most efficient and significant role of Urbanism is that of suggesting possible scenarios and programs rather than prescribing a single and unrealizable fixed rendition of a future reality. Therefore, Urbanism as a discipline must focus on discovering hidden potentials, testing physical possibilities, and delineating paths toward concrete objectives. Subsequently, urban architecture and its field of action must execute the objectives in a manner that is well attuned to the social, political, and economic contexts of the territory in question.

Urbanism must establish the importance and relevance of the project in the city in order to reclaim its well-accepted social presence. Urbanism cannot continue to hide behind laws and guidelines that establish basic forms of order but do not fully exploit the city's operative potential. Formally, urbanism and its forms of project were up to par in the formation of modern cities such as Paris, with the introduction of boulevards that signified a major urban innovation in the nineteenth century. We must now, however, seek innovative urban systems that can engage our newly extended urban and territorial dynamic. We must define the role of flows, increased movement, and new kinds of territorial symbolism in a context where projectual decisions are the contested result of multiple voices – voices which depart from the processes and sequences associated with traditional city-making.

As Aidan Southall has shown us, given the scope and ambition of recent urban transformations, it is important to place such changes and to understand them in a more extensive historical context. Throughout history we clearly can find processes of transition and rupture that occur in a continuous manner, and establish a dialectic relationship between them. Given this, we must not only redefine the role of the traditional city today but also define its role in relation to newly emerging urban forms, such as sprawl and territorial growth.

Cities have proven to be instrumental in establishing forms of exchange among their dwellers since their structure fosters different forms of collective living. Newer forms of urbanity present a less rigid socio-physical structure, where forms of exchange occur more in terms of life-style rather than in terms of specific places of labor. However, new forms of mobility can cause an increasingly nomadic condition which dilutes – to a lesser or greater degree – the urban and collective experiences that for years have served as the backbone of our cities.

The issues presented above bring to the table a series of work initiatives that redefine the way in which the constructed environment is being intervened today. Rather than prescribing specific ideologies about the city, they present an introduction to some of the most effective methods and techniques being currently deployed in a broad variety of contexts. More than being definitive, these compositional lines of work should be animate and generative, since only through understanding and questioning already established lines can we continue to enlarge this growing pool of fine-tuned urban strategies.

[5]

[6]

1 Bercy Front de Park, Paris, France. Jean Pierre Buffi Associates.
2 Bercy Front de Park, Paris, France. Jean Pierre Buffi Associates.
3 View of Bilbao Riverfront, Bilbao, Spain.
4 View of Bilbao Riverfront, Bilbao, Spain.
5 Fresh Kills Competition (winning entry), Staten Island, NYC. Field Operations (James Corner).
6 Caen Unimetal Park, Caen, France. Dominique Perrault.

7 Donau City, Vienna, Austria. Hans Hollein / Dominique Perrault.
8 Donau City, Vienna, Austria. Hans Hollein / Dominique Perrault.
9 Blur Building, Yverdon-les-Bains, Swiss Expo. Diller Scofidio + Renfro.
10 Stepping Stones (competition entry), Bucharest, Romania. Chora (Raoul Bunschoten).

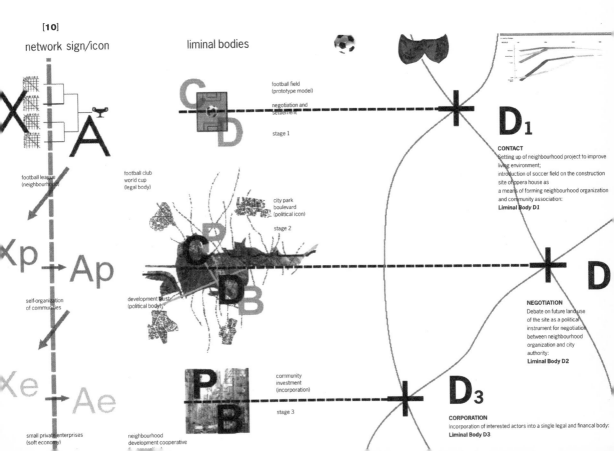

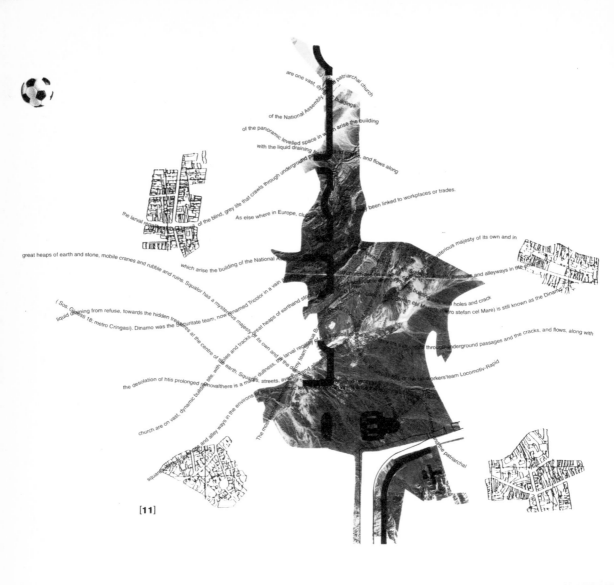

[11]

[12]

[14]

[13]

11 Stepping Stones (competition entry), Bucharest, Romania. Chora (Raoul Bunschoten).
12 Paseo Atlantico, Oporto, Portugal. Manuel de Solà-Morales.
13 Privately Owned Public Spaces, Manhattan, NYC. Various authors.
14 Paseo Atlantico, Oporto, Portugal. Manuel de Solà-Morales.

[15] [16]

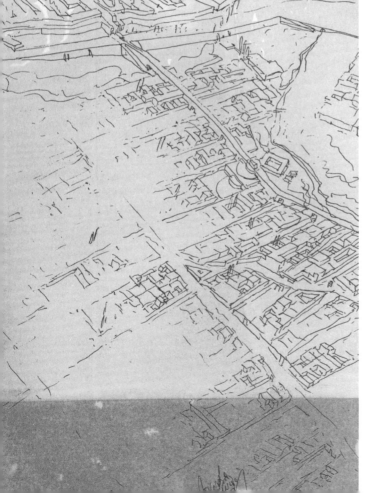

[17]

15 Malagueira Residential Complex, Evora, Portugal. Álvaro Siza.
16 Malagueira Residential Complex, Evora, Portugal. Álvaro Siza.
17 Malagueira Residential Complex, Evora, Portugal. Álvaro Siza.
18 Toledo City Core Restructuring, Toledo, Spain. Bau-Barcelona.
19 Toledo City Core Restructuring, Toledo, Spain. Bau-Barcelona.
20 Genova City Core Restructuring, Genova, Italy. Genova City Hall.

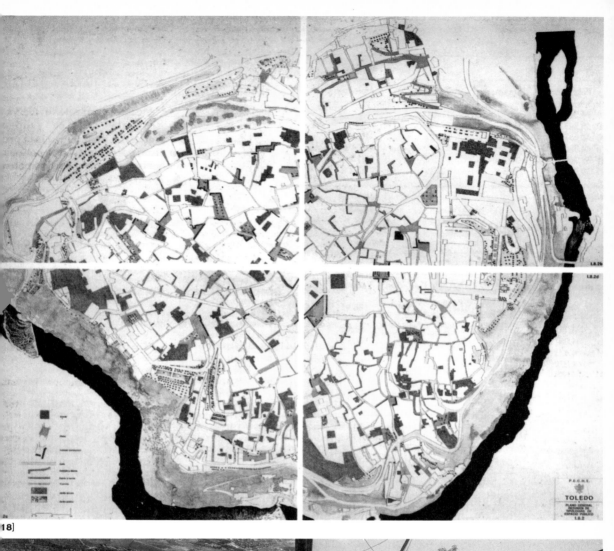

18]

19] [20]

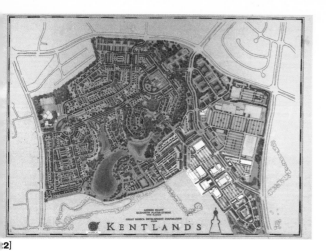

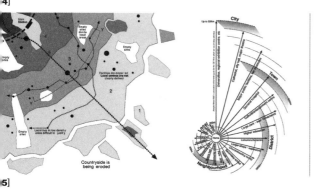

21 Genova City Core Restructuring, Genova, Italy. Genova City Hall.
22 Kentlands Residential Complex, Gaithersburg, Maryland (U.S.A.). Duany Plater-Zyberk.
23 Pujiang Village, Puijang, China. Vittorio Gregotti.
24 Satellite Imagery, City of Los Angeles.
25 Urban Task Force, London, England. Richard Rogers Partnership.

Situations in an Inhabited Landscape

Kees Christiaanse

Art(ificial) Landscape   The man in the photo is farmer Gietema from Lelystad-South, the Dutch version of Broadacre City, designed by KCAP-ASTOC in the Flevopolder, land that was reclaimed in the sixties. His farm was built in the style of a classical manor house. The term 'farm' has taken on a broader meaning over the past few decades and has come to mean a combined place of work and residence in suburban areas with moderate population density, where various types of professional activities are pursued. Similarly, a residential house with an office or a studio can now be referred to as an 'urban farm'. In the gardening centre of the farm, flowers are cultivated in the style of a sumptuous French garden. The gardening centre not only sells gardening products and plants, but the hothouse complex also features a restaurant and a conference centre. After discovering that exclusive flowers have the same distribution points in the US as European vintage cars, he started exporting his flowers by flying them to Atlanta stored in vintage cars, and from there the further distribution takes place. On one of his business trips he met his Taiwanese wife. Their son is half-Frisian and half-Chinese. Gietema's work clothes are custom-designed. In a barn on his land, eels are bred in big water tanks and then smoked. Gietema gave up cattle breeding and traditional farming a long time ago, after taking over the farm from his parents. Due to the milk quota, many Dutch farmers have immigrated to former East Germany, where they can produce more, at lower cost.

As a true Dutchman, Gietema has immediately turned the Dutch government's obsessive environmental protection policies into trade, by planting a forest on his land which earns him a CO_2 subsidy. Besides timber production on a modest scale, the forest can be used for camping, hiking and hunting – for a fee. His land borders on the lake area Oostvaardersplassen, a nature reserve that owes its existence to the hydraulic engineers' failure to drain this part of the reclaimed land. Paradoxically, this area – even though just 50 years old – now is the only little piece of Holland whose landscape still reflects the original appearance of the Rhine and Meuse delta. In this area Gietema now runs a research centre on migratory birds. For some time now, the area has come to develop into one of the most important resting places of migratory birds in Europe. In the nature reserve, Scottish Highland cattle were released that, through cross-breeding with other breeds, have developed into an animal similar to the aurochs, that has been extinct since the Middle Ages. Unfortunately, those prospective aurochs have yellow plastic earmarks, just like normal cows, and just like normal cows, a large number of them were preventatively taken to the slaughterhouses during the BSE crisis. Besides these primeval cows, Przewalski horses have been released over the past few years. This species of horse, that then didn't exist anywhere but in zoos, grew into a herd of six hundred horses in the Oostvaardersplassen area. Three hundred horses have been flown to Ulan Bator, where they were released in the Mongolian steppes. While initially many of the horses died, now a wild herd is gradually growing in a place where they had disappeared 'for good' a hundred years ago.

Gietema has about 60 employees. Besides a few Eastern Europeans from the new EU member countries, about 20 of them are Moroccan young offenders, employed with a subsidy, who serve their sentences by doing educational work as environmental protection workers. They are equipped with electronic tags tied to their feet that register their movements through the GSM network. The facility management on the farm is handled by a new hard- and software system capable of regulating the fully electronic household, from water sensors for the flowers to 380 V three-phase current, all through a single two-wire cable. Gietema's hobbies are gliding, breeding Frisian pedigree cows and doing mechanical work on his father's old John Deere tractor, together with his son. Gietema also owns a farm in Spain that is currently being extended with a successful combination of a fruit plantation and a holiday resort. In the photo we can see him looking at the model together with his father. Through this project Gietema came across urban planning issues. Just in time he bought a large plot of land east of Groningen whose price had come down a lot following the decline in agriculture. To keep the surrounding landscape 'productive' and stem the exodus of economic activity, two large areas of building land were flooded and turned into 'farm' landscapes with lots of water, called Meerstad and Blauwe Stad. Gietema is a shareholder and co-designer of Meerstad. It's paradoxical that the most densely populated country in Europe – that time and again had to reclaim land from the sea – has spare land to flood to meet its people's demand for living by the water.

Almost everything in this – seemingly fictional – story is true. Gietema really exists. He doesn't have a farm in Lelystad, but he is a farmer's son. He is a partner in KCAP-ASTOC and together with his wife Shiuan When, he has already designed three 'new towns' in China, one of which has been built. This story tells us of the unimaginable transformation our world is going through. On the one hand this explains the complete lack of orientation, the banality, and the escape to tradition, in urban planning as well as architecture. But on the other hand it also illustrates the unlimited possibilities of creatively leading a new kind of existence and it gives us a chance to replace our cynicism towards our environment with inspiration.

Landscape as a Carrier '… Since what we call modern is the consequence of incomplete modernisation and must necessarily define itself against a nonmodern residuality that no longer obtains in postmodernity…'[1]

We speak of postmodernism when everything has become modern and modernity can no longer define itself as the dialectical counterpart of what is nonmodern. New currents arise that set themselves apart from the ubiquity of modernity. As early as around 1940, Salvador Dalí writes, after entering an elevator in a New York skyscraper: 'I was surprised by the fact that instead of electricity it was lighted by a large candle. On the wall of the elevator there was a copy of a painting by El Greco hung from heavily ornamented Spanish red velvet strips – the velvet was authentic and probably of the fifteenth century... In Paris, on the other hand, the modern architects à la Le Corbusier were racking their brains to find new and flashy, utterly anti-Parisian materials, so as to imitate the supposed "modern sparkle" of New York...'[2] Clearly, it's hopelessly dogmatic to hold on to Modernism as an expression of modernity. The essence of modernity is the complete freedom of life(and)style, in freely chosen combinations of archaic and modern elements. Postmodernism is accompanied by a trend towards compensating for the lack of 'location' and orientation resulting from the over-supply of information and speed, with an escape to tradition, without however wanting to give up the comfort of modern technology.

Rob Krier's residential quarter in the urban extension Brandenvoort near Helmond is a carefully designed, traditional little village. The houses and the public space were built with real stone-like materials. The details look authentic, but the strips dividing the window frames were glued onto thermopane. Brandenvoort – in itself a project carried out in a conceptually harmonious and consistent way – definitively puts paid to the modernism of Dutch residential style. Liberated from the straitjacket of modernism, project developers and their 'bread architects' have started building weak copies of Brandenvoort all over the place, leaving a trail of banality and populism.
The turning point was the planning of the urban extension Schuytgraaf near Arnhem. What do we, as urban planners, do when we can no longer control the exact programming, the speed of realisation and the quality and style of the architecture? We certainly don't stop!

This fact is exemplary of the current circumstances and should be seen as a challenge for the invention of new strategies. Since we were no longer able to influence what is private, we focussed on strengthening what is public. We are ultra-conservative in the way we deal with the landscape. The spatial divisions – all ditches, rows of trees, dikes and historical 'anchors' – were preserved, to prevent the new parts from forming an autistic enclave. A forceful design of the landscape by West 8 used dense rows and groups of trees to create three-dimensional landscape rooms, behind which the various neighbourhoods thoroughly 'vanish'. Between the settlement areas the old meadows are preserved and will be grazed by animals – there are no parks or artificial greens. For the division of the land within the settlement we used self-developed scatter software, so that the building will be arranged quite randomly, but within the defined constraints. This is not new – in many villa areas from around the turn of the century it's precisely this kind of anarchic arrangement forming a high-quality collection of heterogeneous elements that

---

1 Fredric Jameson (1994), quoted in **Ladders** by Albert Pope (New York: Princeton Architectural Press, 1996) p. 20.
2 The Secret Life of Salvador Dalí, New York, Dial Press 1942, quote by Rem Koolhaas in **Delirious New York** (NY: Oxford University Press, 1978) p. 217.

neutralizes the mediocrity of the various architectural styles. There are also uniform garden-city like settlements in a historical style, from a specific region or simply old-modernistic, where sub-urbanites were able to choose their favourite shelters from the globally interlinked world.

Urban Country Road The B1 is one of the oldest main roads in Europe. It runs from Aachen, via the Ruhr district and right through Dortmund, to Berlin, where it used to be the main link between Potsdam and the central residence. After Berlin, it crosses the Oder River near the town of Küstrin and then runs all across Poland to end in Kaliningrad. Before the existence of rail and motorway, it was literally the vital artery of the area it crossed. Just as in our time urban development is created in 'vista locations' near motorway exits, a linear stretch of economic and representative structures was created in the cities along the B1: the 'urban country road'. The moment of the creation of urbanity is marked by the transition from a mere linking function of the road to an urban-functional complex with the diverse relationships characteristic of cities.[3] In further stages of urbanisation, the urban country road turns into a high street with a central function in the surrounding urban fabric. These developments take place everywhere in the world. One place where they can be observed particularly clearly is the main roads between the Flemish cities in Belgium.[4]

The B1 has repeatedly undergone profound transformations during the past century. In Berlin it crosses Renzo Piano's Potsdamer Platz complex (a peripheral development in the garb of a metropolitan ensemble), and in Dortmund it forms a European version of Venturi's strip typology, consisting of a four-lane road with trees and a tramline in the middle, lined by farms converted into gardening centres or car showrooms and supplemented by residential quarters, hospitals, supermarkets, petrol stations and other urban facilities.[5]

Today, the long-distance transportation function of urban country roads has largely been taken over by motorways, which have profoundly changed the urban space. While the urban country road is a component of an open urban space, multiply interlinked with the surroundings parts of town, the motorway, with its limited number of widely spaced connection points, produces isolated development enclaves, so-called 'ladders'[6], which due to their closed character, are incapable of creating multiple urban relationships, no matter how finely meshed a motorway network is in a polycentric agglomeration and how small the intervals between the connection points are. In urban areas they create a huge mobility problem: as a result of their simple method of providing access – manageable through minor capacity increases – congestion problems arise at the connection points, while once enclave-to-enclave 'short-circuits' are built, congestion arises within the urban fabric.

And also the present-day suburbanisation exhibits more or less closed structures, reinforced by the largely monofunctional character of detached single-family homes which are equally incapable of forming multiple urban relationships. The residents of 'greenfield' enclaves reach their homes from the motorway, via simple access roads carrying separate traffic, widely separated from buildings and obscured by noise protection structures. At the transition point from the access road to the residential area, there will be a 'facility unit', consisting of a minimum concentration of shops, schools and a health-care centre. This fatal constellation, irreversible due to its logic, arouses a yearning for the return of the urban country road, a hybrid of a motorway and a street, whose finely meshed grid, interacting with the surrounding buildings, forms a network of multiple connections with the open landscape: Broadacre City 'revisited'.

In the socio-cultural and economic circumstances that are reflected in the atomisation of the landscape by Suburbia and Edge City, which therefore inevitably elevate the issue of mobility into a paradigm, a search for less hierarchical, more integrated access structures seems sensible.

The access road to the residential area of Langerak, in the new town Leidschenrijn near Utrecht, was deliberately given a slightly curving path. Through its asymmetric position between two string-shaped farming settlements, Oude Rijn and Groene Dijk, a wide zone for residential development was created on one side, while on the other side a 'strip' is formed, in which the existing development can be complemented with a mixture of residential buildings, a school, shops, a housing complex for senior citizens and other facilities forming a heterogeneous skyline of various forms and heights. Some of the houses have glass rooms, similar to shop windows, with 3.2 m high ceilings, whose use as residential space is prohibited thanks to noise protection laws. For that reason, they are officially called 'circulation space' or 'storage space'. Between the houses and the street there is a so-called 'flex strip' that can be used by the residents as a front area, for placing advertising signs or displaying goods.

The residential development on the other side is oriented perpendicularly to the street, and its soundproof front walls go up to the edge of the canal that runs parallel to the street. Through this type of structuring, the conditions for a modest type of urban country road were created. The residents do not drive blindly to their neighbourhoods, but they drive *through* their neighbourhoods. In the strip along the street a heterogeneous zone of communal uses is forming, while the crossroads permit the minimally required linkage to the surrounding area. It is a bizarre phenomenon that the development of sustained urban structures had to be accomplished in such a grotesque way; but the success did not take long to materialise: within a few months after the completion, an animal clinic moved into one of the 'circulation rooms', complete with

3 'Roads no longer merely lead to places, they are places' (J.B. Jackson).
4 De Geyter, Xaveer; **After Sprawl** (Rotterdam: NAI Publishers, 2002) pp. 51, 53.
5 M. Koch, H. Sander, K. Wachten; **Stadtraum B1, Visionen für eine Metropole**, 2002.
6 Pope, Albert; **Ladders** (New York: Princeton Architectural Press, 1996).

advertising and a true surgery theater, where the sub-urbanites – who comprise no less than 80% of the population – can have their pets treated.

Urban Railway Track? A local train or a stopping train will be a supporting factor for an open urban structure in its catchment area. The short distances between the stops ensure a high degree of accessibility and interaction between urban concentrations. While the intercity train has similar effects between urban centres on an interregional level, it lacks any kind of interaction with the area in between these centres. To the contrary, similar to the way motorways cause diverse urban structures to vanish, as described by Albert Pope in 'Ladders'[7], a high-speed rail link condemns those areas in-between to an existence as a 'hinterland'. By itself it wouldn't be so bad to have places where life isn't as hectic, if the intercity train did not, just like the motorway, come with an immense, impenetrable barrier – its railway track, that mutilates the qualities of the surrounding landscape up to a large distance. Firstly, it is questionable – just as in the case of the motorway – whether the huge investment in noise protection could not be used more efficiently, and secondly, the necessity of very wide curves and very small slopes for the train leads to destructive 'prostheses' along the whole railway: hills are razed and enormously heavy concrete bridges are constructed in order to keep the track as level as possible. This is not in line with the concept of the 'bundled infrastructure corridor', which was meant to minimise disruption to the landscape by concentrating motorway, railway and other types of technical infrastructure. In practice, this thought – wise in principle – is more often used as a 'path of least resistance', as it simplifies the decision-making process. Due to the sharper curves and steeper slopes of motorways, the train cannot follow the path of the motorway. In conjunction with topographic and land policy constraints, this leads to an endless chain of left-over areas and no man's land, that can't be used for anything but private gardening plots or car junkyards. These corridors were so wide that it is questionable whether this kind of land use brings sufficient returns to society.

The magnetic levitation train 'Transrapid' combines the advantages of local trains and the regular intercity train ICE. It barely produces any noise, accelerates and brakes extremely fast, and reaches a higher maximum speed than the ICE. It can handle much tighter curves and steeper slopes, so that it could easily be built alongside a motorway. This kind of train could travel from Schiphol Airport to Groningen in about 50 minutes while stopping ten times on the way – the ICE takes more than one hour, with only 3 to 4 stops. With the Transrapid, fewer trains could run at a higher frequency. We have examined two variants of the Schiphol-Groningen rail link: one based on the Transrapid in combination with the existing motorway route and the other based on an ICE railroad track separate from the existing corridor. The Transrapid track can fairly easily be mounted on supports. The girders are slim and elegant, and in terms of their appearance similar to generally accepted structures in the Dutch landscape, like the sluices of the Delta Works or the closure dike of the IJsselmeer. Building the railroad track alongside the existing motorway would eliminate the need to buy land, and the car drivers as well as the train passengers would have a free view. If desired, it would be possible without much cost, to pass the railroad track through the centre of a city. Long-term we can expect to recover the additional cost of the Transrapid through the indirect economic returns of the low demand for space and the higher capacity and number of stops. The latter factor would bring a huge development potential. The reduced travel times as a result of the high speed and the high frequency of stops would bring all cities with Transrapid stops into commuting distance of the economically important Randstad area, which would support the trend towards 'sprawl' (not only of residential locations, but also of economic functions) – a trend that is now hampered by rigid mobility policies – and new development opportunities would arise.

This is one aspect of the general discussion on spatial planning in the Netherlands. This discussion covers a range of options, from concentration in urban areas at one end to a complete dissolution in the landscape at the other. In reality, however, both a concentration in urban areas and a further atomisation across the country is taking place. Through the possibility of reaching many places quickly, the Transrapid can contribute to creating a structure of 'bundled deconcentration' of residential uses, work and recreation.

The ICE railway track, separated from the existing infrastructure corridor, has a different idea at its basis. Without the need to more or less adapt to the path of an existing motorway, the (relative) freedom is created to choose a track that allows to revitalise areas that have come to be inefficiently or poorly used as a result of the crisis in agriculture. The investments into the railway track could then, for example, also contribute to processes of reallocation and consolidation of agricultural land or the creation of nature reserves. This opens up the possibility of a better integration of the railway into the landscape, without a danger of 'wastelands' that might result from bundling tracks. A track that is chosen intelligently in this way, can generate huge additional benefits to society in the long term.

Bathtub Urbanism[8]

The northern part of the Watergraafsmeer district in Amsterdam, originally a polder, is an urban bathtub. It is surrounded by water and dikes, a train station on an embankment and the ring motorway – all elements of closed systems of the present time – and hence the area does not relate to anything but itself. Whether a residential area is located there, an industrial area or a univer-

---

7 ibid.
8 The term 'bathtub urbanism' was invented by Klaus Overmeier from Berlin.

sity campus, barely concerns the rest of the city. This remarkable, shocking observation is actually nothing special. It is characteristic of many other suburban enclaves as well, but in their case it is simply not perceived as acutely, due to their softer edges. On the one hand it seems attractive that apparently the city can consist of replaceable 'patches', but on the other hand their autistic character due to their very concentrated type of land development and their monofunctional nature creates problems: a lack of social control, no day-and-night rhythm (only day or only night), no multiple functional and cultural interconnections that stimulate mobility... a primitive, one-dimensional constellation. In this area, which contains a few complexes of existing buildings, orchards, a farm and wetlands, the new campus of the University of Amsterdam, the Science Park, will be located. The area will be made accessible via three points, through a double cross of axes along the old grid of the polder. More wasn't possible. The area is divided into building zones in an east-west direction, with green bands in between. The building zones follow a special scheme. The buildings in the area are not planted on the lawn like cakes, but are – in a manner of speaking – 'folded' around communal courtyards and form a network with neighbouring buildings. In this way, not the form of the buildings themselves is emphasised, but the web of public and semi-public spaces between the buildings: a labyrinth, an anti-hierarchical network, that is 'infinitely' extensible. A network for slow traffic that winds through the courtyards and atriums like a beaten path through bushes, malleable as 'chewing gum', expanding and shrinking in sync with the day-and-night rhythm of the opening hours. The green areas, besides providing open space, also serve as logistics zones, for example for deliveries or as a 'cable channel', where anything from glass fibre cables and nitrogen pipes to a district heating system can be placed, as needed. In this way, all future laboratory buildings have flexible access to the technical infrastructure.

The bathtub effect in that area is reinforced by the separation of functions imposed by regulations, preventing a mix of residential and work uses. Since the university buildings and research institutes produce certain emissions, housing development is projected in the western part of the area, adjacent to an existing urban residential area. This situation illustrates the dilemma that, while it is now generally considered desirable to create a mix of different functions, this possibility is severely limited by regulations, technical constraints and most of all the structure of the society.

The reality of today is a radical separation of the various components of the city – a spatial expression of its social, economic and cultural fragmentation. Large-scale operations and a separation of functions are inherent features of our society and are fundamentally irreversible. Even a layman will observe that airports, container terminals or industrial zones are indispensable, monofunctional components of our cultural landscape. Their presence is so prominent and far-reaching that they dominate all other developments within their sphere of influence. An airport is hard to reconcile with a finely-meshed fabric of a city. And also the 'positive ghettos' in the suburbs are reluctant to be mixed up with adjacent industrial areas that protect their residences from the noise of the motorway. A mixing of functions can therefore only occur on the level of clusters. A strategic handling of the grain size and the relative positioning is essential to a successful, sustained urban structure in which these clusters are of mutual benefit.

In the Science Park this strategy is expressed by the labyrinthine structure that, analogous to Nolli's map of Rome, puts a coherent web of public and semi-public spaces over these clusters. By placing communal facilities near the connection points, an interaction between different functions is allowed to take place on a communal level. Hopefully the Science Park will be able to use its isolated position in a way that creates an interaction with its surroundings.

Mixed Bathing In Rotterdam, the city and the port authority have jointly set up a 'City Ports' limited company that holds all port areas within the ring motorway. Its aim is to bring about a sustained development of the harbour basin that is in harmony with the city. In the past, port areas whose functions had become obsolete, had been transferred to the city and were used for the development of housing, offices, restaurants and/or entertainment places. From then on, they were no longer called 'port areas', but 'former port areas'. The founding of 'City Ports' meant a significant change of paradigm. The aim was no longer to develop ready-made plans and designs for the port area, but strategies of transformation that take the status quo as their point of departure. The underlying idea is that the existing economic activities, before completely being relegated to second-class uses (car junkyards, recycling facilities etc.) and setting in train a process of financial and social devaluation, are supplemented by permanent and temporary urban activities. It is hoped that allowing these lively, diverse functions to seep in will enable the port areas to maintain a stable value to society and, if possible, turn into attractive features of the city in the long term. This strategy also offers an opportunity to test new forms of mixing of functions or the effects of experimental uses, like, for example, the concept of the 'city farms' (see under 'Art(ificial) landscapes'). The authorities have expressed their willingness, in principle, to apply a lenient interpretation of emission and noise protection regulations, since otherwise it would be impossible for this kind of radical mix of, for example, small-scale residential use and large-scale industrial use to exist.

Waalhaven and Eemhaven also belong to the 'City Ports'. Located between these two biggest harbour basins from pre-war times is the peninsula Heijplaat. Heijplaat consists of a garden city settlement from the thirties, built for people working in the docks, and is surrounded by dockyards – among others the RDM area – container terminals and trans-shipping facilities. A very characteristic part is the 'quarantine area', a wonderfully idyllic place of half-overgrown brick buildings, where an artists' colony has been residing for over 20 years, an 'urban catalyst' avant-la-lettre. There are young adults who were born there in the seventies. Jointly with several Rotterdam organisations, KCAP has taken the initiative to set up a 'Waalhaven Research Group', in cooperation with ETH-Zürich, that will document, interpret and evaluate the existing structures as well as develop various development scenarios and investigate and visualise their consequences. They are putting different, quite extreme models side by side. These models range, for example, from a cautious evolution of the current situation to gradually filling up the harbour basin and building a new suburban part of town on the reclaimed land. Furthermore, they examine what critical mass of new apartments and facilities would be necessary to preserve Heijplaat as a viable garden city and attractive residential enclave. They also investigate the options for further consolidating port buildings worth preserving with cultural events or centres for starting businesses.

The difference between this area and other port areas closer to the city lies in its size and its distance from the urban area. This makes it impossible to control the development. Also, these areas are not in crisis. Given a total hands-off policy, these areas would still successfully attract all sorts of businesses, but this would lead to hectares and hectares of monocultures without urban qualities, which is no longer acceptable in a modern agglomeration with substantial structural problems.

The situation described above shows that the profession of the urban scientist is no longer a cut-and-dried skill, but a discipline that must be almost completely defined and invented by the person exercising it and is nearly as complex as the 'intermediate landscapes' they are active in. The most important part may be that they can no longer simply sit in their office and wait for the assignments to come, but have to identify, define and conquer projects by their own initiative.

1 Lelystad-south: Urban landscape, Warande.
2 Schuytgraaf Arnhem: Villages in green area.

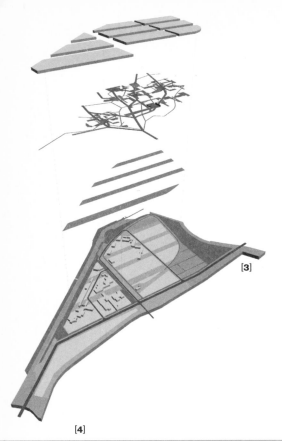

3 Heijplaat: Simultaneous chess.
4 SciencePark, Amsterdam.
5 Langerak, Leidsche Rijn.
6 Zuiderzeeline: bundle of HSL-north with A6.

[3]

[4]

[5]

[6]

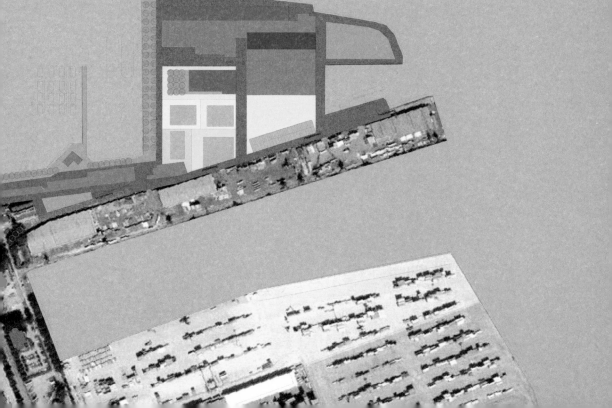

The Polder as a Design Atlas
Clemens Steenbergen and Wouter Reh

Introduction

Design research as carried out in the Department of Landscape Architecture at Delft University of Technology puts the emphasis on three research areas, each closely related to contemporary design issues. The first area is concerned with revealing the design tools and typologies of villa architecture in the widest sense of the word, that is to say both Architecture and Landscape; the second is the urban landscape; and the third is the landscape of the Dutch lowlands. An important aspect of Research by Design is the checking and updating of the range of tools used in landscape architecture, and in doing so, providing a clearer picture of the specificity, legitimacy and utility of those tools and of the development and dynamics of architectural landscape design. This updating makes it possible to establish and to make known clearer limits to the contribution made by the discipline and what can be expected of it in the light of contemporary design tasks.

This article is concerned with the theory and technology of the transformation – urban or otherwise – of Holland's polder landscape generally, and the reclaimed lake landscape in particular.[1] It goes without saying that this is one of the major tasks in the urbanisation of the west Netherlands. The question that we have posed is whether there is such a thing as a neutral 'grammar of landscape architecture', capable of mediating between the program and the form of the country. In this sense, has Holland's reclaimed lake landscape any significance for contemporary design tasks? A necessary first step towards making visible the instruments used in landscape architecture is the analysis of reclaimed land as a designed landscape, describing its form more accurately and revealing its formal grammar.[2]

A major role in the urban transformation of reclaimed land is played by what is known as *polder grammar*, the compositional system used in the design of the man-made landscape in different reclaimed lakes, combining the most rational division possible with the most effective implementation possible, while at the same time leaving room for the site's natural features.

We have examined the significance of this *polder grammar* for later architectural treatment or transformation of the landscape with the help of three examples. Watergraafsmeer involved a reconstruction of the way in which urban form was determined by the form of the polder. Purmer, on the other hand, involved finding a way to make consistent use of the polder grammar to design a prototypical model of a polder town. Haarlemmermeer involved reconciling polder grammar with the stratification of the contemporary urban landscape.

Polder grammar helps the nature of the thought process involved in the architectural treatment

---

[1] Here and in what follows 'Holland' refers to the area occupied by the present provinces of North and South Holland rather than the Netherlands as a whole.
[2] cf. W. Reh, C.M. Steenbergen, with D. Aten, **Zee van Land. De droogmakerij als atlas van de Hollandse landschapsarchitectuur** (Amsterdam: 2004).

or transformation of the landscape to become transparent, avoiding any contingencies as much as possible. Urban landscape, then, is given an objective basic form, creating varied programmatic and situational processes.

The paradox of objective form The question of the spatial qualities of the Dutch polder landscape dominated the urban design and planning debate in the first half of the last century. With the layout of the Zuider Zee polders, for instance, beauty and 'experiential value' were considered to be 'added value' for the landscape. Van Eesteren however, in his inaugural lecture as professor at the Faculty of Architecture in 1948, wondered whether this beauty was not the direct consequence of a commodious layout: 'When an arrangement satisfies the requirements of suitability, should one not expect that the landscape will automatically become beautiful?'[3]

He did however appreciate more than anyone else, that suitability is not achieved automatically, and that in a deeper sense the timeless beauty of the polder landscape and the monumental quality of human use, transcending physical construction, are in fact the product of the architectural control of the task: 'The science of urban design teaches us functions and prescribed aims, while the art of urban design gives us the ability to group things spatially and give them form.'[4]

The question is, however, how to precisely organise and give form to an urban design problem. The essential elements in a land development scheme are not the architectural additions or accents, but rather the art of leaving things out, of achieving an organisation of the elements which make up the technical scheme that is pure enough to make it both useful and beautiful. According to Van Eesteren, beauty comes into being when the designer allows the functional program and the distinctive natural and geographical features to speak for themselves. He saw this as related to the attempt to achieve spatial quality without demarcation, an unbounded, 'cosmic' spatial coherence whereby the breadth of the space extends everywhere.[5]

Subsequently, what can the reclaimed lake landscape teach us in this context? It is interesting to consider how that landscape achieved its delicate balance of form, skirting the boundaries of formal objectivity. At what point does leaving things out transcend certain invisible but important limits, causing the landscape to lose its animation and changing serenity into vulgarity and nihilism, a formal technocratic diagram, no longer expressing any search for beauty?

1 Landscape architecture – research methods

The following table divides architectural design research into different areas of research according to the variability of object and context. The following differences emerge:

Design research analyses a fixed object in a fixed context

Research by design varies either the object or the context

Typological design research considers a fixed object in varied contexts

Identifying study investigates a fixed context on the basis of varied objects

Designing study varies both object and context.⁶

		OBJECT	
		Fixed	Varied
CONTEXT	Fixed	Design Research	Identifying study
	Varied	Typology	Designing Study

For research in landscape architecture, the following distinctions apply: In *design research*, existing examples are selected and treated as architectural landscape compositions for the purposes of analysis. In *research by design*, the researcher looks for relationships between interesting compositional properties of the example analysed and a new task or situation, with the aim of investigating the possibilities offered by those properties. This method should not be confused with a theory of composition based on the application of design rules.

In contrast, the procedure could be compared with *heuristics*, a discipline which when applied methodically leads to new discoveries and inventions. Thus the difference between the creative application of compositional ability and the application of rules laid down by a particular theory of composition can be reduced to the old antithesis between *heuristics*, 'the theory of discovery', and *hermeneutics*, 'the theory of exegesis'.⁷

In what follows, the research into the way in which the town plan was accommodated into the polder Watergraafsmeer should be considered a piece of pure *design research*, wherein both the program and the context are known. The design experiments for Purmer and Haarlemmermeer should be seen as *research by design*, since in these cases both the town plan and the landscape were transformed stage by stage. The aim is to achieve a critical development of the internal logic of the experimental composition of the polder town. From a scientific point of view, the results constitute a *design hypothesis* that can be both discussed and tested.⁸

---

3 C. van Eesteren, **De conceptie van onze hedendaagse nederzettingen en cultuurlandschappen, hun verschijningsvormen en uitdrukkingen** (Inaugural lecture, Delft 1948).
4 ibid.
5 ibid.
6 **Cf.** T. M. de Jong, D. J. M. van der Voordt (eds); **Ways to study and research. Urban, architectural and technical design** (Delft: 2002). See also T. M. de Jong, **De wetenschappelijke rol van Bouwkunde** (Lecture, Delft University of Technology 10–19–2001).
7 C.M. Steenbergen, H. Mihl, W. Reh; 'Design Research, Research by Design', in C.M. Steenbergen, H. Mihl, W. Reh, F. Aerts (eds); **Architectural Design and Composition** (Bussum: 2002) pp. 12–25.
8 Clemens Steenbergen, Wouter Reh; 'The composition of new landscapes', in Clemens Steenbergen, Henk Mihl, Wouter Reh, Ferry Aerts (eds); **Architectural Design and Composition** (Bussum: 2002) pp.192–207.

Composition analysis   How should existing examples be analysed to ensure that a new design can make creative use of any insights gained? Despite their typological similarities, landscape architecture compositions are in many respects unique and resistant to facile reduction. If compositional analysis is to be productive, the analysis must be carried out in a specific way, beginning totally independently of the original motives for or aims of the design. The search is for instruments that can be constitutive as well as generative for the development of new design proposals.

A landscape architectonic composition can be analysed as a conceptual scheme or a diagram of the physical elements active in the spatial operation of the design (the *active compositional elements*) together with a number of formal principles regulating the relationship between the elements that make up the composition, elements such as volumes, rhythm, axiality, zoning, colour and texture.

The term composition is always linked to a particular example and refers to the specific way in which the elements are combined in the design. We call this reduction of the design example the *compositional scheme*. Thus the 'internal logic' of a composition shows up in a compositional scheme as a consistent system of 'internal rules' or 'laws', as an individualised architectural system. In this sense, each example has its own compositional scheme.[9] A compositional scheme can not be directly linked to a new design task, but will undoubtedly contain hidden design tools and experimental skills which will be usable, provided they are properly isolated from their historical concept, as an *invention*, retaining their magic, their originality.

Polder grammar   In the first half of the 17th century, a new scientific and technological approach to land reclamation made possible the draining of the large lakes which made up the Holland lagoon. The development and ingenious linking together of new types of windmill, and the creation of the catchment basin system determined the overall character of Holland's newly reclaimed lake landscape. The Beemster, with its unsurpassed technical and architectural mastery of the polder plane, has become a classic example.[10] The landscape architecture of reclaimed lakes has specific characteristics and a specific compositional scheme. Each polder field embodies a completely individual polder architecture, and indeed could be said to have its own *polder grammar*, consisting of a clear subdivision in an ideal geometrical polder division, surrounded by a more irregular 'polder mantle' and a 'polder fringe', the adjustments necessary to the natural shape of the edge.

The *polder plane*, the largest geometrically regular area in the polder, comprises a number of modules. The smallest module is the *model farm plot*. Farm plots were combined to form *polder blocks* with shared drainage and access arrangements. Groups of two or four polder blocks forming a square or rectangle make up *polder modules*, the units in turn making up the polder plane.

Even the spatial image of the reclaimed lake landscape was permeated by rational characteristics. The farmhouse, with its ingenious building system, was in keeping with the physical properties of the polder landscape, the arrangement of the property was in keeping with the layout of the reclaimed land and useful crops were grown along the polder drainage ditches. Land reclamation often went hand-in-hand with the layout of a *landscape of country estates*, to mark a piece of urban territory. Thus what was developed as a man-made landscape was transformed into a Holland Arcadia, an idealised mirror image of urban life.

2 Design research – the form of Holland's reclaimed lake landscape

Plan analysis of Watergraafsmeer The natural shape and orientation of the lake, with its 'head' on the Nieuwe Diep and a 'tail' on the Amstel, was determined by the prevailing south-west wind. This shape was translated into the egg-shaped polygon of the ring dike. The vast Watergraafsmeer basin was divided into sections of four by two straight avenues, making the dimensions of the reclaimed land decipherable and reflecting its position with respect to Amsterdam. The intersection formed the centre of the reclaimed land. Middenweg, the most important thoroughfare, was lined up with Amsterdam's Zuiderkerk. Around it extended a rectangle of water canals, running on both sides to series of linked windmills. This staggered bayonet shape formed the polder's diagonal, so that the egg-shape of the polder was incorporated into the geometry of the plot division.

The central rectangle, the largest area in the polder capable of regular plot division, formed a polder plane area consisting of four smaller polder blocks, each containing four agricultural plots. Between the *polder plane* and the Braak by the Nieuwe Diep, a *polder mantle* was delineated, at right angles to the plot division. Around the edge of the plot, a division was made that co-ordinated the depth between the edge of the polder and its surface area.

The orientation relative to Amsterdam led to the zoning of land, a consequence of which was the pattern of urbanisation of the reclaimed land. In the north-west corner of the polder, near the present site of Amstel Station, were vegetable gardens. The same area, running into the polder from the ring canal, also contained small garden houses and recreational gardens. Larger country houses and pleasure gardens were located round the intersection between Middenweg

---

9 C.M. Steenbergen, W. Reh; **Architectuur en Landschap. Het ontwerpexperiment van de klassieke tuinen en landschappen** (Bussum: 2003).
10 W. Reh, C.M. Steenbergen, assisted by D. Aten; **Zee van Land. Op. cit.**

and Kruisweg, which was also the site of an open green mall and a country inn. The farms, on the other hand, were spread round the entire area of the polder.

Many smaller independent garden areas developed in the farmyards, kitchen gardens, playgrounds and country estates along the drainage ditches and the ring canal. Together these garden areas divided up the large space occupied by the polder into smaller landscape elements, increasing in scale as they progressed from the Amsterdam side to the Nieuwe Diep, creating a series of spaces. The large green mall in the centre of the reclaimed area was a direct translation of the plot depth into the dimensions of a public space. These characteristics gave the polder a park-like appearance.

Watergraafsmeer, in contrast, had a more intimate character than the fertile, large-scale, agricultural Beemster. Its good accessibility from Amsterdam led to the creation of a landscape of public urban gardens, analogous to the 16th- and 17th-century pleasure gardens outside London and Paris. These urban recreation facilities were fully integrated into the polder layout. In this sense Watergraafsmeer was a forerunner of the urban park, Holland's first polder park.

Plan Analysis of Purmer The Purmer Lake had the shape of a narrow ellipse. The reclaimed land was divided according to a simple system, with plot divisions extending all the way from a centreline to the edge. The structure was not modular, but rather an assembly of standard plots grouped into blocks. The *polder plane* could be said to be reducible to the 'mill run' [*Middeltocht*], the geometric division down the centre that was formed by the elliptical-shaped lake.

The polder roads were situated about halfway along these long plots, mid-way between the line of the Middeltocht and the gently sloping edges. The area between the polder roads is to be seen as a *polder mantle*, the remainder as a *polder fringe*, at which the border treatment varies from place to place. The Nes, at the southernmost point of the polder, is the most striking deviation in the polder geometry.

The polder consisted of three long areas running lengthways. Polder roads with rows of farmhouses bordered the central area, about 1.5 km wide by 10 km long. The adjoining areas were bordered by a row of farmhouses on one side, and the winding edge of the polder on the other. Between the Nes and the ring canal at the southernmost point of the reclaimed land, a smaller scale area was developed. Because of the transparency of the rows of farmhouses, the entire polder area could be seen in its entirety from various points along the roads, a view that normally could only be obtained lengthways from the ring dike in clear weather. The special relationship between the view across the polder and the continuity of the space along the polder is typical of this style of plot division, yet provides a very special spatial quality.

In conclusion, the Purmer system was worked out in less detail than the rational system used for Beemster. In Purmer, efficiency was sought not in a pure geometry with repeatable components, but rather the reverse – minimal adjustment of the geometry to the dimensions of the natural shape of the ground, maintaining the plot depth to the full extent permitted by the irregular edge of the polder. In Purmer, a rational polder layout was reduced to a few well-chosen lines, dividing the whole polder as economically as possible, and providing a more precise adjustment of the basic geometry to the natural conditions.

Plan Analysis of Haarlemmermeer　The ring canal closely follows the natural shape of the lake as it originally was. A co-ordinate system divides the area with the Hoofdvaart as the central axis, and the Kruisvaart as a dominant axis at the two-thirds line. The Hoofdvaart, which provides the axis of symmetry, is formally independent; the two roads running alongside it give it a special profile. The orientation of the grid was partly chosen to improve the protection of the land against flooding, and was accommodated as efficiently as possible into the natural shape of the lake, a funnel-shape fanning out from the south-west to the north-east.

The agricultural division of Haarlemmermeer was based on a 2 × 3 km *polder module*, and accessed from the centre. This module was bordered on the short sides by a polder road and a small lane, and on the long sides by a drainage ditch. Consequently, the rectangular polder plane could be comprised from twelve polder modules with the Hoofdvaart as the central axis of symmetry. Around the polder plane lay several more complete polder blocks, together forming a *polder mantle*. Where the grid irregularly joins the Ringvaart, the natural form of the land reflects in this infill the remnants of the original peat bog landscape.

The geometry of the polder mainly comes to life in the main lines of the grid, the views over the waterways, the rows of farmhouses, and the side roads. A special connection is made at the point where the polder roads come out by the three pumping stations on the dike. The Stelling van Amsterdam, forts which still exist at the far ends of the flood protection quay and along the Hoofdvaart, were not located at special points in the geometry of the polder, but rather serve to anchor the Westeinder lake and the Spaarne to the polder edge.

The motorways and the railway embankments are included in the grid, except where they enter and exit the polder. Each runs along the border of a polder block, which puts them at two-kilometre intervals. The two motorways run along the inner edge of the dunes to the south, in the fork between the A4 and the A44, and come together in the peat bog landscape between the sandbars. The runways of Schiphol airport, however, have an autonomous alignment system.

The spatial structure of the Haarlemmermeer suggests a fusion of elements from three earlier reclamation projects: the parallel rows of farmhouses from Zijpe, the uniform grid from Beemster and the spinal column of mutually perpendicular waterways from Schermer. The rows of farmhouses in Haarlemmermeer are twice as far apart as those in Zijpe, creating little if any spatial interference.

The rows of farmhouses are planted with a single row of trees on either side of the road, which is bordered by a water ditch. The average size of a farm was originally forty hectares, i.e. two plots, which meant that the farmhouses were sometimes grouped two by two, or four by four. Wherever the side roads were planted, they formed 2 × 3 km polder sections. Through the course of time, the width of the roads broadened such that their relationship with the space occupied by the polder rooms has changed.

The three Neo-Gothic steam-powered pumping stations, emblems of technology, were monumental both in terms of design and location. The Lijnden pumping station on the north side of the polder and De Leeghwater by the Kagerplassen, were situated on the axis formed by the Hoofdvaart. The Cruquius pumping station was a feature of the Kruisvaart. The new villages of Hoofddorp and Nieuw-Vennep were carefully planned in advance. The situating of the two places at two important intersections of the polder system led to a street plan that accommodated the polder plot division. The design of the public space that abutted the polder edge, however, was not given any special attention, either in regard to landscape architecture or urban design.

Few people appreciated the significance of the polder in the development of the region. Connections with the surrounding country still remained at the level of a 17th-century reclamation project. Similarly, no one but the pioneer J.P. Amersfoordt appreciated the significance of the polder as an urban expansion area, an area on which to project urban prosperity and cultural aspirations. Instead the polder was seen as a sort of colony, an opportunity for land speculation, needing no particular attention from either town planners or landscape architects.

The result was a meagre landscape, almost completely bare, totally lacking in Arcadian inspiration except for a few private initiatives hardly worth mentioning. The technological body was sparsely clad. Quantitatively, planting remained at the level of a 17th-century reclamation project, although the polder plane was much greater and the spatial task structurally changed. This reclamation project was an example of the polder scheme in its most neutral and flexible form. Even today the polder remains a landscape of colonisation.

In a deeper sense, the colonisation of Haarlemmermeer is still incomplete. The coarseness and spatial indeterminacy of the agricultural grid combined with the location and extent of the reclaimed land are in some sense advantageous as an urban matrix. The continuing process of urban transformation has yet to reveal the latent formal quality of this man-made agricultural prairie.

3 Design Research – the Landscape Architecture of Holland's Polder City

Watergraafsmeer – Fitting a Town into the Polder Layout   An important subject for research is the way in which the agricultural polder pattern was 'translated' into an urban street plan during the urban transformation of Watergraafsmeer. We must therefore make a distinction between the way in which this urbanization was foreseen in the 1939 Watergraafsmeer Extension Plan (as part of the General Extension Plan [the AUP]) and the way in which it was actually put into practice after the Second World War, because of the interesting nuances which can be observed.[11]

In the 1939 Extension Plan, the surface of the agricultural polder was transformed into an *urban polder plane*, with a slightly amended boundary. Within this area, Watergraafsmeer gained three new garden villages – Amsteldorp, Frankendael and Middenmeer. An *urban polder mantle* was marked out between the built-up area, which had by then extended across the border of the polder, and the polder plane itself, roughly defined by the intersection of the access road over the Berlage bridge (a continuation of the plot division of the peat bog on the far side of the Amstel) with the polder grid. The Amstel station was planned to go in a corner left open by the Omval.

In its basic structure, the AUP was simpler than the plan eventually put into practice. The AUP still showed the return of the original polder plane, but with staggered adjustments to the drainage ditches on either side. The layout of new city roads had the effect of distorting the original polder plane. In addition, the articulation of the *polder mantle*, the edge where the polder grid met the natural landscape, disappeared.

Furthermore, a boundary made to the west by the Gooiseweg was not completed according to the polder geometry. The A10 interstate road follows the southern edge of the polder. The Amstel station and the A10 make use of the difference in height between ground level in the peat

11 C.M. Steenbergen, W. Reh, E. Dekker; **De Watergraafsmeer. De stedelijke transformatie van een landbouwpolder** (Internal report, Delft 2002).

bog landscape along the Amstel and ground level in the reclaimed land. The Amstel station stands at dike level, like a patrician's house on the Amstel with a view at one time over the entire polder landscape. In the original design by Schelling in 1939, more details were given. However, the overall effect was lost when the area in front of the station was also built-up. The A10 is also up at dike level, but has no visual connections with the lower level of the reclaimed land.

The new urban plot divisions reflect the plot division of the polder. If the original plot division is projected on the urban plot division it becomes obvious that part of the division into plots was directly translated into the urban street plan, and indeed it probably served as the basis for that plan. Another part was erased.

About half of the Watergraafsmeer was zoned as public green space, with sport parks the size of complete *polder blocks* and garden allotments along the edges, subsequently creating a neutral pattern of regular urban units located next to one another. The Frankendael country estate remained. The *polder edge* was divided into different functions, including the present Amsterdam Science Park. All this served to functionally anchor the reclaimed land as a whole in the urban network.

Purmer – The Geometrical Processing of a Polder Layout   In the Purmer, the urban expansion was carried out as part of the northward enlargement of the Randstad that was provided for in the second policy document on town and country planning prepared by the Dutch government (1965). The western section of the polder was filled in between 1980 and 1990. The built area spilled over the edge of the polder, as it were, yet was contained by a strip of wooded polder plots. The layout of the urban plot division was potentially formal, making the relationship with the centre of Purmerend quite weak and the spatial quality of the polder landscape almost non-existent.

The experimental design could be said to have been a 'rehab' of the original design process, but this time with a consistent creative application of the polder grammar of the agrarian polder systematic, using the program of existing urban expansion as a starting point. This design experiment was carried out in a number of steps.[12]

The first step was to define an *urban polder plane* as opposed to an agrarian polder plane. In Purmer this area could not be derived directly from the geometry of the reclaimed land but had to be created by a geometrical correction to the *polder mantle* and to the edge west of the Middeltocht. The area was then comprised from a number of urban polder blocks along the Ilpendammer road where the row of buildings in the village forms a diagonal across the space left free at the front side of the plots. If a reflection is made of this pattern along the Middeltocht, an image of the urbanization of the land reclamation area as a whole occurs with the row along the Oosterweg in the mirror position.

This correction made it possible to allocate an *urban polder mantle* and *polder edge*, whose spatial function was still to be determined. The demarcation of the *polder edge*, which included not only the ring canal but also a few groups of linked windmills and the reclaimed piece of peat bog landscape near Ilpendam, established the original natural shape of the reclaimed land.

The third step was the allocation of the urban program and the distinction of special landscape characteristics – those parts of the *polder plane*, the *polder mantle*, and *polder edge*. This distinction crystallised the spatial structure of the Purmer polder town on the plot level.

The entire *urban polder edge* was transformed into a lake to be used for storing water (from the east side of the 'proto-urban' agricultural area to the area of water reeds). In doing so, a certain distance could be maintained from the centre of Purmerend and the extension into the peat bog. The *urban polder mantle* was laid out into the woodland, just as was the small piece of reclaimed peat bog landscape by Ilpendam, whereby an ecologically worthwhile connection was established between water, woods and reedfields.

The existing farmhouses along the Ilpendammer road were retained. Housing, commercial and industrial buildings, and parking were spread over the urban polder plane in such a way as to create as many whole *urban polder blocks* as possible. Each of the remaining plots was allocated a special function. The division of the polder block into *urban plots* was carried out on the basis of existing plot division or combinations of two or more agricultural plots. Similarly the ditches round the plots could be upgraded to form component parts of the urban water system.

Entry and exit roads, and side drainage ditches were included in strips of reedland, or located on the borders between different areas. A number of agricultural plots along the thoroughfare leading to the centre of Purmerend were left free as a reminder of the earlier agricultural use of the land, thus retaining an immediate visual connection with the agricultural part of the polder. The horizon was left unobstructed, ensuring complete transparency. The application of polder grammar led to the creation of a straightforward urban street plan, in which the landscape qualities of the reclaimed land were established on a continuing basis.

12 C.M. Steenbergen, W. Reh, E. Dekker; **De Purmer als polderstad. Onderzoek naar de ontstaansgeschiedenis en de groei van Purmerend en een verstedelijkingsalternatief voor de droogmakerij** (Internal report, Delft 2003).

Haarlemmermeer – Stratified Processing of the Polder Layout Haarlemmermeer, sometimes referred to unkindly as the 'rubbish dump of the Randstad', is a prime example of a Randstad urban transformation area. What it lacked was a convincing image of the combination of town and country that constitutes the metropolitan polder city. The aim of the design experiment was to develop this image systematically. The starting point was provided by the current interpretation of the urban design of Haarlemmermeer, the presence of Schiphol, and plans for future infrastructure. Here too the design experiment was guided by a consistent, creative application of polder grammar, as established in the structure of the agricultural polder, taking into consideration both the surrounding landscape and the natural landscape as it existed before reclamation. The aim was to discover whether the stratification and spatial magnitude of the metropolitan polder city could be encapsulated in a version of the polder grammar.[13]

The complexity of the design task becomes obvious when the already realised buildings as well as the current and future infrastructure are projected upon the divisions of the agricultural polder. In contrast to the village in the Purmer, the *stratification* and complexity of the metropolitan polder city is already indicated by this image. The oldest layer is that of the vanished natural landscape and the old reclamation works. The man-made landscape of reclaimed land forms a second layer. The regional town, in which urban functions are dispersed in an infrastructure network, provides a third layer. The task here can be understood to be a confrontation between three different *formal systems*: the organic/natural system, the technical system of the man-made landscape, and the functional system of the regional urban network. Apart from the spatial zoning, the montage and the way the three 'qualities' of the polder interacted with one another served as a means of structuring the design.

Within the polder plane there still exists no more than one complete polder block, which gives that block a special status as the only area in which the land development system used for the reclaimed land is immediately visible. The landscape of the block is marked as an architectural void, a window in time, capturing the polder's agricultural past.

The zoning of the agricultural polder provided a basis for the first zoning of the street plan of the polder city. The boundaries of the agricultural *polder plane* were exceeded here and there, but then could be delineated as an *urban polder plane*, an area for increased density. The area as a whole could be seen as a sort of urban plantation. The urban plot division could in principle have been a modular subdivision of the 200 × 1000 m agricultural plot, for example, or a 100 × 200 m building block.

The *polder mantle* was particularised into a woodland scenery with country estates, capable of accepting special urban functions. Their characteristics were introduced at the edges of the spatial tension between the urban polder plane and the polder fringe.

The variation in the ground level at the edge was so large as to encourage the dynamics of the natural landscape to again play a role, whether or not along the lines set by the agricultural plot division. The dominant theme chosen here was the creation of marshland, which could develop into lakes, bogs or reedland, particularly along the western edge where seepage from the dunes found its way to the surface. The fragments of peat bog landscape contained as part of this operation took on the role of a landscape reserve, establishing a connection with the former landscape outside the reclaimed area.

The presence of the airport meant that it was not possible for the entire urban polder plane to be built up. Accordingly it was zoned thematically into residential areas, industry to the south of Hoofddorp, and a polder park around Schiphol's runways. The drainage system was converted into a new water purification system for the polder town, providing the different stages required to clean polluted water.

The projection of these layers on top of one another gives a picture of the landscape architecture dimension of the metropolitan polder city and positions the design with respect to the *time series* of landscape development.

The polder town can now be developed spatially as an open *lowland series* within a 'landscape envelope'. The situation of the polder town between Amsterdam, the coastal landscape, and the grassed area of peat bog landscape determine the spatial catalogue of this metropolitan polder landscape.

4 A New Urban Polder Typology

Significance and Testing of Design Experiments As in any technical science, the ultimate test of an experimental model is found in its implementation. However, an important difference is that an experiment in one of the natural sciences can be tested empirically because it can be considered a repeatable closed (or context free) system. In the case of landscape architecture design research, however, it is less a matter of testing the end product in use, but more a question of checking compositions and spatial models against an example or a type that has already proven its utility, and checking the rules and conditions used in the transformation against anticipated new functions and programs. Possible areas to be checked include the economical use of resources, social admissibility, and administrative 'momentum'.

13 C.M. Steenbergen, W. Reh, E. Dekker; **Proefontwerp van de Haarlemmermeer als polderstad** (Internal report, Delft 2003).

The function of an experimental design, then, is to suggest a proposal or potential solution that can lead to a model. It follows that when checking design experiments, the internal logic of the conceptual experiment is more important in the first instance than any attempt to reconcile it with practical requirements. In this respect, experimental design differs fundamentally from practical design but comes close to empirical research. This process assumes a continuous series of experimental compositions until ultimately a new and balanced composition is found.

From the analysis of the Watergraafsmeer garden city, the Purmer, and Haarlemmermeer design experiments, a number of design instruments were uncovered, which could well assist in guiding the development of a new typology of urban polders.

In the first place, it seems that the way a piece of reclaimed land is divided can be directly translated into an urban street layout. The essential tools are provided by the polder grammar. The spatial quality of the polder town turns out to depend on a consistently creative application of polder grammar, taking account of the essential characteristics of the landscape architecture of the reclaimed land. In a careful transformation, an elementary way can be found to keep these landscape architecture characteristics permanently anchored in the urban plan. The form of the polder town emerges from the form of the land.

From the Haarlemmermeer design experiment, polder grammar can apparently also be applied to complex issues of urban design. A fully urbanised polder landscape provides the setting for a violent collision between the structure of the network city and the spatial characteristics of the original landscape. In such a case, polder grammar can serve to play off the *genius loci* and the technical formal logic of the agricultural polder in a stratified structure against the predominantly functional formal logic of the city. The nature of the experimental design shifts from end image to process image.

Landscape architecture seems to possess three keys to the visual coordination of the formal system of the metropolitan polder city. The first is the articulation of the traffic network into the landscape architecture, sometimes referred to as the *landscape of flows*. The second is the zoning and rational division of the urban patterns of utilisation to follow the pattern of a *polder plantation*. The third is the

14 W. Reh; 'De derde ontginning', in W. Reh, D.H. Frieling, C. Weeber et al.; **Delta Darlings** (Delft 2003) pp. 10–39.
15 S. Marot; 'Voorwoord', in C.M. Steenbergen, W. Reh; **Architectuur en Landschap. Het ontwerpexperiment van de klassieke tuinen en landschappen** (Bussum 2003) pp. 9–13.

montage and setting of the natural and visual qualities of the polder landscape and the wider surroundings into the metropolitan topography, a so-called *landscape theatre*.

The Tenet of the Polder   Holland's man-made landscape has always had a rational and 'proto-urban' character. Conversely, the urbanisation of the Netherlands should be seen as a continuous transformation of the landscape, in fact a succession of reclamations. Holland's reclaimed lake plays a leading role in the iconic image of the lowlands. This paradigmatic status is due not only in a technical sense in that the lowlands set their own requirements for hydraulic engineering and water management, but also as a compositional principle. In the creation of the landscape of Holland's reclaimed lake, a struggle for priority is always under way between nature, technology and art. The end result is a balanced landscape which does justice to all the essential characteristics of Holland's lowlands.[14]

The simple, decipherable polder layout, with its elementary use of lines and areas, is so neutral and so strong that it is capable of accepting a wide range of different programs and compositions while still continuing to make explicit reference to the powerful images of the Dutch landscape. The relationship between plot, polder module, polder plane and polder edge provides a basis for countless new developments in landscape architecture, urban design and architecture.

The *polder grammar* and polder typologies also have an operational significance. Taken together, Holland's areas of reclaimed land provide a design atlas which, as Sébastien Marot put it, contains 'a rational catalogue of situations', establishing and securing the *genius loci* of the lowlands.[15] This design atlas contains possible starting points for landscape architecture in an urban transformation. However, the key question is still how to preserve and to renew the unity of the polder landscape as an architectural construction.

In the reclaimed lake landscape, theatrical form is completely dismantled. The formal elements are distributed over the landscape in a homogeneous pattern of identical, reiterating models. In reclaimed land, the classic scheme of landscape architecture composition is reduced to a neutral topographical framework, an 'objective' form with its own monumentality. In the reinvention and development of this form lies the challenge which constantly enables Dutch landscape architecture to confirm its virtuosity in contemporary design issues.

Bibliography

C. van Eesteren, **De conceptie van onze hedendaagse nederzettingen en cultuurlandschappen, hun verschijningsvormen en uitdrukkingen** (Inaugural lecture, Delft 1948).

J. Heeling, V.H. Meyer, J. Westrik, **Het ontwerp van de stadsplattegrond** (Amsterdam 2002).

T. M. de Jong, **De wetenschappelijke rol van Bouwkunde** Lecture, Delft University of Technology 19-10-2001.

T. M. de Jong, D. J. M. van der Voordt (eds.), **Ways to study and research. Urban, architectural and technical design** (Delft 2002).

S. Marot, 'Voorwoord', in C.M. Steenbergen, W. Reh, **Architectuur en Landschap. Het ontwerpexperiment van de klassieke tuinen en landschappen** (Bussum 2003) pp. 9–13.

M. Reints, **Nacht- en dagwerk** (essays) (Amsterdam 1998).

W. Reh, C.M. Steenbergen, P.H. de Zeeuw, **Landschapstransformaties. Stedelijke transformaties van het Nederlandse landschap** (Internal publication, Delft 1995).

W. Reh, C.M. Steenbergen, 'Ontwerpend onderzoek en landschapsvorming. Methode, kritiek en perspectief'. In: **Ontwerpen aan de zandgronden van Noord- en Midden Limburg. Drie over Dertig** (Maastricht 2003).

W. Reh, C.M. Steenbergen, assisted by D. Aten, **Zee van Land. De droogmakerij als atlas van de Hollandse landschapsarchitectuur** (Amsterdam 2005).

W. Reh, D.H. Frieling, C. Weeber et al., **Delta Darlings** (Delft 2003).

C.M. Steenbergen, H. Mihl, W. Reh, 'Design Research, Research by Design', in: C.M. Steenbergen, H. Mihl, W. Reh, F. Aerts (eds.), **Architectural Design and Composition** (Bussum 2002) pp. 12–25.

C.M. Steenbergen, W. Reh, 'The composition of new landscapes', in C.M. Steenbergen, H. Mihl, W. Reh, F. Aerts (eds.), **Architectural Design and Composition** (Bussum 2002) pp. 192–207.

C.M. Steenbergen, W. Reh, **Architectuur en Landschap. Het ontwerpexperiment van de klassieke tuinen en landschappen** (Bussum 2003). Published in English under the title **Architecture and Landscape. The Design Experiment of the Great European Gardens and Landscapes** (Basle, Berlin, Boston 2003).

C.M. Steenbergen, W. Reh, E. Dekker, **De Watergraafsmeer. De stedelijke transformatie van een landbouwpolder** (Internal report, Delft 2002).

C.M. Steenbergen, W. Reh, E. Dekker, **De Purmer als polderstad. Onderzoek naar de ontstaansgeschiedenis en de groei van Purmerend en een verstedelijkingsalternatief voor de droogmakerij** (Internal report, Delft 2003).

C.M. Steenbergen, W. Reh, E. Dekker, **Proefontwerp van de Haarlemmermeer als polderstad** (Internal report, Delft 2003).

E. Terlouw, 'Een Eigen Huis', part I, 'De markt, de stad en suburbia', part 2, 'De mogelijkheden van de kavel', **OASE** 52 (1999).

P.H. de Zeeuw, C.M. Steenbergen, E. de Jong, 'De Beemster, een arena van natuur, kunst en techniek', in T. Lauwen (ed.) et al., **Nederland als kunstwerk. Vijf eeuwen bouwen door ingenieurs** (Rotterdam 1995) pp. 153-167.

Spatial model of Purmer

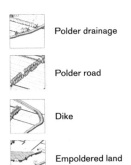

- Polder drainage
- Polder road
- Dike
- Empoldered land

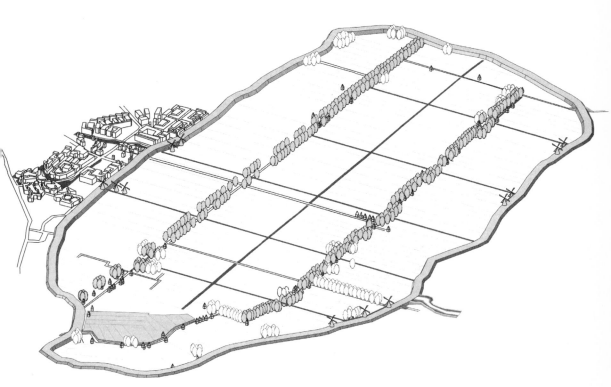

Urban polder grammar

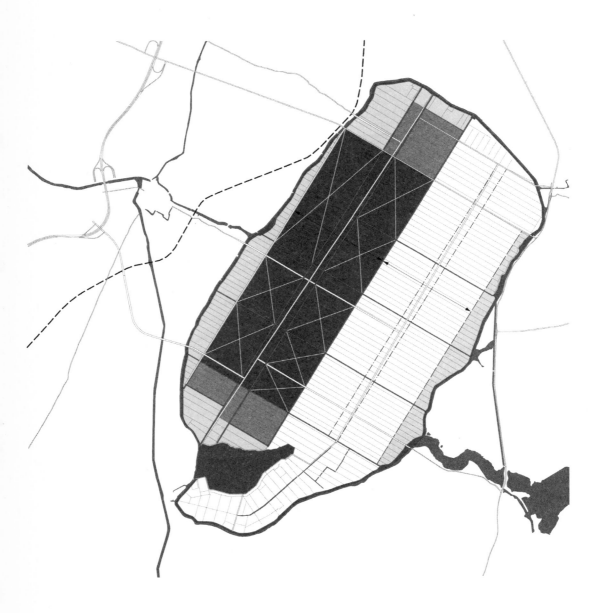

Spatial model of Haarlemmermeer

- Pumping station
- Empoldered land
- Dike
- Military dike
- Ribbon of farms
- Canal

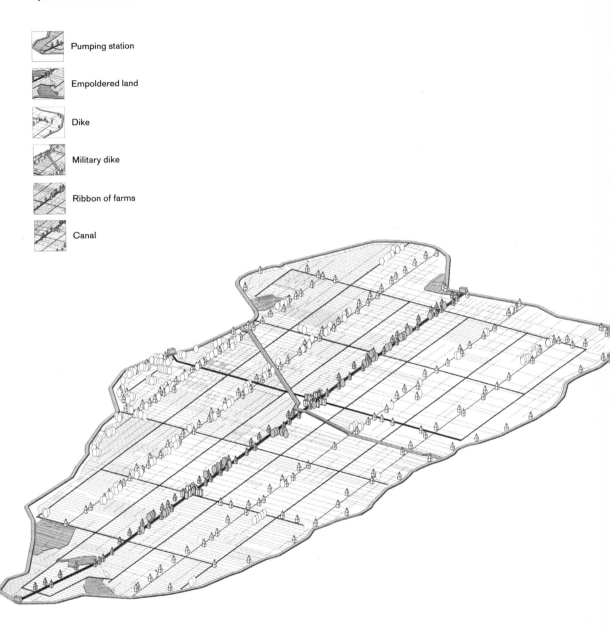

Urban polder grammar

The metropolitan polder town as a formal problem

Field Building and Urban Morphology: A Challenge for the Next Generation of Urban Designers

Anne Vernez Moudon

Introduction

The role of research in design has long been a contentious subject opposing those who see design as a creative art and those who see it as social practice (Boswell, Duany et al. 2004). Experience has shown, however, that the dichotomization of design as art versus science, and of creativity versus social responsibility leads to conflicts and prevents from thinking positively about the future of the design professions and the role of research in their evolution.

In this article, I argue that urban design, specifically, can and should contribute significantly to current and future societies. The unprecedented growth in the urban population worldwide makes it obvious that city design and building are of paramount importance to our future. City design and building demand large monetary and non-monetary investments, and involve a large set of tasks. So far, however, urban design has contributed to these efforts primarily on a project by project basis. It could and should contribute more broadly to the policy and regulatory frameworks that shape individual projects and urban development in general. Many policy makers and regulators are now conscious that their actions affect urban form and development. They are seeking help from urban designers, not only for specific projects, but also for data on and analyses of how urban form affects people's behavior. Urban transportation and environmental planning are prime examples of sectors that seek to work closely with urban form makers and developers. To work with these sectors, however, urban designers need to develop new ways of communicating their knowledge.

Second, I argue that the urban morphological approach provides a useful conceptual and theoretical framework for researching the design characteristics of urban form. Urban morphology focuses on the spatial and material attributes of the built landscape and thus uses a language that is familiar to designers. At the same time, its strong and tested theoretical framework speaks to social scientists. Urban morphological approaches can therefore integrate, and perhaps even transcend, the disciplines of design and the social sciences. They may be a vehicle to build a field of research in urban design.

Finally, I show that new spatial databases and GIS software offer exciting opportunities to research in the built landscape, and to improve our understanding of how this landscape affects behavior and quality of life. Such an understanding will help guide the creation and the transformation of future life-enhancing environments.

Urban Design Then and Now

Urban design developed as a complement to the professions of architecture and urban planning in the 1960's (Kahn and Crawford 2002) when the world had 2.5 billion people. Modernist urban design and development principles primed at the time, proudly stating that 'Design' (or 'design') was to serve the 'masses' and to provide them with a living environment worthy of the technological and scientific revolutions (Newman 1980). Modernists spelled out the three basic 'functions' that living environments had to fulfill: *residing, working, and recreating* (Newman 1980; Mumford 2000). This introduced a strikingly new set of tasks for professions that had previously catered to the very wealthy. It greatly expanded their clientele from previously small governing elites to increasingly large bureaucratic societies. In retrospect, the position of creating environments to 'reside, work and recreate in' might have been not altruistic, but coveting the immense and untapped potential of a very fast growing population. On the other hand, however, the Modernists displayed a communistic streak that, although it has been forgotten (Newman 1980), commanded them to address the needs of *all* people, and specifically those who had not enjoyed the benefits of professional services before.

Modernist design principles did contribute to changing environments since the mid 20th century. This was a time when European cities were in need of reconstruction and American cities began their irreversible suburban growth. New towns and large housing projects restructured the European landscape. In the US, downtowns and office environments sprouted in most cities, ostensibly following a diluted version of the Modernist recipes.

The ensuing decades saw startling growth in population, which led to the building of more cities than ever built in human history. Fifty years after the emergence of urban design, the world population grew almost three-fold to 6.4 billion people, half of whom live in cities (Abu-Lughod 1991; Verlag 1992). Eighty percent of the world population is expected to be urban in the near future. It follows that modernist approaches to urban design are spreading into non-Western cultures: directly so, as individual designers from Europe and the US export their ideas and build more and bigger structures than they have ever built in their lifetime; and indirectly so, as bureaucracies worldwide emulate the regulatory systems of Western planning and development.

How much contemporary urban development worldwide derives from Modernist principles, and how much Modernist principles embody broader capitalistic streaks is subject for debate (Castells 1983; Relph 1987). The fact remains that, to date, Modernist urban design principles have hardly been revisited. For example, theories of urban design continue to refer to and contrast the Garden City and the Modernist movements (Dunham-Jones 1997; Dunham-Jones 2004). Postmodern approaches did make some attempt to inform design at the scale of buildings, but they by and large remain silent with respect to urban form. Further, post-modernism is, almost by definition, 'formless', focused on design process and intentions behind the form making process (Chase, Crawford et al. 1999; Ellin 1999). If it projects forms, the latter are in dialogue with and, more often than not, in opposition to Modernism (Koolhaas 1998; Koolhaas, Boeri et al. 2000).

Worth mentioning are the primarily us-based debates about 'new urbanism' and 'smart growth' (Environmental Protection Agency and International City Management Association 2001; Congress for New Urbanism 2004). However valid or useful these theories prove to be, they do not easily apply to urban development elsewhere. us cities and metropolitan areas have spread at densities that are a fraction of those in other parts of the world (Los Angeles and New York Metropolitan areas have densities that are 10 to 30 times lower than those of Hong Kong or São Paolo), and they have relied on rates of automobile ownership that are extensively greater than those of other countries (Newman 1999). Attempts to broaden these movements and address issues of urban sustainability have yet to proceed effectively (CEU 2004).

The dearth of design principles addressing or even adapted to the numerous contemporary new towns springing up in all parts of the developing world, points to the curious inability of the urban design profession to catch the opportunities of a milestone moment in human history. If there ever was a time to revisit *De Re Aedificatoria* (Alberti 1988) or the *Leyes de las Indias* (Orejon 1961), it should be now.

Worth noting is that while *urban* design remains silent in the wake of the largest production of new cities in history, the *design* of buildings and objects has gained significant momentum. Societies of the developed world show tremendous interest in the value of environmental and industrial design. Many historic buildings and districts are being cleaned up and indeed gentrified. Everyday objects are also subjected to design 'refinement', and 'basic' housing projects of the early and mid 20th century systematically are undergoing redesign in Europe and in North America.

From Practice to Research: New Clients for Shaping Urban Form

The combination of the impact of tremendous growth in population on the production of built environments, and of the developing taste for designed 'things' is an opportunity to direct urban design to a new set of design 'clients', who make policies and regulate how people live, work and recreate in urban environments. In contrast to traditional clients (land owners, developers, and communities), who seek 'projects', or designed 'objects', the 'new' clients want to understand how urban form affects people's behaviors, and conversely, how people's behaviors can affect urban form. They are people in education, transportation, health, public safety, community development, and so on, who see how their policies are linked to the way environments are shaped (Neubauer 2004). For them, the design of environments has an increasingly broad societal value as an integrative approach to community building, where improvements can be made not only in their own specific sectors, but also to the overall quality of life.

The 'new' clients seek, for lack of a better word, a field of *research* in urban design, while traditional clients shaped and catered to the professional *practice* of urban design. Policy makers and regulators want critical 'descriptions' of extant environments, and 'analysis' of simulated future environments. Analyses and simulations need to be systematically carried out, focusing on distinct and definable aspects of environments (building height, sidewalk width, and so on), in order to facilitate the evaluation of alternative strategies to modify environments and to select effective interventions.

Public or quasi-public agencies and institutions house the majority of the new research-oriented design clients. Such agencies have shaped the development of many professions and fields of research over the past century – engineering, medicine, and education, to cite a few. They already ventured into design and building, addressing, for example, the role of urban design in promoting public safety and deterring crime (Newman 1996), as well as the effects of land use and environment on travel behavior and mode choice (Dunphy 1997; Ewing and Cervero 2001). Lately, public health professionals are revisiting the potential relationships between built environment, physical activity, and access to healthy foods (Frumkin, Frank et al. 2004; Moudon 2005).

The Need For Field Building

Any field in the process of development must look for ways to build and to disseminate knowledge. Like many other professions, urban design developed as a form of practice, which has been at the center of early student training and the focus of its journals. Urban design has also had its share of advocacy, articulating theories of 'good' urban design as well as trying them out in practice. To fully flourish, however, a field typically builds a research component that focuses on the development of knowledge and on testing the value of this knowledge in practice.

A Research Component to Urban Design

Knowledge building can be thought of as a three-part cycle (adapted from Lang, Rittel, Schön, and others; Kunz, Rittel et al. 1977; Schon 1983; Lang 1987) (Figure 1). The cycle is a closed loop, linking practice, advocacy, and research.

It includes three types of knowledge: substantive, applied, and normative. The cycle works as follows: substantive knowledge is built in research; it is interpreted and transformed into normative knowledge in advocacy; normative knowledge is then applied in practice; applied knowledge is thereby tested and returned to the reassessment and development of substantive knowledge. There is no hierarchy in the cycle, and participants can enter the process in any of these three interrelated arenas of knowledge. Typically, research is done in academia, and practice in the 'real world', while advocacy is shared by both professionals and academics. However, individual and institutional boundaries between the three elements can be blurred (this has taken place in many fields, such as medicine and the health sciences), and participants can be involved in all three parts of the cycle. This knowledge building cycle is integrative, and involves many different ways of thinking and acting.

Urban form as the socio-physical product of urban design

The practice of urban design leads a process that culminates with the creation of urban forms. As the direct product of practice, urban forms embody urban design's professional territory, and in turn define the territory of its discipline. To create urban form, however, urban designers do not operate in a lone vacuum. Their designs reflect more of what 'can' be done, rather than what 'should' be done (Barnett 1995), because they synthesize social, economic, and cultural forces at work at the time of creation. Therefore, the forms created through the urban design process have both a physical/spatial dimension and a social one. Indeed, the spatial dimension of urban form can only be understood through social processes at work, making urban form a socio-physical product. In this sense, urban design operates in a multi-disciplinary context, which relates directly to the social sciences.

Furthermore, urban forms have a significant temporal dimension. Once in place, they are quickly used, and become a social tool serving those who inhabit or own them. Sociologists claim that urban forms are experienced and interpreted, and indeed 'practiced' by people (Lefebvre 1984 [1971]), and economists cast people as both consumers and producers of neighborhoods (Galster 2001).

Research in urban design needs to consider the continuum of, first, creating urban form (an initial phase that has been the focus of urban design professional practice); then, the use of urban forms, including the various types of transformations made to them (add-ons, demolition, and rebuilding). It also needs to link up to a range of social-science-based disciplines. Table 1 depicts a simple epistemological map for urban design (Moudon 1992).

Theory and Method: The Urban Morphological Approach

Urban morphology, a field dedicated to the study of urban form, can logically provide a theoretical and methodological basis in creating a research arm for urban design. The morphological approach seeks to represent, measure, and analyze basic elements of urban form and the processes of city building. Because the approach focuses on built space – buildings, open spaces, and so on – it uses the spatially constructed language of the designer, yet adapts this language to the epistemological tradition in the social sciences (Conzen 1978; Moudon 2000).

Urban morphology provides a clear theoretical framework to classify the spatial attributes of the built environment in two and three dimensions, and to understand the relationships between these attributes. The approach encompasses several levels at which the spatial structure of the built landscape can be understood. These levels are commonly called 'scale', as in the architectural and the building scales, the site scale, the neighborhood scale, the city scale, and so on. Also, and importantly, it includes a temporal dimension acknowledging the dynamics of these spatial systems (Moudon 2002).

Three classes contain the basic elements of urban form: (1) the land subdivision, (2) the built spaces, and (3) the uses attributed to land and buildings (M.R.G. Conzen's three elements of the town plan) (Whitehand 1981). First, the smallest element is the individual lot (parcel or plot), which, as the principal spatial unit of ownership/use, defines the limits of territories for initial building or development and subsequent transformations of built space. The lot has both a spatial and a temporal aspect, and is therefore an important basic element of urban form. The second element consists of the building(s) and its (their) attendant open spaces, which define space in the third dimension. Buildings typically sit on individual lots, each of which is controlled by the social entity that owns or uses the lot. Lot and building constitute the smallest level of two- and three-dimensional space attribution in the city. They are the basic unit of urban space (Moudon 1986; Cadastral Template 2003). The third element is the street-block, which groups lots and buildings into the next level of spatial attribution.

The street-block establishes the spatial boundaries of a group of lots. In most cases, these boundaries define the collective space of streets. Street-blocks separate public from private space, as well as settlement from circulation space. Lot/buildings and block/streets define space by ownership or use. They are the spatial territories of small social units such as families, and small groups of families. The territories they define are physically recognizable (at the perceptual and experiential levels): lots and buildings may be differentiated one from the other by open space, fences, doors, and windows of different types. Similarly, streets and blocks are also readily distinguishable, streets being used by many different types of people, while lots and their buildings retain a singular identity. Lots and their buildings can often be perceived as modules of space along streets.

The lot is the key element of urban space with three important attributes: it is the smallest unit of (1) socially and (2) temporally defined space at which urban form is created and transformed; it acts as (3) the link between streets and blocks – each lot has at least one side or part of one side facing and connected to a street, while its other sides are shared with other lots in the block.

The next level of spatial aggregation of urban form groups street-blocks. Criteria used to define this level typically highlight the homogeneous characteristics of basic urban form elements, such as similar lot sizes (in the case of a subdivision); consistent architectural character of the buildings (their style or color, reflecting one design and perhaps one designer/developer), and complementary uses on the land – residential, mixed, and so on. Conzen defines this next level as a *morphological unit* – a discrete set of street-blocks, lots, and buildings, usually developed at one point in time, that share homogeneous spatial and architectural characteristics. Italian morphologists call this level the *tessuto*, also referring to the architectural homogeneity of the buildings (Moudon 1994).

Up to the morphological units or *tessuti*, the spatial aggregation of lots into street-blocks is based on physical and spatial criteria. Beyond this level, however, urban space is aggregated based on socially defined criteria, referring typically to such jurisdictional or legal boundaries as cities, counties, metropolitan regions, and so on. These spatial aggregations tend to be very heterogeneous in terms of their urban form characteristics.

Of note, the level of the neighborhood commonly used by planners is socially rather than physically defined. Neighborhoods may have consistent urban form characteristics (age of most buildings), and may be bound by natural or man-made physical features (hills and rivers, major roads, and so on). By and large, however, the singularity of a neighborhood comes from its dominant social and economic status.

Urban form elements and levels of spatial aggregation serve to structure inquiry and research. They provide a conceptual framework similar to that used in other spatially grounded fields such as geography and landscape ecology (Turner and Garner 1991; Hartshorn 1998). The morphological approach also provides strong theoretical guidance for analyzing urban form dynamics, by highlighting the lot or the parcel as the unit of 'urban form at its most elemental level'; and as a 'consistent' element to '[classify] spatial and temporal relations' (Batty 1999).

A strong conceptual and theoretical framework from which urban form can be classified and analyzed is essential to structure the dialogue with allied disciplines (sociology, economics, psychology, transportation, real estate development, public administration, environmental sciences, and so on) and to consolidate the inter- or trans-disciplinary basis of the field of urban design (Figure 2).

Data: New Databases in GIS

Field building and research depend on good data. Most cities now have data in geographic information systems (GIS), which use the parcel as the smallest polygon to which different attributes are assigned (Chrisman 2002; Cadastral Template 2003). Data available for the Seattle Puget Sound region, for example, include more than one million parcels covering a territory of approximately 1,000 square miles. Originally built for tax collection purposes, the databases attach to each parcel such attributes as land use, number of buildings, parcel size, zoning, and assessed property value. Today, these databases also serve to monitor land supply and capacity (Moudon and Hubner 2000).

The potential of parcel-level databases in advancing morphological research is virtually limitless. Whereas past research relied on manually drawn maps and extensive and exhaustive field work, which necessarily limited the extent of areas and cities being considered, it is now possible to examine entire metropolitan areas for which the data are available. Also, the large territories for which geo-spatial data are now available correspond to the urbanized areas that have replaced the small cities and towns that existed until the 19th and early 20th centuries. They therefore permit the application of morphological approaches to contemporary urban settings and their corresponding life styles.

Research Applications

Morphological approaches to urban form using GIS databases come in three basic types: studies in representation, measurement, and analysis.

- **Representation of urban forms (Figures 3, 4)** GIS produces maps that serve the same visual purpose as hand-drawn maps. The advantages of GIS are the speed with which the maps can be produced and their accuracy. Furthermore, the extent and resolution of the maps can be readily and precisely selected, and they can be checked visually. This allows for experimentation in adjusting the level of map detail with the extent of the area being depicted, hence suggesting that innovative ways to map urban form will ensue as more research is performed. The GIS capability of zooming in and out of an area extent puts the researcher in a position to compare almost instantly conditions in many specific spots of a large region. It also begins to link what has been termed the architectural scale with the planning scale – researchers can read maps of the same areas at different extents.

- **Measurement studies (Figure 5; Table 2)** GIS databases greatly facilitate the measurement of elements of urban form. For the first time, it is feasible to give values to the myriad of pieces of urban form, from block sizes, to street length, building coverage, and so on. Net and gross measures can be precisely defined based on the area considered in the denominator. To date, the sizes of urban form elements have been treated in a fairly cavalier fashion by the design and planning professions, resulting in many difficulties and errors. In addressing the now contentious subject of residential density, for example, planners and community groups typically use different reference areas and therefore different measures of density to argue their case.

- **Analyses (Figure 6)** This is the most powerful function that GIS data brings to design and planning. The quantitative dimension of GIS allows researchers to use mathematics and statistics to complement the traditional visual language of design. This new dimension has scared many designers away, leading some to think that parcel GIS has nothing to do with urban form… Yet each new generation of students brings a steadfastly increasing number of young professionals who understand the power of parcel GIS and acquire the skills necessary to use them. Students in urban design and planning not only quickly learn the software, but also dive into basic statistics and modeling. (In the graduate class I teach on urban form, the number of students using ESRI ArcView programs has increased from some 20% in 2000 to 80% in 2004, with the current students organizing a special course for 2005). Data in GIS can be subjected to hundreds of operations, which, applied to urban form, can include the following:
- *Inventories* of all types of forms and uses, from vacant land, street trees, sidewalks, historic structures, and various building types.
- *Spatial clustering* of all types of forms and uses, to investigate how forms or uses agglomerate in space – apartments, neighborhood retail, and so on. Spatial clustering helps identify special environments such as walkable environments, food environments, sports environments, and so on.
- *Neighborhood analyses*, to simulate possible influences of land uses and forms on proximate areas, or, conversely, to define 'catchment areas' for specific activities and related facilities (public transit systems, retail shops, community centers, schools, parks, and so on).
- *Proximity analyses*, or studies of distance between activities and facilities, which can be measured as airline ('as the crow flies') and network (using formal circulation routes).
- *Sampling*, which, when done randomly, allows researchers to reduce the number of observations and thereby facilitates analyses. It is possible to random sample any type of facility and carry out the research on the sample rather than on the entire inventory of facilities (Lee, Moudon et al. 2005).
- *Modeling estimations* can probe and predict the interactive effects of urban form and all types

of behaviors from driving, walking, shopping, and so on (Moudon, Lee et al. 2006). Of particular interest are comparative analyses of subjective measures of urban form (how people perceive the forms) with objective measures (actual measures of urban form), which help designers calibrate the forms they create to how people will sense or interpret them.

Conclusions

Population growth and migration to urban environments over the next few decades can help propel urban design into an essential social task. Entire metropolitan areas are emerging that will define ways of life for future generations. Making these areas supportive of human and other life now and later is critical. And when the population stops growing, economic development will, it is hoped, bring further demand for improving environments for people. In order to serve people, and to provide places for *residing, working and recreating* as the Moderns foresaw, urban design needs to expand as a field of both practice and research. The urban morphological approach offers a simple yet rich conceptual framework, and a set of theories and methods to structure research about livable urban space. New detailed data in powerful Geographic Information Systems now greatly facilitate the tasks ahead, including inventorying existing urban forms, monitoring their use and functionality, assessing ways of changing or transforming them, modeling and simulating future environments.

References

Abu-Lughod, J. L. (1991). **Changing Cities: Urban Sociology**. New York, HarperCollins Publishers.

Alberti, L. B., 1404-1472 (1988). **De Re Aedificatoria. On the Art of Building in Ten Books.** Cambridge, MA, MIT Press.

Barnett, J. (1995). **The Fractured Metropolis, Improving the New City, Restoring the Old City, Reshaping the Region**. New York, Icon Editions.

Batty, M. (1999). 'A Research Program for Urban Morphology.' **Environment and Planning B: Planning and Design**: 475-476.

Boswell, S. E., A. M. Duany, P. J. Hetzel, S. W. Hurtt and D. A. Thadani (2004). **Windsor Forum on Design Education**, Miami, FL, New Urban Press.

Cadastral Template (2003). A Worldwide Comparison of Cadastral Systems. Cadastral Country Reports Based on a Jointly Developed Pcgiap/Fig Template.

Established under UN mandate by Resolution 4 of the 16th unrcc-ap in Okinawa, Japan in July 2003. http://www.cadastraltemplate.org/. January 1, 2004.

Castells, M. (1983). **The City and the Grassroots: A Cross-Cultural Theory of Urban Social Movements**. Berkeley, CA, University of California Press.

CEU (2004). Council for European Urbanism. http://www.ceunet.de/newurbanism.htm.

Chase, J., M. Crawford and J. Kaliski, Eds. (1999). **Everyday Urbanism**. New York, Monacelli Press.

Chrisman, N. R. (2002). **Exploring Geographic Information Systems**. New York, Wiley & Sons.

Congress for New Urbanism (2004). Land Development Regulations. http://www.cnu.org/pdf/code_catalog_8-1-01.pdf. November 1, 2004.

Conzen, M. P. (1978). 'Analytical Approaches to the Urban Landscape.' **Dimensions of Human Geography**. K. W. Butzer. Chicago, IL, The University of Chicago Department of Geography. Research Paper 186, Chapter 8: 128-165.

Dunham-Jones, E. (1997). 'Real Radicalism: Duany and Koolhaas.' **Harvard Design Magazine** Winter/Spring.

Dunham-Jones, E. (2004). A Modernist Education. **Windsor Forum on Design Education**. S. E. Boswell, A. M. Duany, P. J. Hetzel, S. W. Hurtt and D. A. Thadani. Miami, FL, New Urban Press: 79-90.

Dunphy, R. (1997). **Moving Beyond Gridlock: Traffic and Development**. Washington D.C., Urban Land Institute.

Ellin, N. (1999). **Postmodern Urbanism**. New York, Princeton Architectural Press.

Environmental Protection Agency and International City Management Association (2001). Getting to Smart Growth: 100 Policies for Implementation.

Ewing, R. R. and R. Cervero (2001). 'Travel and the Built Environment – a Synthesis.' **Transportation Research Record**(1780): 87-114.

Frumkin, H., L. Frank and R. J. Jackson (2004). **Urban Sprawl and Public Health: Designing, Planning, and Building for Healthy Communities**. Washington, DC, Island Press.

Galster, G. C. (2001). 'On the Nature of Neighbourhood.' **Urban Studies** 38(12): 2111-2124.

Hartshorn, T. A. (1998). **Interpreting the City: An Urban Geography**. New Work, Wiley & Sons.

Kahn, A. and M. Crawford, Eds. (2002). **Urban Design: Practices, Pedagogies, Premises**. New York and Cambridge, MA, Urban Design Program, Columbia University: Van Alen Institute, Projects in Public Architecture; Harvard Graduate School of Design, Harvard University.

Koolhaas, R. (1998). **Oma Rem Koolhaas Living, Vivre, Leben**. Bordeaux, France, Basle, Switzerland, Boston, MA, Arc en reve centre d'architecture, Birkhauser Verlag.

Koolhaas, R., S. Boeri, S. Kwinter, N. Tazi and H. U. Obrist (2000). **Mutations, Harvard Project on the City, Multiplicity**, Barcelona, Spain and Bordeaux, France, ACTAR; Arc en reve centre d'architecture.

Kunz, W., H. W. J. Rittel and W. Schwuchow (1977). **Methods of Analysis and Evaluation of Information Needs: A Critical Review**. Munich, Verlag Dokumentation.

Lang, J. T. (1987). **Creating Architectural Theory: The Role of the Behavioral Sciences in Environmental Design**. New York, Van Nostrand Reinhold Co.

Lee, C., A. V. Moudon and J.-Y. Courbois (2005). 'Built Environment and Behavior: Spatial Sampling Using Parcel Data.' **Annals of Epidemiology** Jul 5;[Epub ahead of print].

Lefebvre, H. (1984 [1971]). **Everyday Life in the Modern World**. New Brunswick, N.J., Transaction Books.

Moudon, A. and M. Hubner, Eds. (2000). **Monitoring Land Supply with Geographic Information Systems, Theory, Practice, and Parcel-Based Approaches**. New York, John Wiley & Sons, Inc.

Moudon, A. V. (1986). **Built for Change: Neighborhood Architecture in San Francisco**. Cambridge, MA, MIT Press.

Moudon, A. V. (1992). 'A Catholic Approach to Organizing What Urban Designers Should Know.' **Journal of Planning Literature** 6(4): 331-349.

Moudon, A. V. (1994). 'Getting to Know the Built Landscape: Typomorphology.' **Ordering Space: Type in Architecture and Design**. K. A. Franck and L. H. Schneekloth. New York, Van Nostrand Reinhold: 289-311.

Moudon, A. V. (2000). 'Proof of Goodness: A Substantive Basis for the New Urbanism.' **Places** 13(2): 38-43.

Moudon, A. V. (2002). 'Thinking About Micro and Macro Urban Morphology.' **Urban Morphology** 6(1): 37-39.

Moudon, A. V. (2005). 'Active Living Research and the Urban Design, Planning, and Transportation Disciplines.' **American Journal of Preventive Medicine** 28(Supplement 2): 214-215.

Moudon, A. V., C. Lee, A. Cheadle, C. Collier, D. Johnson, T. L. Schmid, R. Weathers and L. Lin (2006). 'Operational Definitions of Walkable Neighborhood: Theoretical and Empirical Insights.' **Journal of Physical Activity and Health** Supplemental Issue in press.

Mumford, E. P. (2000). **The CIAM Discourse on Urbanism, 1928-1960**. Cambridge, MA, MIT Press.

Neubauer, D. (2004). Mixed Blessings of the Megacities, Yale Center for the Study of Globalization. http://yaleglobal.yale.edu/index.jsp. December 27, 2004.

Newman, O. (1980). 'Whose Failure Is Modern Architecture?' **Architecture for People**. B. Mikellides. New York, Holt, Rinehart, and Winston**:** 45-58.

Newman, O. (1996). Creating Defensible Space. US Dept. of Housing and Urban Development, Office of Policy Development and Research. Washington, D.C.,

Newman, P. W. G., J.R. Kenworthy (1999). **Sustainability and Cities: Overcoming Automobile Dependence**. Washington, DC, Island Press.

Orejon, A. M., Ed. (1961). **Las Leyes Nuevas De 1542-1543; Ordenanzas Para La Gober Nacion De Las Indias Y Buen Tratamiento Y Conservacion De Los Indios**. Sevilla, Spain, Publicaciones de la Escuela de Estudios Hispano-Americanos de la Universidad de Sevilla.

Relph, E. C. (1987). **The Modern Urban Landscape**. London, Croom Helm.

Schon, D. A. (1983). **The Reflective Practitioner: How Professionals Think in Action**. New York, Basic Books.

Turner, M. G. and R. H. Garner, Eds. (1991). **Quantitative Methods in Landscape Ecology, the Analysis and Interpretation of Landscape Heterogeneity**. New York, Springer-Verlag.

Verlag, C. (1992). **World Urbanisation, Crowding into the Cities**. Cheltenham, England, European Schoolbooks Publishing Ltd.

Whitehand, J. W. R., Ed. (1981). **The Urban Landscape: Historical Development and Management, Papers by M.R.G. Conzen**. Institute of British Geographers Special Publication No 13. New York, Academic Press.

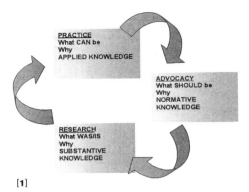

1 The knowledge-building cycle
2 Urban form scales, disciplines, and client groups
3 Comparing Rotterdam and Puget Sound regions at the same scale (Puget Sound region in 1930)

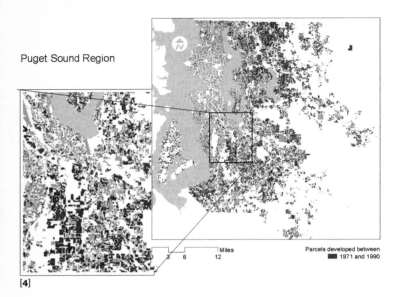

4 Zooming in and out of area extent from the same parcel-level database (Puget Sound region in 1990)
5 Puget Sound region map showing King County, higher density Cluster at one scale, and Downtown Kirkland at another scale (see Table 2 for data associated with these area extents)
6 Surface model depicting the likelihood of walking sufficiently to enhance health, based on the age of the population and on the detailed characteristics of the built environment

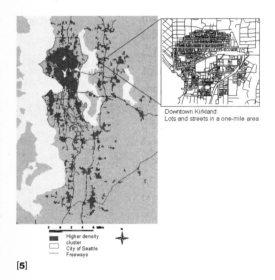

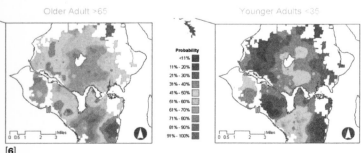

Epistemological Map for Urban Design

Eleven Domains of Inquiry	Methods of Research and Inquiry					Impact on practice
	Dates	Strategies [1]	Modes [2]	Focus [3]	Ethos	
Urban history	On-going	Lit, Phen	Hist-desc	Obj, subj	Etic	Critical assessment of past design, forces on the BE
Picturesque	1950-60s	Lit, Phen	Emp-ind, hist-desc	Obj	Etic	Visual attributes of cities
Image	1960-70s	Pos	Emp-ind	Subj	Emic	How people see and understand cities
Environment-behavior	1950-on	Pos	Emp-ind	Obj, subj	Emic	How people feel, use, and interact with the BE
Place	1970-on	Phen, Lit, Pos	Emp-ind, hist-desc	Obj, subj	Etic, emic	How people feel, use, and interact with the BE
Material culture	1920-on	Lit, Pos, Phen	Emp-ind, hist-desc	Obj	Etic	Object qualities of the cultural landscape
Typo-morphology	1950-on	Lit, Pos	Emp-ind, hist-desc	Obj	Etic	Processes and products of city building
Space-morphology	1950-on	Pos	Theo-ded, hist-desc	Obj	Etic	Urban space geometric attributes
Nature-ecology	1980-on	Pos	Emp-ind	Obj, subj	Etic	Natural forces and the BE
Economic development	1990-on	Pos	Emp-ind, theo-ded	Obj, subj	Etic	Economics of design and development
Regulatory frameworks	1990-on	Pos	Hist-desc, emp-ind	Obj, subj	Etic	Public controls

Adapted from Moudon **Journal of Planning Literature** 1992

1 Lit=literary; Phen=phenomenological; Pos= positivist
2 Hist-desc= historical-descriptive; Emp-ind= empirical inductive; Theo-ded=theoretical deductive

Distribution of housing type and location in King County, the Higher Density Cluster, and Downtown Kirkland

	Total Residential		Single Family		Multi-Family		City of Seattle		Outside of City	
	Units*	%	Units	%	Units	%	Units	%	Units	%
King County	669,385	100	410,210	61.3	259,175	38.7	247,294*	36.9	423,740*	63.1
Higher Density Cluster	335,277	100	97,961	29.2	237,316	70.8	199,737	59.6	135,540	40.4
Kirkland	2290	100	608	26.6	1,682	73.4				

Concluding with Landscape Urbanism Strategies

Kelly Shannon

This essay presents a synthesis of the conclusions of my recently defended doctorate thesis, Rhetorics & Realities, Addressing Landscape Urbanism, Three Cities in Vietnam [KU Leuven, May 2004]. To a certain degree, it deals with all four of the so-called 'big questions' as put forth by the session's chair, Han Meyer, and moderator, Louisa Calabrese. First, the 'urbanizing landscape' of rapidly modernizing contexts (Southeast Asia and Vietnam) frames the scope of the research. Second, the essence of 'centrality' was fundamentally questioned at both the national scale (hence the concentration on three secondary cities) and city scale (resulting in the strong support for the notion of organized dispersal instead of the more commonly advocated 'compact city'). Third, the importance of infrastructure and mobility was stressed as a particular sector where – nearly globally – the State continues to exert control and therefore can be essential in guiding future development. Fourth, the investigation was premised upon an understanding and modern-day reinterpretation of traditional urban morphologies and their relationship to productive landscapes. The research fundamentally readdressed present-day practices in urbanism and attempted to expand the knowledge-base of the profession.

Dialectical Readings The research was premised on a dialectical reading, a zig-zag weaving back and forth between global issues, challenges and a call for new operational tools, on the one hand, and regional and local traditions and realities, specific problems and possible solutions, on the other hand. A multitude of inter-related rhetorics and realities were confronted: the myths of rapid urbanization/modernization and the transition from 'tradition to modernity' with the role of sites and the local-global nexus; the discourse of 'development' and sustainability with the lack of spatial considerations; the theory and biases of contemporary landscape urbanism practice with the historical and non-designed formation of productive landscapes and urbanism; the modern paradigms of urban planning with counter-voices and peripheral resistive practices; the view from 'above' with the experience from 'below'.

Across the globe, conceptual redefinition and reformulation of the urban – even to the point of dissolving it – have dominated the discourse as the imageability, legibility and particularity of intervention in cities has become less clear. The speed and intensity of changes regarding urbanization and modernization have been overwhelming while simultaneously providing new sources of inspiration for urbanity. Of late, there has been a contemporary revival-of-sorts of the tenacious 1950's spirit (the counter-voice of Team X to CIAM), the mid-1960's populist movement concerns, and the traditionalist urbanism and post-modern relativism of the 1970's and 80's. As well, 'other' realities – caused by shifting ideologies and national boundaries, economic development, migration, war and urban survival strategies, ecological challenges – and vernacular and transitional contexts often subvert the Western urban paradigms of the last five decades. They question the universal validity of tried-and-true urban models. Observable urban realities and response to canonic urban rhetoric and models have pointed to alternative descriptive methods and operative strategies which engage in a critical interplay between global and local models of urbanity.

At the same time, a merely social and ecological definition of the enigmatic term 'sustainability' has undermined its potential structuring capacity of the built environment. Few (if any) of the myriad of global sustainability campaigns explicitly address spatial issues – either at the larger territorial scale or specific interventions at the scale of urban projects. Instead, they tend to champion the environmental, social and governance aspects of development. Sustainability has entered practice through legislation more than by way of concrete conceptual grounding. Without a doubt, the broadening of an ecological agenda (inclusive of many of UN-Habitat's programs) towards the built environment and urbanism could both (re)inform the sustainability discourse and substantially revise urban design tools.

Fieldwork was an essential component of the research; an empirical descriptive method of urbanism as critically mapped from 'below' was augmented by the view from 'above'. Although quantitative, descriptive and consensus data abounds in Vietnam, it tends to be either very general or edited to reflect government policies and emphasis. Due to the general unavailability of precise information, fieldwork took on heightened significance, while, at the same time, the on-site observation and analysis of three Vietnamese cities and their peripheral landscapes was not merely descriptive. It served as both 'ground-truthing' of productive tensions in urban and rural phenomena that have been alluded to among various literature sources and as a base for the discovery of unspoken/unwritten realities.

The fieldwork should also be understood as a sort of critical realism (critical in the process of selection and what to map). A critical reading of urban fabrics and morphologies and patterns of functioning (inhabitation, mobility, production, etc.) was made in both a diachronic and a synchronic sense. 'Layered narratives' revealed the cities' urban histories as complex spatial translations of different eras and ideologies – from feudal, to Chinese-influenced imperial dynasties, to French colonial exploits, to American military influences and dependent capitalism, to Soviet-block policies and socialism and finally to *doi moi*, market-based socialism and spontaneous development. The resilience and potency of the cities' multi-layered narratives questions the popular assertion of a linear development path from 'tradition' to 'modern' whereby a next phase replaces a previous one.

On the whole, the landscape urbanism discourse that has developed in Europe and the United States has emerged less as a theory but more as a way to innovate at the level of design practice [Hight 2003]. As Mohsen Mostafavi has recently commented, the relationship between landscape and urbanism, on the one hand, has witnessed a literal transposition of techniques and vocabulary of one to the other and, on the other hand, has displayed an affinity as a metaphoric and metonymic register [Mostafavi 2003:7]. The latter primarily addresses the 'dynamic' rhetoric of landscape and remains, despite claims to the contrary, fundamentally embedded in aesthetics. Landscape and nature are often referenced in contemporary architecture, urban and landscape urbanism theory as ever-changing, temporal and indeterminate. Yet, landscapes can also be understood to be in a continual state of slow, predictable evolution. To a certain degree, nature is reliable, a source of continuity which is able to adapt to different imposed (and temporary) realities. Indeed, landscapes exist as tensions between the dynamic/stable and permanent/impermanent. The appeal of much writing on landscape urbanism lies in its abuse of post-modern rhetoric.

Following the thesis's critical reflection on the limits and applicability of landscape urbanism's theoretical paradigms and a representative selection of built work, it can be concluded that a number of biases continue to constrain the potentially larger meaning of the field's emergence. Within the field is an obvious infrastructural bias. In most instances, infrastructure projects tend to solve engineering problems while missing a more intimate relation with numerous layers of the landscape. Although landscape urbanism projects have attempted to remedy this perspective, a number of projects go to the other extreme, whereby aesthetics take precedence. The creative marriage of civil/hydrological engineering feats to landscape and design of the public realm could lead to more convincing integration of infrastructure and the landscape.

A second dominant bias within the field is ecological – fed by the rich West's concerns for the environment – and contains two perspectives. The first reflects an attitude in existence since the 1970's in which technical aspects of ecology drive territorial development. The second relates to shifts in ecological thinking which have been reflected in landscape urbanism's prominent consideration of dynamic, non-deterministic processes. Ecology, both as a formal model and even more as a metaphorical mode, is a powerful component in the thinking and design of landscape urbanism projects. Rural agricultural environments are often described with terminology from the discipline of ecology. Concepts of porosity, density, heterogeneity, matrix, corridor, edge and patch are terms rooted in landscape ecology. However, in large territories where productive landscapes still play a vital role, they are also relevant terms in regard to urbanization. The interrelations of networks of infrastructure, landscape and settlement support different habitats and social landscapes across territories.

Finally, there exists a narrow interpretation of urbanism in the prevailing rhetoric. Rapid transformations in the countryside are largely excluded in lieu of focus on urban cores and *ex-* and *peri-*urban areas; there is a definitive urban bias to the landscape urbanism discourse. No doubt this is due to the fact that landscape urbanism in Europe and North America has primarily been concerned with post-industrial 'brownfield' sites – many of which are presently undergoing a process of reclamation and re-use. At the same time however, 'greenfield' sites are raising new challenges. The more dominant stance towards greenfields remains squarely in the orbit of conservative strategies – whether embedded within agricultural, forestry and natural reserve policies or in a more limited extent, when new development comes into the picture, within movements such as New Urbanism. In Europe and North America, post-agricultural sites and shrinking cities are realities, the former the result of over-exploited large-scale domains with no economic sustainability and the latter related to both post-industrialization (particularly in entire regions in the former East Europe) and urban sprawl. Landscape strategies are increasingly employed as part of environmental cleanup and the process of restoring imbalanced ecologies. The natural landscape has also been revalidated as an organizing force – protective, productive and/or regenerative landscapes – in containing urban sprawl.

Until the mid-1990's, the dominant tendency in the architectural/urbanism discourse towards the notion of dispersion, was that of negative associations. However, during the last decade, there has been a noticeable degree of sympathy towards the periphery and dispersed landscape – as witnessed in Pope's writing on 'ladders' and the American city [Pope 1996], Neutelings' 'ring culture' [Neutelings 1989], Boeri's addressing of European peripheries [Boeri 1998], Secchi and Viganò's writing and projects addressing the 'città diffusa' [1984, 2001], De Geyter's work on 'after-sprawl' [De Geyter 2002] and De Meulder and Dehaene's studies on the fragmented Flemish landscape [Dehaene and De Meulder 2003]. Dispersion of urban form – in all of the terms for which it has been described (of which the vagueness of the terminology reveals its inherent weakness) – has resulted in a basic dissolution of the modernist metaphor of the functionalist city. Innovations in transportation and communication networks have given the term proximity an entirely new meaning. The periphery and dispersed landscape is now a subject that is accepted within theoretical debate and viewed as an interesting theme, yet there remains a lack of strategies both to effectively describe the reality and to qualitatively intervene.

Learning from Vietnam A central hypothesis was that both the theory of landscape urbanism, as well as its practice, could be substantially reformulated by case studies from a region of the world where the relation of landscape to urbanism continues to hold deep-rooted pragmatic and symbolic meanings. It was postulated that examples from the non-West could inform landscape urbanism's broader theory and contribute to the expanding – but as yet Euro-American-centered – archive of urban and regional research. Roots for a paradigm shift, as evidenced by Team x investigations, can be embedded in 'other' realities. In the non-Western world, both brownfield and greenfield sites present enormous challenges and opportunities for the profession. The pressures of urban expansion and indeed of new conurbations can profit from the possibilities afforded by the comprehensive and strategic development of urbanism with the landscape – be it productive or symbolic. Indeed, in traditional landscapes and in non-Western cultures, productive landscapes and their surrounding settlements have historically dealt explicitly with landscape urbanism – long before the 'emerging' discipline was named as such. Southeast Asian wet paddy civilizations, coco, coffee and fruit plantations in Latin America, European vineyards and Dutch delta-works provide valuable lessons for the new field. The continual interaction of human settlements with productive landscapes for necessity and survival has evolved over the millennia into complex systems of balance. In many contexts with strong agricultural roots, landscapes are often neither city nor country, but simultaneously both. Urban and rural are insufficient terms to describe the visual chaos of the hybrid 'second nature' patchwork territories, as landscapes are shaped neither by aesthetics nor symbolic aims but defined in pragmatic terms. At the same time, the design strategies developed for the case studies have drawn upon not only a number of notions of the contemporary landscape urbanism discourse, but also from the existing logics of landscape and urbanization in the case study cities.

Urbanization in the region of Southeast Asia, in general, and Vietnam, in particular, has been marked by their relation to productive landscapes – either water- or land-based: terraced mountain slopes for agriculture (rice, tea, coffee and rubber), wet paddy cultivation in the low-lying territory and fish- and shrimp-farming in the coastal zones. The flip-side of the productive landscape has been the symbolic landscape, where ancestors are worshipped and spirits of the earth inspire deep-rooted legends; ritualized attachment to the territory is embodied in the Vietnamese landscape.

The case studies of three secondary cities in Vietnam have shown the historical and contemporary importance of the country's symbolic and productive relationship to human settlements in the landscape. Water and mountain genies remain deeply rooted in Vietnamese mythology and beliefs and the socio-culturally codified traditions of the predominantly rural society are evolving, but not being lost, as rapid urbanization progresses. Although settlement changes can still be observed against the relative inertia of the landscape, the intense pressures exerted by such rapid urbanization processes are taking their toll on the landscape – as the system of interlocking eco-systems are upset and the balance of consumptive to productive land dramatically mutates.

Vietnamese cities have been ideal testing grounds for theoretical territorial scenarios to confront contemporary conditions.

Urbanistically, upon first glance, Vietnam appears to be under-developed and under-infrastructured. However, fieldwork revealed that there exists a density of small-scale support systems, formal and informal, which spatially structure settlement in a logical manner. The historically distinctive settlement patterns of the country remain visible – whereby the north has a strong tradition of autonomous village/urban entities while the south has always displayed a perpetual reliance on networks of communication and trade. And although Vinh, Hue and Can Tho all display a relative absence of large-scale infrastructure linking them to the rest of the world, they are themselves densely infrastructured from the point of view of productivity of the land. Irrigation systems for the cultivation of rice and flood protection have created a tightly woven system of waterways, dykes and access routes. The productive land systems in the low-lands have resulted in the overlapping of spatial networks across vast territories – permitting habitation nearly everywhere. Intensification of urban development has been where networked systems overlap. Dispersed settlements in the productive territory are evidence of the interdependence of urban and rural. At the same time, the three cities each reveal unique geographical and morphological systems.

As well, Vietnam's 'layered narratives' strengthen each city's particularity. Vietnam's rich crossbred urbanity coupled with extreme weather (a hot, humid tropical climate, devastating typhoons and extreme flooding) challenges the contention between the cultivated (civilization) and the wild (untamed nature). Fragments of the richly layered urban paradigms (feudal, imperial, colonial, socialist and capitalist) remain and it is proposed that they need not be rejected but creatively re-appropriated. For example, although all that remains of Vinh's 1831 citadel are two crumbling gates, the space could be reconfigured as an open space – working into the proposed system of interconnected public places which could also serve as a hydrological system of absorptive surfaces. Similarly, the city's historical twinning with the former German Democratic Republic – most spatially visible in the Quang Trung housing estate and the broad expanse of Quang Trung Street (highway one as it passes through the city center) – uniquely marks the city. Without a doubt, strategies are necessary for improving the housing typologies and the estate's open spaces (for which existing urban agriculture and informal markets provide precedents), however it would be a shame to 'beautify' and densify the stretch along the boulevard to the extent where this spatially powerful historical layer is no longer present. In Hue, the paradox of UNESCO recognition is that only fragments of the imperial urban patrimony remain. War and naturalization of the monuments (where tropical vegetation has colonized structures that were not maintained for decades) have eroded the physical structures, while, at the same time, the monumentalization of nature continues through the establishment of new sacred sites (for burial sites and clan houses).

The spatial reflections of multiple world-view narratives in Vietnamese urbanism challenge any one particular all-embracing ideological construction and typo-morphological paradigm; there remains a necessary ambiguity. Nature and wild, urbanity and civilization themselves are only understandable and perceptible to a certain degree. Nature is an elusive system unto itself. Similarly, there exists a profound wildness in cities that no single world-view narrative has been able (or should be able) to conquer. Neither is smooth, single-layered nor controllable. The unique historical relation that Vietnamese cities have maintained with their surrounding landscape provides insights into the continued symbiotic relationship between the urban/rural, man-made/natural and accommodating/resistive forces, typical of Vietnam's wild cities and urbane wilderness.

As has become apparent in the case studies in Vietnam, ecological constraints and opportunities provide a rationality hidden in reality for seemingly indiscernible patterns of urbanization, and these same logics can guide design strategies for new layers of development. The language of landscape ecology offers a means to rethink urbanism; its terminology could be translated into strategies and techniques for analysis and design in landscape urbanism. Landscape ecology's notions of structure, function and change can be read in terms of urbanism – wherein structure is the formal structure of a city, function is the day-to-day working of a city and change is the dynamic restructuring of the city over time. Landscape urbanism can provide new (infra)structure for programmed and non-programmed activities that evolve over time.

For example, the landscapes within landscapes of Can Tho – an infrastructured urban system embedded within a larger productive and inhabited territory premised on an intricate water network – contrast sharply with the governmental-approved city extension plans. If the inherent potential of the existing land mosaic were exploited, alternative patterns of urban growth could feasibly replace the land-use-based system that is presently being pursued. In the case of Vinh and Hue, the issue of flooding could to a much greater degree inform future urbanism; the historical relation of agricultural land to settlement and an accord between surfaces that shed water and those that absorb or retain water could be more explicitly addressed in the creation of infrastructure and areas for urban expansion (which requires either the filling of lowlands to create dry, higher lands for settlement or the development of typologies that work with seasonal changes in water levels).

The case study fieldwork in Vietnam confirms a positivist interest in patterns of dispersal. The ancient spatial structuring of settlement in relation to productive paddy fields continues to exert its logics on the landscape. The dispersal of the population across the plains has been continued by the country's post-war urban policies of decentralization and reclamation of unproductive landscape through its formation of 'new economic zones'. More recently, the central government has been pursuing a strategy of further decentralization and dispersal by developing a series of satellite cities. The satellite cites are intended to relieve development pressure on the existing urban centers, to increase accessibility to public services and to boost local economies. Unfortunately, however, the government-approved plans for the satellite cities tend to follow tried-and-true urban models and end up as impositions in the territory. They do not build upon the existing logics of the multivalent system of dispersed urbanity that exists.

In Vinh, Hue and Can Tho, 'contested territories' are numerous and promise to increase as the Vietnamese economy is more integrated into the global system. However, the market is fickle and constantly subject to economic and political reorientation and therefore comprehensive, long-term planning is no longer a feasible option. Instead, however, strategies can be devised which protect a public realm for these cities and guide development to follow existing logics and daily-use tendencies. It is a 'mission impossible' to attempt to achieve balance of any sorts in Vietnamese cities, yet intelligent strategies could create new narratives which could capitalize on the existing ambivalent urbanity to allow that chaotic and layered richness to prevail. As the omnipotent forces of global capitalism begin to pervade, single-use zoning and banishing of the informal need be avoided. There exists an opportunity to exploit the ambiguities inherent in the city's structuring of urban/rural, public/private, productive/reflective/consumptive landscapes in an alternative strategy of modernization and urbanization.

Landscape Urbanism Strategies for Vietnam A number of landscape urbanism strategies for Vietnam can be distilled from the possible scenarios of the case study proposals. The first deals with the increasingly problematic issue of flooding. Presently, as development forges ahead and mineral surface areas cover previously absorptive areas of land mass, the incidence and severity of flooding rises exponentially. Alternatively, in the proposals, the marriage of civil engineering, urbanism and landscape could be developed as operative devices that realistically deal with the hydrological regime – not unlike the ingenious solutions of Frederick Law Olmsted in the 19th century. In Hue, the Vietnamese city which suffers from the most extreme flooding, a combination of 'hard' and 'soft' approaches could be developed in a possible scenario; a by-pass canal, doubling as a large public park (and with public buildings), could be combined with strategies of accommodating excess water in flood pockets, reservoirs and low-land areas. In Vinh and Can Tho the primary approach of the possible scenarios stressed the natural tendency of surface absorption and drainage (by large surfaces of productive, low-lying paddy fields) and irrigation (by a complex network of canals, channels and ditches). In Vinh, the present practice of filling-in low-lying areas (preparing the ground for future urbanization) could be redirected to allow the city to maintain the more natural process of acting as a sponge. In Can Tho, the covering of canals by roads could be combated by the enhancement of the water network as not only a productive and absorptive necessity within the system but also as a viable transport system for the 21st century.

The second strategy regards that of the urban/rural interface. The landscape urbanism scenarios proposed to enrich the existing discontinuous, heterogeneous and multi-polar agglomerations which may be termed either urban countryside or rural metropolis, as opposed to developing an artificial disparity between city and country. The consumptive spaces of the city could be counterbalanced by productive spaces. In Vinh, spaces for urban agriculture could be strategically located to serve multiple functions simultaneously – working as low-land absorptive surfaces for excess water, acting as spaces of decompression in dense fabrics, creating desirable microclimates for adjacent areas and providing an enlarged public realm through a series of interconnected urban parks. At the same time, as the scenario proposed for Hue suggested, urban parks along the rivers (which are partially working as expansive territories for floods, productive and also recreational fields) could also serve to create an enlarged public realm and delimit sprawl. The initial public investment necessary to create such a by-pass canal/urban park would be arguably substantial, however, if the Hue government could retain property rights of the immediate urban plots and only lease them as property values increased; costs could be recuperated in the long run. As well, experimental agricultural fields (related to the proposed university buildings) and productive areas of the lagoon could be integrated into the urban park which stops the sprawling growth of the southern portion of the city. In Can Tho, the existing richness of the urban/rural interface could be enhanced by the proposed scenario. The fruit tree orchard plantings along waterways could be extended directly into the urban fabric (to the Hau Riverfront) not only defining a network of flowering public promenades but also directly tying the surrounding territory into a continuous ecological mesh with urbanity. Although not fully elaborated, a significantly revised urban/rural interface would demand new typologies of housing and buildings in order to more meaningfully relate to productive landscapes and low land areas (requiring a reinvestigation of the traditional house on stilts typology). As well, urban agriculture could provide a direct link between urban and rural daily activities. Urban agriculture parks could also be envisioned not merely as productive landscapes but also as meaningful public space.

The third strategy dovetails into that of open productive space. The contemporary market system of agriculture demands an economy of scales. In order to increase the size and therefore cost effectiveness and competitiveness of Vietnam's productive landscapes, it is recommended that existing urbanization is consolidated, leaving large expanses of territory available for production. Linear development, along water-ways and rail/roads could be strengthened by investment in infrastructure improvements and public transport systems (with boats, railroads and buses). It is proposed to simultaneously densify existing agglomerations along linear infrastructure lines (ribbon development), at nodes, or within well-defined urban massing – interrupted by productive and ecological corridors. Densification is suggested as a device of 'traditional' urbanism to create a stronger sense of urbanity as evidenced in the proposed interventions in Hue and Can Tho where linear development follows the existing logics of urbanization – along roads as well as waterways.

The fourth and fifth strategies are inter-related. The former concerns the creation of micro-climates and new natures (as opposed to protection of a larger number of small, less productive and less ecologically meaningful patches), while the latter addresses the development of new energy models and infrastructure. In Vinh, a possible scenario is developed with a productive (fruit and bamboo tree) forest which could be planted on the city's western edge – protecting the city from the winds, creating a microclimate, providing a limit to urbanization and operating a part of the proposed public park system. Similarly, the scenario for Hue's by-pass canal/urban park and Can Tho's orchard/waterway systems create favorable microclimates. In Vinh, the problematic, seasonal dry, hot 'Laos' wind is harnessed for energy as wind turbines are proposed along the National Highway No. 1 bypass and off-shore. The intensive solar radiation of Can Tho could eventually be channeled through solar collectors and photovoltaic modules in buildings, and less expensive initial cost methods such as systems for the reduction of summer energy requirements (sun screens, natural ventilation, etc.). In addition, water collectors (retention reservoirs) would make sense in all three cities.

The sixth strategy reveals what must be considered a core principle of landscape urbanism – namely defining operative mechanisms/devices to accent the basic structuring characteristics capable of guaranteeing the existing diversity and quality of the landscape and to counteract negative tendencies, while concurrently providing a sustainable means for further urbanization. Multiple readings of the existing landscape were necessary, not only from above (as read from aerial photographs and plans, when available), but also as experienced and mapped from 'below' by way of fieldwork. In the three case study cities, detailed and diversified articulation of the structuring of the territory by geographical formations as well as by the modified nature of man-made agricultural landscapes proved fundamental to revealing the identity of each case study city. Landscape urbanism strategies were proposed to stress the existing qualities in the landscape – to make the invisible visible, to reveal the hidden logics of the territory, to develop the rationality hidden in reality. From 'above' a number of readings were translated into proposals. In Vinh, the arcing land masses or islands of higher land in the low flood plain (caused by erosion and water run-off from the mountains to the north-northwest) were accentuated in the new buildable territorial formation. This reading 'from above' was thereafter nuanced in a possible scenario whereby articulation of the western and eastern edges of the land masses responded to the climatic and everyday uses experienced 'from below'. Similarly, in Hue, the parallel and powerfully structuring ecologies of mountains, plains, lagoons and sea – all traversed by a network of rivers – were underlined by development of the city and its surroundings along infrastructural routes that bordered the various natural systems, while the Perfumed River was recognized as the prime organizer of the territory. Densification along National Highway No. 1 marks the transition from the mountains to the plain; the expected linear developments structure the transition from plain to lagoon to sea (with road networks merely being the improvement and completion of existing systems). The scenario's inclusion of an organizing central spine (public platform for access and activities) resulted from fieldwork. In Can Tho, it was indeed the view 'from below' that informed the possible scenario. By boat, the city was experienced as a porous double mesh of interconnected waterways (rivers, canals and an industrialized grid of irrigation channels), orchards, paddy fields and roadways (highways, urban boulevards, roads paralleling waterways and paths atop dykes), with settlements, where rural and urban environments wove into one another. In the proposed scenario, the 'natural' system could be extended into the urban core and investment into infrastructure (waterways, roadways), thereby establishing likely, linear sites for new settlement.

Finally, there is a strategy of developing alternative transport systems which take advantage of the existing logics of the landscapes. For all three cities, a network of both public bus routes and water-based vaporetto/taxis has been proposed by merely enhancing the existing infrastructures. In many instances, systems can operate more as networks than they presently do by the simple addition of missing-links in the road system or by the strategic addition of bridges and/or ferries. In Vinh and Hue, the under-utilized railroad could be reactivated to become a vital element in the public transport system. The advantage of water-based systems in the Vietnamese context need also be stressed. Water transport not only provides benefit in terms of ecology but also stresses an inherent identity issue. Water-based transport accommodates multiple and not exclusive uses – allowing slow/fast, traditional/modern and small-scale/large-scale craft to co-exist. The most important aspect of this last landscape urbanism strategy has indeed to do with its 'strategic' aspect. The formative capacity of the territory through infrastructure remains one of the most powerful devices for directing future urbanization. At the same time, high costs render the process of building infrastructure as slow – particularly in contexts of severely limited resources. The Asian Development Bank and the World Bank, in addition to a host of bi-lateral cooperation projects, are presently pursuing the development of new highways systems and large bridges/tunnels throughout the country. The landscape urbanism proposals work more from the perspective of improving existing networks and strategically accenting a hierarchy of systems across the territory.

In concluding on Vietnam, it must be mentioned that there remain a number of pressing issues/questions that were made visible in the fieldwork and during the elaboration of the landscape urbanism strategies which demand greater study and attention. Indeed, fieldwork is never finished and merely opened up other issues to study. The sacred landscape was historically very important in Vietnam, and as in the instance of Hue, it appears that this tradition is continuing with renewed vigor. Yet, contemporary planning and urban design take neither the cosmological ordering of imperial tombs nor the sites of ordinary tombs and clan-houses into serious consideration. Further anthropological research is necessary to reveal the hidden logics and opportunities in spatial, cultural and even economic structuring of the contemporary additions to the sacred landscape. Another aspect that was brought to the fore in fieldwork and proposed interventions concerned water management. Obviously, in-depth hydrological investigations are necessary for feasible solutions to the increasing problem of flooding and saline inundation. Although the landscape urbanism strategies developed are naïve from a technical standpoint, they do work with the common-sense logic of territories operating as sponges with elastic thresholds between land and water. As well, the strategies reveal a commitment to the notion that hydrology and civil engineering need not be developed sectorially in order merely to solve technical problems, but that they can merge with urbanism and the making of public space. And finally, there a cost issue pertaining to the extent of infrastructure works proposed. Of course, a number of the strategies proposed build upon planned changes by the Vietnamese government for the road network, but at the same time, the financing for many such projects has yet to be determined. Presently, there are comprehensive studies that are being developed by traffic engineers, donor agencies and the Asian Development Bank and these would have to be considered more significantly in further development of landscape urbanism strategies for Vietnam.

Landscape Urbanism Principles: Another Taxonomy

The taxonomies of landscape urbanism that have been made by various landscape architects and urbanists in Europe and North America have focused either on defining the steps to meaningfully intervene in sites (Marot, Girot), categorizing the processes by which the landscape serves as a metaphor for urbanization (Corner, Geuze) or classifying terms by which projects may be related to the landscape (Smets). However, a taxonomy beyond the programmatic, graphical, aesthetic and metaphorical has yet to be elaborated.

This research offers a possible beginning for the development of a series of operative landscape urbanism principles. The specificity of the Vietnamese case studies has led to the abstraction of yet another taxonomy for the globally emerging discipline. The first principle of landscape urbanism concerns the careful reading and understanding of the existing landscape and creating urbanity that simultaneously stresses the assets and solves the threats particular to the location. The view from 'above' includes the reading of ecosystems, watersheds and geographical/topographical formations, etc. which are critical to a comprehensive analysis of the larger territorial setting of cities and their peripheries. The view from 'below' is complementary and requires extensive fieldwork and an understanding of the territory from a haptic and experienced sense. Spatial structures can be designed so as to be tied to both the logics of the landscape and its everyday use. Stemming from this reasoning, it can be deduced that landscape urbanism strategies need to develop through addressing multiple issues simultaneously. For example: engineering feats and infrastructure provision need to marry with enlargement of the public realm; settlement patterns need to clarify the legibility of the landscape; expanded productivity of landscape needs to overlap with provision of public amenities, etc.

The second principle stresses the unavoidable bias of infrastructure, whereby the creative marriage of civil engineering, urbanism and landscape can simultaneously serve multiple purposes, including the solving of hydrological and other ecological challenges, qualitatively defining an enlarged public realm and blurring or clarifying the boundaries between the urban and the rural land-uses, morphologies and typologies. In developing and developed countries, infrastructure works remain one of the most important building activities left of the State (if not the only one) and this fact must not be underestimated. Beyond merely being an essential component of the expanding communications/accessibility network, infrastructure's relationship to urbanization can consciously be managed. Alongside and in conjunction with infrastructure, places of either congestion or decompression, of urbanity or of a rural nature, of productivity or of recreation can be consciously suggested and/or designed. Existing urban fragments can be strengthened and interconnected via infrastructure – infrastructures not only of transportation but also of open space, communication, services, etc.

The third principle underscores the interdependence of productive and consumptive land and suggests a renewed emphasis on no-nonsense productive landscapes for both national and subsistence economics; it proposes the replacement of the urban/rural dichotomy by the notion of hybrid urban countryside/rural metropolis. Productive pockets/parks in cities not only provide an additional subsistence/income base for the less wealthy but, if configured as a connective system of open spaces, can also create ecological corridors and spaces of decompression which revalorize urban fabrics. At the same time, urban servicing of rural settlements decreases rural poverty, while densification of existing rural settlements can address the issue of economies of scale and the need for larger agricultural plots for mechanized farming.

The fourth principle is tied strongly to the third and is one of creating desirable micro-climates and protecting strategic public areas from inclement and undesirable conditions (noise, wind, flood, etc.). The public realm can be qualitatively up-graded by the creation of interlocking landscapes which also serve as wind/noise shields, productive parks, flood and excess run-off reservoirs.

Finally, landscape urbanism, as defined by its synthesizing of landscape, architecture and urbanism is uniquely placed to create ambiguous and rich thresholds – not only of built and unbuilt, urban and rural but also more nuanced overlappings of public/private realms, wet/dry zones, productive/recreational/touristic areas, etc. The proposed landscape urbanism principles are strongly linked to the notion of sustainability in a broadened sense.

From 'Below'; from 'Above' The research concluded with a number of themes that came out of the case study investigation, that are relevant to the broader subject of contemporary urbanity. The analysis and reading of reality from 'below' not only serves as verification, elaboration, and modification of the perspective from 'above'. On the contrary, strategies can be suggested by experience 'from below' – as was the case with the scenarios' development of public platforms for social spaces and rice drying (stemming from an observable use of roads, including National Highway No. 1, as available flat and hard surfaces). From 'below', another scale of issues was experienced through fieldwork and was translated to proposals that dealt more directly with habitation and daily use. In all three cities, the unimaginable proximity of urbanity and productive landscapes was astonishing and perhaps most emblematically visible in Vinh's Quang Trung Estate, along National Highway No. 1, where urban agriculture and small husbandry animatedly occupied the open space; proposals for all three cities built upon the juxtaposition of consumptive and productive territories. As well, the ingenious solutions borne of necessity were noticeable in urban centers as much as in the rural countryside. In spatial terms this translated to the temporary appropriation of spaces – roads (including the highway) which became drying platforms for rice and other agri/aqua cultural products; sidewalks that doubled as private housing extensions, parking areas for motorcycles, café spaces and informal markets; fallow fields serving as sports grounds; waterways as floating markets, etc. There was a conscious effort made in the scenarios' proposals to accommodate the transitory and changing nature of programmatic appropriation of threshold spaces. As well, in each city specific observations from walking, motorcycle and boat trips conditioned the specificity of the landscape urbanism proposals. In Can Tho, the quality of being in the center of the city one moment – in the frenzied urban realm – turning a corner (either on a road or in a canal) and then being in a richly vegetated jungle-like rural world, was recognized as a distinct identity element in the city that could be maintained by a proposed alternative to the officially envisioned urban enlargement. In all three cities, but most powerfully revealed in Hue, was the incredible variety of landscape and urban/rural typo-morphologies within relatively small distances. Not only is the maintenance of difference important for the cities in order to compete on multiple platforms and offer their citizens choice in lifestyle and livelihood, but also to retain the identity that is inherent in the landscape (both natural and as it has been shaped by man over the millennia). As well, as tourism becomes a larger contributor to the Vietnamese economy, access to the diverse territories, landscape types and the local customs can become a major asset setting Vietnam apart from the region's mass tourism.

The tradition of descriptive urbanism has yet to be fully enveloped by landscape urbanism. However, as has been revealed in the case study investigations, there is an obvious advantage in doing so. The strategies developed for the case study cities were based on strengthening the existing; proposals were always based on pre-existing elements in the landscape. The translation of descriptive urbanism into descriptive landscape urbanism could substantially reinform the emerging discipline.

Towards Another Voluntarism?

Although upon first glance the landscape urbanism strategies proposed for Vietnam in the three case study cities could be criticized for being overly voluntaristic, it must be recognized that they build upon contemporary tendencies – both in the existing planning context as well as in daily reality. The country's tradition of urbanism has been founded upon an unequivocal belief in planning and a certain power to impose radical spatial configurations upon the territory – beginning from the country's first organized development of irrigation systems and spanning until the present-day practice of forming new satellite cities upon productive paddy land. Indeed, it is for this very condition – whereby the country's centrally-controlled system (pre-imperial, imperial, colonial and Communist-endowed ideologies and heritage) continues to exert 'public' control to a degree which has for the most part disappeared from the world – that Vietnam is a context in which such broad, structuring strokes of landscape urbanism remain feasible. The proposed large-scale interventions are (infra)structural and aim to protect and enlarge the collective, public realm of rapidly urbanizing cities, while at the same time acting as supports for appropriation by unprogrammed activity, mobility/transport and as platforms for investment. The landscape urbanism strategies proposed are indeed feasible because fifty percent of what is proposed already exists – the fieldwork has summarized tendencies of informal urbanism and everyday use.

Of course, it is recognized that there exists a danger and paradox in the willingness to plan and design tendencies stemming from daily practices. Yet, it could be made to link everyday use to overall proposals which frame structures for the spontaneous to occur. The primary morphology of the landscape could be manipulated at the infrastructural level of reasoning. In Vietnam, there remains a will to plan and in this regard a new voluntarism is advocated which could operate on the level of (infra)structural and strategic planning. The voluntaristic and normative traditions of planning need not be rejected a priori, but can instead be focused (as a neo-modernist approach) on the strategic and structuring of an enlarged public realm.

A New Phase of 'The Modern Project'

On the one hand, the thesis contains a critique of the stages of modernization and the faces of modernism – including the rationalities of colonial domination and the basic logic and pragmatism of overall state planning supported by master planning. On the other hand, the fieldwork could suggest a 'post-modern surfing on the waves attitude', modified by a populist sympathy for daily practices. But, it could also be seen as espousing landscape urbanism as a new phase of 'the modern project' – as a new paradigm to spatially, socially, culturally and ecologically correct the omnipotent forces of world capitalism. It is fundamentally grounded in the belief that architecture and urbanism are socially-based and have a responsibility to enlarge and to protect the public realm. Despite the erosion of the Welfare State in the Western world, there do remain strong layers of resistance to global capitalism and its spatial consequences. Vietnam is in the midst of an important transitional period, whereby it is loosening its socialist grip on power and opening itself to the forces of the market. This unstable condition could become a powerful ambivalence: during the coming years, there is the great opportunity for the country to limit and concentrate its capacity for central control for the purposes of providing basic services, infrastructure and facilities. Indeed, this may be read as a continuation of the modernist project through the creation of uncompromising dedication to an expanded public realm while, at the same time, opening up, in a selective way, to the realities of the global economy. Landscape urbanism could become an operative tool to make this goal feasible.

Potentially, the international appeal of landscape urbanism could lie in its commitment to protecting and increasing the public realm.

The Power of Sites within the Global/Local Nexus

Although the approach of landscape urbanism has eluded precise definition, it appears to be generally agreed that landscape urbanism is essentially rooted in a belief in the intelligence and power of place – not so much in the conservative sense of Martin Heidegger's and Christian Norberg-Schultz's *genius loci* (Norberg-Schultz 1979) but more in Elia Zenghelis' contemporary interpretation of uncovering existing logics of reality and finding the capacity of sites by distinguishing the junk from the potentials (Zenghelis 1993). Landscape urbanism could be an urbanism strategy that gives voice to the restorative and resistive social and cultural formation of territories – and the evocative power of landscapes. Landscape urbanism deals with sites too large and too complex for definitive, one-off solutions. The overlaying of ecological and urban strategies can offer a means by which projects may create new systems of interconnected networks that complement the existing structures.

In a landscape urbanism strategy, the site becomes the controlling instrument of the interface between culture and nature; site phenomena are generative devices for new forms and programs. Integrating the principles of ecology's positive feedback system results in landscape urbanism's appreciation of/working with larger regional scales (watersheds, ecosystems, infrastructures, and settlement patterns). A dynamic interrelationship of urban and rural works to create an urban countryside/rural metropolis. The form and character of landscape urbanism strategies derive from the social and cultural formation of the physical fabric.

A descriptive landscape urbanism could evolve from the careful reading of layered contested territories and 'designerly' investigation of potentials. The existing logics of landscapes (including its historical layers and ad-hoc daily appropriations) could be reorganized at different scales and connected to new (infra)structures. Specific logics from the 'junkyard' of existing landscapes could be stressed and new interventions with structural capacities could reformulate reality. Landscape urbanism strategies could become powerful tools for negotiation between different actors and within the 'contested territories' of 21st-century cities.

References

Boeri, Stefano (1998) **Italy, Cross Sections of a Country: An Eclectic Atlas of Contemporary Italian Urban Landscape**; Milan: Scalo.

Boeri, Stefano (2003) **USE Uncertain States of Europe: Multiplicity USE – A Trip through a Changing Europe**; Milan: Skira Editore.

De Geyter, Xaveer (2002) **After-Sprawl: Research for the Contemporary City**; Rotterdam: NAi Publishers.

Dehaene, Michiel and De Meulder, Bruno (2003) 'Hybrid Figures in the Dispersed City: Towards a Retroactive Urbanism for the Flemish Urban Landscape', **Architecture and Urbanism 393**, Tokyo, June 2003, pp. 126-131.

Hight, Christopher (2003) in **Landscape Urbanism: A Manual for the Machinic Landscape**; London: AA Publications.

Mostafavi, Mohsen and Najle, Ciro (2003) **Landscape Urbanism: A Manual for the Machinic Landscape**; London: AA Publications.

Neutelings, Willem Jan (1989) 'The Antwerpen Ringzone', **Casabella** vol. 53, no. 553/554, Jan./Feb. 1989, pp. 42-45.

Pope, Albert (1996) **Ladders**; Houston and New York: Rice University School of Architecture and Princeton Architectural Press.

Secchi, Bernardo (1984) 'New Technology: New Forms of the City being Created by the Electronic Revolution', **Casabella** vol. 43, no. 501, Apr. 1984, pp. 16-17.

Viganò, Paola (ed.) (2001) **Territories of a New Modernity**; Naples: Electa Napoli.

Zenghelis, Elia (1993) Interview by N. Dodd and N. Joustra in **Berlage Cahier 2**; Rotterdam: 010 Publishers.

1 The Yen river and Truong Son Mountain landscape
2 The interdependence of urban and rural is visible in Vinh
3 The large-scale rebuilding of Vinh began in 1974 and the Quang Trung Estate is emblematic of the new 'city of socialist man' — a gift from the German Democratic Republic

[4a]

[4b]

[5]

[6]

4a-b A paradoxical situation arises in Hue, where monuments are being naturalized (after destruction in war) and new ad-hoc monuments are conquering nature in the form of burial tombs and family clan houses
5 Indigenous water-based urbanism in Can Tho
6 A proliferation of large-scale industrial zones and export processing zones
7 The Vinh-Cua Lo approved master plan to 2020
8 Productive landscape of Vinh
9 Edge between the high-land urbanized islands and the low-land productive land
10a-b Hue's approved master plan to 2020

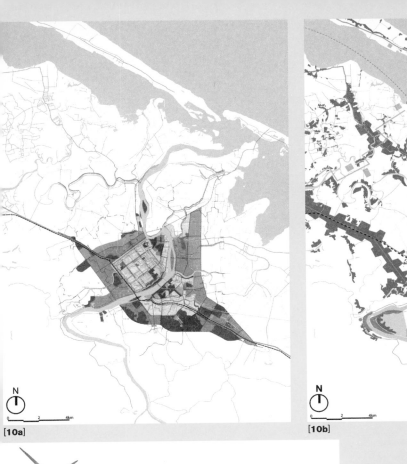

[10a]

[10b]

[8]

[9]

[11a]

[11b]

[12a]

[12b]

[4a]

[4b]

[13a]

[13b]

1a-b Proposed by-pass canal in Hue
2a-b System of water transport hubs in Hue
3a-b The approved master plan to 2020 in Can Tho
4a-b The potential of water-based urbanism

Section VII:

Project & Research

The history of Western architecture is intimately bound up with the development of the European city. From Antiquity to the Medieval, through the age of the Renaissance, the Baroque and Neo-Classicism, and well into the industrial era the subsequent urban architectures have determined the characteristic composite form of the European city.

This section wishes to investigate the role and impact of the architectural project on the formal identity of the European City. In what way have architectural interventions contributed to and catalyzed the process of transformation and renewal of existing urban areas, both in the past and present? And, can the architectural idea of a 'European city' still persist, in a time of ongoing globalization, or has it by now become an anachronism? These questions will form the general frame of this section that includes papers by Leen van Duin, Adalberto Del Bo, and Henk Engel.

Further, such specific research themes as the following provided an outline for that which we invited authors to address:

• *Typo-morphological research*: plan-analytical studies of urban areas in European cities that investigate the coherence between urban morphology and building typology, both currently and historically.
Sub-questions:
-Which are the typological and morphological elements that characterize the specific form of the European City?
-How do transformations in urban morphology effect changes in building typologies, and vice versa?

• *Research by Design*: design studies for urban areas in European cities that investigate the spatial potential for transformation and renewal by means of concrete design proposals: architectural interventions.
Sub-questions:
-Which building typologies, programs and architectural languages can contribute to the renewal of urban areas?
-How to relate new architectural interventions to the existing urban and built structures?

• *Theoretical research*: investigations into the theories, methods and techniques of typo-morphological research and architectural design, i.e. the architectural discourse on the city.
Sub-questions:
-Why and how should typo-morphology be a prerequisite for architectural design?
-Which are the innovative ideas and techniques in the field of design methodology and design studies?

Traditional Architecture and the Swinging Market

Leen van Duin

Some towns in the Netherlands have grown gradually; others were developed on a predetermined pattern. The typical historical Dutch 'water town' is an example of both: its development is mainly based on peat land reclamation patterns. The associated engineering problems, such as the control of surplus water, required the construction of a broad encircling moat and a series of narrow drainage canals. Over a period of time, the parcels along the canals were filled in.

The nineteenth and twentieth centuries saw the development of a string of urban districts on the fringes of Dutch towns. This type of urban extension is still commonplace today, as can be illustrated by the development of large urban projects on what are known as Vinex locations. These banal, mass-produced projects are hidden behind a mask of exotic architecture. They are the exponents of the current situation in the Netherlands, where every project has to be newer than new and can literally be anything. Architecture has gone completely wild. Even a shovelful of traditional methods with a touch of retro is considered new.

It was, paradoxically enough, the Dutch functionalist par excellence, Van den Broek, who argued in 1948[1] that the traditional approach of the *Delftse School* was based on the fundamental rules of form, construction and function, which are significant not only in the tradition of building, but should always be regarded as essential to architecture – in the sense of art as well as science: the protective nature of the house, the enclosure of the wall, the bordering of space. Naturally the traditionalist wants to provide a counterbalance to the all-too-rapid technical and social changes. However, in his search for the essence of architecture, he certainly does not resist innovations and changes at any price. But he tries to connect the most fundamental questions of architecture and architectural design with (local) traditions. Does such a tradition exist in the Netherlands? Is it in fact possible to identify the features of a specific architectural approach?

It is obvious that cultural expressions are not strictly confined within national borders. From the later Middle Ages onwards, a number of Dutch towns for instance were already part of the Hanseatic League – a corporation of merchants around the Baltic Sea – and with the trading contacts this brought also some sort of cultural alliance. A special Northern European tradition developed with a preference for simplicity, the unadorned use of façades of bricks and mortar. The architecture associated with this tradition emanated a quiet and organic quality. The interior of buildings – churches, town halls, cloisters – consisted of simple brickwork, in which columns and pilasters were distinctive only through the incidence of light on the surfaces. The interiors are both natural and pure, forming places within which the hectic everyday life has slowed down.

It might be said that in the flat Dutch landscape, with its wide vistas, rational thought could

---

[1] J.H. van den Broek, **Creatieve krachten in de architectonische conceptie**, Inaugural lecture, Delft 1948.

easily develop from the seventeenth century onwards. The concept of nature and society as exact and manageable categories according to the Newtonian model became immensely popular and an integral part of the Dutch tradition. In his travel impressions from the 1930's, the English writer Aldous Huxley characterised the Dutch polder landscape accurately: 'In a landscape that is the ideal plane surface of the geometry books, the roads and canals trace out the shortest distances between one point and another. In the interminable polders the road-topped dykes and gleaming ditches intersect one another at right angles, a criss-cross of perfect parallels. [When Huxley sees a farm on the other side of the canal, he says] How perfectly it fits into the geometrical scheme! On a cube, cut down to about a third of its height, is placed a tall pyramid. That is the house... Geometry calls for geometry. ... Delightful landscape! I know of no country that is more mentally exhilarating to travel in. No wonder Descartes preferred the Dutch to any other scene. It is the rationalist's paradise.'[2]

Besides urban architecture, we can find in the Netherlands rural and industrial buildings, farms and brick kilns for instance, which were built in a functional way in accordance with ancient traditions. These were buildings without any kind of pretension, without a striking façade, where traditional skills were predominant instead of academic considerations. It is precisely this quality that the Austrian architect Loos described in his plea for a natural architecture.[3]
The unaffected simplicity of the German architect Tessenow, with his sober, refined and highly personal buildings, had a huge influence on the development of a traditionalist movement in Northern Europe from the 1920's onward. It is certainly not the case, though, that tradition provides a ready-made recipe for an architectural design. According to the Italian architect Grassi[4], it takes time and effort to rediscover the mastery encapsulated in quite ordinary things. The house for instance, the most common built object, has survived the centuries virtually unchanged – with the exception, of course of the formal experiments that engulfed us in the twentieth century. Yet, however much Grassi emphasises the continuity of the profession, he puts the personal vision of the architect first and foremost. He clarifies the various positions one can adopt from a traditionalist point of view; i.e., Schmitthenner's or Tessenow's. Both point to a regional tradition, but where Schmitthenner repeats himself with largely nostalgic considerations of the order and ritual of the traditional bourgeois house, Tessenow with a free and virtuoso reworking of the same features, manages to highlight aspects which seem to be forgotten within the tradition: the honest character and value of the forms, their *raison d'être*. His designs surprise us time after time because of their authenticity.

Traditionalists emphasise the tectonics of building, human dignity, the joy of producing something by hand, and resist the blurring of differences in the process of internationalisation and globalisation of architecture. The brick architecture of Kromhout, Kropholler and Granpré Molière illustrates this Dutch traditionalism in the first quarter of the twentieth century. Granpré gave a fine definition of the term, when he compared tradition with the roads in a country area: once a road has been made, no one thinks any more about making yet another one through the open fields. It appears not always necessary to ask why an approach is the way it is. The fact that the road has already been there for a long time makes it worthy of respect and one that should be followed, naturally within certain limits. In Granpré's view this doesn't mean that the existence of traditional form makes issues of functional and structural organisation superfluous, but experimenting with form for the sake of the experiment itself is absurd. Not everything needs to be different all the time, sometimes something is so beautiful that it needs to be repeated. Granpré described three undivided but distinctive angles of approach for the architect from a philosophical and theological viewpoint: the spontaneous inspiration of the artist, the passing down of ideas (read: tradition) and the laws of nature, i.e. the everlasting that is embodied in nature by the Creator himself. An architect should strive for a harmonic unity between these three angles of approach. Granpré searched for an eternal architecture – which he wanted to anchor in the Catholic religion and the associated social ideals – and in doing so he acquired quite a few followers grouped in the Delftse School. Hans van der Laan was one of them.

The Delftse School wanted to achieve harmony, the unity of man and his environment, a situation in which man, according to the doctrine of Thomas Aquinas, can find himself: the unity of body and soul. Only if the architect was to become a craftsman again, could society be protected against uprooting by mechanisation and far-reaching division of labour in the nineteenth and twentieth centuries. One studied medieval architecture, in particular the Romanesque, in which the relationship between matter and soul was supposed to be still intact. This study was not so much intellectual and theoretical in its intent. One believed in the importance of seeing history with one's own eyes and organised study trips, particularly to Italy. Siena above all was the focus of interest, because old architecture should be experienced. On sketching excursions, one sharpened one's own viewpoints and capabilities by observing historic buildings in their context. Sketches were made over and over again, in order to understand even better what one in fact already knew, and one thus become expert in the language of past form and its potential use in the future.

Under the influence of Van der Laan, the attention among traditionalists shifted in the 1950's from ethical and philosophical issues to a system of intrinsic relationships, and thus the specific laws of architecture. Just as many architects before him, from Vitruvius to Le Corbusier, Van der

2 Quoted from: M. Reints, Nacht- en dagwerk, Amsterdam 1998.
3 A. Loos, 'Architektur', in: **Trotzdem**, Vienna 1932.
4 G. Grassi, 'Een mening over het onderwijs en over de voorwaarden van ons werk', in: **Oase** no. 28, 1990.

Laan was interested in the connection of figures with architectural ideas, which resulted in his theory about a proportional system The Plastic Number (1960).[5] In this study, he placed a range of quantities of a building one on top of another as coefficients, so to speak. The dimensions of the elements of a building – columns, walls, floors – are related to each other, i.e. they are traced back to a system of dimensions from which the relationship between the elements and the whole can be identified.

Even the architect Rietveld expressed his admiration for Van der Laan's work: 'Two aspects appeal to me in particular... The first is that the ordinance does not want to be any more than a lesson in the correct and consistent use of dimensional units and forms, leaving the architect totally free in the way he designs a scheme for these dimensions and forms as a result of the function and construction; the second is the confidence that speaks from this work, i.e. that the correct use of this ordinance necessarily leads to the desired architectural expression and harmonic unity.'[6] With this second characterization in particular, Rietveld emphasized the rationalist aspect in Van der Laan's approach. In doing so he made a direct connection between this form of traditionalism, known as the *Bossche School*, and the geometry and proportional systems which had such a long tradition in the Netherlands, and in which Berlage, and later on the architects of De Stijl, were so interested.

Under the influence of the morphological studies of *Tendenza* – Rossi, Grassi, Tafuri[7] – the 1980's saw a resurgence of interest in the Netherlands as elsewhere in tradition and historical continuity. One doesn't need to believe in eternal values, but searching for possibilities of continuity and durable building, one has to see Van der Laan's work. I myself became interested in his work through a brilliant article by Padovan in 1984.[8] He compared Van der Laan's theories with those of the Italian rationalists of *Tendenza*, who enjoyed such international interest after Rossi's exhibition *Architettura Razionale* of 1973. For the Rationalists as well as Van der Laan, architectonic form doesn't originate from a minute response to specific functions, but from the general idea of 'being housed', and the universal logic of building.

In the same period, the Luxembourg architects Rob and Leon Krier tried to resist the indifference and demolition of cities in the second half of the twentieth century. In the exhibition *Rational Architecture* of 1975, they argued unequivocally for a traditional approach to architecture and for the reconstruction of the European city. According to the Kriers, the city as a system of spaces – streets, squares, city parks – has been replaced by the modernists with an object-orientated collection of buildings, which can only be explained from administrative regulations and with no other relation to each other than a transport system:

mobility and the contemporary network city. Just as *Tendenza*, they asserted that morphological studies should form the basis for any architectonic intervention.
In my view, only typological studies can produce knowledge about opportunities for buildings that are characterized by continuity, clarity and exactness, and can be used for a variety of functions. In doing so, we have to redefine a number of categories in order to explore possibilities of architecture today: the link between form and function, the meaning of dimension, scale and proportion of hybrid buildings, the role of conventions and architectural history.

As stated above, under the influence of the postmodern movement architecture has gone wild. Among Dutch architects, it was above all Sjoerd Soeters who, around 1980, championed heterogeneity as an antidote for the rigidity of the modern movement. To be sure, this resulted in spectacular designs, but the 'anything goes' principle was also to blame for the unfortunate practice of randomly throwing together a variety of style quotations. The architect came to resemble a knowledgeable conversationalist, who cynically and without a hint of analytical reflection, is never at a loss for the right fashionable words. His first priorities for a building were that it please the eye and appeal to the world of advertising, with a corresponding look of total comfort and luxury. This new architecture has a cinematic character; its intention is to offer a total experience, providing real-time sensual stimuli against the backdrop of the city as a kind of better-than-life, illusion-filled theme park. Market principles play a dominant role in such architecture – the customer is always right – and the architect has to supply whatever the client wants.

Much is being said about the confusion of this development and, parallel to it, an absence of appropriate architectural criticism. In countless publications, a range of attempts at opinion-forming on the matter are currently being made, but – with the exception of project developers and, naturally, Prince Charles himself – no one knows what to make of the unabashed longing for good old-fashioned architecture. Based on an analysis of Europe's oldest architectural magazine, Britain's *Architectural Review*, the historian Ed Taverne anticipated today's internet chat when he observed in 1978 that falling back on the architectural movements of the past, with their easily recognisable symbols and rich, associative potential, had a soothing effect on a powerful group in society: the comfortable middle-class consumer. In the meantime, the architect's role as satisfier of society's longings would seem to have attained general acceptance. The architect renders his services to administrators, land owners and clients desiring to realise a romantic, picturesque setting contrasting with the dominant (but ground-losing) architectural style in the spirit

---

5 Van der Laan, **The Plastic Number** (1960).
6 G. Rietveld, 'Boekbespreking Le Nombre Plastique', in: **Bouwkundig weekblad** no. 18, 1960.
7 The studies of **Tendenza** are focused on the rules which regulate the city as an artefact. It creates a framework for architectonic interventions. **Cf. e.g.**: A. Rossi, **L'architettura della città**, (Padua 1967) Dutch edition: **De architectuur van de stad** (SUN Nijmegen, 2002). G. Grassi, **La construzione logica dell'architettura** (Padua, 1967). Dutch edition: **De logische constructie van de architectuur** (SUN: Nijmegen, 1997). M. Tafuri, **Teorie e storia dell'architettura**, (Rome, 1968).
8 R. Padovan, 'The Rational Architecture of Hans van der Laan', in: **Transactions** vol. 3 no.1, 1984.

of the query: 'What is it like to live in a picture?' Now that everyone can buy his own home, residential areas are turning into theme parks: 'to your left the Trojan dwellings, to your right a baroque district'.

Brandevoort, the new retro-neighbourhood being built in the Dutch town of Helmond, is already the subject of international attention, especially in England of course. A few hundred of its planned 6,000 dwellings are now finished. The plan includes a fortified centre, encircled by picturesque ramparts and a moat, in turn surrounded by the suburbs Brand, Schutsboom and Stepekolk. As in Almere or on the banks of Amsterdam's IJ, where the city's canals were copied, in Helmond – the middle of nowhere – medieval defences are being built, telling us the story of a Dutch city that never existed before now. The architecture has the look of an ensemble of old Dutch houses which has gradually developed in the course of time, and which has been given a historic but affordable appearance. The originator of the urban plan is Rob Krier. One might ask oneself why a retro architect like Krier enjoys such popularity in the Netherlands. The answer is not very hard to find: a housing association – in the Netherlands often the commissioning party – will just as easily opt initially for Mecanoo or MVRDV only to switch a week later to Krier or one of his co-religionists. Sometimes home consumers want a modern lifestyle, other times a traditional one, and the customer is always right – right? Today's home trends are heading in the direction of an invented past. They have nothing to do with the preservation of traditional values; it is all about market: what sells sells.

This way of thinking rapidly gained ground from the mid-nineties onwards and explains how for example Almere got a Belgian castle, why Coevorden is also planning one, as well as the fact that in the town of Haverleij near Den Bosch 1,000 dwellings in the form of compact castles are presently under construction. And it comes as no surprise that most of these developments are the brain-child of Soeters again. With only five dwellings per hectare, the building density in Haverleij is so low that each house could easily have been freestanding. But instead, the architect concentrated them in castles and a small fortified town with 450 dwellings, which Rob Krier is designing. A seventeenth-century fortified town, English country houses, Loire castles – it is all possible, as long as the architecture is recognisable by today's Digital and Hypermobile Man. The example of Haverleij demonstrates pre-eminently how retro architecture can express a past which never existed. In 400 years' time, when archaeologists uncover the remains of Haverleij, they will face a special challenge when it becomes clear that the buildings which once stood there were concrete replicas of not existing buildings, with a thin skin of brick, designed by the same architect who gave the Vinex location Brandevoort its historical touch.

In high-brow circles, opposition to the retro movement is growing. Architects who still cling to their crumbling status as an intellectual and artistic avant-garde sneer at the tastes of the masses. The trend, however, is clear: the norm is no longer a quest for modernity. A medieval castle, or any retro fantasy can all be contemporary. Nevertheless, it would be a grave mistake to confuse the retro architecture of today with traditionalism. At the time it was an approach which placed great importance on durability and defined transformation in terms of continuity and gradual development. Retro now swings with the market. I see no connection at all between the traditional architecture of, for example, the Delftse School, and the retro machine of today. Retro is a condition of our time. I confess I allow myself to be seduced by the sense of security and light-heartedness in the creations of the Kriers and company. But what I want to dwell in, though, is the apparent everydayness in the traditional architecture as Loos described.

For the future Large Urban Projects in the Netherlands, research is needed to explore the durable development of the European City. Not to copy them, but to move with caution between historical knowledge and today's questions. Such a research project should be included in schools of architecture, because only there can students and professionals work free from the pressure of the market. It is this approach I want to explore in the faculty of Architecture in Delft in the coming years. At first we shall focus on the Dutch city. Should our studies lead us back to Siena we will let you know.

[1]

[4]

[2]

[3]

[6]

1 Traditionalism in the Netherlands: housing in Vreewijk, Rotterdam, c. 1920.
2 Brandevoort: example of a suburb, c. 2002.
3 Haverleij: maquette of the small fortified town.
4 Brandevoort: the canal, c. 2002.
5 Brandevoort: the market place, c. 2002.
6 Haverleij: example of a castle.
7 Haverleij: maquette of a castle.

Next page:
8 Brandevoort: retro-neighborhood of the Dutch town of Helmond, c. 2002.
9 Brandevoort: the moat.
10 Brandevoort: plan of fortified center, encircled by suburbs.
11 Haverleij: plan of a castle.
12 Haverleij: plan of a small fortified town surrounded by castles in a golf course.

[8]

[10]

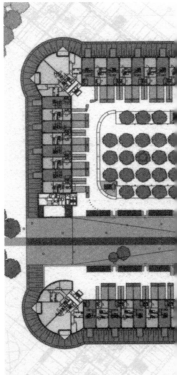

[11]

[9]

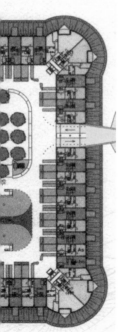

[12]

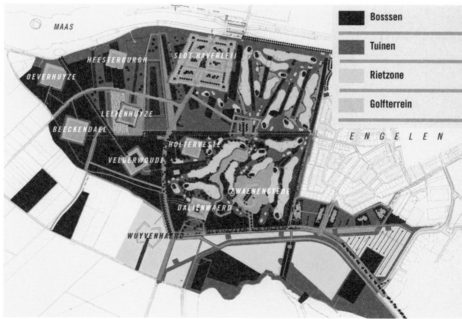

To Create Order out of the Desperate Confusion of Our Time

Adalberto Del Bo

In this paper, I will try to address the various issues raised in the situation in Italy, and the contribution of Italian studies to urban design and the idea of city. Specifically, I will speak about the results of research on the works that Hilberseimer and Mies van der Rohe conducted together, as well as work that has been carried out in the PHD program in Architectural Composition.

The unifying concern of cities and architecture in Europe today, is the series of problems that are affecting its territories – those heavy and common issues related to the process of transformation brought about by a strong and widespread settling pressure for which adequate responses have not been identified which could be satisfactory in terms of architectural order and organisation. The most common words used to describe a city are: *complex, manifold, fluid, metamorphic, a-local, ephemeral, random, chaotic, fragmented, indefinite, open, scattered*. I would also add the term uncontrolled – even if the concept of control is today out of fashion. All of these concepts are seen by trustworthy sectors of architectural culture in positive terms; perhaps because only *chaos* is believed to be an effective expression of the accelerated and incoherent contemporary condition. Yet it is necessary to pause in order to screen the situation in a severe and, at the same time, ingenuous way. We can try to understand the origins of such a deep crisis and evaluate those examples to be taken as a starting point in the search for possible alternatives among the general formulations defined by architectural culture up to now.

For example, the current phenomenon in Italy mainly has to do with a process of residential expansion which took place in the absence of plans whose dimension would be adequate to the size of the issues to be tackled. Contemporary planning is unfortunately still based on the undersized administrative dimensions that were established in the past. These processes of expansion, by indiscriminately extending the outer urban areas, have caused nuclei that used to be separate, to be now joined together, thus characterising vast areas once rich in history, memory, identity, and form with a strong indeterminacy. This amalgamation has caused, besides a disorder that cannot be easily cured, a widespread sense of disorientation.

The phenomenon, especially in the Northern areas, has to do with a significant migration of people leaving the city perceived as being too polluted, costly, and dangerous, and moving to outer areas. These persons mostly dwell in one-family houses with a garden. Besides this phenomenon and within this altered framework of relationships, the settlements of the past which used to be cornerstones in the historical constitution of the territory, are now transfigured and tend to lose their identity as well as their strategic role as regulating elements of the settlement. These demographic phenomena are today widely known, especially thanks to studies conducted by sociologists, geographers, artists, and men of letters. On the other hand, urban planning is unfortunately absent from this debate, protagonist of a serious failure in the functional and qualitative control of cities and regions.

Italian culture has not yet been able to find an adequate answer to this situation, both from a building-typological and from a morphological point of view, nor has it attempted to orient the work in this direction by promoting studies, researches and public competitions on the topic. European architecture should hold a vast debate on these common issues that deal with the shape of settlements and the subsequent idea of city that should be taken as a reference. Schools of architecture should especially be involved in this debate since they are entrusted with the task of training young people and of orienting, through research, the choices made by society towards the future.

The urban problems we are facing today in Europe are more or less still the same ones listed in 1933 in the Athens Charter, though unfortunately worsened by time. The Athens Charter is an extremely interesting text that shows very concisely the clarity with which, seventy years ago, the architects belonging to the Modern Movement were capable of evaluating the state of cities, of pointing out necessary solutions and of identifying the potential role of architecture. As everyone knows, the document was compiled by Le Corbusier who published the 95 Points in 1942 when architects from the European cities in ruins looked beyond the war toward reconstruction. The first decade after the Second World War was a time of intense thinking on the city and on architecture (and not only in Europe under reconstruction). It is a crucial period of time that has been little considered by critics and not yet well investigated. During that time, some very significant buildings were built and there was an extraordinary concentration of studies and formulations on the new city made by some of the masters of the Modern Movement who were then in their full maturity.

In 1946 Le Corbusier, responding to a survey of a magazine, foresaw with far-sightedness the transformations that American cities would undergo and the similar risks the French territory could run. 'To the endless suburbs of New York and Chicago will be added new and endless suburbs. An increased urban Catastrophe! ... The great waste of the garden city will keep ruling...' At the same time he warned France of the danger of getting this problem in its own territories. 'France should do the opposite of the United States and shouldn't waste the things nor the existence of its children. It must do so compelled by its size, by its current poverty, by the spiritual balance that distinguishes it.'[1] Fifty years later it can be said that the cry for the French situation and culture did not come to anything.

The situation in Italy was quite different. Those specific issues did not seem to have especially interested the architectural culture of that time; and also today the interest is more in poetics and in language than in issues dealing with the shape of the city, of the territory, and of the settlements. Continuity as a characteristic element of the dialectic relationship with the transformed reality, and hence the need of a deep knowledge of reality itself, accounts for a feature that spanned architecture over time, with the awareness that the nourishment of architecture is life and architecture itself.

To this purpose, it is significant to recall Le Corbusier's note[2] in which he puts six points, starting from an initial act of looking, tracing the order of an extraordinary sequence of growing analytical knowledge through a process of deep understanding that arrives at knowledge and comprehension in seeing: *'La clef, c'est: regarder… Regarder, observer, voir, imaginer, inventer, creer'*.[3] According to the sequence, *seeing* and *imagining* are the (somehow inseparable) contact points between analysis and project, the crucial point of the dialectical passage in the design process between reality and its transformation (the latter is somehow already present in reality itself, since *understanding* is already *imagining*). *Invention* and *creation* are, then, very architectural operations (typological and constructive); they rely upon decisions that are aimed at identifying and establishing forms more directly.

The six points above are probably no less solid, enduring and eloquent than the celebrated five points that Le Corbusier entrusted to the history of modern architecture. Besides being a useful key to understanding the architecture of the Swiss master, the points are a general direction and hint on the procedure in the relationship between analysis and project, being a tool with which one can measure the most specific contributions that historical-typological analyses can yield.

Researches on the founding characters in architecture are undoubtedly precious contributions for knowledge, given the input they are capable of offering to a project and to the construction of a theory of design. Italian research (the Milanese and Venetian ones especially) has been able to build around the mid-sixties within the university a true school of architecture, based upon the search for general principles that were shared and recognizable. This position has its origins in the reaction that developed in Italy after the Second World War against the idealistic culture (which especially flourished with fascist provincialism): a neo-enlightenment position that aimed to assign to reason and to rational survey techniques a positive cognitive function.

The writings of Aldo Rossi, Carlo Aymonino, Giorgio Grassi and others, that started from these political-cultural assumptions, were mainly based upon the thoughts of European culture and in particular upon the French, British and German studies dating back to the early 1900's and between the two wars. In this vast and profound operation of renewal of urban studies and in

---

1 Le Corbusier; **Propos d'urbanisme** (Paris: 1946).
2 Autograph note by Le Corbusier, Cap Martin, 1963.
3 Le Corbusier, **op. cit.**

establishing a solid position on the city and on the project, lies the construction of theoretical positions that had a wide international
echo and that promoted a strong diffusion of Italian architectural culture. As Massimo Scolari recalls:

> Urban analysis and the concept of typology were the signal of a different way of considering urban facts and marked a more conscious project. There was a commitment to a design that wouldn't merely be superimposed on the city but would rather get into its historical and cultural roots, thus rebuilding a dialectic relationship with the pre-existing typological, morphological and monumental elements.[4]

This project is a fine tuning of the characters related to the typological-morphological analysis of urban facts founded upon scientific methodologies applied to historical materials using a comparative procedure and techniques based upon the fact-finding inquiries of cartography, topography and building surveys. If we set apart the sterile quarrel on the scientific nature of architecture (the question of whether it is a science or not, this naming issue is effectively solved with the definition of architecture as 'art and science at the same time'), the issue of the dialectic relationship between transformations and the knowledge of local characters has been a problem not easy to solve for the many naive people who believe that the relationship between analysis and project is mechanical and that the project can directly follow the results of the analysis. After all, the discourse on the presumed objectivity of the analysis and on the supposed subjectivity of the project (as well as on the respective opposites) cannot but stay open, given the fact that the analysis, though based on facts, is nonetheless and inevitably an interpretative operation.

Given the failure of the control of the city (let's think about the huge problems of traffic and pollution, besides the self-evident disorder), the main topic that architectural culture has to face and find possible solutions for is the idea of city; as it has always happened in history, it is important to point to general alternative solutions for urban and territorial transformations, since it is not allowed to witness the continuous deterioration and worsening of the situation without attempting to define some alternative solutions.

The studies on the idea of city advanced pretty significantly in the early years after the Second World War. One can think of Hilberseimer's projects for Chicago, and the built hypothesis of the *Unité d'Habitation*; this is a chapter which should be deeply reconsidered and with no prejudices looking at the experience had up to now (which is not at all negative). The theme of the idea of city is today neglected and considered anachronistic by two opposite positions: on one side, the widespread mistrust in the ability of architecture to contrast the current transformation processes; on the other side, the fear (an ideological fear that originates from the frequent confusion between the idea of city and the idea of ideal city) that architecture tends to force upon the city a highly-demanding vision. These are the positions of those who surrendered to chaos and of those who believe chaos is an inevitable phenomenon and is thus a full and accepted expression of the contemporary condition.

In his first speech as Director of the school in Chicago, Mies van der Rohe points out as an objective the need 'To create order out of the desperate confusion of our time'.[5] And he adds, always speaking of order in architecture:

> Order is more than organization. Organization is the determination of function. Order, however, imparts meaning. If we would give to each thing what intrinsically belongs to it, then all things would easily fall into the proper place; only there they could really be what they are and there they would fully realize themselves. The chaos in which we live would give way to order and the world would again become meaningful and beautiful.[6]

This passage is especially significant from a theoretical point of view which was confirmed by Hilberseimer himself in his writings, and was at the basis of their collaboration, dealing with the identification and the distinction between the components that contribute to defining the architectural project. Besides the obvious dispute with the functionalist positions (that expected to mechanically identify order in functional organization), Mies states a methodological issue that deals with the theory of architectural design. He identifies the greater complexity and the autonomy of the ordering activities aimed at shaping the whole as well as the crucial role of organizational aspects as an integral part of the project, as a complex of decisions that are intimately bound to the project itself.

It is crucial to reassert this position especially today when architecture tends to fulfil the task of clothing surfaces; a sort of decorative role, not far from the stylism of fashion. It is curious, to this purpose, to notice that this happens when in electronics – which is an activity that is far from architecture (from a disciplinary point of view) though central in the contemporary world – the word *architecture* is granted an ordering and structuring role by means of the introduction of terms such as *Systems Architecture, Architecture of the Web* or *Information Architecture*.

The Lafayette Park project in Detroit (in the second half of the 1950's) is here presented as an excellent example of an interpretation of the characters of the new city starting from the Euro-

---

4 Scolari, Massimo; 'L'impegno tipologico', **Casabella** no. 509, 1985.
5 Mies van der Rohe; Inaugural address as Director of the Department of Architecture at Armour Institute of Technology, Chicago, November 20, 1938.
6 Mies van der Rohe; 'Miscellanea, notes for conferences', in Neumeyer, Fritz; **Mies van der Rohe. Das kunstlose Wort. Gedanken zur Baukunst** (Berlin: 1986).

pean experience. This project is an extraordinary, though little known, piece of work that derives from Hilberseimer and Mies van der Rohe working together in their full maturity. The collaboration between the two German masters, besides being a fundamental step for understanding the history of architecture in the Modern Movement and afterwards, is also a clear example of the proposed distinction between order and organization. The project shows almost didactically the organizing principles expressed by Hilberseimer in his theoretical work, which Aldo Rossi properly indicates in the preface to the second edition of *The Architecture of The City* (1970) as a significant example of a research '… where the analysis of the city and the construction of the architecture, one strictly depending on the other, are aspects of a general theory of rationalism in architecture'.[7] The organizing principles are the following: building within nature; elimination of car traffic; easy pedestrian access to parks, schools and shopping centers, etc. without the need to cross any roads; construction of mixed-use buildings (*Mischbebauung*); attention to the best sun exposure and to the shadows cast by tall buildings onto the land occupied by small houses, elegant row houses or houses with patio (in the park) designed according to the best Western tradition.

The Lafayette Park project, inserted in the urban grid of downtown Detroit (among the most troubled Western cities nowadays victim of a devastating process of implosion), replaced a slum that was entirely demolished in application of a federal law that was strongly opposed after the Second World War. The project, surrounded by throughways, consisted of two residential parts connected by a wide central area for parking, and immersed in a green surrounding. The project is a complete replacement of the previously existing situation, of which no traces are left; a true new beginning, a return to the natural original condition with no reference whatsoever to the past. The streets entering the settlement (which are oriented like the grid of Detroit) are alternate and this allows easy access to the houses from the parking areas which, like all streets, are located one level below the houses.

The shape of the Lafayette Park settlement is defined by a *planimetry* that was conceived by Mies van der Rohe with an extraordinary attention to the elements and to their relationships. The spaces defined by the low houses are identified by an orthogonal distribution that can locate, within the variety of the alignments, a complex and open structure of interior spaces, of successions and of perspectives that find in the masterly arrangement of the green areas, a spatial dimension that is domestic and collective at the same time. The low houses are designed according to a clever typological and structural study that is aimed at achieving a direct relationship with nature by means of glass façades that also contribute to making the buildings look like extremely light structures. The rooms of each house allow open perspectives onto the natural surroundings thanks to a careful use of technology. The painted steel structure and the large aluminium window frames are the elements that link together, in a unitary though measured way, the various buildings. They represent, in their relationship with the new landscape of the area, an artificial perfection.

The design and the location of the tall buildings identifies a double and articulate level of spatial relations: those defined within the settlement as a whole (a relationship with its own surroundings at the ground level of the low houses) and those related to the city where each tall building takes part in the overall spatial dialectic. In the end, the scheme is a singular, tight relationship ruled by Composition, and inside which buildings are spatially related by height and, at the same time, are connected with their surroundings at the ground level for which they are the most significant reference. This type of organization, of course, refers to an idea of life that is different from that of the current city, and seems to match a mixed and intermediate character between countryside habits and urban life. This condition is not alien to those inhabitants who, for a series of reasons, decided to live in the outer areas of the city to pursue the idea that the success of the garden city, in the several and conflicting versions defined over time, had partly decreed as an ideal of life. It must be recalled, to this purpose, that after 50 years of experimentation, by now Lafayette Park is protected by the association of residents and that in the 90's it was included in the *National Register of Historic Places* of the United States. A study is currently investigating the nature of the relationships between the elements and their shape in the Lafayette Park project. This study has pointed out the decisive role played by the use of the golden section in defining the architectural order and in creating spaces and volumes that are connected in one single process of formal definition.

This process is applied to urban construction according to those systems that were present in the layout of cities founded in the Middle Ages; a quintessential process that has been carried out with absolute precision (and it couldn't be otherwise) by means of an extremely complex system of relationships that has to do with the control of the overall design (usually geometric) and the use of golden sections for the spatial relations and for the design of the surfaces. This kind of attention has to do with artistic practice and the theoretical construction on art. These aspects were very much alive in the avant-garde debates between the two World Wars – at the Bauhaus where both Mies van der Rohe and Hilberseimer taught along with various artists, among which Wassily Kandinsky and Walther Peterhans, who were engaged in the theory of figuration. Towards the end of his life, Mies said in an interview:

> The problem of the building art has really always been the same. The qualitative is achieved

7 Rossi, Aldo; Introduction to the second edition of **L'architettura della città** (Padova: 1970).

through proportions in the building, and proportions do not cost anything. For the most part these are proportions between things… It is of course much work for the architect to articulate the in-between spaces. The artistic is almost always a question of proportions.[8]

Mies's words are extremely clear; there are no trade secrets here. Even though the words are different from Le Corbusier's, the call for the employment of the traditional elements of architectural composition is considered critical. The plan form of the IIT project in Chicago and its extraordinary construction confirm the indicated line of work and point out, in the analysis of the various subsequent solutions, the presence of a deep ordering structure and a high degree of complexity in the relationships that were established.

As with Le Corbusier's Modulor's construction, Mies van der Rohe's adoption of systems of golden proportions is a very significant point. By means of a plea for order and for classical tradition the German master tends to oppose any possible form of subjectivism, arbitrary act and gratuitousness in defining architecture (because the 'authentic quality' we are talking about cannot but be architecture itself). He points out the need to search for stable and shared forms, for elements that find their very explanation in themselves and in the relationships they establish among themselves. One finds in this plea the constant quest for objectivity, for that *Sachlichkeit* that, along with the adhesion to classical qualities – the very essence of the European city – was the constant reason, first of all in ethical terms, for a research that represents a precious patrimony to be investigated more in depth, especially today in times of hard subjectivity.

This research deals with topics that are very close to the concept of *Concinnitas*, a rhetorical notion that in Alberti's theory becomes a true aesthetical category, the supreme rule for the perfection of the work of art. A research, in the end, that allows a positive answer to the questions posed by this Conference on the persistence of the architectural idea of a 'European city' and on the innovative ideas and techniques in the field of design methodology and design studies.

The work, developed in a design workshop as part of the Ph.D. in Architectural Composition and presented here[9], deals with an alternative version to the project they are going to build in the outer area of Malpensa's international air-

---

8 Mies van der Rohe; Video Interview, 1968.
9 The description of the Design workshop in the Ticino area (2003) refer to the images 6, 7, 8.
The students in the Ph.D. workshop were: Sara Biffi, Francesco Bruno, Stefano Cusatelli, Paola Galbiati, Andrea Palmieri, and Francesca Scotti.

port in the northern part of the Milanese metropolis. The airport is located in the park of Ticino, a wide territory, historically and environmentally rich. Old and famous settlements have been built on the canal's network, beginning in the 13th century in order to put Ticino's river in connection with Milan, a huge and clever engineering and architectural work. The project looks forward to the relations between the settlement's structures (houses, hotels, Research Centers) and the territory's shape, with a particular regard to the residential settlement's thematic and its connection within the local typological character, particularly with the great rural courts of the Lombard tradition.

[1]

POWER PLANT	8. GYMNASIUM AND NATATORIUM	15. HUMANITIES (LEWIS BUILDING)
METALS RESEARCH	9. INSTITUTE OF GAS TECHNOLOGY	16. CHEMISTRY
ENGINEERING RESEARCH	10. LITHOGRAPHIC TECHNICAL FOUNDATION	17. METALLURGY AND CHEMICAL ENGINEERING
AUDITORIUM AND STUDENT UNION	11. RESEARCH LABORATORY	18.
ELECTRICAL ENGINEERING	12. ARMOUR RESEARCH FOUNDATION	19. FIELDHOUSE
CIVIL ENGINEERING	13. ARCHITECTURE AND APPLIED ARTS	20. ATHLETIC FIELD
LIBRARY AND ADMINISTRATION	14. MECHANICAL ENGINEERING	

[2]

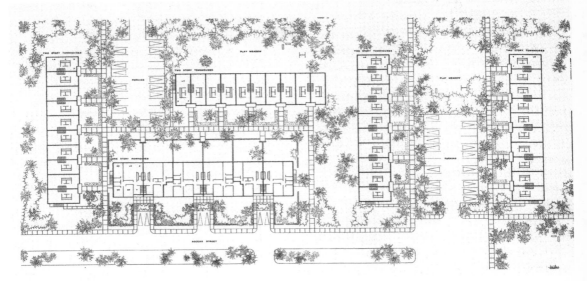

[3]

[4]

1 Photomontage of Gratiot Redevelopment (Lafayette Park); site model, Detroit, 1955.
2 Ludwig Mies van der Rohe, photomontage of aerial photograph and model of site plan, IIT Campus, 1940.
3 Detail of the plan with townhouses and courtyard houses, Lafayette Park, Detroit, c.1956.
4 View from Lafayette Park.

[5]

[7]

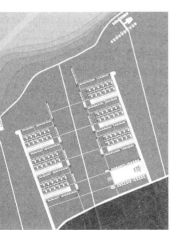

5 PhD Design workshop. Model view.
6 PhD Design workshop. New settlements in the Ticino area.
7 PhD Design workshop. The plan with patio houses and great courts.

Merz City

Henk Engel

The tremendous historical need of our unsatisfied modern culture, the assembling around one of countless other cultures, the consuming desire for knowledge – what does all this point to, if not to the loss of myth, the loss of the mythical home, the mythical maternal womb? Let us ask ourselves whether the feverish and uncanny excitement of this culture is anything but the greedy seizing and snatching at food of a hungry man – and who would care to contribute anything to a culture that cannot be satisfied no matter how much it devours, and at whose contact the most vigorous and wholesome nourishment is changed into 'history and criticism'?
Friedrich Nietzsche, *The Birth of Tragedy*, 1872.[1]

In 1992, Ed Taverne asked me to write an article for *Stedebouw, De geschiedenis van de stad in de Nederlanden van 1500 tot heden* (Urban design, The history of the city in the Netherlands from 1500 to the present day). The subject assigned to me was 'The shape of the city – the Netherlands post-1945', to be covered in 3,000 words and a few illustrations. The result, in cooperation with Endry van Velzen, was little more than a first introduction; ever since, this subject has never left my mind.

The initial opportunity to approach this problem in a wider context came with an entry to the international competition *Inside Randstad Holland*, held by the E.O. Wijers Foundation in 1995.[2] The second opportunity came with the study *The Post-war City, a Contemporary Design Task* commissioned by the Stimuleringsfonds voor Architectuur (The Netherlands Architecture Fund) in 2000.[3] Besides these two macro-studies, mention should also be made of a micro-study, an analysis of the historic centre of Gouda, carried out in support of a development study for the Bolwerk site, commissioned by the local authority and supported by the Stimuleringsfonds voor Architectuur.[4]

In the last of these studies it became clear how much typo-morphological urban research could benefit from the construction of a proper macro-level which would provide a structure for micro-level analyses. In fact, the issue was a re-evaluation of the 'socio-economic survey' which typo-morphological urban research had shown in a bad light.

The 1920's saw the introduction of urban design research in support of town planning, with pioneering work being done by T.K. van Lohuizen in particular.[5] Urban design research made a dis-

---

1 Nietzsche, Friedrich; **Basic Writings of Nietzsche**, 'The Birth of Tragedy', translated, edited, and commentaries by Walter Kaufmann (New York: Modern Library, 1992) p. 136,§23.
2 Motto: 'Overholland', Henk Engel, Endry van Velzen (De Nijl Architecten), in cooperation with H+N+S Landschapsarchitecten. Publication: van Blerk, Henk (ed.); **Inside Randstad Holland, Designing the inner fringes of Green Heart Metropolis**, appendix to **Blauwe Kamer Profiel**, no. 3, (1995) pp. 28–29 and pp. 58–62.
3 Hereijgers, Ad and van Velzen, Endry et al.; **De naoorlogse stad, een hedendaagse ontwerpopgave** (Rotterdam: Nai, 2001).
4 De Nijl Architecten, **Het Mierennest, bebouwingsstudie voor de Bolwerk-locatie** (research report, internal publication, Rotterdam 2000). Publication design: **Landschapsarchitectuur en stedenbouw '01–'03** (Bussum: Thoth, 2004) pp. 82–85.
5 van der Valk, Arnold; **Het levenswerk van Th. K. van Lohuizen 1890–1956** (Delft: Delft UP, 1990).

tinction between a 'technical survey' and a 'socio-economic survey'. A 'technical survey' was concerned with recording the physical state of the planning area, with particular attention to soil conditions, drainage, existing ground utilisation, building stock, and the ownership situation. A 'socio-economic survey' chiefly involved research into factors determining the pattern of population distribution.

The urban design research field quite soon found itself overshadowed by a single objective, the objective of the 'socio-economic survey', that is to say the prediction of future developments such as increases in population and traffic. In particular the development of scientific planning research in support of the government's spatial planning led to the total disappearance of the 'technical survey'. It should therefore not be surprising that the subsequent research and design still being advanced by Van Lohuizen and Van Eesteren as forming 'the unit of urban design activity' began to drift apart.

Since the 1980's numerous attempts have been made to bring design and research back together. Inspired by foreign examples, increasing use has been made of typo-morphological urban studies to support design.[6] With a little goodwill it could be said that in contemporary urban research the accent has shifted in favour of the 'technical survey'. However, the most important objective of this type of research is not always the gaining of a more profound insight into the immediate physical context. In a more general sense, the objective of the study of formal urban structures is to derive 'design instruments' which will give some guarantee of improved urban quality.

I see my contribution to this anthology as an opportunity to reconcile this pursuit of 'urban architecture' – which occurs in the Netherlands in such diverse work as that of Aldo van Eyck, Rem Koolhaas and Rob Krier – with the 'process of urbanisation' in order to provide a framework for more detailed urban studies. The object to be analysed in this case is Randstad Holland, the last century's most remarkable creation of Dutch urban design, with guidance provided by the map of the reconstruction of the Randstad since 1850. This map shows the urbanisation of the Randstad area at four stages, 1850, 1940, 1970 and 2000.[7] An important inspiration for this map was the map showing the development of towns in the Netherlands prior to 1795, as prepared by J.C. Visser for the second edition of the *Atlas van Nederland* (Atlas of the Netherlands), volume 2: *Bewoningsgeschiedenis* (History of population distribution, 1984). The method used by Visser was borrowed from the socio-economic survey, applied retrospectively.[8]

The aim of the map of the reconstruction of Randstad Holland was to provide an overview. Cartography is an excellent way of making urbanisation understandable as a physical and spatial phenomenon. The map was based on one simple parameter, the expansion of the built-up urban area. To make it possible to understand the time dimension of the process of urbanisation, use was made of a 'morphological periodisation' based on four kinds of urban fabric which even today can still be recognised in various Dutch towns as clearly distinct districts.

Randstad Holland   *Only as creators! – This has given me the greatest trouble and still does: To realize that what things are called is incomparably more important than what they are. The reputation, name, and appearance what it counts for – originally almost always wrong and arbitrary, thrown over things like a dress and altogether foreign to their nature and even to their skin – all this grows from generation unto generation, merely because people believe in it, until it gradually grows to be part of the thing and turns into its very body... We can destroy only as creators. – But let us not forget this either: it is enough to create new names and estimations and probabilities in order to create in the long run new 'things'.* Friedrich Nietzsche, *The Gay Science*, 1882.[9]

With its 6 million inhabitants, the Randstad is currently the most densely populated region in the Netherlands. Forty per cent of the Dutch population is concentrated in this area. The total area of the Randstad is comparable to that of the urban agglomerations of London, Paris or Milan. The difference is that the Randstad is not centred round a single dominant city. The four largest cities in the Randstad – Amsterdam (727,053 inhabitants), The Hague (440,743), Rotterdam (592,665) and Utrecht (232,718) – are all relatively small. Administratively speaking, the Randstad is a hodgepodge of some 35 local authorities, distributed over three provinces and between eight and ten district water boards. In contrast to classic metropolises like London, Paris and Milan, the Randstad is a group of towns, a network of towns and villages, of which some 25 originated as historic market towns.

The operational significance of the Randstad lies in the field of national spatial planning. In fact the Randstad is a recent invention, devised with the aim of putting Holland on the map as a metropolis. In 1750, Amsterdam was fourth in size on the list of European cities, after London, Paris and Naples. By 1850 the capital of the Netherlands had dropped to sixteenth place, subsequently ending up, in 1950, at number 25[10], at which time the Randstad would have come seventh. The Netherlands would suddenly have counted again, a point that is not without importance when it comes to attracting multinational companies and institutions. But at the

6 Engel, Henk and Claessens, François; 'Massawoningbouw, object van stadsanalyse en architectuur', in Komossa, Susanne (ed.); **Atlas van het Hollandse bouwblok** (Bussum: Thoth, 2002) pp. 266–275.
7 A first version of sections of this map was prepared for Hereijgers, Ad and van Velzen, Endry (eds.); **op. cit.**. For the complete atlas (23 pages) cf. Engel, Henk and Pané, Iskander and van der Bogt, Olivier; 'Randstad Holland in kaart', in **OverHolland 2** (Amsterdam: SUN, 2005).
8 Thurkow, A. et al.; **Atlas van Nederland. Vol. 2: Bewoningsgeschiedenis**, 'The Hague 1984', map 16. For a comprehensive account see Visser, J.C.; 'Dichtheid van de bevolking in de laat-middeleeuwse stad', in **Historisch Geografisch Tijdschrift** 3 (1985), pp. 10–21.
9 Nietzsche, Friedrich; **The Gay Science**, trans. by Walter Kaufmann (New York: Vintage, 1974).p. 122.§58.
10 Hohenberg, P.M. and Lees, L.H.; **The Making of Urban Europe 1000–1950**, (Cambridge, Mass./London: Harvard University Press, 1985) table 7.2, p. 227.

time, the name 'Randstad' meant nothing, internationally speaking. This situation only began to change in the 1960's.

The story goes that the name 'Randstad' was introduced by Albert Plesman, the founder of KLM. Around 1930, flying over Holland, Plesman was the first to recognise the potential of Holland's urban landscape. What he foresaw was a 'great horseshoe-shaped city of more than 3 million inhabitants', consisting of 'centres of population in Utrecht, the Gooi, Amsterdam, Haarlem, the Bollenstreek, Leiden, Wassenaar, The Hague, Delft, Schiedam, Rotterdam and Dordrecht, each bleeding into one another'.[11] It's a great story. Arnold van der Valk has expressed the suspicion that Plesman did not really get the idea from the air, but rather from seeing maps showing the 'Urban sphere of influence of Holland-Utrecht' which Van Lohuizen had prepared for the 1924 *International Urban Design Conference* in Amsterdam. After that conference, these maps put in an appearance on various occasions.[12]

If this modified myth of the origin of the Randstad is correct, it convincingly confirms the idea that in urban design work maps are not only a tool, but also serve 'mainly as a means of presentation'. As De Casseres put it, 'Urban design is as closely tied to the *representation of the Earth's surface* as to *the Earth's surface itself*.'[13] Yet Van Lohuizen's maps were mainly intended to show the dangers of unbridled urbanisation. Their purpose was to support the case for regional planning, which it was hoped would keep urbanisation on the right lines and avoid the negative consequences inherent in the development of a metropolis.

In fact the name 'Randstad' which Plesman linked to a view from an aircraft or a view of Van Lohuizen's maps, emphasised the positive side of the process of urbanisation, for example the possibility of setting up a world-class airport. It was the name that gave the image a permanent place in people's imagination, so perhaps Plesman, rather than Van Lohuizen, is to be thanked for the fact that in the 1960's the Randstad was thrust into the role of providing an alternative model for the development of a metropolis.

In *The World Cities* (1966), the English geographer Peter Hall sang the praises of the polycentric Dutch metropolis by comparison with the traditional monocentric metropolises: 'It seems virtually certain that at the present time the Dutch solution offers the correct model for most metropolises still undergoing growth.'[14] In spite of the spatial planning policy that until quite recently effectively sabotaged the actual creation of a Randstad metropolis, the subsequent virtual reality of the Randstad has been a factor which has significantly contributed to Randstad, Holland currently being granted the same status as other leading European urban regions.

Subsequently, it has also been common to associate the special characteristics of the Randstad with the structure of the earlier pattern of urbanisation. In 1525 Holland was already the most urbanised part of the Netherlands. Forty-four per cent of the population lived in towns, albeit small towns. The shape that we know from the maps of the geographer Jacob van Deventer is based on the towns as they developed between 1400 and 1550. If we look at the way that towns were distributed at that time, it seems that the foundation for the development of the Randstad was laid in that period. In 1560 the area now known as the Randstad contained all of Holland's towns (the towns in the area presently occupied by the provinces of North and South Holland) and two-thirds of all the towns in the Netherlands containing more than 10,000 inhabitants. Just as today, the urban population was distributed over many towns, large and small, none of which enjoyed a dominant position.[15]

De Vries and Van der Woude have however pointed out that this situation was only typical of the first half of the 16th century: 'The striking thing however is that in the later Republic, around 1525, the area could still by no means be called a focus of international economic activity.'[16] During this period, Antwerp was the centre of international trade. Henry Pirenne hit the nail on the head when he said that when Antwerp stood at the centre of European trade, the Netherlands was 'a suburb of Antwerp'.[17] A hundred years later the situation was totally changed. The centre of world trade had moved to Amsterdam. The level of urbanisation increased to 61% (1675) and the network of Dutch towns displayed a distinct hierarchy, at the top of which stood Amsterdam, the metropolis of the 17th century.

With this in mind, we can now take a fresh look at the polycentric structure of the Randstad that so many authors have praised. To get some kind of grasp of the historic development of the Randstad agglomeration, it is important to understand it as a network of towns, large and small. This idea requires some explanation. In the 1960's and 1970's, 'structure' was a vogue word; since the 1990's the same can be said of 'network'. People see networks everywhere, indeed 'network' has even become fashionable as a verb. In the present context, the expression 'network of towns'

---

11   Musterd, Sako and de Pater, Ben; **Randstad Holland. Internationaal, regionaal, lokaal**, (Assen: Van Gorkum, 1992) p. 1.
12   van der Valk, Arnold; **op. cit.,** pp. 50–62. About Plesman in particular, p. 60.
13   de Casseres, J.M.; 'Stedebouw en kaartenwetenschap', in **Tijdschrift voor volkshuisvesting en stedebouw**, volume 8, no. 4 (April 1927), pp. 85–86.
14   Hall, Peter; **Zeven wereldsteden,** trans. M.G. Schenk in co-operation with E.F. van den Berg-Brouwer (Baarn: Wereldakademie, De Haan/Meulenhof, 1966) p. 97 and pp. 120–121. Originally **The World Cities** (London: Weidenveld en Nicholson, 1971). See also Burke, Gerald L.; **Greenheart Metropolis** (London: Macmillan, 1966). **Cf.** van der Cammen, Hans and de Klerk, Len; **Ruimtelijke ordening, van Grachtengordel tot Vinex-wijk** (Utrecht: Het Spectrum, 2003) p. 225.
15   See note 8 and van der Cammen, Hans (ed.), **Four Metropolises in Western Europe** (Assen/Maastricht: Van Gorkum, 1988) pp. 120–121.
16   de Vries, Jan and van der Woude, Ad; **Nederland 1500–1815, de eerste ronde van moderne economische groei** (Amsterdam: Balans,1995) p. 86.
17   Pirenne, Henry; **Histoire de Belgique** III (Brussels: Lamartin,1923/1907) p. 259. Cited in Braudel, Fernand; **Beschaving, economie en kapitalisme (15de–18de eeuw), Deel III, De tijd en de wereld** (Amsterdam: Contact, 1990) p. 37.

has a strict definition. Recent literature about urbanisation processes has contrasted this designation for a collection of towns which maintain connections with one another with 'system of central places', a theoretical model for the distribution of towns devised in the 1930's by the German geographer W. Christaller.[18]

The basis of the 'system of central places' is the role played by a town as the centre providing such services as marketplace, administration, etc. for a more or less sizeable hinterland. Services may be provided at various levels, so creating a hierarchical system of towns headed by a large town acting as the 'central place' for all 'central places' at lower levels. Such a system of towns, with its associated service areas, constitutes a region. According to Christaller, if one disregards geographical differences and differences in population density, all systems of towns display the same structural characteristics, i.e. the same regular geometric pattern of distribution of towns and an order of precedence between those towns which in every case bears a fixed relationship to their respective sizes.

The differences which systems of towns display in real life was explained by Christaller by reference to disturbances in the regular pattern of distribution caused by specific geographical conditions, and to the population density in the relevant region, which determined the actual number and sizes of the towns and the distances separating them from one another. Nonetheless, further research disclosed remarkable deviations, even in the system of towns in South Germany from which Christaller derived his theory. In today's research these deviations are explained in terms of inter-regional trade, and this led to the introduction of the term 'network system'[19], a term used to refer to a system of towns not tied to a particular territory. Networks of towns are by nature unstable: trade routes shift and the dominant economic centres, the metropolises, change places.[20]

Which of the two systems deserves to be accepted as the primary model of urban creation is controversial. Hoppenbrouwers has observed: 'It is extraordinarily difficult to work out what precisely gave the decisive thrust to the definitive urbanisation of Holland, and when that thrust took place. There is undeniably some connection with the reclamation of peat bogs and the associated structural changes in agriculture, which in the long term shed much more labour than it took on.'[21] The reference here was to the subsidence caused by the reclamation of peat bogs, which had the long-term effect of making agriculture impossible and forcing the changeover to cattle breeding. These changes made Holland dependent on imported cereals, which in turn made it necessary to start producing for export.

The effect on the further development of the towns in the area covered by the present Randstad is clear: a matter of decisive significance was that in successive periods these towns formed part of network systems extending far outside the area. So far indeed that the system of Holland towns must itself be thought of as a network, in which the relationships between different towns were by no means fixed. Changes in long-distance trade and the relocation of trade routes through the area frequently caused internal changes in the system of towns. None of this detracts from the fact that the towns in the area covered by the present Randstad also functioned as central places and indeed still fulfil this role. However, this no more gives a basic explanation for the pattern of distribution of those towns than it does for their sizes. Such an explanation requires the urbanisation of the Randstad area to be looked at in a wider perspective.[22]

De Vries and Van der Woude distinguish four major waves of urbanisation for the areas by the North Sea in the second millennium. In the first wave, the most urbanised area was in Flanders and Brabant (Bruges, Ghent and subsequently Antwerp, in the 14th to 18th centuries). In the

---

18 Christaller, W.; **Die zentrale Orte in Süddeutschland** (Jena, 1933). In the Netherlands the system of central places mainly became known as a normative planning instrument for the distribution of settlements and services. Christaller presented his results at the 1938 International Geographical Conference in Amsterdam as a planning principle pre-eminently applicable to the layout of newly reclaimed land. See **Congrès Internationale de Géographie** (Amsterdam, 1938). See also Bosma, Koos and Wagenaar, Cor (eds.); **Een geruisloze doorbraak. De geschiedenis van architectuur en stedebouw tijdens de bezetting en de wederopbouw van Nederland** (Rotterdam: NAi, 1995) pp. 165–166. For the application of the system of central places to the layout of the IJsselmeer polders, see Hemel, Zef; **Het landschap van de IJsselmeerpolders, inrichting en vormgeving** (Rotterdam: NAi, 1994) pp. 164–165. Christaller's method of calculation has also been applied in the programming of urban extensions by van der Cammen, Hans and de Klerk, Len; **Ruimtelijke ordening, van Grachtengordel tot Vinex-wijk** (Utrecht: Het Spectrum, 2003) pp. 135–136.
19 Hohenberg, P.M. and Lees, L.H.; **op. cit.**, pp. 47–73.
20 A point of clarification: in the present context a system is an articulated and ordered whole. The structure is the description of the internal build-up of the system, the ordering principle, often long-lasting if not permanent. The structure ensures that the system remains intact whatever changes might occur. A network is the least complicated way of referring to a collection of linked objects. Yet networks of towns also have a structure: 'hierarchies of centres', 'nodes and junctions', 'gateways and outposts', 'cores and peripheries'. **Cf.** Hohenberg, P.M. and Lees, L.H.; **op. cit.**, p. 5. See also Braudel, Fernand; **op. cit.**, pp. 37–39.
21 Hoppenbrouwers, P.C.M.; 'Van waterland tot stedenland', in: de Nijs, Thimo and Beukers, Eelco; **Deel I Geschiedenis van Holland tot 1572** (Hilversum: Verloren, 2002) p. 120. The analysis by De Vries and Van der Woude comes to a similar conclusion: 'The fact that in this area urbanisation translated itself into a large number of towns rather than a smaller number of larger towns, speaks volumes about the relative importance of the push-factor provided by the crisis in agriculture and the pull-factor exerted by a booming urban economy'. De Vries and Van der Woude, **op. cit.**, p. 35.
22 van Engelsdorp Gastelaars, Rob; 'Verstedelijking in Nederland tussen 1800 en 1940', in Taverne, Ed and Visser, Irmin (eds.); **Stedebouw. De geschiedenis van de stad in de Nederlanden van 1500 tot heden** (Nijmegen: SUN, 1993) pp. 30–38 and 174–179. Engelsdorp Gastelaars even distinguishes four subsystems in the urbanisation of the Netherlands between 1800 and 1940: '(a) the network of towns involved in international trade in the 17th and 18th centuries via primary markets, (b) the network of towns and villages that became involved in industrial production during the 19th century, (c) the system of central places that was involved in caring for the country's population and (d) the system of central locations involved in the increasingly closely-knit administrative organisation of the Kingdom of the Netherlands between 1800 and 1940.' (p. 175) It appears from his further account that in many towns these subsystems overlapped.

second wave, the centre moved northwards to Holland (Amsterdam, in the 17th century). Subsequently, in the third wave, the centre moved to England (London, in the 18th and 19th centuries); the centre of the fourth and last wave of urbanisation so far, from 1870 onwards, lay in Germany (the Rhine-Ruhr area).[23] The formation and development of Holland's towns was completed within the framework of these four waves of urbanisation.

An Anatomy of the Randstad

If mankind had never built any places of worship, architecture would still be in its infancy. The tasks that mankind set itself on the basis of false assumptions (as for example that the soul could exist apart from the body), provided the stimulus for the development of the highest forms of culture. 'Truths' are incapable of supplying such stimuli.
Friedrich Nietzsche, *Posthumous fragments*, 1876-1877, §23 p.167.

It is tempting to see the formation and development of Holland's towns in the light of these four waves of urbanisation. During each such wave the system of towns in Holland occupied a different position within the more comprehensive system of towns round the North Sea. At the same time, of course, a change took place in the relationships between, and rankings of, the towns in the area covered by the present Randstad. Looked at in this way, it is possible to locate the formation of the Randstad in the fourth wave of urbanisation, the last so far.

The starting point for Holland's system of towns around 1800 was conveniently charted by J.C. Visser in his contribution to the *Atlas van Nederland, Vol. 2: Bewoningsgeschiedenis*, which also shows the effects of the earlier waves of urbanisation. Table 1 shows that on a number of occasions in the five centuries between 1300 and 1800, the ranking of the nine most important towns in the Randstad by number of inhabitants changed significantly.[24] The ranking in 1400 corresponded to the first wave of urbanisation. Their favourable location between four economic centres formed the starting point for the first blossoming of Holland's towns. The largest town was Utrecht, followed at some distance by the other towns of Holland, none of which was significantly different in size from the rest.

The ranking in 1670 corresponded to the second wave of urbanisation. Amsterdam had become a centre of world trade and stood at the head of a network of towns which had developed a hierarchy and specialisation. It is impossible to give the precise date between the first and second waves of urbanisation on which the ranking in the first system of towns changed into the new ranking of the second period. If we go on to look at the coming into being of the Randstad's present system, we see that it began with a clear break resulting from the third wave of urbanisation.

For the Netherlands, the third wave of urbanisation was a period of economic stagnation. Many of the towns in Holland went through a period of serious de-urbanisation. The low point occurred at the beginning of the 19th century. Although Amsterdam managed to maintain the size of its population during this period of stagnation, the economic structure of the urban network lost coherence. The de-urbanisation of Holland was accompanied by a contraction in the economic activity in the centre, with the result that by the end of the 18th century, Amsterdam, with 221,000 inhabitants, had become even more dominant, relatively speaking, than in the period of its greatest growth.[25]

By 1800, all that remained of Holland's urban network was the system of waterways that was to remain the sole transport system until the late 19th century. The region of what would become the Randstad also remained a territorial entity for the purposes of military defence from the 17th century onwards. The 'Hollandse Waterlinie' (a defensive strip of flooded land) formed the cornerstone of 'Fortress Holland'. Although by nature invisible, the Waterlinie was seen as a reliable national boundary until 1940, when the German bombers simply flew over it on their way to Rotterdam.[26] The notion of 'Randstad Holland' was born at just the right time to take the place of the outdated concept of 'Fortress Holland'.

The course of the fourth wave of urbanisation was supported by massive population growth. At first sight it seemed that the fourth wave of urbanisation would see no significant change in the 1795 ranking, established in the period of stagnation after the second peak urbanisation period (see table 2).[27] The structure of Holland's urban network appeared to be stable, the only growth being quantitative. However a glance at the precise population figures makes it clear that the 18th-century monocentric urban network was vanishing rapidly.

There were two major changes. First of all, some towns changed place in the bottom half of the list. This was not really significant. More important was that the five towns in the Randstad area with populations between 60 and 150 thousand inhabitants were joined by 15 more towns of a similar size, averaging 90,000 inhabitants, and they in turn were joined by a further 20 towns of

---

[23] De Vries and Van der Woude, **op. cit.**, p. 87. For developments in a European context see Braudel, Jan de Vries, Hohenberg and Lees, and Lesger, Clé; 'Stedelijke groei en stedensystemen' and 'De dynamiek van het Europese stedensysteem', in Taverne, Ed and Visser, Irmin (eds.); **op. cit.**, pp. 30–38 and 104–111.
[24] The population sizes used for **Table 1** for the period prior to 1300 were derived from Visser, J.C.; 'Dichtheid van de bevolking in de laat-middeleeuwse stad', in **Historisch Geografisch Tijdschrift**, 3 (1985), pp. 10–21, and for later dates from Lourens, Piet and Lucassen, Jan; **Inwoneraantallen van Nederlandse steden ca. 1300–1800** (Amsterdam: NEHA, 1997).
[25] Wagenaar, M. and van Engelsdorp Gastelaars, R.; 'Het ontstaan van de Randstad, 1815–1930', in **K.N.A.G. Geografisch Tijdschrift** XX (1986) no. 1, p. 16.
[26] Heesen, Willem and van Winden, Wilfred; 'Het strategisch landschap', in Brand, Hans and Brand, Jan (eds.); **De Hollandse Waterlinie** (Utrecht/Antwerp: Veen, 1986).
[27] The population figures used in **Table 2** were derived from Smook, Rutger A.F.; **Binnensteden veranderen** (Zutphen: Walburg Pers, 1984), apart from the figures for Rotterdam and the year 2000, for which use was made of CBS data.

between 30 and 60,000 inhabitants (averaging 40,000 inhabitants). At that time almost half the population of the Randstad, 2.6 million inhabitants in all, lived in these 40 relatively small towns. The other half lived in the four major cities, Amsterdam, Rotterdam, The Hague, and Utrecht. These cities, at the top of the list, provided the setting for the second great change from the situation prevailing at the beginning of the 19th century.

Amsterdam had lost its dominant position. Amsterdam and Rotterdam had become almost equivalent in size, and the difference between the sizes of the populations of The Hague and Utrecht and the two top cities had become much more modest. In short, the network of Holland towns had become polycentric. Supporters of the Randstad model see in this the original configuration of a new kind of Metropolis. However, the polycentrism of the network of Holland towns can also mean that in time Holland, to paraphrase Pirenne, will once more function as a suburb – this time not of Bruges or Antwerp, as in the 14th and 15th centuries – but of the economically powerful Rhine-Ruhr area.

As in the 16th century, this situation could have formed the starting point for metropolitan development in the immediate future.[28] It is however doubtful whether the large urban extensions since the 1970's provide any indication pointing in this direction. These urban extensions no longer had much to do with increases in numbers of inhabitants. The course of further urbanisation underwent a significant change in direction after 1970. By the end of the 1960's the forecast that the population of the Netherlands would rise to 20 million inhabitants by 2000, had had to be reduced significantly. After 1970, the rate of increase in population began to fall, although the fall was masked by the enormous production of housing which took place after that date. In the last 30 years of the 20th century, the size of the housing stock in the Netherlands almost doubled.

At this point we come up against one limitation of a purely cartographic approach. One of the most important characteristics of the urbanisation which took place during the 20th century, the increase in land utilisation by towns, can not be seen from a study of the maps. The situation is much the same for another factor, the revolution in traffic and transport. The maps do show the development of the various infrastructures required by this revolution, but give no insight into the decrease brought about in the time taken to get from one place to another. Other factors that could be mentioned include increased prosperity and changes in the composition of the population and the way that people spend their time. All these factors have radically altered both urban living and the way the town is perceived, not only by town dwellers in general, but also in particular by people professionally involved in the production of towns.

To correct the image created in professional circles, I shall only focus attention here on the increase in land utilisation by towns. Table 3 gives figures for the increase in population and the size of the built-up urban area for the nine most important historic towns in the Randstad area between 1850 and 2000.[29] It appears that in the last 150 years the total size of the built-up area occupied by the nine towns has become 22 times larger, while the population has only increased by a factor of 4.6. The increase in the size of the built-up area seems to be as much determined by population growth as by expansion, i.e. reduction in the density of urban land utilisation by a factor of 4.7 (see diagram 1).

The phenomenon of expanding land utilisation is difficult to fathom. It is to be hoped that detailed investigation of individual towns will lead to greater understanding. In any event, three factors play an important part: the development of specialised work areas, more generous plot division for housing and a decrease in the average occupancy level. The first two factors are not new. The idea that a key factor in all this was the invention of modern functionalism in town and country planning is based on a misunderstanding. De Pater mentions 'processes of segregation, dispersion and reduction in density' as playing a role in all periods of urban development.[30] The introduction of new means of transport – trains, trams, bicycles and cars – undoubtedly gave a new dimension to these processes.

The fall in the average occupancy level, to the extent that we know today, is a new phenomenon. To discover the effect on the course of the urbanisation process during the past century and a half we need to establish the share played by the fall in average occupancy level in the increase in housing stock, and the only reliable housing stock figures available are for the Netherlands as a whole. Diagram 2 shows that the average occupancy level only started to fall after 1910.[31] Until that year there was even a rise, from 4.5 in 1850 to 4.9 in 1910. In 1910 and 1940 the level fell to 4.2. This fall was however partly made good during the war years. Thereafter the pre-war trend continued: by 1970 the average occupancy level had fallen to 3.4. By then the fall had become a well-known phenomenon in professional circles. The immediate cause was the reduction in the number of people living together in the same house, brought about by the decreased housing shortage and smaller family sizes.[32]

For the Netherlands as a whole, half of the increase in the housing stock in the period from 1850

28 Reh, Frieling, and Weeber; **Delta Darlings** (Delft: Internal Faculty Publication, 2003).
29 The population numbers in **Table 3** were derived from the same sources as above for **Table 2** (see note 26). The sizes of the built-up urban area were calculated with the help of the map of the reconstruction of the Randstad.
30 de Pater, Ben; 'Van land met steden tot stedenland. Een kleine historische stadsgeografie van Nederland', in **Historisch-geografisch Tijdschrift**, volume 7 (1989) no. 2, pp. 51–53.
31 **Diagram 2** is based on population and housing stock figures derived from van der Cammen, Hans and de Klerk, Len; **op. cit.**
32 In 1930, the average occupancy level in Amsterdam was 3.74. The A.U.P. had already taken into account a further decline to 3.34. **Algemeen uitbreidingsplan van Amsterdam, nota van toelichting** (Amsterdam: Publication from Gemeente Amsterdam, 1934) pp. 78–79. See also appendix IV.

to 2000 was needed to house the growth in population. The other half was necessitated by the fall in average occupancy level. If this fall had not occurred, the entire increase in housing stock in the period 1970-2000 would have been superfluous. As far as I know, nobody around 1970 foresaw that the rate of decrease in occupancy level would go on increasing. Since 1970 the decrease in the rate of population growth has gone hand-in-hand with the fall in the average occupancy level.

It is however only too easy to miss the point that the effect of the fall in average occupancy level is related to the size of the total housing stock. The capacity of the existing housing stock declines, but as the size of the stock increases, the demand for housing also increases. The substantial increase in the size of the housing stock after 1970 provided a complete code to the cycle of population growth which started at the beginning of the 19th century and is now nearing its end.

In the period 1970-2000, 70% of the increase in the size of the housing stock was required to satisfy the need produced by the reduction in occupancy level; only 30% was necessitated by population growth. At the same time the number of inhabitants per hectare in existing urban areas fell by 30%, so eroding support for the provision in those areas of services such as schools and shops. These are at any event some of the ingredients of what people sometimes refer to as the urban crisis of the 1970's, which followed a period of reconstruction and economic prosperity.[33]

The shape of the town   *Stone is more stony than it used to be. – In general we no longer understand architecture… Everything in a Greek or Christian building originally signified something, and indeed something of a higher order of things: this feeling of inexhaustible significance lay about the building like a magical veil. Beauty entered this system only incidentally, without essentially encroaching upon the fundamental sense of the uncanny and exalted, of consecration by magic and the proximity of the divine; at most beauty* mitigated the dread *– but this dread was everywhere the pre-supposition – what is the beauty of a building to us today? The same thing as the beautiful face of a mindless woman: something mask-like.*
Friedrich Nietzsche, *Human, All Too Human*, 1878.[34]

Unlike those who were involved in keeping the urbanisation process on the right track in the last century, we today find it easier to appreciate that even the most recent wave of urbanisation is not one of unlimited growth. According to the most recent estimate, an unchanged policy on immigration will mean that within the foreseeable future the size of the population of the Netherlands will reach a maximum, and then begin to decline.[35] Even the fall in average occupancy level will reach its limit. The result will be the disappearance of the most important spur to expansion of the built-up urban area.

This of course says nothing about the further economic development and the functioning of the Randstad, both of which are beyond the scope of the present analysis. But the present study is undoubtedly important in gaining a picture of the development of the Randstad. The factors determining the form of urbanisation of the area covered by the Randstad are not exclusively structural. It is impossible to get round the fact that since the 1901 Housing Act, the production of towns in Holland has gradually come to be led on all levels by design activities and so by architectural and urban design proposals relating both to towns that people want and towns that people want to avoid.

Originally the special value of the Randstad Holland concept was that it provided an alternative to the spectre of unbridled urbanisation. As a model of decentralised urbanisation 'Randstad Holland' has certainly helped to avoid the negative consequences of the classic development of a metropolis. Nobody has ever paid much attention to the fact that it also destroyed a number of the positive opportunities that such a development could offer. Town and country planning in the form in which it made its first appearance in the Netherlands in 1924 at the International Urban Development Conference in Amsterdam, was guided by the vision of the town laid down at the time, in no uncertain terms, by Raymond Unwin.

As Unwin put it, 'If a town wants to be the home of a real community, it has to possess the following characteristic properties: a properly defined shape, cohesion between its parts and a limit to the size to which it can grow as a healthy, independent municipality.'[36] This statement is not so much anti-urban as anti-metropolis. The metropolis was contrasted with the ideal of the small European town. At that time, town and country planning was seen as an instrument to improve the living conditions of the population. The key issue was health, mental and physical. And health here meant the health of the social body, the community.

This was the vision that to a significant extent determined the pattern of distribution of urbanisation in the Randstad area after 1945. In the 1960's a policy of 'bundled deconcentration' was introduced. Centres of urban growth were designated to divert the pressure of urbanisation away from major cities. When shortly thereafter, in the 1970's, doubt began to be cast on this policy, changing direction was no longer the work of a moment, which explains why the Randstad has remained a conglomeration of separate towns and substantial villages. Although the volume of

33 van der Cammen, Hans and de Klerk, Len; **op. cit.**, pp. 233 ff.
34 Nietzsche, Friedrich; **Human, All Too Human** (Cambridge: Cambridge UP, 1996) §218 p. 101.
35 The most recent estimate by the CBS gives a population of 17 million people by 2035; thereafter a population decline.
36 Quote from Unwin on the cover of Engel, Henk and van Velzen, Endry; **Architectuur van de stadsrand, Frankfurt am main 1925–1930** (Delft: Publicatie Bureau Bouwkunde, 1978).

daily commuter traffic suggests that these settlements no longer function as independent units, in the foreseeable future hardly anything could change the present pattern of distribution of the built-up urban area. For now this is something we have to live with, like it or not.

In the face of this reality, every typical idealistic approach to the city or the metropolis risks looking rather pathetic. Typo-morphological urban research will continue to run this risk as long as it persists in deriving 'design instruments', intended to be capable of guaranteeing urban quality, from a single stage of urban development. In *The Architecture of the City*, the book more responsible than any other for the great popularity of typo-morphological urban research, Aldo Rossi distances himself from this kind of 'urban architecture'. In the book, Rossi makes clear that the physical structure of a town can not be reduced to a single principle. A townscape is not a 'unit'; rather, it displays breaks and contrasts, all of which have something to say about the town's use and history.[37] Rossi refers to Frits Schumacher in support of this view.

In this connection it is important to reread what Schumacher himself has to say on the subject. In 1951, Schumacher wrote:

> In essence, today's 'metropolis', indeed even today's large town, is no longer a construction which can be reduced to a single basic principle. It is composed of individual districts, each with its own very different sociological characteristics. This differentiation can even be seen as a character trait. The administrative district and the business district, the industrial district and the residential district, all of which occur in a variety of types, distinguish themselves from one another with ever-increasing clarity. It would be totally wrong to want to force them to conform to a single formal law. The dominant geometric spirit in the administrative district is utterly different from that in the business district, and is expressed differently again in the industrial district. We can however easily recognise in the different kinds of residential district the characteristics which determine the type, whether it be 'medium-sized town', 'small town', 'garden city', indeed even 'village'.[38]

By comparison with Unwin's organic concept of the town, the observation of Schumacher, a member of the same generation as Unwin, is truly refreshing. At the same time it is clear that Rossi goes a step further. On the basis of further examination of the historic town, what Schumacher saw as characteristic of today's town was designated by Rossi as an integral component of the concept 'town'. In Rossi's words,

> A town is by its very nature not a creation which can be traced back to a

single basic idea. This is true of the modern metropolis, but equally true of the concept of the town itself as the sum of many parts, neighbourhoods and districts, which differ significantly from one another and are differentiated by their formal and sociological characteristics. Indeed this differentiation is one of the typical characteristics of the town. It is senseless to want to subject these different zones to a single kind of explanation or a single formal law.³⁹

Looked at in this way, it is understandable that Rossi did not see the latest urban developments as a new phenomenon. The 'new scale' of urbanisation is no reason to introduce a different concept of the town.⁴⁰ If the term 'town' stands for a 'durable entity', then the term applies to the physical quality of the town, the town as artefact. There is an essential distinction between what might be called the *living town* and the *concrete town*. In urban research the physical town is generally considered the resultant of social forces. The 'outline urban theory' which Rossi included in *The Architecture of the City*, was not intended to provide a generalising urban concept. With many outflanking movements the importance was demonstrated of 'a theory which comprehends the town as architecture'.

The town can be analysed in many ways, but in Rossi's view only an architectural approach offers the possibility of penetrating into the unique phenomenon of towns as they are. When Rossi speaks in this connection about 'the idea of the town as a synthesis of all its qualities' he means a concrete town: 'Athens, Rome, Constantinople and Paris are urban ideas.'⁴¹ Morphological urban research shows the complex relationship between architectural forms and history. The town is where architectural forms outlive the original reason for their construction. This is precisely what makes them open to changing functions and meanings. In this connection, Rossi could also have referred to Kurt Schwitters, the Merz artist.

The point of view that Schwitters advanced in the architectural debate in the 1920's, a debate heavily charged with Utopias, bore witness to an unprecedented realism. In Bruno Taut's magazine supplement *Frühlicht*, the architect wrote:

> Of all the arts, architecture is by nature the most geared to Merz thinking. As is well-known, Merz means using any old works of art that happen to be available as material for new works of art. For architecture, the recalcitrance of the materials used for building houses means nothing more than reusing old materials, over and over again, including them in new designs. This results in the creation of extraordinarily rich and beautiful buildings, because for an architect it is not the style of the old component that is normative, but the idea of the new '*Gesamtkunstwerk*'. This is the way our towns, to take one example, should be dealt with. By

37 Rossi, Aldo; **De architectuur van de stad** (Nijmegen: SUN, 2002) p. 63 note 1.
Translated from the Italian.
38 Schumacher, Fritz; **Vom Städtebau zur Landesplanung und Fragen der Städtebaulicher Gestaltung** (Tübingen: Wasmuth, 1951) p. 37.
39 Rossi, Aldo; **op. cit.**, p. 62.
40 **ibid.**, p. 8.
41 **ibid.**, p. 152.

carefully demolishing the most disturbing parts, including houses both ugly and beautiful in a single comprehensive rhythm and distributing accents correctly it should be possible to transform the metropolis into an enormous work of Merz art.[42]

Both Schwitters and Rossi took as their starting point the appearance of the town, the townscape. Neither fought shy of using concepts derived from the idealistic urban aesthetics introduced by Camillo Sitte in his exposition on the discipline of architecture produced at the end of the 19th century.[43] Both Schwitters and Rossi agree with Sitte in seeing the town as a work of art, a collective work of art that only comes into being over time. Thus both award the town the status of 'quasi-subject'.[44] But it is precisely here that Schwitters and Rossi find starting points for the radical repositioning of concepts drawn from idealistic urban aesthetics. By definition the townscape is not homogeneous but polymorphous, fragmented, full of contrasts and contradictions. A townscape is a collage.

In Rossi's view, the image of the town is an expression of collective imagination and memory: '… just as a memory is linked to facts and places, so is the town on the *locus* of a collective memory, of which architecture and landscape are the expression. And just as new facts are constantly being added to the memory, so new facts are constantly fusing with the town.'[45] The image derives its coherence from the memory. Rossi took this as the starting point for urban analysis and design. At least, that was the explicit aim, for his designs also display a different kind of coherence. Things forgotten and neglected are at least as important to the structure of a memory as what is remembered.

Schwitters made clear that unity and coherence are always the result of a *coup d'état*. The central category here is style. In his collages references made by the fragments of reality are forced to the background, suspended, by a process of stylisation. The unity and coherence of a Merz picture are based totally and

---

42 Schwitters, Kurt; 'Schloss und Kathedrale mit Hofbrunnen', in **Frühlicht** no. 3, 1922. Re-published in Schwitters, Kurt; **Das literarische Werk. Band 5 Manifeste und kritische Prosa** (Cologne: DuMont, 1981) p. 96.
43 Sitte, Camillo; **Der Städtebau nach seiner künstlerischen Grundsätzen**, (Vienna: Greiser, 1889). Dutch translation **De stedebouw volgens zijn artistieke grondbeginselen** (Rotterdam: 010 Publishers, 1991).
44 Jansen, Harry S.J.; **De constructie van het stadsverleden** (Groningen: Wolters-Noordhoff, 1991) esp. chapters VI and VII.
45 Rossi, Aldo; **op. cit.**, p. 156.
46 Wiesing, Lambert; **Stil statt Wahrheit. Kurt Schwitters und Ludwig Wittgenstein über ästhetische Lebensformen** (Munich: Wilhelm Fink, 1991) and Elderfield, John; **Kurt Schwitters** (London: Thames and Hudson, 1985).
47 Schwitters, Kurt; 'Watch Your Step', in: **Merz** 6, 1923. Re-published in Schwitters, Kurt; **Das literarische Werk. Band 5 Manifeste und kritische Prosa** (Cologne: DuMont, 1981) p. 168.

completely on the maker's personal sense of style.⁴⁶ Applied to the town this technique introduces total arbitrariness. This is why Schwitters' view of the town can not do without a concept of style that does justice to the town as 'quasi-subject'. In conclusion, the way that Schwitters put this can be helpful to the further study of the architecture of the town:

> Style is the expression of the common will of the many, at best of everyone, the democratic will to give things shape. But since most people – and even a few artists, now and then – are mainly idiots, and since these idiots are most convinced about things they know about and universal agreement can only be found somewhere in the middle, style is generally a compromise between art and non-art, between play and utility.⁴⁷

	1300	1400	1560	1670	1735	1795
1	Utrecht	Utrecht	Amsterdam	Amsterdam	Amsterdam	Amsterdam
2	Dordrecht	Dordrecht	Utrecht	Leiden	Leiden	Rotterdam
3	Haarlem	Haarlem	Haarlem	Rotterdam	Rotterdam	Den Haag
4	Delft	Delft	Delft	Haarlem	Haarlem	Utrecht
5	Leiden	Leiden	Leiden	Utrecht	Den Haag	Leiden
6	Gouda	Gouda	Dordrecht	Delft	Utrecht	Haarlem
7	Amsterdam	Amsterdam	Gouda	Den Haag	Gouda	Dordrecht
8		Rotterdam	Rotterdam	Dordrecht	Dordrecht	Delft
9		Den Haag	Den Haag	Gouda	Delft	Gouda

[1]

	1850	1880	1910	1940	1970	2000
1	Amsterdam	Amsterdam	Amsterdam	Amsterdam	Amsterdam	Amsterdam
2	Rotterdam	Rotterdam	Rotterdam	Rotterdam	Rotterdam	Rotterdam
3	Den Haag	Den Haag	Den Haag	Den Haag	Den Haag	Den Haag
4	Utrecht	Utrecht	Utrecht	Utrecht	Utrecht	Utrecht
5	Leiden	Leiden	Haarlem	Haarlem	Haarlem	Haarlem
6	Haarlem	Haarlem	Leiden	Leiden	Leiden	Dordrecht
7	Dordrecht	Dordrecht	Dordrecht	Dordrecht	Dordrecht	Leiden
8	Delft	Delft	Delft	Delft	Delft	Delft
9	Gouda	Gouda	Gouda	Gouda	Gouda	Gouda

[2]

Reconstruction map of the Randstad: expansion of the built-up urban area in the Holland Randstad for the years 1850, 1940, 1970 and 2000.
1 Ranking of the nine most important historic towns in the Randstad from 1300 to 1795.
2 Ranking of the nine most important historic towns in the Randstad from 1850 to 2000.
3 Population, extent of the built-up urban area and density of the nine most important historic towns in the Randstad from 1850 to 2000.

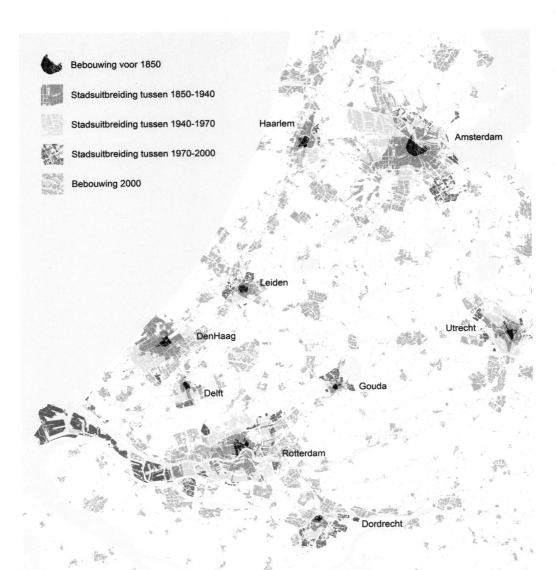

Verdunningsdiagram van negen historische steden in de Randstad
schaal ca. 1:700.000

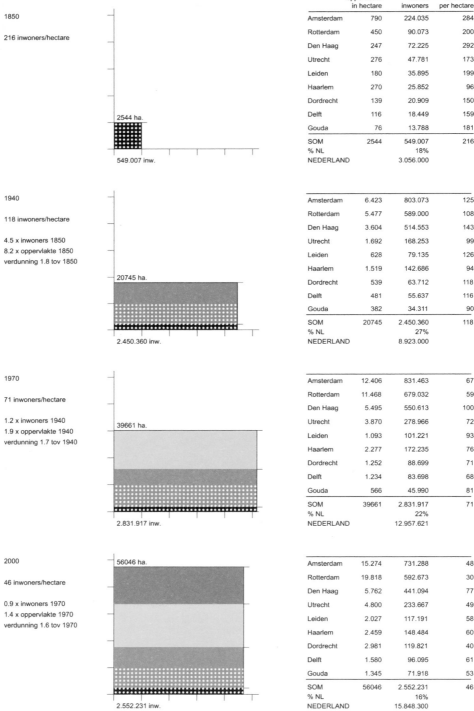

1850
216 inwoners/hectare

2544 ha.
549.007 inw.

	oppervlakte in hectare	aantal inwoners	inwoners per hectare
Amsterdam	790	224.035	284
Rotterdam	450	90.073	200
Den Haag	247	72.225	292
Utrecht	276	47.781	173
Leiden	180	35.895	199
Haarlem	270	25.852	96
Dordrecht	139	20.909	150
Delft	116	18.449	159
Gouda	76	13.788	181
SOM	2544	549.007	216
% NL		18%	
NEDERLAND		3.056.000	

1940
118 inwoners/hectare
4.5 x inwoners 1850
8.2 x oppervlakte 1850
verdunning 1.8 tov 1850

20745 ha.
2.450.360 inw.

Amsterdam	6.423	803.073	125
Rotterdam	5.477	589.000	108
Den Haag	3.604	514.553	143
Utrecht	1.692	168.253	99
Leiden	628	79.135	126
Haarlem	1.519	142.686	94
Dordrecht	539	63.712	118
Delft	481	55.637	116
Gouda	382	34.311	90
SOM	20745	2.450.360	118
% NL		27%	
NEDERLAND		8.923.000	

1970
71 inwoners/hectare
1.2 x inwoners 1940
1.9 x oppervlakte 1940
verdunning 1.7 tov 1940

39661 ha.
2.831.917 inw.

Amsterdam	12.406	831.463	67
Rotterdam	11.468	679.032	59
Den Haag	5.495	550.613	100
Utrecht	3.870	278.966	72
Leiden	1.093	101.221	93
Haarlem	2.277	172.235	76
Dordrecht	1.252	88.699	71
Delft	1.234	83.698	68
Gouda	566	45.990	81
SOM	39661	2.831.917	71
% NL		22%	
NEDERLAND		12.957.621	

2000
46 inwoners/hectare
0.9 x inwoners 1970
1.4 x oppervlakte 1970
verdunning 1.6 tov 1970

56046 ha.
2.552.231 inw.

Amsterdam	15.274	731.288	48
Rotterdam	19.818	592.673	30
Den Haag	5.762	441.094	77
Utrecht	4.800	233.667	49
Leiden	2.027	117.191	58
Haarlem	2.459	148.484	60
Dordrecht	2.981	119.821	40
Delft	1.580	96.095	61
Gouda	1.345	71.918	53
SOM	56046	2.552.231	46
% NL		16%	
NEDERLAND		15.848.300	

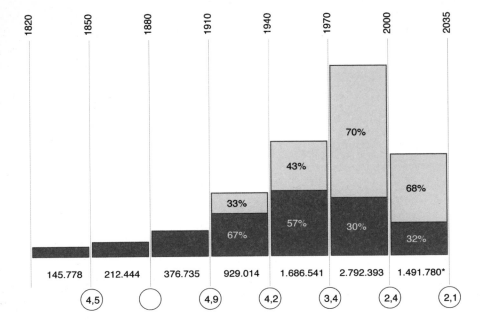

Toename van de woningvoorraad geheel nederland

▪ aandeel t.g.v. de bevolkingsgroei

▫ aandeel t.g.v. de dalende woningbezetting

*prognose

[5]

4 Expansion (reduction in density) of urban ground utilisation in the nine most important historic towns in the Randstad, from 1850 to 2000.
5 Increase in the housing stock in the Netherlands as a whole, showing how much was due to population growth and how much to a decrease in average occupancy rate.

Section VIII:

Theory and Praxis

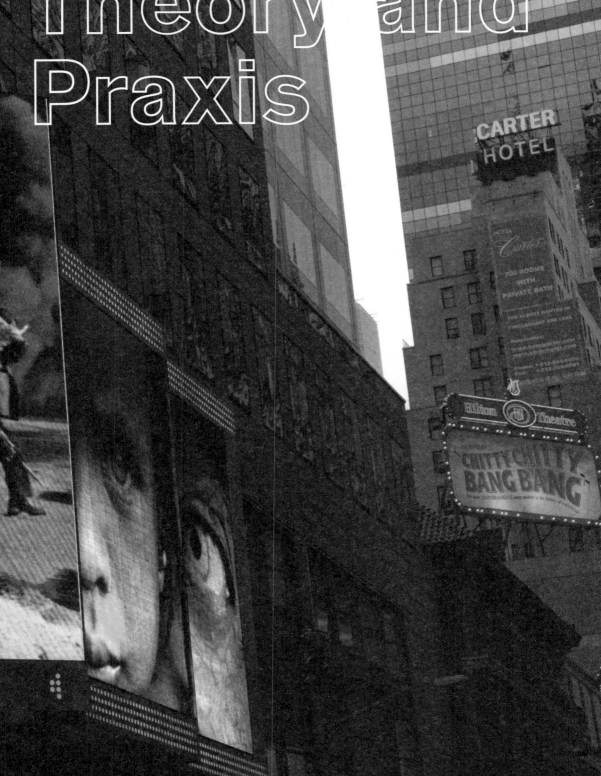

This final section deals with the relation between doctoral research and design practice. In many cases, especially in schools of architecture, the conventional view of research and design theory is that of a set of ideas and concepts that are supposed to be relatively independent of a specific field of design. The practice of design in that case is understood as the object of the theory, the implication being that 'theory knows better'. Theory is supposed to guide and steer the design process. Subsequently, theory takes the overarching hand in this game. The idea is that practice is 'informed' by 'theory', theory being understood as 'scientific' and practice being relatively 'blind'. The problem this dichotomy raises is twofold: practice is hollowed out and reduced to the implementation of rules that come from the outside, and theory and research are allowed only as a cover up for the messiness of daily practice. Consequently, a certain idea of 'instrumentalism' is implied – theoretical statements are divorced from the possibilities for truth or reference, and theoretical constructs have no ontological import. The problem of *mediation* arises – mediation between the formal analyses of the work, whether built or non-built, and its social ground or societal context.

K.M. Hays, in his book *Architecture / Theory / Since 1968* argues that theory has 'an appetite for modifying and expanding reality, a desire to organize a new vision of a world perceived as unsatisfactory or incomplete'. Fredric Jameson, addressing this problem from the broader perspective of interdisciplinary activity from what he refers to as 'Beyond Critical', has written: 'New theoretical discourse is produced by the setting into active equivalence of two preexisting codes. ... What must be understood is that the new code (or metacode) can in no way be considered a synthesis between the previous pair... It is rather a question of linking two sets of terms in such a way that each can express and indeed *interpret* the other.'

Hays considers both the historical context of architectural production and the object itself as a *text* in the sense that we cannot approach them separately and directly. In Baird, Pérez-Gómez, Moravánszky, Werner, Zimmermann and Mitchell we find a common position; namely that a critical stance is paramount, that research and in particular architecture theory is explicitly related to a critical analysis of society. Although each of these scholars approaches the discourse from varying topics of interest as well as from specific historical perspectives and trajectories, each advance their own discourse within the classical, and often paradoxical, 'theory/praxis' debate.

All the scholars in this section have published extensively on such problems as mediation and interpretation within inter-disciplinary research fields. They also confront similar issues within their own research departments and it is from this, their own position, that their contributions provide insight into the problems confronting researchers within the discipline of architecture and theory today. Specifically, Baird in his paper, '"Criticality" and Its Discontents' describes the present state of affairs in academia concerning issues of what theoretical discourse is applicable to architectural practice today. Certainly, a growing number of younger firms are wishing to shirk off the theoretical baggage of the post-1968 generation and find new tools in which to go forward, into a 'projective practice'. Nevertheless, Baird warns us, this practice must be theorized as well if it is to have legitimacy; the questioning must continue. Ákos Moravánszky, in his essay, 'Power Lines', takes a step into the nineteenth century in order to bring forward issues that are pertinent to contemporary discussions on 'mapping', that is to say spatial representation from a two-dimensional surface into a three-dimensional plane of perspectival experience. Yet, architecture is also seemingly increasingly becoming disassociated from any concern with its social or cultural task, asking the unremitting question of what it means to be human. Alberto Pérez-Gómez, in 'Ethics and Poetics in Architectural Praxis' reminds us that sometimes the just plain old-fashioned – i.e. the poetics of architecture – need not be lost in the struggle for innovation and progression. Finally, Arie Graafland in his contribution to this volume, 'On Criticality', examines what the new technologies have to offer without having to loose sight of political engagement, of a critical theoretical practice, of a contextualizing architecture in a sociological/economic/historical framework. Graafland calls this 'an architecture of the street', and shows that in order to go forward, neither an extreme position of abandoning the past, nor ignorance of innovation need be necessary.

'Criticality' and Its Discontents
George Baird

Shortly before his untimely death, the Spanish theorist and critic, Ignasi di Sola-Morales commented to me that 'if a European architect or architectural scholar wished to study contemporary architectural theory, it was necessary to come to the East Coast of the United States to do so'.

I realize that it is one of the intentions of the Berlage Institute's new doctoral program, and of the creation of the new Delft School of Design to challenge this American hegemony – and recent events suggest that these challenges are meeting with some success. Still, if America does retain some prominence in these matters, then it may be of interest to the reader to learn of a significant divergence that is now appearing in the territory of contemporary theory there, a divergence that is triggering discussions of a growing intensity, such as have not been seen since the beginning of the polemical attacks of the protagonists of deconstructivism on post-modernism, a decade and a half ago. The matter that is now coming into question is the concept of a 'critical architecture' such as has been promulgated in advanced circles in architectural theory, for at least two decades now. The conception can probably be said to have received a definitive early formulation in a text by my Harvard colleague Michael Hays that Robert Somol and Sarah Whiting (two of the prominent recent participants in the discussion) have labeled 'canonical'. The text in question is 'Critical Architecture: Between Culture and Form', published in the *Yale Architectural Journal:* Perspecta 21 in 1984.[1]

Today 'criticality' is under attack, being seen by its critics as obsolete, as irrelevant, and/or as an inhibitor of design creativity. Moreover, the criticisms which are increasingly and frequently being made of it come from an interesting diversity of sources. To start to make sense of this emergent situation, we might try to locate the beginnings of the evident shift of opinion against this once-so-dominant theoretical discourse in architecture. One interesting precursor of current comment was an outburst by Rem Koolhaas, at one of the series of conferences organized by *ANY* Magazine at the Canadian Centre for Architecture in 1994:
The problem with the prevailing discourse of architectural criticism, complained Koolhaas, is the inability to recognize there is in the deepest motivations of architecture something that cannot be critical.[2]

But if Koolhaas' complaint was a harbinger of things to come, probably the first frontal challenge to criticality was a text published by Michael Speaks, the Director of Graduate Studies at SCIARC, in the American magazine *Architectural Record* in 2001.[3] In a startlingly revisionist text, Speaks explicitly abandoned the 'resistance' that he had learned from his own teacher, Fredric Jameson, in favor of a model of a new, alternative, and efficaciously integrated architecture that would take its cues from contemporary business management practices.[4]

[1] See also Somol, Robert and Whiting, Sarah; 'Notes Around the Doppler Effect and Other Moods of Modernism' in **Perspecta 33: The Yale Architectural Journal** (2002) p. 73.
[2] Rem Koolhaas as quoted by Beth Kapusta in **The Canadian Architect Magazine**; Vol. 39; (August 1994) p. 10.
[3] Speaks, Michael; 'Design Intelligence and the New Economy', **Architectural Record** (January, 2002) pp. 72–79.
[4] Upon hearing me present this text at a June 2004 conference at Delft University of Technology,

Before the dust from Speaks' polemic had settled, a subtler challenge was mounted by two other American theorists: Robert Somol of UCLA, and Sarah Whiting of Harvard. Their text, 'Notes around the Doppler Effect and Other Moods of Modernism' appeared in *Perspecta 33* in 2002.[5] In it, Somol and Whiting argued against the conception of a 'critical architecture' that had been long promulgated by Hays. In the place of the hitherto 'critical' architecture, Somol and Whiting proposed one that would be 'projective' instead.

Since the date of Somol and Whiting's publication, the pace of publications on this theme and the number of participants in the discussion have accelerated. At the end of 2002, for example, Michael Speaks followed up his polemic with a longer text published in *A + U*.[6] Since then, additional participants, such as Stan Allen, the Dean of the School of Architecture at Princeton, and Sylvia Lavin, the Chair of the Department of Architecture have joined the fray.

In this essay, I have set myself the task of attempting to briefly clarify how the divergence I have described has unfolded to date, and to summarize what is at stake in it, since it is my view that a great deal is indeed at stake.

Let me begin with a short account of the lineage of 'criticality'. One of its most cogent and internally coherent renditions has been that of the practitioner – and no mean theorist himself – Peter Eisenman, accompanied by the above mentioned Hays. Together, over the past two decades, these two have developed a position that has consistently focused intellectually on concepts of 'resistance' and 'negation'. For Eisenman, the position derives primarily from the work of the Italian historian and critic, Manfredo Tafuri, but it has been fleshed out in Eisenman's own mind by other prominent figures in contemporary thought, including Jacques Derrida, Gianni Vattimo, etc.. For Hays, Tafuri is equally as important a figure as he has been for Eisenman, but he is accompanied by additional figures such as Georg Lukács, Theodor Adorno, and Fredric Jameson.

For Hays, following Tafuri, the paramount exemplar of 'negation' in late modernism was Mies van der Rohe. Like his mentor, Hays has seen the late Mies as embodying a 'refusal' of the terms of contemporary consumer society, in the very surfaces of his built forms. In this regard, the Seagram Building on Park Avenue in New York City is as important a case study for Hays as it is for Tafuri himself. For his part, Eisenman has, over his career as a designer and thinker, welded precepts from Tafuri to others derived from the theories of minimalist art practices, as they have been articulated by such figures as Rosalind Krauss. In his hands, this has produced not so much a series of built forms embodying refusal or resistance; but rather, a design method which is for Eisenman more important as a process than it is for the architectural products resulting from it. Still, notwithstanding these differences in nuance, Eisenman and Hays have formed a formidable pair of advocates of 'resistance' in contemporary architecture and architectural theory for the past two decades.

Nevertheless, an enumeration of participants in a recent *Harvard Design Magazine* 'Stocktaking' symposium makes clear that Eisenman and Hays do not exhaust the modalities of 'criticality' that have had influence in recent years.[7] For example, Kenneth Frampton's commitment to 'resistance' to consumer society has been equally as resolute as that of Eisenman and Hays during the period in question, even if his intellectual lineage leads back more to Adorno and Heidegger than to Tafuri. Then there is Michael Sorkin, much more of a New York City 'street-fighter', politically speaking, than any of the figures I have discussed so far. Sorkin is renowned in American design circles for his longstanding courage in having mounted compelling attacks on prominent figures on the American design scene, from Philip Johnson to Daniel Libeskind. But despite widespread admiration for his critical writings, the substantive theoretical form of Sorkin's 'resistance' is not seen to be centrally embedded in his own design production, as Mies's has been seen to be by Tafuri, or Eisenman's has been seen to be by Hays. And this has meant that Sorkin's criticism, powerful as it has been, has nonetheless been limited in its impact on the evolving forms of American design practice.

Perhaps the most distinctive 'critical' American design practice has been that of two figures who did not take part in the *HDM* Stocktaking: Elizabeth Diller and Ricardo Scofidio. For almost as long a span of time as Eisenman's and Hays' collaboration, Diller + Scofidio have produced a remarkable range of projects that have succeeded in embodying 'resistance' in a fashion that bears comparison with the one Tafuri admired in the late work of Mies. Still, most of the work in question has comprised museum and gallery installations, rather than buildings. And the museum has continued to be a more receptive venue for critical work than the street has been during the period in question, as witnessed by the parallel art practices of the Belgian Marcel Broedthaers, the German transplant to New York Hans Haacke, etc.. How interesting then, that the first major exhibition curated by the recently appointed architecture curator of New York's Whitney Museum – Michael Hays – should be of the work of Diller + Scofidio: surely a recent, and can we perhaps say 'late' triumph of American 'criticality'? What is more, it is interesting to note that in their Whitney show, Diller + Scofidio chose to exhibit many of the museum gallery

---

Stan Allen suggested that the Joan Ockman/Terry Riley pragmatism symposium held at MoMA in November of 2000 might be an earlier challenge to criticality than Speaks' polemic of early 2002. This may be so, but since I did not attend the conference and since it was devoted to pragmatism per se rather than to criticality, there is some question in my mind as to whether this challenge can appropriately be described as 'frontal'.

5 cf. Somol, Robert and Whiting, Sarah; 'Mining Autonomy' in **Perspecta 33: The Yale Architectural Journal** (2002) pp. 72–77.

6 Speaks, Michael; 'Design Intelligence: Part I, Introduction', **A + U** (December, 2002) pp. 10–18.

7 cf. 'Stocktaking 2004: Nine Questions About the Present and Future of Design'; **Harvard Design Magazine** (Spring Summer 2004); Number 20; pp. 5–52.

projects that have made them famous, and almost none of the building projects on which their recent design practice has focused, which will have to meet the more difficult test of being critical 'in the street' as it were. It will be interesting to see how successful this firm will be in sustaining its compelling stance of 'resistance' in buildings rather than in installations, and in the now changing climate of American architectural theory.

What then can we say about this changing climate? What reasons can we adduce for the increasingly threatened state of 'criticality'? And what are the key features of the approaches to architectural theory that are being offered up in its place? It seems to me that there are a number of strands to the story, most of them interesting, if not all of equal historical consequence.

One of them, as far as I can tell, is a purely biographical – not to say generational – predicament. It is a commonplace to note that Peter Eisenman has been a major influence in American architectural education since his founding of the New York based Institute for Architecture and Urban Studies – and of its in-house Journal *Oppositions*, some three decades ago. By now, I think it can be said that Eisenman's influence on his protégés can be compared with that of his own mentor Colin Rowe on his. But Rowe, it would seem, was an easier-going mentor than Eisenman has been able to be. As a consequence, getting out from under the influence of the master has been a much greater challenge for the protégés of Eisenman than it was for those of Rowe. I do not think it is a coincidence that so many of the protagonists of the currently proffered alternatives to 'criticality' are former protégés of Eisenman. For example, Stan Allen, Robert Somol and Sarah Whiting all fall into this category. To the extent then that Eisenman himself has maintained such obdurate loyalty to 'criticality' over a long span of time, he has produced a corresponding tension amongst his followers, in respect to their understandable career efforts to cut loose from him. I suspect that we could go even further, and speculate that to the extent that he has also maintained a stance of continuing contempt for what he, following Rowe, has called 'décor de la vie', he has opened the door to a revived interest in surface, and texture – and even decoration – on the part of some of his revisionist followers? Whatever the final answers to these intriguing biographical questions turn out to be, it is clear that an effort to transcend a certain Eisenmanian hegemony in the upper echelons of American architectural culture, is one of the more personal current tendencies evident to observers such as myself.

What is more, it is probably no less co-incidental, that an alternate referent frequently turned to in the various discourses of 'post-criticality' is the same Rem Koolhaas I have already quoted. For example, Koolhaas performs a crucial bridging role in Somol and Whiting's 'Doppler Effect' text, enabling them to shift from a 'critical' stance to a 'projective' one. Then too, it is almost inconceivable that the post-Utopian pragmatism so pervasive in leading Dutch architectural circles nowadays (and which one suspects, to some extent, has been imported to the United States by Speaks), cannot be tracked back to the influence of the stormy young Koolhaas himself, on the early generations of his own protégés, in the late 1970's and 80's.[8]

But this reference to Koolhaas brings me back to the short quotation from him that I cited early in this text, and which will serve to move us from biographical and generational considerations to more substantive ones. For, in the commentary in question, Koolhaas went on to speculate as follows:
> Maybe some of our most interesting engagements are uncritical, emphatic engagements, which deal with the sometimes insane difficulty of an architectural project to deal with the incredible accumulation of economic, cultural, political but also logistical issues.[9]

Here we see Koolhaas the ambitious *Real-politiker* once again exhibiting his intense belief in the necessity of a professional, architectural efficacy. And for him, if it turns out that 'criticality' constrains efficacy, then to that extent, 'criticality' must give way.

What is more, as the 1990's wore on, as it became apparent that Eisenman's own design interests would increasingly focus on process rather than products; and as it became apparent that the putative tropes of the then-ascendant 'deconstructivism' were much less 'critical' than many had expected them to be; all this contributed to a dissipation of the robust energy that had earlier been embodied – theoretically at least – by the project of 'critical architecture'.

Then too, it is probably the case that the trajectory of the life of Manfredo Tafuri from his retreat from contemporary criticism in the mid-1980's, to his death in 1994, contributed further to this shift of mood. After all, Tafuri had been the most assertive contemporary advocate of an architecture which would not accept the terms of reality as they were presented. Indeed, in an extended series of essays over the span of the 1970's, he formulated an utterly distinctive conception of the architectural 'project', one which would at one and the same time, propose a new architectural form; would do so on the plane of the entire urban entity in which it was to be located; and would, by inference, transform that entire urban entity itself into something new. Needless to say, there were not too many successful historical examples of this bold and ambitious method that he could point to – Le Corbusier's Plan Obus for Algiers being one of the few. Given that before he had retired to Venetian history in the Renaissance, he had already dismissed the American architectural avant-garde as 'architecture in the boudoir', it cannot be

---

8 A number of forms of a post-Utopian, and post-critical European (and mainly Dutch) practice are described in an essay by Roemer van Toorn, published in the **Harvard Design Magazine** (Fall 2004/Winter 2005).
9 Kapusta; **op. cit**.

denied by the time of his death, his overall theoretical stance was a somewhat disheartening one – especially for American audiences, who had had the dystopic side of Tafuri's sensibility so predominantly emphasized to them in earlier years.

In any event, with Speaks' polemic of 2001, there commenced a stream of 'counter-critical' texts that has continued up to the present day. Speaks' *A+U* text of late 2002 is probably the most developed argument that he has contributed to the ongoing discussion to date. Entitled 'Design Intelligence', it starts off with a calculatingly particular usage of the term 'intelligence' – that of the American Central Intelligence Agency – and then moves on to argue that in the contemporary design world, 'visionary ideas have given way to the "chatter" of intelligence'.[10] Speaks then went on specifically to dissociate himself from a whole series of design tendencies he saw as obsolete, disparaging the influences of both Derrida and Tafuri along the way:

> Post-modernism, Deconstructivism, Critical Regionalism and a host of other critical architectures in the late 1980's and 1990's posed … as false pretenders to Modernism. Whether effetely Derridian or ponderously Tafurian, theoretically inspired vanguards operated in a state of perpetual critique. Stuck between a world of certainty whose demise they had been instrumental in bringing about, and an emergent world of uncertainty into which they were being thrown headlong, theoretical vanguards were incapacitated by their own resolute negativity.[11]

Instead, Speaks argued for what he called a 'post-vanguard' professional practice, defined by 'design intelligence, and not by any formal, theoretical or professional identity'. He went on:

> Accustomed in ways that their vanguard predecessors can never be to open source intelligence (OSINT as it is called by the CIA), gathered from the little truths published on the web, found in popular culture, and gleaned from other professions and design disciplines, these practices are adaptable to almost any circumstance almost anywhere.[12]

Compared to the strongly pragmatic – even anti-theoretical – stance advocated by Speaks, Somol and Whiting's 'Doppler Effect' remains a model of enduringly 'theoretical' – not to say 'philosophical' – nuance. They summarize their concern about what they label 'the now dominant paradigm' by observing that in their view, in recent years, 'disciplinarity has been absorbed and exhausted by the project of criticality'. They employ the design production of Peter Eisenman, together with the theory of Michael Hays to attempt to demonstrate this. In this respect, perhaps their most important claim is that:

> … for both [Eisenman and Hays], disciplinarity is understood as autonomy

(enabling critique, representation and signification), but not as instrumentality (projection, performativity, and pragmatics). One could say that their definition of disciplinarity is directed against reification, rather than towards the possibility of emergence.[13]

And they conclude this part of their argument by observing that 'as an alternative to the critical project – here linked to the indexical, the dialectical and hot representation – this text develops an alternative genealogy of the projective – linked to the diagrammatic, the atmospheric and cool performance.'[14] Perhaps not surprisingly, this schema leads them in turn to propose as an alternative to the precedent offered by Eisenman, the one they ascribe to Rem Koolhaas. In doing so, they contrast the 'two orientations towards disciplinarity: That is, disciplinarity as autonomy and process, as in the case of Eisenman's reading of the Domino, and disciplinarity as force and effect, as in Koolhaas' staging of the Downtown Athletic Club'.[15] And they conclude: 'Rather than looking back or criticizing the status quo, the Doppler projects forward alternative (not necessarily oppositional) arrangements or scenarios'.[16]

So even though they eschew the extreme polemical stance taken up by Speaks, they do not, in the end, differ all that fundamentally from the position he espoused. For his part, in his contribution to the *HDM* 'Stocktaking', Stan Allen offered up a commentary which is broadly parallel to the two just cited. Like Speaks, Allen identified a need to 'go beyond avant-garde models' and to make use of 'popular culture and the creativity of the marketplace'.[17] Indeed, he explicitly endorsed both Speaks, and Somol and Whiting's arguments, citing both texts in his own.

Most recently, in parallel presentations at Princeton, Harvard and Toronto, Sylvia Lavin has entered into the debate, and has made a distinctive contribution to it, calling for a new appreciation of, and consideration for 'the provisional' and the 'ephemeral' in the world of contemporary architecture and design. Characterizing modernism as excessively preoccupied with the 'fixed' and the 'durable' in the world, she argued that reconsideration of such qualities in the environment could be both liberating and productive of new design possibilities.[18]

What should we make of this unfolding divergence of opinion between two important generations of thinkers on the scene of American architectural theory? Let me conclude by offering a few observations of my own. First of all, let us step back a little from the front lines of this battle,

10 Michael Speaks, **A+U; op. cit.**, p. 12.
11 **ibid.**, p. 16.
12 **ibid.**, p. 16.
13 Somol and Whiting; **op. cit.**, p. 73.
14 **ibid.**, p. 74.
15 **ibid.**, p. 74.
16 **ibid.**, p. 75.
17 Allen, Stan; contribution to 'Stocktaking 2004: Nine Questions About the Present and Future of Design'; **Harvard Design Magazine** (Spring Summer 2004); Number 20.
18 Sylvia Lavin, in lectures delivered during the spring of 2004 at Harvard, Princeton, and University of Toronto.

and take a closer look at figures that lie in the background. From my comments thus far, I think it is clear that Manfredo Tafuri looms large behind American formulations of a 'critical architecture', and that having long exhibited discomfort in regard to its implications, Rem Koolhaas has served as a model for some of the orientations to practice that have been proposed as alternatives to it.

But there is an additional *eminence grise* looming in back of a number of the members of the camp who have criticized the influence of 'criticality'; this is the American art critic and commentator, Dave Hickey. Not yet as well known in architectural circles as Tafuri once was, Hickey is a recently appointed MacArthur Fellow, who has written on a wide range of social and cultural issues in the United States and elsewhere. A keen observer of a wide range of popular cultures – and an art critic with a decidedly skeptical view of the continuing pertinence of the artistic tradition of minimalism (this alone places him far from Eisenman) – Hickey is cited by Somol and Whiting as the author of an interpretation that opposes the performing styles of two American film actors, Robert Mitchum and Robert di Niro. Interpreting Hickey on the two American actors, Somol and Whiting contrast the styles as 'hot' and 'cool'. 'While cooling suggests a process of mixing [and thus the Doppler Effect would be one form of cool], the hot resists through distinction, and connotes the overly difficult, belabored, worked, complicated. Cool is relaxed, easy.'[19] Thus it is clear that Somol and Whiting are eager to employ Hickey as part of their effort to dispel the American legacy of Tafuri in our field.

For me – and let me say that I share the two authors' fascination with Hickey – this possibility is not so clear. I shall return to this in a moment, but first, I want to review a few interesting paradoxes that arise within the overall spectrum of opinion I have outlined above.

To start with, let us take the matter of the design avant-garde (as Allen calls it) or vanguard (as Speaks does). Both commentators dismiss it as obsolete and irrelevant. This is clearly a rebuff to Eisenman, who has always seen a certain American cultural avant-garde as being the embodiment or resistance. Yet, interestingly enough, Tafuri himself declared the avant-garde in architecture obsolete and irrelevant, long before the new critics of criticality did. So strong is the tendency of American theorists to see Tafuri through an Eisenmanian lens, that they fail to take note of the fact that the American architects and planners he most admired were not avant-gardists at all, but rather, such figures as Eliel Saarinen, Clarence Stein, Henry Wright – not to mention such figures as the New Deal creators of the Tennessee Valley Authority. So preoccupied are Tafuri's American readers with 'architecture in the boudoir', that they fail to pay comparable attention to '*socialpolitik* and the city in Weimar, Germany' where

Tafuri's impatience with avant-gardism, and his strong commitment to professional 'engagement' indisputably lie. So, in the first of our paradoxes, we may observe that were he still alive, Tafuri would align himself with the disenchantment of the younger Americans with their own avant-garde, and would support their desire for 'a form of practice committed to public legitimacy, to the active engagement of new technologies, and to creative means of implementation' as Stan Allen advocates.[20]

Then, there is the matter of 'instrumentality'. At one point in their text, Somol and Whiting present 'instrumentality' as the definitive opposite to 'autonomy'. And in doing so, they summarize under this term, three of the key features of the new approach they are recommending: projection, performativity and pragmatics. But of course, Tafuri was also deeply committed to the idea of projection; indeed, as I noted earlier in this text, his highly activist conception of the architectural 'project' lay at the heart of his theoretical position. Similarly, to the extent that we can read 'pragmatics' as having at least partly to do with architectural programs, Tafuri was clearly as interested in program as a medium of design innovation as Koolhaas has ever been. But like '*Socialpolitik*', Tafuri's powerful commitments to projective efficacy, and to programmatic innovation, are hard to see through an Eisenmanian lens. Parenthetically, I concede that I do not find a comparable commitment to 'performativity' in his writings, but I do note that the political stance with which he was associated in the days of his journal *Contropiano* strategically distinguished itself from that of the Italian Communist party, on account of its commitment to active participation by workers in the ongoing formulation of party positions, as opposed to the top-down party line control advocated by the party leaders – and for this reason was labeled an '*operaista*' political tendency.

Let me return now to Hickey, and to the possibility of his being enlisted for the polemical purposes of the younger generation disillusioned with criticality. To be sure, there is in his heteroclite sensibility, a startling and engaging openness to distinctiveness which has engaged Somol, Whiting, and Lavin, and to which they in turn all seem to be committed. He is even willing to engage the 'decorative' in ways that would seem to lend support to some of the speculative comments Sylvia Lavin has made in recent academic discussions. His characterization of Mitchum's acting style, and his interests in jazz reinforce further his association with cultural stances that can be called 'relaxed' or 'easy'. But of course, it also remains the case that when all is said and done, Hickey himself continues to be engaged by an obdurate – if implicit – quest for 'authenticity'. At the first lecture I ever heard him give, he delivered an extended descriptive comparison of two southwestern American cities he knows well, Santa Fe (where his 2002 exhibition 'Beau Monde' was held) and Las Vegas (where he lives). Summing up his critical assessment of the two

19 Somol and Whiting, **op. cit.**
20 It is interesting to note at this juncture in my argument that in his contribution to the **HDM** 'Stocktaking', Stan Allen himself observed that such forms of practice as he is endorsing here, can be found in a number of locations in Europe, but have 'so far resisted translation to the US'. So perhaps the collectivist European legacy of Tafuri may remain stronger than has been acknowledged?

cities he made the dazzling observation that he prefers Las Vegas to Santa Fe because he prefers 'the real fake to the fake real'.[21] Indeed, to the extent that the protagonists of any version of a post-critical project want to enlist him to challenge the legacy of Eisenman, a comment from his introduction to the 'Beau Monde' show will give them pause. Discussing his selection of artists and works to be included in the show, he observed: 'Rather than asking the post-minimalist question "How rough can it get and still remain meaningful?" I found myself asking the cosmopolitan question: "How smooth can it get and still resist rationalization?"'[22]

So, we can now see, even the 'cool' and speculative Hickey continues to be engaged by a form of 'resistance'. It seems to me that the provocative question he suggested he had asked himself about his Santa Fe show is one that one could easily imagine being asked in regard to a work of Diller + Scofidio such as their 'Soft Sell' 42nd Street installation of 1993.[23] And speaking of 42nd Street, is it not interesting also to recall that the very figure with whom I began my account of the erosion of the dominant discourse of criticality was Rem Koolhaas. For Koolhaas himself, notwithstanding his interests in 'creative means of implementation', and in other key parts of the post-critical agenda, has nonetheless participated in more than a few recent episodes of vigorous critical engagement. I could start by recounting the fascinating episode during which, having been brought to Harvard to attack Andrés Duany and the New Urbanism, he declined to do so, waiting only for an appropriate moment to chastise Duany severely for his failure 'as a prominent American architect' to speak out against the destruction of the distinctive street culture of Manhattan's 42nd Street as a result of its sweeping Disneyfication. And I would probably end with his recent attack on the Chinese authorities for their lamentable and all-too-pragmatic approval of the destruction of extensive historic residential districts of the city of Beijing.

Thus it seems to me that the political alignments, and the theoretical complexities that this interesting divergence of opinions has brought to the surface to date, do not so much constitute the conclusion of a story, but rather, only the beginning of one. A number of important questions remain to be asked, it seems to me, before a truly robust and durable new professional stance will be able to be achieved. For example, while it is probably true that 'relaxed' and 'easy' cannot

21 Hickey, Dave; 'Dialectical Utopias'; **Harvard Design Magazine**; Winter/Spring 1998, pp. 8–13.
22 Hickey, Dave; **Beau Monde: Towards a Redeemed Cosmopolitanism** (Santa Fe, New Mexico: Site Santa Fe, 2001) p. 76.
23 'Soft Sell' is documented in Diller + Scofidio and Teyssot, George; **Flesh: Architectural Probes: The Mutant Body of Architecture** (Princeton, N.J.: Princeton Architectural Press, 1994) pp. 250–253.

be reconciled with 'difficult', it is not so clear to me that they cannot be reconciled with 'resistant' either. And it is equally clear to me that a much more developed pursuit of social and political parallels between architecture and cinema would be one potent way of articulating such subtle distinctions further.

Then too, I am very curious to see to what extent the putatively 'projective' forms of practice being advocated by the new critics of criticality will develop parallel models of critical assessment with which to be able to measure the ambition and the capacity for significant social transformation of such forms. Without such models, architecture could all too easily again find itself conceptually and ethically adrift. For example, while it is clear from a multitude of cultural perspectives that the 'decorative' as a formal category, can be integrated within new forms of practice, it is also clear that those forms run some risk of being reduced to the 'merely' decorative. Enough architectural episodes of the 'merely' decorative have occurred recently to serve as a warning. Most fundamentally in my view, it is clear that a newly projective architecture will not be able to be developed in the absence of a supporting body of projective theory. Without it, I predict that this new architecture will devolve to the 'merely' pragmatic, and to the 'merely' decorative, with astonishing speed.

May I conclude, then, by calling for much more careful reflection from us all, before the respective roles of critique, innovation, authenticity, and expanded cultural possibility can be integrated into an 'operative' new theory of praxis for our times.

Power Lines

Ákos Moravánszky

Alajos Landau, a painter and art teacher in the late 19th century in Budapest, published in 1882 his *Drawing Geometry*[1] (Rajzoló geométria), a book on geometry and perspective for students. One of its illustrations shows a *csikós*, the horseman of the Hungarian flatlands, the *puszta*, clad in his traditional dress, staring at a mysterious point situated somewhere on the horizon, and his eyes are guided toward it by the long straight road with electric power lines running along its side. It seems strange to find this figure familiar from romantic painting and poetry in Landau's book. In the Hungarian culture of the 19th century, the *csikós* was an important symbol of national virtues. His semi-nomadic life (often also associated with an outlaw attitude toward the Austrian-controlled central government), his free movements on the roadless *puszta* were expressions of an idealized identity very different from the realities created by the rapid modernization of the country. The uninterrupted vistas of the *puszta* create the illusion of unimpeded freedom: '*My spirit soars, from chains released, when I behold the unhorizoned plain…*', wrote the poet Sándor Petőfi.[2]

Landau's horseman seems to observe the constructed perspective image rather than being part of it; the horse is not standing on the road but seems to hover outside its space. We don't see the vanishing point on the horizon because the back of the horseman's head blocks our view. In order to construct the new road which cuts through the many possible paths of the *puszta*, a map is necessary. This map projects its logic onto the land and puts the vanishing point on the horizon. The straight lines of modernization appear in this mythic region of resistance as a result, and as a sign of a violent and oppressive power. Our horseman must marvel, looking at the point where no visible building or object appears: 'What is there?' The point seems to be strongly present, different from all the other points on the horizon. Yet this question is precisely the one which Alberti was asking when he looked at the vanishing point as a kind of kingpin of the newly-invented perspective: 'What is there?' Yet he gave the answer only hesitantly: '*quasi l'infinito*'.

Theoria   The Greek origin of the word theory, *theoria*, is worth considering. *Thea* is an occurrence which wants to be understood, and *theoros* is an observer, who looks at what is occurring and tries to grasp and understand it. The Greek idea of theorizing was connected with the oracles. A *theoros* was sent by a *polis* (city) to a place of oracle like Delphi, to be present at the oracle and report with authority; that is, without altering it: 'for neither adding anything would you find a cure, nor subtracting anything would you avoid erring in the eyes of gods'[3], as the Greek poet Theognis has warned the *theoros*. The decision of the Athenians whether to start a war against the Persians, or take instead a defensive stance, depended on the report and interpretation of divine utterances by the *theoros*. His task was to create a narrative to bridge the gap between human intelligence and divine interaction. The narrative of the *theoros*, however, had to

---

[1] Landau, Alajos and Wohlrab, Flóris; **Rajzoló geométria** (Budapest: Franklin Társulat, 1882).
[2] Petőfi, Sándor; 'Az Alföld' (1844), in **Petőfi Sándor összes versei** (Budapest: Osiris, 2001).
[3] Maurizio, Lisa; 'Delphic Oracles as Oral Performances: Authenticity and Historical Evidence', in **Classical Antiquity** Vol. 16 No. 2 (Oct. 1997), pp. 308–332.

be negotiated: in cases when the Athenian ambassadors rejected an oracle, they refused to confer authority to the *theoros*.

Let us focus, for the time being, on the visual aspect of *theoria* as observation and on the image as a visual access to the world. Martin Heidegger explains in his essay 'The Age of World Picture' that an important prerequisite for the world to become picture was man's anthropological transformation. For the Greek philosopher and poet of the fifth century BC, Parmenides, man was essentially a 'hearer of being', but in Heidegger's view, Plato's concept of the *eidos* (form as an object of knowledge) constitutes the precondition for the shift to the modern concept of the world as picture.[4]

Landau's perspective demonstrated the use of perspective as a primary means of world-constructing. His illustration, compared with other, picturesque representations of the *puszta* in romantic painting, shows how the certainty of the constructed perspective is now claiming a rational, even technical access to former mythical territories. In Heidegger's words, 'it is not simply that we have a different picture of the world, but we now have the world as a picture'.[5] This means that *theoria* as observation of the world (*Weltanschauung*) is indeed a supreme practice, the conquest of the world as picture.[6]

Heidegger saw in Descartes' interpretation of man as subject the metaphysical precondition for all future sciences ('anthropologies'). That is to say, the subjectification of being takes place in representation (*Vorstellung*, in the sense of imagination as well), as the creation of an image. The Cartesian equation between truth and the 'certainty of imagination' will also be evoked by Michel Foucault.

Foucault, in a chapter of his book *The Order of Things*[7] entitled 'The Limits of Representation' discusses the liberation of the order of our knowledge from the discourse of representation. He is clearly influenced by Heidegger's description of Descartes' triumph, which is the break with pre-classical epistemology and the introduction of the discourse of representation. He focuses in this chapter, however, on the second break, which he situates at the end of the 18th century. From this moment on, the order of things will no longer be anchored in the space of representation, but shifted to the realm of the invisible:

> In order to find a way back to the point where the visible forms of beings are joined – the structure of living beings, the value of wealth, the syntax of words – we must direct our search towards that peak, that necessary but always inaccessible point, which drives down, beyond our gaze, towards the very heart of things.[8]

However, in Foucault's view, only the empirical sciences freed themselves from the burden of representation; human sciences remained firmly situated in their space.

We now return to Alajos Landau's *csikós* to find out more about the ties between the 'world picture' and architectural representation around 1900. The hegemony of the straight lines of projection, clearly demonstrated by Landau's constructed view, became an important theoretical basis for the architecture of Otto Wagner. The Viennese architect drafted his first monumental perspective views of ideal cities (e.g. *Artibus*, 1880) at the time when Landau was working on his *Drafting Geometry*. A later architectural perspective by Wagner, the well-known bird's eye view of the central park (*Luftzentrum*) of the future twenty-second district of Vienna shows the power of projection in a similar sense as in the Landau drawing: the street lines of the urban grid were just as alien to the existing circular-radial system of Vienna as the power lines in the *puszta*. The perimeter of the district reveals the difficulty in accommodating the rectangular grid within this section of the city, located between two radial streets.

Wagner was invited to participate in a city planning conference in New York in 1910, an occasion to develop this ideal project, published as *'Die Grossstadt. Eine Studie über diese'* (1911) in German, translated as 'The Development of the Great City'.[9] Because of the history of the proposal and the similarity to North American urban projects such as the Columbia University campus by McKim, Meade, and White (1894), it is often assumed that Wagner's urban plan was inspired by the grid of the American city.[10] The North American city, however, is just one possible model for the rectangular urban grid. It represents the geometrical projection of a legal/metaphysical space also found in Roman imperial colonies or in colonial towns in South America. The geometry of the mapping shows a high grade of neutrality regarding its methodology as well as its results irrespective of the spatial models, whether Roman law, Spanish colonial legislation (the 'law of the Indies')[11] or Jeffersonian rules of real estate.

Pinakes To discuss the relationship between such models and the urban/architectural form, we have to return to Greek philosophy. Anaximander, a philosopher and speculative astronomer who replaced the closed celestial dome with an open universe, created in the 6th century BC the

---

4 Heidegger, Martin; 'Die Zeit des Weltbildes', in Heidegger, **Holzwege** (Frankfurt am Main: Vittorio Klostermann, 1980), p. 92.
5 **ibid.**, p. 87.
6 **ibid.**, p. 92.
7 Foucault, Michel; **The Order of Things** (New York: Random House, 1973).
8 **ibid.**, p. 239.
9 Wagner, Otto; **Die Grossstadt: Eine Studie über diese** (Vienna: Anton Schroll, 1911). English translation 'The Development of the Great City', **Architectural Record** 31 (May 1912), p. 485–500, reprinted in **Oppositions** 17 (Summer 1979), pp. 103–106.
10 E.g. Sarnitz, August; 'Realism versus Verniedlichung: The Design of the Great City', in Harry Francis Mallgrave, **Otto Wagner: Reflections on the Raiment of Modernity** (Santa Monica: The Getty Center for the History of Art and the Humanities, 1993) pp. 101**ff.**
11 The law of the Indies, issued by the Spanish king Philippe II in 1573, regulated the urban form of Spanish cities in America. **Cf.** Gasparini, Graziano; 'The Law of the Indies. The Spanish-American Grid Plan: The Urban Bureaucratic Form', in **The New City** 1 (Fall 1991), pp. 6–33.

first *pinax* – a 'cognitive map', a philosophical diagram of the earth, at the same time a material object, like a bronze tablet with engraved surface. Anaximander's map was circular, describing the shape of the earth as a column-drum.[12] The circular form was originally a form of assembly and warfare, and from this practice it was transferred to the *polis*, the urban political space.

Otto Wagner's aerial perspective, however, shows a different origin. It can be related to another *pinax*, that of Hippodamos from Miletos, who was not only the first urban theorist but the first political theorist as well. He codified the orthogonal urban grid as a proposal that served Athenian political intentions well. The defensive contraction of the circular city lost its significance, as the grid was able to connect, creating a large trading area, a 'common market' for the Greeks and Barbarians. The birth of this proposal marked the zenith of Athenian political aspirations, not only in terms of military expansion and territorial gains but also spreading the ideal of *hellenikón*, a Greek way of life, a product of Athenian culture. This dissemination was based on the exportation and imposition of exemplary models – including that of the city. 'By nature we are all equal, foreigners and Greeks'[13] – wrote Antiphon, a friend of Pericles. The democracy of Pericles was the consequence of this new relationship between the *pinax* as ideal model and the urban form which the *pinax* shapes.

In the form of the historic center of Vienna the old ring of military fortifications is still inscribed. The vast circular space of the *Ringstrasse* occupies the place of the former belt of fortifications. For the new urban space, alternative models as 'competing visions' did exist – Carl E. Schorske discussed these diverging proposals in his classic study *Fin-de-siècle Vienna*.[14] Camillo Sitte, who as urban theorist was more influential than Otto Wagner, commented with sarcasm on the so-called 'modern systems'. He charged the large open spaces with causing *agoraphobia*, a modern mental illness. Wagner, on his part, insisted on the opposite, arguing in his *Modern Architecture* that urban dwellers underwent an anthropological change: 'The modern eye has ... lost the sense for a small, intimate scale; it has become accustomed to less varied images, to longer straight lines, to more expansive surfaces, to larger masses, and for this reason a greater moderation and a plainer silhouetting of such buildings ... seems advisable.'[15] Both Wagner and Sitte studied contemporary theories of visual perception, and interpreted urban place using the observations of such authors as Hermann von Helmholtz and Hermann Maertens.[16]

A further argument for the straight line was its economic efficiency in modern Capitalism: '... the busy man, whenever possible, moves in a straight line, and that the person in a hurry is surely annoyed by the smallest time-consuming detour. The last decades have even carried the banner "Time is money".'[17]

Power Points, Power Lines The transformation of the ideal model into a perspective view is a projection, a process of utmost importance for Wagner who selected his employees and students on the basis of their drawing skills: 'When composing, the architect has to place great importance on the effect of perspective', Wagner wrote. The architect 'must organize the silhouette, the massing, the projections of the cornice, the distortions, the sculptural line of the profile and ornaments in such a way that they appear properly emphasized from a *single vantage point*. This point will ... be that location where the work can be viewed most frequently, most easily, and most naturally...'[18] Again, part of the explanation for the dominance of the vantage point is the need for a moment of stasis in the restless metropolis:

> One of the attributes peculiar to human perception is that in examining any work of art the eye seeks a point of rest or concentration; otherwise a painful uncertainty or aesthetic uneasiness occurs. This will always prompt the architect to design a focal point where the rays of attention combine or organize themselves.[19]

Wagner was very aware of the power exerted on the body by architecture, and the role of visual perception in this process. He emphasized that some works of architecture are composed for two viewing distances, for instance buildings with domes and towers. The façade in such cases has to satisfy with its details the observer on the square or the street, 'while the high, richly silhouetted superstructure was either an integral part of a *veduta* or resonated harmoniously with the cityscape in order to become a characteristic landmark visible from afar...'[20] Wagner turned such observations directly into design methodology. When designing a church for the mentally ill (St. Leopold am Steinhof, Vienna 1901-1907), he stressed the issue of avoiding back light that would disturb the patients while attending mass and emphasized the unobstructed view of the altar as an achievement that distinguishes his building from many well-known examples of church architecture.

More significantly, the construction of the perspective image determines the built object even in its smallest details. The lines of projection leave their traces on the skin of the buildings. The

---

12 Hahn, Robert; **Anaximander and the Architects: The Contributions of Egyptian and Greek Architectural Technologies to the Origins of Greek Philosophy** (Albany: SUNY, 2001).
13 Farinelli, Franco; 'Squaring the Circle, or the Nature of Political Identity', in Franco Farinelli, Gunnar Olsson, Dagmar Reichert; **Limits of Representation** (Munich: Accedo, 1994), p. 23.
14 Schorske, Carl E.; **Fin-de-siècle Vienna: Politics and Culture** (Cambridge: Cambridge University Press, 1961).
15 Wagner, Otto; **Modern Architecture: A Guidebook for his Students to this Field of Art**. Introduction and translations by Harry Francis Mallgrave (Santa Monica: The Getty Center for the History of Art and the Humanities, 1988), p. 109.
16 von Helmholtz, Hermann; **Handbuch der physiologischen Optik** (Hamburg, Leipzig: Leopold Voss, 1856–1866) and Maertens, Hermann; **Der Optische-Massstab oder Die Theorie und Praxis des ästhetischen Sehens in den bildenden Künsten** (Bonn: Carl Georgi, 1877).
17 Wagner, **Modern Architecture**, p. 110.
18 ibid., p. 86.
19 ibid., p. 87.
20 ibid., p. 87.

façade of Otto Wagner's apartment block on the Neustiftgasse in Vienna (1909-1910) shows a fine grid of horizontals and verticals, lines cut into the plastered surface, locating the building within the grid of the big city. Wagner wrote: 'Architectural treatments that seek their motives in the architecture of palaces are completely inappropriate to such cellular conglomerates, simply because they contradict the interior structure of the building.'[21] The significance of the project lies in its closeness to the *pinax*, to the endless grid of the 'cellular conglomerate' that generated the form even if the house is a mere fragment of his visionary bird eye's view. The big letters of the street name and number are the only identifying mark of a 'prototype' which could be projected anywhere.

The Adoration of the Perspective In the architectural drawing, theory and practice intersect, the place of transformation where the abstract lines of projection turn into building mass. The significance of presentational drawings, the large perspective views of Wagner, is revealed by their almost devotional treatment, as if they were relics, with flowers as an additional layer covering the frame of the drawing. Wagner, in his *Moderne Architektur*, also discussed the question of how architectural works are to be presented graphically:

> A picture, a sculpture, a room, a building, or any other artistic object directly affects the senses of the viewer through the eye, thereby greatly facilitating understanding and judgment. Yet the understanding of plans and elevations requires an intellectual concentration for which the viewer usually lacks the desire, more often the capability, thus making judgment more difficult… Also, many architects are content to present drawings in an unimaginative way, not consistent with the demands of modern taste.[22]

Otto Wagner's *Moderne Architektur* appeared in four editions between 1896 and 1914. He made some changes to updated editions. In the 4th edition, Wagner added two new chapters and changed the title to *Die Baukunst unserer Zeit* (The Building-Art of Our Times). There are no architectural drawings in the four editions, only small photographs, mostly details, and vignettes arranged in pairs, which also reflect playfully on Wagner's favorite dichotomies. Wagner commented: 'I wanted to avoid [graphic illustrations] because my earlier publications illustrate what is said here. They show clearly how the views expressed ripened in me.'[23] In stressing the specific character of architectural perspectives, Wagner emphasized:

> … the architect's task must always be to put his ideas down on paper in the clearest, most distinct, neatest, most purposeful, and most convincing way possible. Every architectural drawing has to document the taste of the artist, and it must never be forgotten that what is to be presented is a FUTURE, not

an existing work. The craze to offer the most deceptive possible picture of the future is an error, if for no other reason than because it involves a lie…[24]

He demanded 'a personal and impressionistic presentation…'[25] No wonder that Wagner was convinced that 'we should see in architecture the highest expression of man's ability, bordering on the divine'[26] and his architectural perspectives became objects to be presented on iconostasis-like scaffoldings. Both architect and his drawing are objects of adoration, the former because of his divinatory talent, 'with his happy combination of idealism and realism … [he] has been praised as the crowning glory of modern man.'[27]

Practical and Theoretical Cultures

Wagner's praise was rooted in his conviction that the architect's sensitivity to read symptoms of modernity – from fashion to changing social habits and mentalities which he carefully registered in his writings – makes him uniquely capable to create environments that communicate the new values. While various institutionalized practices from statistics to medicine can be regarded as modes of management of the society, architecture and urbanism exert a much more direct power by establishing boundaries, regulating flows and opening accesses. Many new changes are introduced in the name of rationality, but Wagner insisted that the works of the engineer, even if rational, would be rejected:

> The engineer who does not consider the nascent art-form but only the structural calculation and the expense will therefore speak a language unsympathetic to man, while on the other hand, the architect's mode of expression will remain unintelligible if in the creation of the art-form he does not start from construction.[28]

Wagner emphatically referred to modern life as the main source of the architect's experience: 'The main reason that the importance of the architect has not been fully appreciated', he wrote in 1896, 'lies … in the language he has directed to the public, which in most cases is unintelligible.'[29] To make the language of the architect understandable again, Wagner urged the student of architecture to visit the centers of modern life, to experience 'where modern luxury may be found, and there he might train himself completely by perceiving the needs of modern man.'[30] The legibility of architecture requires that it be rooted in the 'needs of our time':

> All modern creations must correspond to the new materials and demands of the present if

21 ibid., p. 109.
22 ibid., p. 101.
23 ibid., p. 124.
24 ibid., p. 102.
25 ibid., p. 102.
26 ibid., p. 62.
27 ibid., p. 61.
28 ibid., p. 94.
29 ibid., p. 65.
30 ibid., p. 69.

they are to suit modern man, they must illustrate our own better, democratic, self-confident ideal nature and take into account man's colossal technical and scientific achievements, as well as his thoroughly practical tendency...[31]

This 'truth' of modern life must be the basis for the new architecture and urbanism. But can these principles, this abstract idea of modernity translated into a new aesthetics, be a 'practical culture', to use Friedrich Schiller's distinction?

For Schiller, the re-evaluation of practice was a response to the disillusion after the failed revolution of 1789. Schiller was convinced that political changes could not achieve the hoped-for results unless they were preceded by a change in the individual's way of perception (*'in seiner ganzen Empfindungsweise'*). Enlightenment as a theoretical culture, operating in the realm of language, cannot be successful without a practical culture which requires aesthetic education: '... if man is ever to solve that problem of politics in practice he will have to approach it through the problem of the aesthetic, because it is only through beauty that man makes his way to freedom.'[32]

Wagner was convinced that the modern architect as 'the crowning glory of modern man' is particularly qualified to achieve this transformation by the means of the architectural project and projection. Just as architecture and urban design based on rationality will be 'unsympathetic', opposed by inhabitants and observers, a projection founded on the rational procedures of perspective will meet resistance if not transformed into an artwork, 'a personal and impressionistic presentation'. The hegemony of the perspective in the studio and school of Otto Wagner was, however, already rejected by Adolf Loos, and in the Bauhaus, isometric projection (axonometry) was regarded as a new optical access to the object of design.

Projection is not the representation itself but the process – turning theory into practice, turning design into anthropology, and turning urbanism into the organization of society. Therefore it is imperative that the models, the *pinakes* which regulate the process of projection are constantly called into question. Today, with the keyword of *globalization* a new (and at the same time, very old) cosmological image, a new *pinax* seems to control the reorganization of socio-economical space. The case study I presented here shows that there is no reason to justify the legitimacy of one singular model as a fateful morphology that determines the management of resources and human bodies, but to point out its geometrical simplifications and political distortions, and compare it to other possible and even co-existent models.

31 ibid., p. 78.
32 Friedrich Schiller; 'Letters on the Aesthetic Education of Man', in **Essays** edited by Walter Hinderer and Daniel O. Dahlstrom (New York: Continuum, 1993), p. 90.

1 Illustration from Alajos Landau and Flóris Wohlrab, **Rajzoló geométria** (Budapest: Franklin-Társulat, 1882).
2 Otto Wagner, 'Artibus', bird's eye view, 1880. From Otto Antonia Graf, **Otto Wagner** (Vienna, Cologne, Graz: Böhlau, 1985).
3 Otto Wagner, 'Die Groszstadt', center of the XXII. District of Vienna, bird's eye view, 1911. From Graf, op. cit.
4 Otto Wagner, 'Die Groszstadt', center of the XXII. District of Vienna, 1911. From Graf, op. cit.

Abb. 883, 159, 1 Zentrum des XXII. Bezirkes, Vogelschau

[3]

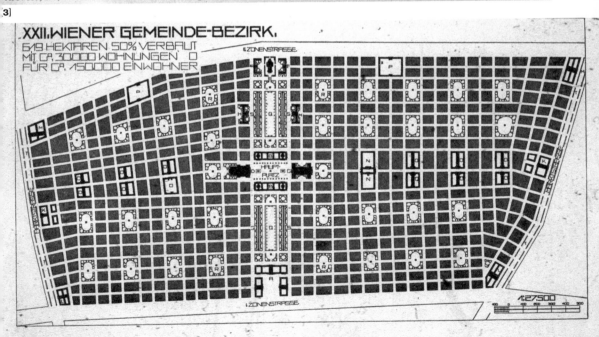

[4]

[5]

[6]

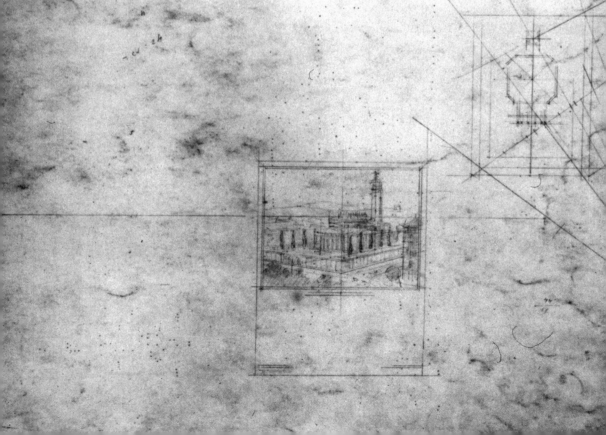

STUDIE ZUR BERLINER DOM·FRAGE

[7]

[8]

5 Plan of the foundation of the city of San Felipe of Santiago in Cuba. From **The New City** 1, 1991.
6 Otto Wagner, Friedenskirche (Church of Peace), perpective sketch, 1917. From Graf, op. cit.
7 Otto Wagner, Project for Berlin Cathedral, 1891. From Graf, op. cit.
8 Otto Wagner, Project for Berlin Cathedral, detail.

[9]

[10] [11]

[12]

[13]

[14]

9 Presentation of Otto Wagner's project for the cemetery church in Währing on the 5. exhibition of the Secession, 1899.
10 Otto Wagner, cemetery church in Währing, 1898. From Graf, op. cit.
11 Otto Wagner, Künstlerhof, perspective sketch, 1917. From Graf, op. cit.
12 Otto Wagner, Apartment house Neustiftgasse 40, Vienna, 1909. From Graf, op. cit.
13 Otto Wagner, Apartment house Döblergasse 4, detail of the façade. Photo Á. Moravánszky.
14 Page from Otto Wagner, **Die Baukunst unserer Zeit** (Vienna: Schroll, 1914).

Ethics and Poetics in Architectural Praxis

Alberto Pérez-Gómez

Today, whether we are living at the end of Progress, or simply enduring the continuing crisis of modernity, we may recognize that the reality of our discipline is infinitely complex, both shifting with history and culture, and also remaining the same, analogous to the human condition which demands that we continually address the same basic questions to come to terms with mortality and the possibility of cultural transcendence opened up by language, while expecting diverse answers which are appropriate to specific times and places. Architecture, like other forms of making traditionally associated with the Fine Arts, is an ontological mutant. Its historical being cannot be reduced to a species of works. It is a naïve prejudice to identify the tradition of architecture with a chronological collection of buildings, understood as useful 'creations', whose main significance was to delight through aesthetic contemplation. This popular modern conception was a product of the Enlightenment that only came to fruition in the 19th century, particularly after the dissemination of J.N.L. Durand's *Recueil et Parallèlle des Edifices de Tout Genre, Anciens et Modernes*[1] and its rendering of architectural history as a progressive sequence of rational building types. Such a reduced and objectified understanding leads us to a dead-end. In the wake of the failures of functionalism and heroic formalism to create significant spaces for the majority of mankind to dwell, it becomes difficult indeed to defend the cultural legitimacy and importance of our discipline, particularly if identified exclusively with the efficient *production* of buildings. A more careful appraisal of our architectural traditions and their changing political and epistemological contexts, suggests a different way to understand architecture's 'universe of discourse' – operating in the realm of what Giambatista Vico called in the early 1700's 'imaginative universals': A discipline which over the centuries has seemed capable of offering humanity, through widely different incarnations and modes of production, far more than superfluous pleasure or a technical solution to pragmatic necessities.

Our technological world often encourages skepticism about architecture having any meaning *at all*, other than providing for shelter. This is not surprising, for in our technological world, increasingly synonymous with the global village, 'universal' truths obtain legitimacy by association with the technical achievements of applied science, ultimately issuing from a mathematical language: the one thing that seems to 'stand' for all, regardless of local languages or cultures. Our technological world is one driven almost exclusively by efficiency of means. Efficiency stands as an absolute value, not only in economics or industrial production, but in all orders of life, demonstrable through mathematical argumentation. Thus the means can claim to be unaffected by the social consequence of its ends. Furthermore, the technological world is one in which specialization is deemed as the only solution to the proliferation of information, disregarding real *knowledge*, that is our ethical capacity to respond for our actions, in view of our total life experience, here and now. Within this framework, meaningful architecture is nothing other than efficient building.

---

1 Durand, Jacques-Nicolas-Louis; **Recueil et Parallèlle des Edifices de Tout Genre, Anciens et Modernes** (Paris, 1801).

Yet the mystery at the origins of human technique prevails. Our reason may be capable of dismissing the quality of the built environment as central to our spiritual well-being, yet our dreams and our actions are always set *in place*, and our understanding (of others and ourselves) could simply not *be* without significant places. Our bodies can recognize and understand, despite our so-called 'scientific' common sense and its Cartesian isotropic space, the wisdom embodied in a place, in a culture; its profound, untranslatable expressive qualities. With little effort we may recognize how architecture, in those rare places that speak back to us and resonate with our dreams, incites us to real meditation, to personal thought and imagination, opening up the 'space of desire' that allows us to be 'at home' while remaining always 'incomplete' and open to our personal death, this being our most durable human characteristic. Even binary spaces could not appear to resemble reality if we were not first and foremost mortal, self-conscious bodies *already* engaged with the world through orientation and gravity. We don't merely 'have' a body, we 'are' our bodies. It suffices to try to think in a totally dark room for more than a few minutes to convince oneself of the 'reality' of this unarticulated, pre-conceptual 'ground' of being which depends so much upon 'architecture' as the external, visible order, one primarily responsible for making our limits present.

So, where does it inhabit, this architecture? Architecture communicates to us not *a* particular meaning, but rather the possibility of *recognizing* ourselves as complete, in order to dwell poetically on earth and thus be wholly human. In the Western tradition, the products of architecture have ranged from the *daidala* of classical antiquity to the sun-dials, machines and buildings which Vitruvius names as the three manifestations of the discipline, from the gardens and ephemeral architecture of the Baroque period to the built and unbuilt 'architecture of resistance' of modernity such as Le Corbusier's La Tourette, Gaudí's Casa Batlló, or Hejduk's 'masques'. This *recognition* is not merely one of semantic equivalence, rather it occurs in experience, and as in a poem, its 'meaning' is inseparable from the experience of the poem itself. It is embedded in culture, it is playful by definition, and is always circumstantial. These artifacts, *thaumata*, convey wonder, a form of beauty grounded in *eros* (*Venus-tas*), clearly understood by Marsilio Ficino and his alter ego, Francesco Colonna during the Renaissance, as the central issue in Western architecture, distinct from 'formal composition', the late 18th-century alternative to the 'sublime' which modern aesthetics often took for granted. Architectural beauty, like erotic love, burns itself into our soul, it inspires fear and reverence through a 'poetic image', one that affects us primarily through our vision, and yet is fully sensuous, synaesthetic: it is thus capable of seducing and elevating us to understand our embodied soul's participation in wholeness. What differentiates these artifacts from other forms of *ars* or *poiesis*, concerns their intertwining with life itself in the form of significant action, traditionally ritual, ranging from the founding of cities and war, to organized religion and politics.

In other words, good architecture, or simply architecture, to differentiate it from other technical artifacts in our physical environment (such as most buildings in the post-industrial city), offers societies a place for existential orientation. Architecture is the embodiment of a radical orientation, of a metaphysics, in the particular sense given to this word by Spanish philosopher José Ortega y Gasset. When successful, architecture allows for *participation* in meaningful action, conveying to the participant an understanding of his or her place in the world. In other words, it opens up a clearing for the individual's experience of purpose through participation in cultural institutions. When successful, architecture *plays* with power. It is, however, not possible to paraphrase the order it conveys. It is radical orientation in *experience*, beyond words. So, while its theory may be rooted in mythic or poetic stories, philosophy, theology or *scientia*, during different times of its history, architecture is none of these, but an *event*. As such, it is ephemeral, yet it has the capacity of changing one's life in the vivid present – exactly like magic, or an erotic encounter. Thus, it can be said to embody knowledge, but rather than clear logic, it is knowledge understood in the Biblical sense: a carnal, fully sexual and therefore opaque experience of truth. For this reason its 'meaning' can never be objectified, reduced to functions, ideological programs, formal or stylistic formulas. Likewise, its technical medium is open rather than specific (like, say, building typologies), including all artifacts from diverse media that make possible human dwelling and which by definition stand 'at the limits of language', establishing the boundaries of human cultures within which other more properly linguistic forms of expression may take place.

However, in order to appropriately frame and bring about architectural *events*, the architect must necessarily engage language. The main concern of any generative theory of architecture is therefore, *ethical*. Its purpose being to find appropriate language (in the form of stories) capable of modulating a project in view of ethical imperatives, always specific to each task at hand. The practice that emerges from such a theory can never be an instrumental application nor a totalizing operation, one that might be universally applied as style or method. Rather, this *praxis* aims at the production of harmonious, well-adjusted *fragments* that may question, by inducing wonder, the hegemony of the ideological, fundamentalist or technological beliefs embedded in the physical fabric of the global village. This *praxis* may be better grasped as a *verb*, rather than in terms of its heterogeneous products, as a process that is never neutral and should be valorized. The presence of a well-grounded *praxis*, the trajectory of an architect's words and deeds over time, embodying a responsible practical philosophy, is the key element for any architectural criticism, over and beyond the particular 'aesthetic' or 'functional' qualities of a particular work.

The poetic and critical dimension of architecture is not unlike literature and film, addressing the questions that truly matter for our humanity in culturally specific terms, revealing an enigma behind everyday events and objects. The cultural specificity of *practices* in our global village is therefore absolutely crucial. Though technology has already had a homogenizing effect, *praxis* involves much more than technical means and scientific operations – it concerns values, articulated through the stories that ground acts and deeds in a particular culture. This practical wisdom is usually of the order of oral transmission, rather than textual information. The enduring quality of architecture is essential for the perpetuation of cultures. Values, as emerging in the life-world, are preserved by institutions, and embodied in our physical constructions. These diverse practices, like their accompanying dying languages, are valuable endangered species, and must be preserved.

Architectural education cannot be reduced to the instrumental theories that are taken for granted in schools around the world. Interestingly enough, the computer now offers means of accessing ever-changing technical information and its applications, making it possible to liberate valuable school time for face-to-face communication, to make place for a true education through dialogue and debate. It is possible to imagine that we can reverse the tendency towards specialization which also started in the early 19th century, and that has culminated in what Ortega y Gasset called the age of the 'specialized barbarian'. Education does not happen magically with students 'synthesizing' specialized information in the mind. This is a naïve model borne from a Cartesian misunderstanding, for human consciousness is not reducible to mind, and even less, to a computer. Technical courses particularly, should be redesigned with broad questions in mind, teaching the future architect not solutions necessarily, but the origins of issues, their ethical consequences, and tactics for thought.

Design cannot be dictated by functions, algorithms, or any sort of compositional *mathesis*, for the issues of architecture are never simply technological or aesthetic. Architectural design is *not* problem-solving, and formal innovation is not enough, as we are often told by naïve evolutionary theorists. Future architects should be prepared to use their imagination to *make* poetic artifacts rather than *plan* buildings, engaging dimensions of consciousness which are usually stifled by our present educational paradigms. This is not a merely intuitive operation, or unreflective action, but rather the continuation of a practical philosophy and a meditative practice. Making with an awareness of expectations, in a collaborative mode whenever appropriate to the tasks, seeking the magic of coincidence which is the epiphany of order. For poetry, according to Vico, is a kind of metaphysics, the truths that are gleaned from it speak to and through the imagination – consciousness, body and memory all in one – rather than in the language of scientific algorithms. Humanity creates, makes poetry, architecture and institutions, but in a way very different from that of God (or modern technology). I quote: 'For God, in his purest intelligence, knows things, and by knowing them, creates them; but [humans] in their robust ignorance, [do] it by virtue of a wholly corporeal imagination, one liable to perturb in excess.'[2]

In view of these realizations, it is essential to grasp the possible confluence of ethics and poetics in our 21st century. The present excesses of empty computer-generated formalism, with its roots in liberal capitalism, and the far more insidious moral disasters that humanity witnessed during the last century in the name of health and beauty, associated particularly with fascism, but also with communism – all issuing from the ideological extremes of liberty, equality and fraternity, those treasured political values first embraced by the French Revolution – have made us justifiably skeptical. I would like to invoke Plato and argue that beauty, as a form of deeply shared cultural experience, understood as a priori *meaning* in the world of culture, is a fundamental category. The experience of beauty, we may recall our *Phaedrus*, is a vehicle for the soul to ascend towards truth, *(pt)eros* provides the wings. Beauty is truth incarnated in the human realm, it is a trace of the light of Being that mortals can seldom contemplate directly, it is the purposefulness of nature mimetically reflected by the artifact, and the coherence of a scientific theory. Thus multiplicity and contradiction, ourselves included, are resolved in erotic unity. In Gadamer's words: In this 'world below', we can be deceived by what only seems wise, or what merely appears to be good. Even in this world of appearances, however, all beauty is true beauty, because it is in the nature of beauty to appear.[3] This is what makes the beautiful distinct among ideas, according to Socrates. Beauty exemplifies, in Karl Jasper's terms, reason incarnate in existence. This Platonic formulation, reading Plato with Gadamer, Brentlinger and Kosman in an emphatically non-dualistic mode, offers a great challenge, as we must understand it in our epoch of cultural relativism.[4] Does it work?

It is easy to understand taste as participating in local, historically determined norms. Yet, when we move beyond aesthetics, taste takes its place among other forms of *phronesis* (*prudentia* in Latin, soundness of judgment), Aristotle's 'practical wisdom,' grounded in the habits and values which we share with others, and that appear with utmost clarity and certainty. Such self-evidence, manifested in the poetic artifacts and stories of our traditions, can produce judgments that are no less rational for being grounded in *ethos*. These works of art and poetry are indeed capable of moving us, they transform our life and ground our very being.

Eros and the imagination are inextricably linked. This is more than a physiological fact. Our love of beauty is our desire to be whole and to be holy, beauty transcends the *aporia* of necessity

---

2 Vico, Giambattista; **The New Science**, tr. T. G. Bergin and M.A. Fisch (Ithaca: Cornell University Press, 1970), 75.
3 Gadamer, Hans-Georg; **The Relevance of the Beautiful** (Cambridge: Cambridge University Press, 1986). See especially Part 1.
4 See Gadamer; **op. cit**. and the essays by Brentlinger and Kosman in A. Soble (ed.); **Eros, Agape and Philia. Readings in the Philosophy of Love** (St. Paul, MN: Paragon House, 1989).

and superfluity; it is both necessary for reproduction, and crucial for our spiritual *well*-being, the defining characteristic of our humanity. Richard Kearney, among other philosophers in the hermeneutic tradition, has demonstrated the importance of the imagination for ethical action.[5] Contrary to the view of many critical theorists who may believe that there exists an irreconcilable contradiction between ethics (associated with democracy, rationality and consensus) and the poetic imagination, Kearney convincingly shows how it is the lack of imagination which may be at the root of our worse moral failures. Imagination is precisely our capacity for love and compassion, for both 'recognizing' and 'valorizing' the other, for understanding the other as myself, over and above differences of culture and belief. Imagination is both, our capacity for truly free play, and our faculty to make stories and to partake from the language and vision of others.

As architects we are called to make, to build the public realm. The personal imagination is our vehicle. As I have pointed out, what we make could have enormous consequences. This now obvious observation is actually magnified by our historically given condition. We have inherited a great responsibility, for in fact, unlike our ancestors until the 17th century, we effectively *make* history. The world now generally believes in the self-evidence of human-generated change, a particularity of the Western – originally Christian – project that has become universalized. Thus history – our diverse stories, as varied as our cultures – is what we share as a ground for action, together with an indeterminate, somewhat infirm more-than-human world that appears forever fragmented. We don't share, like our more distant ancestors, a cosmological ground, a perception of the universe as a fundamentally changeless totality, limited and straightforward. Only by engaging our imagination's capacity to create compassionately can we negotiate the nearly infinite possibilities for production, in view of our now *real* cultural diversity, and the proliferation of instrumental methodologies and computer software, actually capable of making rational any irrational form, making any 'skin' stable and buildable. The imagination is equally the antidote to the prevailing cynical view about architecture, according to which it matters little what we make, for it will be co-opted by politics and power, its purpose being to exploit, dominate or control the other.

There are, of course, great difficulties. Power in its different guises, as money, politics, and clients' interests, must always be negotiated. Nietzsche has shown us that for this purpose a playful attitude seems to work best. What is important to retain, however, is that despite these difficulties, renouncing innovation is not an ethical option. Our historicity may now reveal the futility of Utopia and the early modern ideal of progress, yet to project inherently means to propose, through the imagination, a better future for a polity; it is inherently an ethical practice, and this is *not* equivalent to a mindless search for consumable novelties disconnected from history.

Throughout human history, architecture has often provided authentic dwelling, allowing individuals to recognize their place in a purposeful natural and cultural context. At times, nevertheless, it has contributed to tragedy. The aesthetic programs underlying Nazi Germany are a case in point. Rather than being underscored by the imagination, the Nazi programs were borne from a rationalized mythology, transformed into the dogma of nationalism. Think as well of the way two very tall yet typical skyscrapers, secular symbols of a triumphant technology, were read as ideological signs by Muslim fundamentalists on September 11, 2001. The sad event of their destruction transformed two largely conventional buildings into architecture, having a nefarious effect on our world civilization. For all these reasons, it is crucial to remember that, however we may deplore the loss of poetic enchantment in the world as it opens itself to nihilism, all we can do is continue to 'weaken' the strong values of all sorts of ideologies and fundamentalist positions, ranging from organized religion to technology, expecting that in the gaps a new, genuine spirituality may emerge. Truly unethical is to pretend that there exists a unique and absolute set of values, articulated in one mythology, dogmatic religion, rational ideology or technology, to the exclusion of others. To attain the goal of weakening strong values, our *praxis* must be fragmented, every problem and question must be carefully contextualized and re-formulated, and the design responses must be specific rather than artificially stylistic or universal. In every circumstance we must be prepared with Nietzsche and Heidegger, to wait patiently for the rustle of the angels' wings that may be passing by, avoiding the planner's dream of total solutions.

Given the dangers at hand, it is crucial to endow ourselves with a language that articulates responsibility, and which may anchor a practice that is, as I have shown, inherently ambivalent. From its inception in human history, technical production, however poetic, carries a dimension that moves 'against nature', it is a necessary danger for *homo sapiens* that can never simply adapt itself to the natural environment, like other animals; a potential curse which is also one of humanity's most precious gifts, narrated by many myths of traditional cultures, such as the stories of Prometheus and Cain. It would suffice to remember here the ambivalent nature of the earliest architectural artifacts in our Western tradition, the Greek *daidala*, both dangerous and wondrous objects, deceitful and yet necessary for the survival of the human spirit, like votive or sacrificial objects, both sacred and polluted. While many of our cultural achievements have been obtained at great cost, like for example the 12th-century urbanization in Europe and the great Gothic cathedrals at the expense of deforestation, or the Greek alphabet at the expense of pictographic, figurative writing, it may be naïve to claim, like some well-intentioned ecologists, that the

5 Kearney, Richard; **The Wake of Imagination** (Minneapolis, MN: University of Minnesota Press, 1988).

answer might be to live closer to nature, following the wisdom of our ancestral cultures. Of course, as Heidegger has beautifully shown, there is a serious danger for humanity, as we live our lives in a world of objects that conceal our finite horizon and impede our access and understanding of the more-than-human world, treating nature rather as a collection of resources to be exploited. For us, nevertheless, reading the landscape like an Australian aborigine, and living at one with nature like our mythical ancestors, are not real options. If something has been lost, like our cultural understanding of *genius loci*, something has been won as well. Our highly artificial culture, through its historical self-consciousness, may embrace the *aporias* of cyclical and linear time and recognize for itself the same mysterious origins as earlier products of *techne-poiesis*. Through historical recollection and our future orientation, we can cultivate both our capacity for stewardship and responsibility, as well as our poetic potential as makers to disclose and celebrate the original mystery as it appears in the primary structure of our embodiment; the meaningfulness of a given world that cannot, however, be reduced to universal categories.

Thus we can reiterate the crucial necessity of language (theory) for an ethical practice. Conversely, we should immediately acknowledge that words and deeds never fully coincide, language is opaque. This is, however, to be celebrated rather than deplored, as it usually happens from a scientistic point of view. This opaqueness characterizes the very nature of human languages, never coincidental with the Word of a god for whom to name is to make. Like the making of technical artifacts, the possession of symbolic, multivocal languages is among the most precious gifts that make us human, perhaps more precious than our approximations to an ideal, scientific or mathematical universal language. As George Steiner has eloquently stated[6], our over three-and-a-half thousand distinct languages for a single species, often in close proximity to each other and mysteriously diverse, and capable of speaking poetically in ways that always enrich our experience of reality, is the ultimate enigma which no evolutionary theory of man can ever reduce.

No matter what we produce as architects, once the work inhabits the public realm, it is truly beyond our control. An expressed intention can never fully predict the work's meaning. The 'others' decide its destiny and its final significance. Despite this logical conundrum, understanding that there is a phenomenological continuity between thinking and making, between our words, in our particular language, and our deeds, is still our best bet. What we control, and must be accountable for, are our intentions. It is usual to dismiss good intentions and award greater value to pragmatic actions. Nevertheless, well-grounded intentions are crucial and rare in the modern world. They imply a whole style of thinking and action, a past life and thick network of connections with a culture, far more than what an individual is capable of articulating at the surface of consciousness, or through one particular product. This is *praxis*, practice in the full Aristotelian sense, comprising its ethical imperatives. In our Western tradition, prior to the early 18th century, this practical philosophy was reconciled with the articulations of reality in mythology, theology, philosophy, or science; it was never dictated, as is often assumed, by aesthetic formulas, functional equations or an ideological program.

In our late modernity, this tradition of architectural discourse, which I will call 'philosophical theory' to mark its difference, is emphatically distinct from 'applied science' or methodology. Theory of architecture is also not synonymous with pure science – however pure science may be seen to inherit the Greek quest for beauty, contemplating the laws that may reduce the plurality of the universe into unity. In fact, the identification of architectural theory with instrumental discourses, even for seemingly sophisticated critical theories of recent times, is at the root of its present 'bad name'. Philosophical theory always involves a double wager. It is always both personal and social. Our making is not only a 'profession', it is our life. Our life is our first reality, and therefore architecture's meaning matters immensely for each one of us – here and now. We may recall Boullée's despair at the roots of his theorizing, particularly poignant at a time when architecture could no longer depend on a scientific discourse for its meaning. Once a modern philosophical theory is understood as primarily driven by ethics, as practical philosophy in the tradition of Aristotle's *phronesis, techne-poiesis* or practice appears as *process*, as a fully embodied, personal engagement with the crafts.

Rather than assuming that 'representations' are neutral, the products of the design process are to be *fully* valorized. It is possible to develop an awareness of the wonders that we can reveal, to ourselves, through human work. This is the nature of Dionysian free play, hopefully revealing, in the sense of Nietzsche and Gadamer, the coincidence of chance and necessity. Gilles Deleuze[7] expresses this very well when he states that play, as affirmation, is reserved for thought and art, where victories are for those that know how to play, how to affirm and to ramify chance, rather than dividing it in order to dominate or win. This characteristic is what makes art real and, according to Deleuze, capable of disturbing the reality, economy and morality of the world. Our imagination effectively constructs the world, an embodied imagination which must be recollected in our era of binary space. This necessitates patience and open-endedness. Every moment of a search is liable to turn up poetic disclosures, through inward vision or projections on the outside world, disclosures that may eventually be translated into an authentic architectural vision.

6 Steiner, George; **After Babel** (London: Oxford University Press, 1975).
7 Deleuze, Gilles; **The Logic of Sense** (New York: Columbia University Press, 1990) p. 76.

In order to speak properly and articulate an ethical position for a particular project, a knowledge of 'philosophical history' becomes crucial. This rhetorical skill is crucial for the development of a coherent *praxis*: philosophical history is capable of making alien artifacts tell us their stories through a hermeneutic process. In the spirit of Friedrich Nietzsche, Hannah Arendt, Paul Ricœur and Hans-Georg Gadamer, this is a history for the future, one meant to enhance our vitality and creativity, rather than one that may immobilize us through useless data, an immoderate respect for the old for its own sake, or unattainable idealized models. The architecture, and particularly the words — now texts — which have articulated the *praxis* of other times and places, must be understood in light of relevant pre-judgments issuing from our contemporary questions, yet in the context of the language (and culture) of their makers, respecting the questions they originally addressed. Thus the process of interpretation, appropriating that which is acknowledged as truly distant, should make it possible to render their voices into our own specific time and politics, rather than assuming a universal language at work, or a progressive teleology. Needless to say, this hermeneutic understanding is equally applicable to our engagement with other synchronic cultures.

Hermeneutics involves the interpretation of architectural intentions, reading the works of other times and places in relation to the epistemological contexts in which they are produced, be they mythical, philosophical, theological or scientific. The aim is to read 'between the lines' and with courtesy, the world of the work, and the world in front of the work; acknowledging the human pursuit of meaning over and above other motivations, yet critically, seeking to understand how these architectural works may respond to the questions of our present humanity. A critical hermeneutics rejects the historical flattening and homogenization of deconstruction and proposes the valorization of experiential content, the mystery which is human purpose and the presence of spirituality. To account for what matters and can change our life.

Within this framework of understanding, ethics appears not through norms or generalities, but through stories that focus on specific works and individuals. In recent critical theory the self, understood as a dangerous inflated *ego*, product of the 18th century, has received a bad name. Feminist and social critiques tend to render art and design as the result of more or less anonymous, more or less

8 See also Richard Kearney; **The Poetics of Imagining – from Husserl to Lyotard** (London: Harper Collins, 1991) for a historical account of this issue in modern philosophy, and my own forthcoming Pérez-Gómez, Alberto; **Built upon Love: Architectural Longing after Ethics and Aesthetics** (Cambridge: MIT Press).

insidious forces. The unmasking of ego-centered interpretations is of course healthy. What is very dangerous, however, is following up this diagnosis with a desire to renounce our personal imagination as if it were some evil, distorting device, in favor of a supposedly objective consensual framework. I have already alluded, with Richard Kearney[8], to the ethical function of the imagination. It is always the *I* who acts, a first person fully embodied and imaginative, caught in a technological and historical world which both endows her with responsibility, and takes it away. The embodied author, fully rooted in language and culture, is also capable of poetic speech, of *making* beyond the confines of a narrow style, ideology or nationality. My claim is that individual architects in history, despite the dangers of the imagination and the opaque relationship between words and deeds, have indeed contributed imaginative answers to our universal call for dwelling. Through their personal reformulation of universal, social questions, in view of their own historically and geographically specific framework of beliefs, truly poetic responses have emerged; answers from which we can learn and develop an ability to act *here and now*.

On Criticality
Arie Graafland

The flat screen of my computer has a name of its own. It's called DIGITAL LIFE. Is there life inside my computer? Well, sometimes it looks like it. It does things on its own. Indeed, it seems to have its own mind. It shows me lively pictures; it communicates for me all over the planet with its equals. It can drive me mad, actually, but mostly I am quite happy with what it does for me. At least it has changed my working life considerably. And isn't that what most life-forms are about? Changing behavior. But where does that leave me as a person? Am I lost as a mere 'interface'?

Quite a few architects and artists will tell me: 'you are not lost, you just have to talk back. Communicate!' Communicate with our (plans for) digital architecture and art forms. I am already part of what William Gibson wrote twenty years ago about cyberspace; I am part of that 'consensual hallucination' called cyberspace. It also seems to be a lot more than that. Certainly, more than the computer itself. It is about networks, compatibility, architectures, access, non-accessibility; it is about power, transfers of money, data and information that flow through it. And it is also about new forms of architectural practices, office organizational strategies, and profiling.

Enabled by the new information and communication technologies, Michael Speaks[1] writes that network practices in the 1990's became communities that are more powerful than any single studio or office. Speaks explores the organizational structures of a few offices that, like academic research groups in the same period of time, became more internationally oriented, using each other's networks and expertise. For example, the Amsterdam based UN Studio (United Network Studio) of Ben van Berkel and Caroline Bos organized their office as a 'network studio'. These network firms proliferated in the 1990's, reflecting the need for small, innovative studios to create working partnerships. Speaks focuses on organization and economic change, indeed, his vocabulary is on 'innovation'. The central idea of this innovation is a highly *organizational* one. Speaks talks about 'knowledge based services' and refers to 'cultural intelligence'. In referring to ShoP (Sharples Holden Pasquarelli), a young firm in New York, he quotes Gregg Pasquarelli who was looking for a new way of practice, in which design was not just handed off, but was part of an entire approach. SHoP to him is unlike most architectural offices – the model is a consultant firm, not the traditional master-builder type office. He stresses the combination of design intelligence with computer design technology and a sophisticated approach to marketing, public relations, and other aspects of the business of architecture in order to create that 'truly innovative' practice.

Speaks is in fact referring to the transition from Fordism to *flexible accumulation*, described by David Harvey and others[2] in the 1990's. This more flexible form of capital and organization emphasizes **the new** as a category, a recurring term in these accounts on the benefits of compu-

---

1 Speaks, Michael; 'Design Intelligence and the New Economy' in **Architectural Record** (01–2003) p. 72.
2 See for example, Harvey, David, **The Condition of Postmodernity: An Enquiry into the Origins of Cultural Change** (New York: Blackwell, 1990).

terization. It is the fleeting, the ephemeral, the fugitive and the contingent in modern life, rather than the more solid values implanted under Fordism[3] that becomes paramount. Harvey discusses three accounts from that period of time, all dealing with the same economic and organizational issues. The rather celebratory account by Halal[4] of new capitalism, emphasizes the new entrepreneurialism. The second is Lash and Urry's *The End of Organized Capitalism*[5], and the third is a book by Swyngedouw from 1986[6] providing in great detail the transformations in technology and the labor process. Many of Speaks' arguments are similar to Harvey's in theorizing this transition.

The period from 1965 to 1973 was one of the inability of Fordism and Keynesianism to contain the inherent contradictions of capitalism. On the surface, Harvey writes, these difficulties could best be captured by one word: rigidity. There were problems with the rigidity of long-term and large-scale fixed capital investments, in labor markets, labor allocation and contracts. The 1970's and 1980's have been a troubled period of economic restructuring, and social and political readjustment. This shift to *flexible accumulation* rests on flexibility with respect to labor processes, markets, products and consumption patterns. Furthermore, the shift is characterized by the emergence of entirely new sectors of production, new ways of providing financial services, new markets and above all, greatly intensified rates of commercial, technological and organizational innovation.

According to Speaks, even the most forward-looking members of the architectural establishment have ignored these innovations. And for sure, for some of the offices their way of working *is* **new**. Greg Lynn (Los Angeles), Douglas Garafalo (Chicago), and Michael McInturf (Cincinnati), mostly architects with smaller offices in different cities, took advantage of electronic imaging and new communication technologies. Together they realized the Korean Presbyterian Church of New York. Stan Allen and James Corner – one in New York, the other in Philadelphia – collaborated in architectural thinking, research and landscape architecture. Also the Dutch firm MVRDV is into research of datascapes, urban design/decision-making models like Regionmaker, and publishing office work as research-related. An important aspect of this innovation is the fact that the principals were far more involved in academic research-related work than the majority of their colleagues ever were. Their colleagues are still more interested in publishing their completed office work. Mostly built work, they do not make that step back, there is virtually no distance from their design production. Probably Rem Koolhaas is the inspiring figure here, his first form of 'practice' being, so to speak, a book on Manhattan. Later publications show more of the tendencies Speaks indicates – writing about architecture and urbanism, a clever mix of proposed and executed office work, critical articles, and a surplus of photographic material combined with an ingenious way of presenting, made possible by advanced digital assembling techniques and the design skills of Bruce Mau and the like. In the 1990's AMO, an offshoot of OMA, was set up when principals Rem Koolhaas and Dan Wood, then project architect for OMA's abandoned Universal Studios Project in Los Angeles, decided to begin this theoretical arm together. At the moment, Reinier De Graaf heads AMO.

Another example of Italian origin is Stefano Boeri[7], exploring urban projects, doing design research, and with an architectural practice in Milan. For some architecture schools, this new methodology meant new forms of architectural research. Koolhaas started his 'Project for the City', an immensely influential and fast paced project at Harvard School of Design. In Rotterdam, the Berlage Institute is always on the cutting edge of design and architectural thinking and has established a name for itself. More recently, the TU Delft established its Delft School of Design, an internationally oriented research laboratory at the Architecture Faculty. They all have in common the exploration of the fields of architecture and critical thinking.

Nevertheless, for Speaks, the times of 'critical theory' are past. At stake here was more than the transition to flexible accumulation in Western economies as proposed by Harvey. Instead, this new flexible accumulation of architectural image and practice, as well as forms of management, is the new successor to the so-called 'exhaustion' of primarily Continental theory. In this depletion, Speaks sees the failure to recognize the important shift in the relationship between thinking and doing that occurred in architecture in the 1990's. Consequently, the more recent focus is on American pragmatism and on these 'newly emerging forms of practice'. For him the new challenge for architecture is to develop forms of practice able to survive the fiercely competitive global marketplace. The idea is that architects use 'intelligence' in a twofold way: as a specific form of practical knowledge characteristic for the profession, and in the practical way the American CIA or military might want to use 'intelligence'. Architects should be able to think ahead and visualize ahead – a form of fore-knowing the effects and, at the same time, the social impacts of their proposals. Yet, in order to be able to do so, they must employ 'intelligence' like the military, be able to work from seemingly endless fragments of 'information', rumours even, and disinformation. The 'chatter' of the outside world should be related to the projective capacity of the profession. The way to do it is just to use your imagination and to play along.

The question is, however, are there other ways to deal with projective practice and how can critical thinking be involved in this procedure? Or is critical thinking, indeed 'exhausted'? The

---

3 Harvey, David; **The Condition of Postmodernity: An Enquiry into the Origins of Cultural Change** (New York: Blackwell, 1990) p. 171.
4 Halal, William E.; **The New Capitalism** (New York: John Wiley & Sons, 1986).
5 Lash, Scott and Urry, John; **The End of Organized Capitalism** (Cambridge: Polity Press, 1987).
6 Swyngedouw, E.; 'The Socio-spatial Implications of Innovations in Industrial Organization'. Working Paper no 20, Johns Hopkins European Center for Regional Planning and Research (Lille: 1986).
7 Boeri, Stefano, Lanzani, Arturo and Marini, Edoardo; 'Ambienti, paesaggi e imagine della regione Milanese' in **AIM**, Associazione Interessi Metropolitani (Milan: Editrice Abiare Segeta spa, 1993).

first thing to be said is that virtually no office in the past has been interested in critical thinking; this investigation has always been solely the domain of philosophy and sociology in the universities. Certainly, critical theory was not a field of interest that played any role in actual office work. The link Speaks makes between modernization of the work process in architectural offices, and 'the exhaustion of Continental theory' is questionable in itself since Continental theory never played any role in daily office practice in Europe and America. Architects get their ideas elsewhere – from confrontation with the specificity of site and program, from work of other architects, from periodicals and professional literature. The discourse on theory, in fact, streams along in all of its convoluted complexity, largely unnoticed by the average practitioner of architecture. If the articles get too complex, simply no one will read them. Indeed, there has always been 'chatter'; and in fact, there will probably always be 'chatter' in the architectural office with its chaotic work processes, and also in the Universities where different interests and ideas have to work together or at least tolerate each other. Speaks' argument runs two ways: firstly promoting flexible and internationally oriented office practices, and secondly announcing the 'exhaustion' of critical thinking in the Universities. I argue that although the two are substantially unrelated practically speaking, they are both relevant questions for architecture and especially for architectural education.

What other ways are there to approach this question? What is needed here, in my opinion, is what I would like to call a '**reflexive architecture**', an architecture addressing its own foundations reflexively, paired with the digitalized work processes on a larger scale than the traditional office practices employed until recently. To be clear, in the end I do not think that the offices can do this reflexive architecture on their own. Most of them will have neither the focus nor the time for extended experimentation. A Studio setting is necessary in order to be able to get the desired focus. The designs in my *Socius of Architecture*[8] were made in our office, but the text came from my work in the University. 'Reflexivity' is an activity mainly in Universities since it relates to 'critique' and to contemporary notions of time and space. If this split between office practice and University research remains, it will leave the historians and critics on the safe side, they won't have to bother with the messy daily practices in the offices where negotiating and adapting are more common than the grand design. However, I think the research groups in Universities cannot do reflexive architectural research without the offices involved. They lack the much needed pragmatic context and client. What is happening in Speaks' discourse is the effort to promote a few American and Dutch offices to the forefront of contemporary architectural practice. His main argument being an organizational and instrumental one, a position of instrumentality covered by a 'pragmatic' stance no longer assessing the outcome of the designs. Not quite a new position.

This 'obsession with instrumentality', as Alberto Pérez-Gómez[9] writes, rages unabated in architectural practice and almost always underscores the 'leading edge' positions. He traced the instrumental obsession in mid-eighteenth-century technical theories in order to probe their myths of rationality. His focus was on pre-modern architecture where essential aspects of architectural knowledge were defined as *techne* founded on *mathemata* that could be transmitted through a 'scientific' treatise. Pérez-Gómez examined the polemic between two instrumental theories in late seventeenth- and eighteenth-century France, the work of Charles Étienne Briseux and his criticism of the earlier writings of Claude Perrault.[10] The contemporary 'obsession with instrumentality' encourages fashionable architectural projects that are oblivious to their cultural context, to their intended programs, to their historical roots, to ethical imperatives, and to our experiencing body. Although Pérez-Gómez is correct, in my opinion the problem is that today's architectural practice is no longer on the level of historical consciousness, or even managerial and organizational levels as Speaks suggests, no longer on a cognitive or historical level, but on the level of a software-driven flattened out *aesthetic reflexivity*.

This kind of aesthetic reflexivity has more recently found an important place in the production and consumption of the culture industries. Architectural books and magazines are also a part of this mechanism. Treatises are no longer an option for an architect, rather a necessity. For the offices it will be hard, if not impossible, to step back from this aestheticization. The conditions under which they work are in a sort of symbolic flow, cultural capital creation and aesthetically cast expert systems that are intrinsic to the current profession. Whereas intellectual property rights are the main form of capital in the culture industries, in architecture what is sold is not the intellectual rights since it is a singular operation, but the 'product', the architectural project, and especially the 'name' the firm or architect has managed to make for himself by way of publishing. Quite a lot of the smaller firms Speaks mentions, do niche marketing, finding holes in a major markets of building practices. What some of them have invented are not so much an economic and managerial innovation as well as a strategic *aesthetic innovation* in *profiling* and *image production*. This invention/promotion of course has been the case for longer periods of time in the twentieth century, and was always followed by a critique from both Marxist and conservative sides. Nevertheless, the critique is also getting more and more complex. In Marxist critique, at least there was always a stronghold, a form of resistance with aesthetic depth, as in for example, Tafuri or Adorno. But what is happening now is the *disappearance* of that subject of resistance in the circulation of images in contemporary information and communication structures. This very disappearance, however, is what Speaks characterizes as the 'exhaustion of Continental theory'.

8 Graafland, Arie; **The Socius of Architecture: Amsterdam, Tokyo, New York** (Rotterdam: 010 Publishers, 2000).
9 Pérez-Gómez, Alberto, 'Charles-Étienne Briseux: The Musical Body and the Limits of Instrumentality in Architecture', in **Body and Building: Essays on the Changing Relation of Body and Architecture** edited by Dodds, George and Tavernor, Robert (Cambridge: MIT Press, 2002) pp. 164 **ff.**
10 **ibid.** p. 164.

We are no longer dealing with reflexive subjects, but reflexive objects, as Lash and Urry[11] argue. They have argued that the current cultural artefacts in the music industry, to use one example, are no longer transcendent as representations, but that they have become immanent as objects amongst other objects circulating and competing in information and communication structures of popular culture. Music has become a lifestyle. Their claim is that with modernization and autonomization, hence differentiation of the cultural, culture became primarily representation. More recently we have seen representations taking up the functional position of objects, objects which only differ from other objects of everyday life in their immaterial form and aesthetic character. Madonna as a star is not just an image, but a representation. She has become a cultural object in the anthropological sense of culture. With the declining significance of social structures and their partial displacement by information and communication structures the aestheticization of everyday life becomes possible.

Our current condition of postmodernity is in effect the generalization of aesthetic modernism to not just an elite, but the whole of the population. Aesthetic modernism, however, presupposed that autonomous subject with depth and reflection. It assumes an aesthetic expressive subject. Lash and Urry argue that the circulation of images in contemporary information and communication structures entails not an aesthetic subject, but these reflexive objects.[12] Although their observations might be too close to Baudrillard's notion of dystopia[13] here, it is true that the subjects tend to be flattened out in the ongoing proliferation of digitalized images. But to me this process is not yet completed, not yet exhausted, there are still critical possibilities left. For sure this flexible accumulation is much more than a merely economic managerial flexibility as suggested by Speaks. In our digital world, contemporary architectural image production is replacing modernistic aesthetics for an 'anaesthetics' as Neil Leach[14] has recently suggested.

Record companies are not so much selling the record, but the artist. For architecture this is not the same situation; architecture does not command that kind of widespread interest in society, although interest is growing rapidly in magazines originally not dealing with architecture. But for the architectural in-crowd, Rem Koolhaas and Frank Gehry are functioning in a comparable way. It is not so much the building, but a 'Koolhaas' as a brand name, reinforced by his own publications and the OMA/AMO office, and even more by the endless publications on his work in books and magazines. Many culture sector firms have become like advertising agencies, and advertising itself has become more like a culture industry, Lash and Urry have argued. For example, the PR firm of Saatchi in London profiles their advertising business as 'commercial communication'. The AMO office is not too far away from the same practice. OMA/AMO's research into Shopping more or less coincided with their Prada account. The office not only designed the shops, but took care of the corporate identity of the company in advertising and publicity.

I am avoiding the already obsolete terms 'innovative' and even the terminology of 'critical architecture' since I am also of the opinion that 'critical' in social theory and philosophy are indeed problematic, and cannot easily be related to a projective aesthetic practice like architecture. Nevertheless, in saying 'critical' we must be precise. Certainly 'critical' can be related to 'retrospective', historical, and critical analyses. Critical itself is either under a lot of pressure, or is fading away completely in social theory and philosophy since for many it seems to have lost the much needed critical subject. There is certainly much more at stake here than the mentioned shift to a new organizational model. It is also a matter of knowledge as Speaks suggests. Philosophical, political, and scientific truths have fragmented into proliferating swarms of 'little truths', appearing and disappearing so fast that ascertaining whether they are really true is impractical if not altogether impossible, he writes.[15]

Yet his altogether too hasty conclusion is a farewell to critical theory; ideas or ideologies are no longer relevant, but *intelligence*. The 'critical architectures' of the 1960's and 1970's had none of the theoretical, political, or philosophical *gravitas* of their early 20th-century predecessors, he writes. On top of that, Post Modernism, Deconstructivism, Critical Regionalism, and many others in the late 1980's and 1990's posed as false pretenders to Modernism. The opposition Speaks is laying out here is about different forms of theoretical and aesthetic practices. His claim is that 'vanguard practices' are reliant on ideas, theories and concepts *given in advance* (my italics), and that 'post-vanguard' practices are more 'entrepreneurial' in seeking opportunities for innovation. That is to say, practices that cannot be determined by any idea, theory or concept.[16]

I think it is here where the misunderstandings are in danger of arising. To my mind, architecture as a projective and creative aesthetic practice can never be both solidly and safely guided by critical theory which is retrospective by definition. The *projection* of architectural thought into a building and its *prospectively* hoped-for aesthetic effects will always be an uncertain stab in the dark, whether it comes from 'entrepreneurial opportunities for innovation', or from 'critical' intentions. It has nothing to do with the idea of 'stable theories' given in advance, or the inventions of entrepreneurial practices. I will stress the aesthetic side of this projective process; it certainly does not mean an 'anything-goes'. There are many stable ways to analyze and organize the context, the program, the construction, the budget, etc. But in many cases, it is this aesthetic effect that is discussed at length beforehand in both educational and practical settings. Philoso-

11 Lash, Scott and Urry, John; **Economies of Sign and Space** (London: Sage Publications, 1994).
12 ibid., p. 132.
13 Baudrillard, Jean; **Simulation and Simulacra**, translated by Sheila Faria Glaser (Ann Arbor: University of Michigan Press, 1995).
14 Leach, Neil; **The Anaesthetics of Architecture** (Cambridge, Mass.: MIT Press, 1999).
15 Speaks, Michael; 'Design Intelligence' in **A+U** (2002:12 December) p. 12.
16 ibid., p. 12.

phy in this context easily leads to a confusion, in many cases coming from Eisenman's writings and especially his linking up with philosophical partners like Derrida and Rajchman where, at least to my mind, there is a false suggestion of *projective* archi-philosophy. Whether this is due to the way Eisenman always publishes his projects, or whether it comes from a genuine philosophical interest in the projects is hard to decide. I think it is possibly both, but what seems to be certain is that it is confusing the American discourse on 'criticality'.

Speaks himself seems to be 'exhausted' by this discourse, but that does not mean that we are completely at a loss here. Certainly it is possible to say something intelligible about economic conditions in design practice, about political choices and decision making in urbanism, about territorial conditions or 'terrestriality', about design ideologies and managerial relations. In the end, architecture and urbanism are about our lives and the way we experience our world. Nevertheless, the comments of Speaks, although I do not agree with his argument, no doubt confront us with the more serious problem, that of critique itself. The argument presents us with yet another opportunity to question the status and usefulness of deconstructionism, critical theory, and our ideas about society and nature.

Indeed, Bruno Latour writes, 'it has been a long time since the very notion of the avant-garde – the proletariat, the artistic – passed away, pushed aside by other forces, moved to the rear garde, or may be lumped with the baggage train.'[17] It looks like we are still going through the motions of a critical avant-garde, but is not the spirit gone, he asks? Staying with the idea that critical thought is a weapon, a '*Waffe der Kritik*' as it was once called, Latour writes that we have to re-think our critical strategies and instruments. The actual threats might have changed so much that we might still be directing our entire arsenal east or west while the enemy has now moved to a very different place. Our critical arsenal with the neutron bombs of deconstruction, with the missiles of discourse analysis, might all be misdirected. And yes, we might be using the wrong arsenal, we will have to go back again to deconstructivist architecture and our newly established digital architectures to see what went wrong. And at the same time come up with alternatives, the latter more imperative than the former.

At first sight there might be a correspondence between Speaks' notions and Latour's. But in fact their positions are very different. Latour's plea is to get closer to the facts, not fighting empiricism, but on the contrary renewing empiricism. The new critical mind for him is to be found not in *intelligence*, but in the cultivation of a stubbornly realist attitude, to speak like William James, a realism dealing with what he calls *matters of concern*, not matters of fact.[18] Instead of moving away from facts, we have to direct our attention toward the conditions that made them possible. For architecture it implies the redirection of our thoughts to what I would call *an architecture of the street*. A reflexive architectural way of proceeding, renewing empiricism, and addressing the sophisticated tools of architectural deconstruction and its inherent construction – or better, the lack of – social construction. The desired outcome of architectural practices discussed at length in architecture schools, books and magazines can never be guided by a rhetoric of 'entrepreneurial architecture', or 'design intelligence'. A discourse which only focuses on organizational questions, and is referring to 'critical' architectures as lost cases, is in fact a hastily post-modern and post-political manœuvre that needs to be addressed. It is still possible to think of critique in other ways. Not so much in the exclusive Marxist way which I will briefly explain a little further on, but in a way Latour and Lash suggest. Not as the critic who debunks, but the one who assembles. The critic in his thinking is not the one who pulls the rug from under the feet of the naïve believers, but the one who offers the participants arenas in which to gather, Latour writes. That is to say, generating more ideas than we have received, not being purely 'negative', but in fact productive.

In order to explicate this progression in the field of architecture, I will relate to an earlier publication where I tried to show both relation and ruptures between architecture and 'critical' theory.[19] In an insightful article in *Harvard Design Review*[20], George Baird sketches out the American discussion on 'criticality', which has its origin in Europe in Marxist and Kantian thinking. The lineage of criticality in architecture more or less starts with Peter Eisenman, accompanied by Michael Hays, who has developed a position consistently focussed intellectually on concepts of 'resistance' and 'negation', Baird writes.[21] Both refer to the Italian historian and critic Manfredo Tafuri. To my mind, one of the crucial notions on criticality is to be found in Hays' book on Hannes Meyer and Ludwig Hilberseimer[22], which is heavily indebted to the work of the Marxist thinker Fredric Jameson.[23] For Hays – following Tafuri[24] – Mies van der Rohe was the paramount exemplar of negation in late Modernism. The work of Mies is examined as critical, or resistant and oppositional. In another writing[25], Hays addresses the surface distortions and formal inscrutability of the 1922 skyscraper project published in the second issue of *G* magazine. Mies insists that an order is immanent in the surface itself and that the order is continuous with and dependent upon the world in which the viewer actually moves. Hays puts

17 Latour, Bruno; 'Why has critique run out of steam? From matters of fact to matters of concern', in **Critical Inquiry**, Winter 2004, p. 226.
18 Cf. Latour, Bruno; **Reassembling the Social: An Introduction to Actor-Network-Theory** (Oxford: Oxford UP, 2005).
19 Graafland, Arie; **op. cit.**
20 Baird, George; 'Criticality and its Discontents', in **Harvard Design Magazine**, Rising Ambitions, Expanding Terrain, Realism and Utopianism (Fall/Winter 2004).
21 Baird, George; **op. cit.**, pp. 16**ff.**
22 Hays, K. Michael; **Modernism and the Post Humanist Subject** (Cambridge, Mass.: MIT Press, 1992).
23 See Jameson, Fredric; **The Political Unconscious: Narrative as a Socially Symbolic Act** (New York: Cornell University Press, 1981) and 'Postmodernism, or the Cultural Logic of Late Capitalism', in **New Left Review** 146 (July–August 1984) pp. 63**ff.**
24 Tafuri, Manfredo; **Theories and History of Architecture** (London: Granada Publishing, 1980).
25 Hays, K. Michael; 'Critical Architecture, Between Culture and Form', in Perspecta 21, **The Yale Architectural Journal** (Cambridge, Mass.: MIT Press, 1984).

the building into the context of the German city at the time, referring to Georg Simmel's ideas on the blasé individual. This sense of surface and volume in fact wrenched the building from the atemporal, idealized realm of autonomous forms, in order to install it in the historical world of that time. The design, then, becomes open to the chance and uncertainty of life in the metropolis. The moment of resistance is that it is not subsumed in the chaos of the metropolis, but rather seeks for another order through the systematic use of the unexpected mirroring of surfaces. Hays addresses the building, not the architect's intentions or œuvre.

Other later works of Mies, like Alexanderplatz in Berlin and the Adam building on Leipzigerstrasse, do not fall under the same category; it all depends on how the buildings can be related to a focused and critical assessment from sociology or philosophy. Instead of Simmel, one could also argue from Walter Benjamin's work. The main question is how the relation is assessed between project and city. Not in the sense most urbanists discuss this relation as a fitting-in with the site, but as a critical reading of both city and architecture at the same time. The later projects abstain from any dialogue with the physical particularities of their contexts; the glass walled blocks could be reproduced on any site. The sameness of the units and undifferentiated order tend to deny the possibility of attaching significance to the arrangements. Yet, Hays argues, it does not mean that these later designs are unrelated to the 1922 skyscraper designs. It is the repudiation of an *a priori* logic as primary focus of meaning that ties them together. Mies' achievement in the Alexanderplatz design was to open up a clearing of silence in the chaos of the nervous metropolis; it is silence that carries the burden of meaning in this project.

Interestingly enough, the new position of 'criticality' seems to be with Elisabeth Diller and Ricardo Scofidio, since Michael Hays' first act as curator of New York's Whitney Museum was to give them a major exhibition. But even more interesting is that they chose to exhibit many of the museum gallery projects that have made them famous, Baird writes.[26] None of the building projects on which their recent design practice has focussed was shown – projects that will have to meet the more difficult test of being critical 'in the street'. To be able to relate to notions of 'lived space', which is part of this criticality in the streets, we will have to relate to the ideas of a critical theory, and especially projective thinking as in *reflexive architecture* (not to be confused with 'critical architecture'). 'Critical' can only be used for theory, not for architecture. In the projects in the *Socius* book[27], it is directly related to an 'architecture in the streets'. Remarkably, the earliest and most severe critique on Eisenman's work came from Tafuri, where Eisenman's work was considered to be fit for the *boudoir*, and *not for the street*.[28] So let us turn to that 'reflexive architecture in the street' and see what it has to offer.

The Human Body and its Ground Before we go back to the architectural discourse, we first have to address more general notions of (human) nature, sustainability, biosphere, and information society. How do we address these questions? The concept of human nature is highly complex; I will not strictly follow the problem of what is called 'the post-humanist subject' as it is already well presented in current cultural discourse or theory.[29] I will address the problem of 'digital worlds' from the problem of *grounding*, and the necessity of a spatio-temporal *'re-framing'* of architectural thought in terms of the organic and inorganic in order to get at ways in which we may rethink the possibility of sustainable action and agency in our times. Cyberspace in particular, Timothy Luke argues[30], forces human beings to re-conceptualize their spatial situation inasmuch as they experience their positions in cyberspace only as simulations in some 'virtual life' form. His argument is that we might need another reasoning to capture these digital worlds. The epistemological foundations of conventional reasoning in terms of political realism are grounded in the modernist laws of second nature, he writes. We might need another epistemic notion on what is *real* and what is *virtual*. In taking up the notions of 'first' and 'second' nature, Luke defines the 'third nature' as informational cybersphere/telesphere.

Digitalization shifts human agency and structure to a register of informational bits from that of manufactured matter. Human presence gets located in the interplay of the two modes of nature's influence. First nature, according to Luke, gains its identity from the varied terrains forming the bioscape/ecoscape/geoscape of *'terrestriality'*. Earth, water and sky provide the basic elements mapped in physical geographies of the biosphere that in turn influence human life with natural forces. Yet a large part of the biosphere is polluted beyond recovery. For example, car and air traffic are jointly responsible for some 40% of the USA's annual energy consumption, but the built environment consumes an equal amount, the rest taken by industry. Urban sprawl in the USA is one of the major problems of energy consumption. Subsidized gas, relatively low taxes on cars, high accessibility by car, and low land prices guarantee more and more sprawl every day. Of course the problem is far more complex than what can be briefly described here in a few lines. My main concern is how to understand our own actions in relation to nature and the possible architectural and urban solutions.

Both architecture and urbanism play an important role in the understanding of digitalized work processes and digital architecture, and the relation to *bioscape*, *ecoscape* and *geoscape*. It is difficult,

26 Baird, George; **op. cit.**
27 Graafland, Arie; **op. cit.**
28 Tafuri, Manfredo; 'L'architecture dans le boudoir', in **The Sphere and the Labyrinth: Avant-gardes and Architecture from Piranesi to the 1970s** (Cambridge: MIT Press, 1987) pp. 267**ff.**
29 cf. Foucault, Michel; **The Archaeology of Knowledge** (Great Britain: Tavistock Publications Limited, 1972), and **The Order of Things** (Great Britain: Tavistock Publications Limited, 1970).
30 Luke, Timothy W.; 'Simulated Sovereignty, Telematic Territoriality: the Political Economy of Cyberspace', in **Spaces of Culture**, editors Featherstone, Mike & Lash, Scott (London: Sage, 1999) pp. 28**ff.**

if not impossible to say where these systems begin or end, where solutions to the environment might be found, what kind of agreement we might reach to solve architectural and urban problems. There is indeed a witches' brew of political arguments, concepts and difficulties that can conveniently be the basis of endless academic, intellectual, theoretical and philosophical debate, as David Harvey writes.[31] Some common language has to be found, according to Harvey, or at least an adequate way of translating between different languages. His common ground is in 'the web of life' metaphor; it might indeed help us to filter our actions through the web of interconnections that make up the living world.

In addition, Luke's definition of the nation state, mass society and global geopolitics as historical artefacts used for constructing and conquering the built environments or social spaces of second nature can help us along this path. It is a domain historically described for architecture by Richard Sennett in his book *Flesh and Stone*.[32] Second nature is discussed in the sense of the technoscape/socioscape/ethnoscape of *territoriality*. Luke might be right that many of the changes today cannot be fully understood with these two concepts alone. The elaborate human constructions become overlaid, interpenetrated and reconstituted with a 'third nature' of an informational cybersphere or telesphere, he argues. As a new concept we might want to see this in a Deleuzian way of a contour, a configuration, a constellation of an event to come. It will also have more and more implications for the way we deal with architecture and urbanism. Architectural and urban design are deeply involved in 'third nature'. Until recently, design was involved in first and second nature, but with digitalization it has entered a third nature. This is not only a question of the 'means' of designing, it has – and will – influence our ways of seeing and experiencing architecture.

On the other hand, Peter Eisenman writes, architecture traditionally was place-bound, linked to a condition of experience.[33] Eisenman refers to the comparable notions Luke is writing about, mediated environments challenging the givens of classical time, the time of experience. Writing about his Rebstock Park project for Frankfurt, Eisenman writes that architecture can no longer be bound by the static conditions of space and place. To his mind architecture must deal with new conditions like the 'event'.[34] Rebstock is seen as an unfolding event – events like a rock concert where one becomes part of the environment. Yet architectural theory has largely ignored this idea. Instead, theory has focussed on notions of figure and ground, according to Eisenman. There seem to be two ways of dealing with this conceptual pair; one leading to contextualism, and one leading to a *tabula rasa* such as the modern movement imagined. With architectural modernism there is no relationship between old and new, or between figure and ground. Ground, or territoriality in Luke's terms, is seen as a clear neutral datum, projecting its autonomy into the future. I think both Luke and Eisenman are right in detecting a 'third nature', but where it will lead is still not clear.

Critical Theory in Brief: the Aesthetic Mode of Writing
Theory has to be grasped in the place and time out of which it emerges. These situations are constantly changing. In that sense, Scott Lash's use of '*allegory*' is an interesting thought that I would like to pursue for a moment. He distinguishes two types of modernism in social theory: on the one hand, positivism, and on the other '*Lebensphilosophie*'. Positivism he understands as structured along the lines of 'system', and *Lebensphilosophie* along the principle of 'symbol'. The forerunner of positivism, whose paradigmatic system building figures run from Rousseau/Condorcet, through Comte, the late Marx, Le Corbusier and more recently Habermas's later work, is French humanist classicism. Lash refers to the not unproblematic opposition of '*Zivilisation*' and '*Kultur*' in Norbert Elias' work.35 Lash's main reference here is Simmel, who worked more in the idiom of symbol than system. Simmel also began to work in a different register, the register of allegory. Lash describes it as a deepening of Goethe's notion of symbol in contrast to French classical allegory which was superficial and ornamental, as it was associated with the salons and the manners of court society. Lash shifts the notion of symbol from the classical to the baroque allegory, from French court society to Spanish absolutism and thus to baroque allegory. For Lash it consists of a completely different register from the original juxtaposition of symbol and allegory, it has to do neither with *Zivilisation* nor *Kultur*. Concepts seem to partially lose their original meaning here, in fact Lash is laying out a different 'plane of immanence' in the Deleuzian sense.

Second and third nature as newly established concepts would need a different laying out of the plane that holds them together. Classical allegory proffers a point for point homology between

---

31 Harvey, David; **Spaces of Hope** (Los Angeles: University of California Press, 2000) p. 215.
32 Sennett, Richard; **Flesh and Stone: The Body and the City in Western Civilization** (New York, London: Norton & Company, 1994).
33 Eisenman, Peter; 'Unfolding Events: Frankfurt, Rebstock and the Possibility of a New Urbanism', in **Unfolding Frankfurt** (Ernst & Sohn, Verlag fur Architektur und Technische Wissenschaften Gmbh.,Germany 1991) p. 9.
34 Eisenman's critique is on architecture theory's neglect of the **event structure** in architecture. He might be right there, but I think it is not only a question of addressing the topic of the event structure, but also the way we write about it. It is not only about an open mind for fleeting events, but very much about a fleeting way of writing about these events. For a large part, architecture history has been focused on what Eisenman calls the figure-ground relationship. Events however go further than just the 'function' of a plan. Events go deeper into the structure of a plan; indeed, they form it for the most part as I tried to show in my Versailles analysis. See my **Versailles and the Mechanics of Power. The Subjugation of Circe. An Essay** (Rotterdam: 010 Publishers, 2003) p. 54, and note 86.
35 Daniel Gordon in his **Citizens Without Sovereignty: Equality and Sociability in French Thought, 1670–1789** (Princeton: Princeton University Press, 1994) draws attention to the influence of Thomas Mann's criticism of French 'civilization'. In his **Betrachtungen eines Unpolitischen** of 1918, Mann makes the same distinction Lash makes between 'civilization' and 'culture', whereby Germany for Mann was more subjected to the latter and France to the former. Gordon's book deals with the to his mind uncritical antithesis as used by Elias to argue about cultural history. Gordon shows that Elias's 'spatial axis' – the difference between France and Germany – is not convincing. See my **Versailles and the Mechanics of Power. The Subjugation of Circe. An Essay** (Rotterdam: 010 Publishers, 2003) p. 54, and note 86.

two narratives; his baroque version posits a significant absence, a 'hole' in the underlying narrative. If the original, 'true' story somehow is not quite right, then the point to point homology between the second narrative and the first is no longer possible. Baroque allegorists such as Nietzsche, Simmel, Benjamin, Adorno and Karl Krauss write in the form of an essay, Lash maintains. The essay might well look '*wissentschaftlich*', but instead emerges in an aesthetic mode – serious and at the same time superficial, light, ornamental.

Baroque allegory is, in fact, opposite to a Marxist explanation. Michael Hays refers to Louis Althusser, the French Marxist, with regard to his idea of 'relative autonomy'.[36] At the other end of the line, then, might be the Frankfurt School of Horkheimer and Adorno. In the American debate on 'critical', the term is used many times by different authors with often divergent meanings, but it might be good to remember that Parisian Marxism in the sixties and seventies was never interested in a 'critical' but in a Marxist 'scientific' way of proceeding. My first two books on the architectural body, were written as an essay in that aesthetic mode to which Lash refers – a seemingly 'light', sometimes 'ironic' architectural critique as in my analysis of Rem Koolhaas' Downtown Athletic Club and Duchamp's Large Glass.[37] Yet the allegorist is, while looking ornamental, simultaneously deadly serious. In Lash's formulation, the allegorist is the father of the illegitimate child of modernity's other.

Conclusion With many of the contemporary architectural electronic imaging techniques and communication technologies, we are in the end loosing all ground. My claim is that we need more ground and permanence in architecture instead of 'folds'. Seen from an architectural perspective, it means that blobs and folds take the city as an additive texture without any coherence; they consume too much space since they want to stand on their own imagined pedestals. They reinforce urban sprawl. Instead of more compact building, they spread out. There is indifference to the environment, grounding is no issue. I think I can agree with Lash's critique. Although I do not think his critique works in the instance of Koolhaas as he suggests, it does work very well for digital architectures: speed supersedes space as indifference supersedes difference.

The source for these digital designs is third nature. Third nature here is largely penetrating first and second nature; it dissolves any notion of ground or context.

36 Hays, Michael K.; 'Ideologies of Media and the Architecture of Cities in Transition', in **Cities in Transition** edited by Arie Graafland & Deborah Hauptmann (Rotterdam: 010 Publishers, 2001) pp. 263**ff.**
37 Graafland, Arie; 'Artificiality in the Work of Rem Koolhaas', in **Architectural Bodies** (Rotterdam: 010 Publishers, 1996) pp. 39**ff.**
38 Castells, Manuel; **The Informational City: Information technology, Economic Restructuring and the Urban Regional Process** (London: Blackwell, 1989).

It is here where my doubts for a possible application to architecture and urbanism begin. Like second nature, third nature is no doubt a social product. Eisenman's Rebstock Park shifts the notion of figure/ground to one of assumed Deleuzian folding. This shift has direct consequences for the grounding of design. We should realize that all spaces are *constructs* and *real*, including our digital worlds. Virtual space in Deleuze's sense is not an unforeseen possibility in the design, to be realized in a certain framing. It is about a question that will open up new uncharted territories. First and second nature do not have more materialized substance; it is indeed more than a collective hallucination restricted to the symbolic domains of social superstructures. It has an immense material base in communication satellites, and fibre optic networks as Manuel Castells has analysed.[38]

In architecture and urbanism, we cannot do without 'ground', nor can we do without critical thinking. I think Deleuze and Guattari are very right in saying that thinking takes place in the relationship of territory and earth. If we lose first and second nature, we lose the very notions of gender, sexuality, ethnic diversity, uneven distribution of wealth, and class. Too easily, the shift from harsh reality into the seemingly endless possibilities of the computer programmes is made, made without much interest for these categories. The location of most of Lynn's constructions is nowhere; they might be anywhere. Just like the complexity of movement in Koolhaas' international airports, they are for the greater part interchangeable. In architecture and urbanism, we can never lose ground; third nature won't be enough. Thinking, in the end, always takes place in relation to territory and earth. We need first and second nature too.

Credits

Delft School of Design Series on Architecture and Urbanism
Series Editor Arie Graafland

Editorial Board
K. Michael Hays (Harvard University, USA)
Ákos Moravánszky (ETH Zürich, Switzerland)
Michael Müller (Bremen University, Germany)
Frank R. Werner (University of Wuppertal, Germany)
Gerd Zimmermann (Bauhaus University, Germany)

Also published in this series:
2 **The Body in Architecture**
ISBN 978 90 6450 568 3
3 **De-/signing the Urban. Technogenesis and the urban image**
ISBN 978 90 6450 611 6

Crossover. Architecture Urbanism Technology
Editors Arie Graafland and Leslie Jaye Kavanaugh
Text editing John Kirkpatrick
Book design by Piet Gerards Ontwerpers (Piet Gerards and
Maud van Rossum), Amsterdam
Printed by Snoeck Ducaju, Ghent

Photo credits
Jeroen Musch p. 31, 32, 33, 34, 35, 36, 37, 38, 39, 484-485
Roemer van Toorn cover, p. 40-41, 136-137, 208-209, 280-281,
386-387, 592-593, 644-645

©2006 The authors / 010 Publishers, Rotterdam
www.010publishers.nl

ISBN 978 90 6450 609 3